LONDON MATHEMATICAL SOCIETY
MONOGRAPHS NEW SERIES

LONDON MATHEMATICAL SOCIETY MONOGRAPHS NEW SERIES

Previous volumes of the LMS Monographs were published by Academic Press, to whom all enquiries should be addressed. Volumes in the New Series will be published by Oxford University Press throughout the world.

NEW SERIES

1. *Diophantine inequalities* R. C. Baker
2. *The Schur multiplier* Gregory Karpilovsky
3. *Existentially closed groups* Graham Higman and Elizabeth Scott
4. *The asymptotic solution of linear differential systems* M. S. P. Eastham
5. *The restricted Burnside problem* Michael Vaughan-Lee
6. *Pluripotential theory* Maciej Klimek
7. *Free Lie algebras* Christophe Reutenaur
8. *The restricted Burnside problem* (2nd edition) Michael Vaughan-Lee
9. *The geometry of topological stability* Andrew du Plessis and Terry Wall
10. *Spectral decompositions and analytic sheaves* J. Eschmeier and M. Putinar
11. *An atlas of Brauer characters* C. Jansen, K. Lux, R. Parker, and R. Wilson
12. *Fundamentals of semigroup theory* John M. Howie
13. *Area, lattice points, and exponential sums* M. H. Huxley
14. *Super-real fields* H. G. Dales and W. H. Woodin
15. *Integrability, self-duality, and twistor theory* L. Mason and N. M. J. Woodhouse
16. *Categories of symmetries and infinite-dimensional groups* Yu. A. Neretin
17. *Interpolation, identification, and sampling* Jonathan R. Partington
18. *Metric number theory* Glyn Harman
19. *Profinite groups* John S. Wilson
20. *An introduction to local spectral theory* K. B. Laursen and M. M. Neumann
21. *Characters of finite Coxeter groups and Iwahori-Hecke algebras* M. Geck and G. Pfeiffer
22. *Classical harmonic analysis and locally compact groups* Hans Reiter and Jan D. Stegeman
23. *Operator spaces* E. G. Effros and Z.-J. Ruan

Operator Spaces

Edward G. Effros
Department of Mathematics at the University of California, Los Angeles

and

Zhong-Jin Ruan
Department of Mathematics, University of Illinois at Urbana-Champaign

CLARENDON PRESS • OXFORD
2000

OXFORD
UNIVERSITY PRESS
Great Clarendon Street, Oxford OX2 6DP

Oxford University Press is a department of the University of Oxford.
It furthers the University's objective of excellence in research, scholarship,
and education by publishing worldwide in

Oxford New York

Athens Auckland Bangkok Bogotá Buenos Aires Calcutta
Cape Town Chennai Dar es Salaam Delhi Florence Hong Kong Istanbul
Karachi Kuala Lumpur Madrid Melbourne Mexico City Mumbai
Nairobi Paris São Paulo Shanghai Singapore Taipei Tokyo Toronto Warsaw

with associated companies in Berlin Ibadan

Oxford is a registered trade mark of Oxford University Press
in the UK and in certain other countries

Published in the United States
by Oxford University Press Inc., New York

First published 2000

A catalogue record for this book is available from the British Library

Library of Congress Cataloging in Publication Data
Data available

ISBN 0 19 853482 5

Typeset by Focal Image Ltd, London
Printed in Great Britain
on acid-free paper by
T. J. International Ltd, Padstow

Dedicated to Rita and Lian

for their support and understanding

Preface

Perhaps the most dramatic shift in our understanding of the physical world occurred in 1925, with the publication of Heisenberg's remarkable paper 'Über quantentheoretische Umdeutung kinematischer und mechanischer Beziehungen'. Heisenberg demonstrated that one could deduce quantum phenomena from the equations of Newtonian physics provided one interpreted the time dependent variables as standing for infinite matrices rather than functions. In contrast to functions, matrices need not commute under multiplication. Heisenberg subsequently showed that this unusual feature of 'quantum variables' may be physically understood in terms of his famous Uncertainty Principle.

Heisenberg's 'matrix mechanics' quickly attracted the attention of a number of leading mathematicians, including Jordan, von Neumann, and Weyl. In particular, von Neumann pointed out that Heisenberg's matrices were more precisely modelled by self-adjoint Hilbert space operators. Nevertheless, despite these early contributions to the codification of quantum physics, mathematicians have, at times, been reluctant to consider the implications of Heisenberg's discovery for their own discipline.

There is now a consensus among scientists that the classical and relativistic notions of measurement and geometry that underlie so much of modern mathematics no longer correspond to our understanding of the real world. Von Neumann was the first to fully appreciate this fact, and he concluded that we should seek 'quantized' analogues of mathematics. He proposed that, as in physics, we should begin by replacing functions by operators.

Von Neumann took the first steps toward mathematical quantization in collaboration with Murray. In a remarkable series of articles, Murray and von Neumann (1936), (1937), and (1943), and von Neumann (1929), (1940), and (1949), they succeeded in formulating an operator version of integration theory. They began by replacing the algebras of bounded complex functions that naturally arise in classical integration theory (or more precisely, the L_∞*-algebras*) by $*$*-algebras of bounded operators on Hilbert spaces.* During the past sixty years, such *operator algebras* have been shown to have a profound structure theory. As von Neumann had anticipated, they provide a natural framework for quantizing other areas of mathematics, including portions of topology, geometry, analysis, probability theory, and algebra.

In this monograph we are concerned with a more recent innovation, the *quantization of Banach space theory.* In retrospect, this development was both straightforward and unambiguous. We recall that a normed space E can always be realized as a *function space*; that is, a linear space of bounded functions on a set S, together with the uniform norm (see §2.1). By analogy, we define a (concrete) *operator space* V to be a linear space of bounded operators on a Hilbert space H. Although such a space V is normed by the operator norm, it actually inherits a more elaborate structure. Owing to the fact that an $n \times n$ matrix of operators on a Hilbert space H may again be realized as an operator on H^n, there is a *distinguished norm* on each of the *matrix spaces* $M_n(V)$. The appropriate morphisms for this structure are the linear mappings which are *completely bounded*; that is to say, which induce uniformly bounded mappings of the matrix spaces.

Long before operator space theory was axiomatized, operator algebraists had used completely bounded mappings to study the structure of C^*-algebras and von Neumann algebras. With this new framework, it has now become clear that some of the most important invariants of operator algebras, such as injectivity, exactness, and local reflexivity, are best understood as being attributes of their underlying operator spaces. In principle, however, operator spaces should have a much wider applicability than these algebraic results might suggest. There are natural operator space analogues for all the 'classical Banach spaces', and in particular, they provide a natural context for studying non-commutative integration theory. The latter subject is playing an increasingly important role in non-commutative analysis. It is now evident that various difficulties that arise in Fourier analysis of non-commutative groups may be overcome once we acknowledge the underlying operator space structures. There are also reasons to believe that this theory will be essential in the study of harmonic analysis on quantum groups. Finally, it seems inevitable that operator spaces and their Frechet generalizations will provide the correct functional analytic settings for other areas of quantized analysis, including the differential systems that naturally arise in non-commutative geometry.

Throughout this work we use the 'classical' theory of *functions* to motivate the 'quantized' theory of *operators*. As in the physical theory, this goal cannot be fully achieved since operator space theory involves phenomena that do not have classical analogues. Nevertheless, function theory has continued to provide our most fruitful guide for the development of the subject. Perhaps one of the most attractive aspects of operator space theory is the manner in which routine notions in Banach space theory re-emerge as deep and beautiful ideas in operator space theory. It is our hope that we shall succeed in communicating the excitement of this subject to our Banach space colleagues.

The monograph is divided into five parts, followed by an appendix in which we have summarized some of the elementary results we shall use from functional analysis.

In a preliminary chapter, we introduce the reader to spaces with 'matrix coefficients'. Although we might have used Banach modules for this purpose, we have adopted a less sophisticated approach, which is closer to that generally used in the literature.

Part I is devoted to the three fundamental results upon which operator space theory is based: the *representation theorem* of Ruan, the Arveson–Wittstock generalization of the *Hahn–Banach theorem*, and the Paulsen–Wittstock *decomposition theorem* for complete contractions. We have attempted to give completely accessible proofs for these important facts. Although there are now elegant tensor product approaches to all of these results, we feel that they may be too forbidding to the newcomer. As an application of this material, we characterize the injective operator spaces in §6.1.

As we explain at the beginning of Part II, tensor products have been crucial in the development of operator space theory. We consider the three most important tensor products. These are the *operator space projective* and *injective tensor products*, which are quite similar to their classical analogues, and the *Haagerup tensor product*, which is quite unlike anything that may be found in Banach space theory. The operator space injective tensor product reduces to the usual spatial tensor product for C^*-algebras, whereas the operator space projective tensor product may be used to construct the predual of the von Neumann algebraic spatial tensor product (see §7.2). The Haagerup tensor product has proved to be especially useful in more algebraic contexts, some of which we briefly consider in Part V.

In his pioneering work on Banach spaces, Grothendieck used a categorical approach that is particularly amenable to quantization. In particular, his study of the links between mapping spaces and tensor products carries over to this new context. In Part III we use the operator space projective and injective tensor products to generalize Grothendieck's theory of approximation properties. We then introduce analogues of the three most important Banach mapping spaces: the *nuclear*, the *integral*, and the *absolutely summing mappings*. It is here that the more subtle behaviour of operator spaces first becomes apparent. Grothendieck's characterization of the dual of the Banach space injective tensor product in terms of integral mappings does not carry over to operator spaces (see (12.1.8) and Proposition 14.2.2). On the other hand, his ingenious use of absolute summing mappings to study the Dvoretzky–Rogers theorem on unconditional summability applies as well to operator space theory (see §13.4).

Perhaps the deepest results in operator space theory are concerned with the unexpected new phenomena that occur in operator space theory. These largely centre around three notions that are key to C^*-algebra theory: *nuclearity*, *exactness*, and *local reflexivity*. These developments are explained in Part IV, which may be regarded as the central portion of this monograph. This is an area of great beauty and depth, and it represents one of the triumphs of the subject. Owing to the work of Kirchberg, Haagerup, Pisier and the recent developments in (Effros *et al.* 1998, 1999), we now have a very precise understanding of how these invariants are related. In particular, we have for any operator space V,

$$V \text{ is nuclear } \Rightarrow V \text{ is exact } \Rightarrow V \text{ is locally reflexive,}$$

and

$$V \text{ is nuclear } \Leftrightarrow V \text{ is locally reflexive and } V^{**} \text{ is semidiscrete.}$$

It has been suggested by Kirchberg that if G is an arbitrary discrete group, then its reduced group C^*-algebra $C^*_\lambda(G)$ is exact. On the other hand, he showed that extensions of exact C^*-algebras need not be exact. Turning to local reflexivity, C^*-algebras need not be locally reflexive, but, surprisingly, *all* C^*-algebraic duals are locally reflexive. This theorem is proved in §15.3.

In Part V we briefly consider some of the algebraic applications of operator space theory. We use the operator space projective tensor product to introduce the notion of a '*quantized Banach algebra*' and we reformulate amenability in this context. The most important examples of such objects are the Fourier algebras of locally compact groups, and more generally the 'L_1-*convolution algebras*' of quantum groups. In contrast to the classical theory of amenability, the Fourier algebra of a locally compact group is amenable in the completely bounded sense if and only if the group is amenable. We have included the proof of this for the simple case of compact groups. We have also included a proof of the strikingly elegant abstract characterization for the unital, not necessarily self-adjoint operator algebras. This 'non-self-adjoint Gelfand–Naimark' theorem illustrates some of the remarkable properties of the Haagerup tensor product.

Our goal in this monograph has been to explain the deep analogy between linear spaces of bounded functions and linear spaces of bounded operators. We have made every effort to provide 'elementary' proofs that will be accessible to readers with a rudimentary knowledge of functional analysis, even when that has restricted the scope of our treatment. This should not be regarded as an encylopaedic summary of the subject, and in particular, the references, notes, and bibliography primarily contain various items that are useful to the topics considered.

It is our conviction that the extraordinary array of techniques developed by Banach space theorists will have many applications in non-commutative analysis, and that conversely, operator space theory will provide Banach space theorists with exciting new vistas for research. More generally, it is our hope that this new formalism will help to unravel some of the difficulties associated with quantization.

Operator space theory has evolved rapidly, and this has made it difficult to write an up-to-date monograph on the subject. Numerous drafts of the book were overtaken by irresistible new approaches, prompting us to repeatedly scrap the text and begin over again. We are indebted to Garth Dales who suggested the project, offered encouragement, and never complained about unfulfilled schedules. We also wish to thank the editors of the Oxford University Press, who gave unflagging support for the project.

Los Angeles E. G. E.
Urbana-Champaign Z-J. R.
December 1999

Contents

1

Matrix and operator conventions

In order to understand operator space theory, one must have at least a passing acquaintance with the algebraic and analytic properties of matrices over linear spaces. The reader who is knowledgeable about these notions may prefer to begin with the next chapter.

It is often best to regard an n by n matrix over a linear space V as a 'linear combination' of the form $v = \sum \alpha_i \otimes v_i$, where $v_i \in V$, and the 'coefficients' α_i are scalar n by n matrices. We expect that this will be a familiar notion to Banach space theorists, who have found it very useful to study norms for linear combinations $v = \sum \alpha_i \otimes v_i$, where the α_i are random variables. Pursuing this line of thought, matrices and operators are the quantum versions of random variables, or in other words, the 'observables' of modern physics. Thus in operator space theory one is interested in computing the norms of linear combinations with 'non-commuting random variable coefficients'.

We introduce the basic matrix manipulations and their matrix coefficient interpretations in the first section. In the following section we review the various natural norms and orderings that can be imposed on matrices, including the Schatten–von Neumann analogues of the ℓ_p^n norms for $p = 1, 2, \infty$. In the final section we briefly discuss the properties of matrices of operators.

1.1 MATRICES AND MATRIX COEFFICIENTS

In its simplest manifestation, the distinction between functions and operators may be seen by comparing (singly indexed) *n-tuples* with (doubly indexed) *matrices* over a linear space. We have found it helpful to notationally distinguish between these two notions, and for that purpose we begin with a brief discussion of vectors.

Given a complex linear space V and an integer $n \in \mathbb{N}$, an *n-tuple* over V is an element of $V^n = V \oplus \cdots \oplus V$. We employ horizontal, and when convenient, vertical displays for an n-tuple $v = (v_j) \in V^n$; that is, we write

$$v = (v_1, \ldots, v_n) = \begin{pmatrix} v_1 \\ \vdots \\ v_n \end{pmatrix}.$$

We let $\varepsilon_i = \varepsilon_i^{(n)} = (0,\ldots 0, 1_i, 0\ldots, 0)$ be the usual basis vectors for $\mathbb{C}^n$. Given a linear space V, we have linear isomorphisms

$$V^n \cong \mathbb{C}^n \otimes V \cong V \otimes \mathbb{C}^n$$

defined by

$$v = (v_i) \mapsto \sum \varepsilon_i \otimes v_i$$

and

$$v = (v_i) \mapsto \sum v_i \otimes \varepsilon_i.$$

Each linear mapping $\varphi : V \to W$ determines a linear mapping

$$\varphi^n : V^n \to W^n : \varphi^n(v) = (\varphi(v_i)),$$

or equivalently,

$$\varphi^n = id \otimes \varphi : \mathbb{C}^n \otimes V \to \mathbb{C}^n \otimes W.$$

We shall often use more general indices. Given a set $\mathfrak{s}$, we define an $\mathfrak{s}$*-tuple* $v = (v_s)_{s\in\mathfrak{s}}$ to be a function from $\mathfrak{s}$ to V. We let $V^{\mathfrak{s}}$ denote the linear space of $\mathfrak{s}$-tuples. To accommodate the previous notation, we take the liberty of letting an integer n also stand for the set $\{1,\ldots,n\}$, writing $j \in n$ if $1 \leq j \leq n$, and we let ∞ stand for the set $\mathbb{N}$. Thus V^∞ is the linear space of all sequences $(v_1, v_2, \ldots)$ with $v_j \in V$.

A bijection of index sets $\theta : \mathfrak{s} \to \mathfrak{s}'$, determines a linear isomorphism

$$\overline{\theta} : V^{\mathfrak{s}} \to V^{\mathfrak{s}'} : v \mapsto v',$$

where $v'_{\theta(s)} = v_s$. When the bijection θ is unambiguous, we shall often not bother to include it in our notation, simply writing $V^{\mathfrak{s}} \cong V^{\mathfrak{s}'}$ to indicate that an 'identification' is being made. We have, for example, that for any two index sets $\mathfrak{s}$ and $\mathfrak{t}$, the 'flip' $\theta : \mathfrak{s} \times \mathfrak{t} \to \mathfrak{t} \times \mathfrak{s}$ determines a natural isomorphism $V^{\mathfrak{s}\times\mathfrak{t}} \cong V^{\mathfrak{t}\times\mathfrak{s}}$.

Given a linear space V and two index sets $\mathfrak{s}$ and $\mathfrak{t}$, we identify $\mathfrak{s}$-tuples of $\mathfrak{t}$-tuples in V with $\mathfrak{s} \times \mathfrak{t}$-tuples by notationally dropping the inner parentheses, or to be more precise, we define isomorphisms

$$(V^{\mathfrak{s}})^{\mathfrak{t}} \cong V^{\mathfrak{s}\times\mathfrak{t}} \cong (V^{\mathfrak{t}})^{\mathfrak{s}}$$

by letting

$$((v_{(s,t)})_{s\in\mathfrak{s}})_{t\in\mathfrak{t}} \leftrightarrow (v_{(s,t)})_{(s,t)\in\mathfrak{s}\times\mathfrak{t}} \leftrightarrow ((v_{(s,t)})_{t\in\mathfrak{t}})_{s\in\mathfrak{s}}.$$

Given a linear space V and integers m and n, we let $\mathbb{M}_{m,n}(V)$ denote the linear space of m *by* n *matrices*

$$v = \begin{bmatrix} v_{1,1} & \cdots & v_{1,n} \\ \vdots & & \vdots \\ v_{m,1} & \cdots & v_{m,n} \end{bmatrix},$$

where $v_{i,j} \in V$, and we write $\mathbb{M}_n(V) = \mathbb{M}_{n,n}(V)$. If $V = \mathbb{C}$, then we let $\mathbb{M}_{m,n} = \mathbb{M}_{m,n}(\mathbb{C})$ and $\mathbb{M}_n = \mathbb{M}_n(\mathbb{C})$.

The *matrix units*

$$\varepsilon_{i,j} = \varepsilon_{i,j}^{[m,n]} = \begin{bmatrix} 0 & \ldots & 0 \\ \vdots & 1_{i,j} & \vdots \\ 0 & \ldots & 0 \end{bmatrix}$$

form a vector basis for $\mathbb{M}_{m,n}$, and we let $\varepsilon_{i,j}^{[n]} = \varepsilon_{i,j}^{[n,n]}$. The row matrices

$$E_j = E_j^n = \varepsilon_{1,j}^{[1,n]} = [0 \ldots 1_j \ldots 0] \in \mathbb{M}_{1,n} \tag{1.1.1}$$

satisfy

$$\varepsilon_{i,j}^{[m,n]} = E_i^{m*} E_j^n. \tag{1.1.2}$$

The *identity matrix* I_n of the algebra $\mathbb{M}_n$ is given by

$$I_n = \sum \varepsilon_{i,i}^{[n]} = [\delta_{i,j}],$$

where, as usual, $\delta_{i,j}$ is defined to be 1 if $i = j$, and 0 otherwise.

The matrix units determine the linear identifications

$$\mathbb{M}_{m,n}(V) \cong \mathbb{M}_{m,n} \otimes V \cong V \otimes \mathbb{M}_{m,n}, \tag{1.1.3}$$

where

$$v = [v_{i,j}] \mapsto \sum \varepsilon_{i,j}^{[m,n]} \otimes v_{i,j} \tag{1.1.4}$$

and

$$v = [v_{i,j}] \mapsto \sum v_{i,j} \otimes \varepsilon_{i,j}^{[m,n]},$$

respectively. In the reverse direction, if $\alpha = [\alpha_{i,j}] \in \mathbb{M}_{m,n}$ and $v_0 \in V$, then the corresponding *elementary tensor* $\alpha \otimes v_0$ is given by the matrix

$$\alpha \otimes v_0 = \sum \alpha_{i,j}\varepsilon_{i,j} \otimes v_0 = \sum \varepsilon_{i,j} \otimes (\alpha_{i,j}v_0) = [\alpha_{i,j}v_0]. \tag{1.1.5}$$

We see from (1.1.3) that any vector $v \in \mathbb{M}_{m,n} \otimes V$ may be represented as a sum of elementary matrices, $v = \sum_{k=1}^{r} \alpha_k \otimes v_k$, or to put it another way, as a *linear combination with matrix coefficients.* This is not a unique representation, and it is often convenient to use other such decompositions (see, for example, §3.4). More often than not, we shall place the matrix coefficients on the left.

There are two natural operations which link the finite matrix spaces $\mathbb{M}_{m,n}(V)$. Given $v \in \mathbb{M}_{m,n}(V)$ and $v' \in \mathbb{M}_{p,q}(V)$, we define the *direct sum* $v \oplus v' \in \mathbb{M}_{m+p,n+q}(V)$ by letting

$$v \oplus v' = \begin{bmatrix} v_{i,j} & 0 \\ 0 & v'_{k,l} \end{bmatrix} \in \mathbb{M}_{m+p,n+q}(V), \tag{1.1.6}$$

or in terms of elementary tensors, if $v_0, v_0' \in V$,

$$\left(\varepsilon_{i,j}^{[m,n]} \otimes v_0\right) \oplus \left(\varepsilon_{k,l}^{[p,q]} \otimes v_0'\right) = \varepsilon_{i,j}^{[m+p,n+q]} \otimes v_0 + \varepsilon_{m+k,n+l}^{[m+p,n+q]} \otimes v_0'.$$

On the other hand, if we are given matrices $\alpha \in \mathbb{M}_{m,p}$, $v \in \mathbb{M}_{p,q}(V)$, and $\beta \in \mathbb{M}_{q,n}$, we define the *matrix product* $\alpha v \beta \in \mathbb{M}_{m,n}(V)$ by

$$\alpha v \beta = \left[\sum_{k,l} \alpha_{i,k} v_{k,l} \beta_{l,j}\right]_{i \in m, j \in n}. \tag{1.1.7}$$

Equivalently, we have from (1.1.5) that if $v_0 \in V$ and $\gamma \in \mathbb{M}_{p,q}$, then

$$\alpha(\gamma \otimes v_0)\beta = \alpha \left[\gamma_{k,l} v_0\right] \beta = \left[\sum_{k,l} \alpha_{i,k} \gamma_{k,l} \beta_{l,j} v_0\right] = \alpha\gamma\beta \otimes v_0. \tag{1.1.8}$$

If $V = \mathbb{M}_r$ and we use the identification $\mathbb{M}_m(\mathbb{M}_r) \cong \mathbb{M}_m \otimes \mathbb{M}_r$, then we have for $\alpha \in \mathbb{M}_m$, $\gamma \in \mathbb{M}_m(\mathbb{M}_r)$, and $\beta \in \mathbb{M}_m$,

$$\alpha\gamma\beta = (\alpha \otimes I_r)\gamma(\beta \otimes I_r). \tag{1.1.9}$$

If $v = [v_{k,l}] \in \mathbb{M}_{m,n}(V)$, then

$$v_{i,j} = [0 \ldots 1_i \ldots 0]\,[v_{k,l}] \begin{bmatrix} 0 \\ \vdots \\ 1_j \\ \vdots \\ 0 \end{bmatrix} = E_i v E_j^*, \tag{1.1.10}$$

where we let $E_i = E_i^m$ and $E_j = E_j^n$. With the identification $\mathbb{C} \otimes V \cong V$, we have

$$E_i^* v_0 E_j = E_i^*(1 \otimes v_0) E_j = \varepsilon_{i,j} \otimes v_0,$$

and thus if $v = [v_{i,j}] \in \mathbb{M}_{m,n}(V)$, then

$$v = \sum \varepsilon_{i,j} \otimes v_{i,j} = \sum E_i^* v_{i,j} E_j. \tag{1.1.11}$$

Given index sets $\mathfrak{s}$ and $\mathfrak{t}$, we let $\mathbb{M}_{\mathfrak{s},\mathfrak{t}}(V)$ denote the linear space of doubly indexed $\mathfrak{s}, \mathfrak{t}$-*matrices*

$$v = [v_{s,t}] = [v_{s,t}]_{s \in \mathfrak{s}, t \in \mathfrak{t}},$$

with each entry $v_{s,t} \in V$ labelled by a *row index* $s \in \mathfrak{s}$ and a *column index* $t \in \mathfrak{t}$, and we let

$$\mathbb{M}_{\mathfrak{s},\mathfrak{t}}^{\mathrm{fin}}(V) \subseteq \mathbb{M}_{\mathfrak{s},\mathfrak{t}}(V) \tag{1.1.12}$$

be the matrices with only finitely many non-zero entries. We do not regard a matrix v as a function on $\mathfrak{s} \times \mathfrak{t}$ since by our conventions such a function determines an $\mathfrak{s} \times \mathfrak{t}$-tuple; that is, an element of $V^{\mathfrak{s} \times \mathfrak{t}}$. To put it another way, we notationally distinguish between functions $v(s,t) = v_{s,t}$ of two variables ('bifunctions'), and functions $v((s,t)) = v_{(s,t)}$ of one variable defined on the product space. We let $\mathbb{M}_{\mathfrak{s}}(V) = \mathbb{M}_{\mathfrak{s},\mathfrak{s}}(V)$, $\mathbb{M}_{\mathfrak{s},\mathfrak{t}} = \mathbb{M}_{\mathfrak{s},\mathfrak{t}}(\mathbb{C})$, and $\mathbb{M}_{\mathfrak{s}} = \mathbb{M}_{\mathfrak{s},\mathfrak{s}}$. For each $n \in \mathbb{N}$, we may identify elements of $\mathbb{M}_n(V)$ with the $v \in \mathbb{M}_\infty^{\mathrm{fin}}(V)$ such that $v_{i,j} = 0$ when i or j is greater than n, and we then have

$$\mathbb{M}_\infty^{\mathrm{fin}}(V) = \bigcup_{n \in \mathbb{N}} \mathbb{M}_n(V). \tag{1.1.13}$$

We can use index set bijections $\mathfrak{s} \to \mathfrak{s}'$ and $\mathfrak{t} \to \mathfrak{t}'$ to determine corresponding linear isomorphisms $\mathbb{M}_{\mathfrak{s},\mathfrak{t}}(V) \cong \mathbb{M}_{\mathfrak{s}',\mathfrak{t}'}(V)$. If we *delete the internal brackets*, then we identify matrices of matrices with simple matrices over product spaces. Thus we use the correspondences

$$\begin{aligned}\big[[v_{(s,g),(t,h)}]_{g\in\mathfrak{g},h\in\mathfrak{h}}\big]_{s\in\mathfrak{s},t\in\mathfrak{t}} &\leftrightarrow \big[v_{(s,g),(t,h)}\big]_{(s,g)\in\mathfrak{s}\times\mathfrak{g},(t,h)\in\mathfrak{t}\times\mathfrak{h}} \\ &\leftrightarrow \big[[v_{(s,g),(t,h)}]_{s\in\mathfrak{s},t\in\mathfrak{t}}\big]_{g\in\mathfrak{g},h\in\mathfrak{h}}\end{aligned} \tag{1.1.14}$$

to define the natural isomorphisms

$$\mathbb{M}_{\mathfrak{s},\mathfrak{t}}(\mathbb{M}_{\mathfrak{g},\mathfrak{h}}(V)) \cong \mathbb{M}_{\mathfrak{s}\times\mathfrak{g},\mathfrak{t}\times\mathfrak{h}}(V) \cong \mathbb{M}_{\mathfrak{g},\mathfrak{h}}(\mathbb{M}_{\mathfrak{s},\mathfrak{t}}(V)). \tag{1.1.15}$$

Once again, we use an integer $p \in \mathbb{N}$ to also stand for the index set $\{1, \ldots, p\}$, and we write ∞ for $\mathbb{N}$. In particular, for $p, q \in \mathbb{N}$ we have

$$\mathbb{M}_p(\mathbb{M}_q(V)) \cong \mathbb{M}_{p\times q}(V) \cong \mathbb{M}_q(\mathbb{M}_p(V)) \tag{1.1.16}$$

and

$$\mathbb{M}_p \otimes \mathbb{M}_q \cong \mathbb{M}_{p\times q} \cong \mathbb{M}_q \otimes \mathbb{M}_p. \tag{1.1.17}$$

We next consider matrices over the tensor product $V \otimes W$ of linear spaces V and W. In contrast to scalar matrices, we shall generally not identify $V\otimes W$ with $W \otimes V$. If we are given $v \in \mathbb{M}_{\mathfrak{g},\mathfrak{h}}(V)$ and $w \in \mathbb{M}_{\mathfrak{s},\mathfrak{t}}(W)$, then we define the *Kronecker product*

$$v \odot w \in \mathbb{M}_{\mathfrak{g}\times\mathfrak{s},\mathfrak{h}\times\mathfrak{t}}(V \otimes W) \tag{1.1.18}$$

by

$$(v \odot w)_{(g,s),(h,t)} = v_{g,h} \otimes w_{s,t}.$$

For finite index sets, it is useful to reformulate this in terms of elementary tensors. If we are given $v = \alpha \otimes v_0 \in \mathbb{M}_m \otimes V$ and $w = \beta \otimes w_0 \in \mathbb{M}_n \otimes W$, then the Kronecker product

$$v \odot w \in \mathbb{M}_{m\times n} \otimes V \otimes W$$

is given by

$$(\alpha \otimes v_0) \odot (\beta \otimes w_0) \cong (\alpha \otimes \beta) \otimes (v_0 \otimes w_0). \tag{1.1.19}$$

If we use the identifications $V \cong \mathbb{C} \otimes V \cong V \otimes \mathbb{C}$, then for $\gamma = [\gamma_{i,j}] \in \mathbb{M}_m$ and $v = [v_{k,l}] \in \mathbb{M}_n(V)$,

$$\gamma \odot v = [\gamma_{i,j} v_{k,l}] \cong [v_{k,l}\gamma_{i,j}] = v \odot \gamma, \tag{1.1.20}$$

where the symbol $\cong$ indicates that we are using the identification of mn by mn matrices with nm by nm matrices. In particular,

$$v \otimes I_m \cong I_m \otimes v = [\delta_{i,j} v_{k,l}] \cong \begin{bmatrix} v & 0 & \dots & 0 \\ 0 & v & \dots & 0 \\ \vdots & & \ddots & \vdots \\ 0 & 0 & & v \end{bmatrix}. \tag{1.1.21}$$

Given a linear mapping $\varphi : V \to W$ and $m, n \in \mathbb{N}$, we have a corresponding mapping $\varphi_{m,n} : \mathbb{M}_{m,n}(V) \to \mathbb{M}_{m,n}(W)$ defined by

$$\varphi_{m,n}(v) = [\varphi(v_{i,j})].$$

It follows from (1.1.5) that

$$\varphi_{m,n}(\gamma \otimes v_0) = \varphi_{m,n}([\gamma_{i,j} v_0]) = [\varphi(\gamma_{i,j} v_0)] = [\gamma_{i,j}\varphi(v_0)] = \gamma \otimes \varphi(v_0),$$

and thus

$$\varphi_{m,n} = id_{\mathbb{M}_{m,n}} \otimes \varphi : \mathbb{M}_{m,n} \otimes V \to \mathbb{M}_{m,n} \otimes W .$$

We let

$$\varphi_n = \varphi_{n,n} : \mathbb{M}_n(V) \to \mathbb{M}_n(W) .$$

It is evident that if we are given v, w, α, and β as above, then

$$\varphi_{m+p,n+q}(v \oplus w) = \varphi_{m,n}(v) \oplus \varphi_{p,q}(w), \tag{1.1.22}$$

and

$$\varphi_{m,n}(\alpha v \beta) = \alpha \varphi_{p,q}(v) \beta . \tag{1.1.23}$$

Given dual linear spaces V and V', we define the *scalar pairing*

$$\langle \cdot , \cdot \rangle : \mathbb{M}_{m,n}(V) \times \mathbb{M}_{m,n}(V') \to \mathbb{C} \tag{1.1.24}$$

and the *matrix pairing*

$$\langle\langle \cdot , \cdot \rangle\rangle : \mathbb{M}_{m,n}(V) \times \mathbb{M}_{p,q}(V') \to \mathbb{M}_{mp,nq} \tag{1.1.25}$$

by

$$\langle v , w \rangle = \sum \langle v_{i,j}, w_{i,j} \rangle \tag{1.1.26}$$

and

$$\langle\langle v , w \rangle\rangle = [\langle v_{i,j}, w_{k,l} \rangle] , \tag{1.1.27}$$

respectively. In some respects, the matrix pairing is the more natural one for operator spaces.

If $V = \mathbb{C}$, then the scalar pairing of $\mathbb{M}_{m,n}$ with itself is given by

$$\langle \alpha , \beta \rangle = \sum \alpha_{i,j} \beta_{i,j} = \text{trace}\,(\alpha \beta^{tr}) = \text{trace}\,(\alpha^{tr} \beta),$$

whereas the matrix pairing of $\mathbb{M}_{m,n}$ and $\mathbb{M}_{p,q}$ is just the Kronecker product

$$\langle\langle \alpha, \beta \rangle\rangle = [\alpha_{i,j} \beta_{k,l}] = \alpha \otimes \beta . \tag{1.1.28}$$

If we let $V' = V^d$ be the dual space of V with the usual duality pairing $\langle v, f\rangle = f(v)$, then the matrix pairing

$$\langle\!\langle \cdot, \cdot \rangle\!\rangle : \mathbb{M}_p(V) \times \mathbb{M}_q(V^d) \to \mathbb{M}_{p\times q} \tag{1.1.29}$$

is given by the useful formula

$$\langle\!\langle v, f \rangle\!\rangle = [f_{k,l}(v_{i,j})] = [f(v_{i,j})] = f_p(v)\,. \tag{1.1.30}$$

Given linear spaces V and W, we identify an $r \times r$ matrix $\varphi = [\varphi_{g,h}]$ of linear mappings $\varphi_{g,h} : V \to W$ with a linear map $\varphi : V \to \mathbb{M}_r(W)$ by letting

$$\varphi(v) = [\varphi_{g,h}(v)]\,. \tag{1.1.31}$$

This provides us with a canonical identification

$$\mathbb{M}_r(\mathbb{L}(V, W)) \cong \mathbb{L}(V, \mathbb{M}_r(W))\,. \tag{1.1.32}$$

1.2 SCHATTEN MATRIX NORMS

Schatten and von Neumann were the first to systematically investigate the matrix analogues of the Banach spaces ℓ_p^n (for the latter, see §A.2). We shall restrict our attention to the cases $p = 1, 2, \infty$. The norm $\|\cdot\|_\infty$ (which we usually denote by $\|\cdot\|$) is determined by the customary identification of $\mathbb{M}_{m,n}$ with $\mathcal{B}(\ell_2^n, \ell_2^m)$, $\|\cdot\|_2$ is the *Hilbert–Schmidt norm*

$$\|\alpha\|_2 = [\text{trace }(\alpha^*\alpha)]^{1/2} = \left[\sum |\alpha_{ij}|^2\right]^{1/2}, \tag{1.2.1}$$

and $\|\cdot\|_1$ is the *trace class norm*

$$\|\alpha\|_1 = \text{trace }(|\alpha|),$$

where $|\alpha| = (\alpha^*\alpha)^{1/2}$. We write $S_\infty(m,n) = M_{m,n}$, $S_2(m,n) = HS_{m,n}$ and $S_1(m,n) = T_{m,n}$ for the linear space $\mathbb{M}_{m,n}$ with the norms $\|\cdot\|$, $\|\cdot\|_2$, and $\|\cdot\|_1$, respectively, and we let $S_\infty(n) = M_n$, $S_2(n) = HS_n$, and $S_1(n) = T_n$ denote $\mathbb{M}_n$ with these different norms. We shall often use the less awkward notation $M_{m,n}$ and M_n even when we are only interested in the underlying linear spaces.

We can use the scalar pairing (1.1.26) to identify T_n with the Banach dual of M_n because we have the relation $\|\alpha\|_1 = \sup\{|\langle\alpha,\beta\rangle| : \|\beta\| \leq 1\}$. Since the alternative pairing

$$M_n \times T_n \to \mathbb{C} : (\alpha, \beta) \mapsto \text{trace}\,(\alpha\beta) = \sum_{i,j} \alpha_{i,j}\beta_{j,i} \tag{1.2.2}$$

is often used to identify T_n with the Banach dual of M_n, a few words are in order. To see that both conventions determine the same norm on T_n, it suffices to show that the transpose mapping

$$\mathbf{t} : M_n \to M_n : \beta \mapsto \beta^{tr}$$

is isometric. This is immediate since if $\xi = (\xi_j)$ and $\eta = (\eta_j)$ are unit vectors in $\mathbb{C}^n$, then

$$\langle \beta\eta \mid \xi \rangle = \langle \beta^{tr}\overline{\xi} \mid \overline{\eta} \rangle,$$

where $\overline{\xi} = (\overline{\xi}_j)$ and $\overline{\eta} = (\overline{\eta}_j)$ are again unit vectors in $\mathbb{C}^n$. Thus for any $\alpha \in \mathbb{M}_n$,

$$\sup\{|\mathrm{trace}\,(\alpha\beta)| : \|\beta\| \leq 1\} = \sup\{|\mathrm{trace}\,(\alpha\beta^{tr})| : \|\beta\| \leq 1\}.$$

The pairing (1.2.2) has the advantage that it is consistent with the trace pairing for mapping spaces (see §A.2). We have chosen not to use it since it necessarily entails some awkward transpositions (this may be seen if one tries to reformulate Lemma 4.1.1).

The intrinsic ordering on the C^*-algebra M_n (see §A.4) is self-dual with respect to the scalar pairing

$$\langle \cdot , \cdot \rangle : M_n \times M_n \to \mathbb{C},$$

in the sense that $\alpha \geq 0$ if and only if $\langle \alpha, \beta \rangle \geq 0$ for all $\beta \geq 0$. To see this let us suppose that $\beta \geq 0$, or equivalently, that $\beta = \delta^*\delta$ for some matrix δ. It follows that $\beta^{tr} = \delta^{tr}\delta^{tr*} \geq 0$, and if $\alpha = \gamma^*\gamma$, then

$$\langle \alpha, \beta \rangle = \mathrm{trace}\,((\gamma^*\gamma)(\delta^*\delta)^{tr}) = \mathrm{trace}\,(\delta^{tr*}\gamma^*\gamma\delta^{tr}) \geq 0$$

(the trace functional is positive on M_n). Conversely let us suppose that $\langle \alpha, \beta \rangle \geq 0$ for all $\beta \geq 0$. Given a unit vector $\xi \in \mathbb{C}^n$, the projection e on $\mathbb{C}\xi$ is positive, and thus,

$$\langle \alpha\xi \mid \xi \rangle = \langle \alpha, e \rangle = \mathrm{trace}\,(\alpha e^{tr}) \geq 0\,.$$

We let $\mathbf{S}_n$ denote the set of *density matrices* in T_n; that is, the matrices α for which $\|\alpha\|_1 = 1$ and $\alpha \geq 0$. Equivalently, $\mathbf{S}_n$ is the set of states on the C^*-algebra M_n. It may be regarded as the non-commutative analogue of $\mathbf{P}_n$, the set of *probability measures* on $\{1, \ldots, n\}$.

Given $\alpha \in M_{m,n}$, $\alpha^*\alpha \in M_n^+$ has a unique square root $|\alpha| = (\alpha^*\alpha)^{1/2}$. We refer to the following as the *polar decomposition theorem.*

Theorem 1.2.1 *For any $\alpha \in M_{m,n}$, there exists a partial isometry*

$$\nu : \mathbb{C}^n \to \mathbb{C}^m$$

such that $\alpha = \nu\,|\alpha|$, ν maps $(\ker\alpha)^\perp$ *onto* range α, $\ker\nu = \ker\alpha$, *and $\nu^*\alpha = |\alpha|$. For any $\alpha \in M_{m,n}$ (respectively, $\alpha \in T_{m,n}$, or $\alpha \in HS_{m,n}$), $|\alpha| \in M_n$ (respectively, $|\alpha| \in T_n$, or $|\alpha| \in HS_n$) has the same norm as α. If $m = n$, then there exists a unitary $\mu \in M_n$ such that $\alpha = \mu\,|\alpha|$.*

Proof For any $\eta \in \mathbb{C}^n$,

$$\|\,|\alpha|\,\eta\|^2 = \langle |\alpha|\,\eta \mid |\alpha|\,\eta \rangle = \langle \alpha^*\alpha\eta \mid \eta \rangle = \|\alpha\eta\|^2.$$

Thus we can define an isometry $\nu : \mathrm{range}\,|\alpha| \to \mathrm{range}\,\alpha \subseteq \mathbb{C}^m$ by letting

$$\nu(|\alpha|\,\eta) = \alpha(\eta)\,.$$

ν can be extended to a partial isometry of $\mathbb{C}^n$ into $\mathbb{C}^m$ by letting $\nu\eta = 0$ for all

$$\eta \in (\text{range}\,|\alpha|)^{\perp} = \ker\,|\alpha| = \ker\alpha.$$

It is easy to verify that $\alpha = \nu\,|\alpha|$. Furthermore, $\nu^*\nu \in M_m$ is the orthogonal projection onto the range of α, and thus $\nu^*\alpha = |\alpha|$. If $m = n$, then we may find an isometry

$$\nu' : \ker\alpha \cong (\text{range}\ \alpha)^{\perp},$$

and if we define the unitary operator μ by

$$\mu(\xi + \eta) = \nu\xi + \nu'\eta$$

for $\xi \in (\ker\alpha)^{\perp}$ and $\eta \in \ker\alpha$, then we have $\alpha = \mu\,|\alpha|$.

If $\alpha \in M_{m,n}$, then we have

$$||\alpha|| = ||\nu\,|\alpha|\,|| \leq ||\,|\alpha|\,|| = ||\nu^*\alpha|| \leq ||\alpha||,$$

and thus $||\,|\alpha|\,|| = ||\alpha||$. It follows from the definitions that

$$||\,|\alpha|\,||_1 = \text{trace}\,(|\alpha|) = ||\alpha||_1$$

and

$$||\,|\alpha|\,||_2 = \left[\text{trace}\,(|\alpha|^2)\right]^{1/2} = [\text{trace}\,(\alpha^*\alpha)]^{1/2} = ||\alpha||_2\,. \qquad \square$$

From the classical Schwarz inequality,

$$|\text{trace}\ (\gamma^*\delta)| = \left|\sum \overline{\gamma}_{i,j}\delta_{i,j}\right| \leq ||\delta||_2\,||\gamma||_2\,. \tag{1.2.3}$$

This may be used to prove the following *non-commutative Hölder inequality.*

Proposition 1.2.2 *For any $\delta, \gamma \in M_n$,*

$$||\gamma^*\delta||_1 \leq ||\gamma||_2||\delta||_2\,.$$

For any $\beta \in T_n$, there exist $\delta, \gamma \in HS_n$ such that

$$\beta = \gamma^*\delta \quad \textit{and} \quad ||\beta||_1 = ||\delta||_2||\gamma||_2\,.$$

Proof Given $\delta, \gamma \in HS_n$, we let $\beta = \gamma^*\delta$. It follows from the polar decomposition theorem that there exists a partial isometry $\nu \in M_n$ such that

$$\beta = \nu|\beta| \ \text{ and } \ |\beta| = \nu^*\beta\,.$$

Therefore, from (1.2.3),

$$\begin{aligned}||\beta||_1 &= \text{trace}\,(|\beta|) = \text{trace}\,(\nu^*\beta) = \text{trace}\,((\gamma\nu)^*\delta)\\ &\leq [\text{trace}\,((\gamma\nu)^*\gamma\nu)]^{1/2}\,[\text{trace}(\delta^*\delta)]^{1/2} \leq ||\gamma||_2||\delta||_2\,.\end{aligned}$$

On the other hand, if $\beta \in T_n$, then it follows from the polar decomposition theorem that there exists a partial isometry $\nu \in M_n$ such that

$$\beta = \nu|\beta|$$

and $\nu^*\nu$ is the projection onto the range of $|\beta|$. If we let $\gamma = (\nu|\beta|^{1/2})^*$ and $\delta = |\beta|^{1/2}$, then

$$\beta = \gamma^*\delta\,.$$

Since $\delta^*\delta = |\beta|$ and $\gamma^*\gamma = (\nu\,|\beta|)^{1/2}(|\beta|^{1/2}\,\nu^*) = \nu\,|\beta|\,\nu^*$,

$$\|\delta\|_2\|\gamma\|_2 = [\mathrm{trace}\,(|\beta|)]^{1/2}\,[\mathrm{trace}\,(\nu\,|\beta|\,\nu^*)]^{1/2} = \mathrm{trace}\,(|\beta|) = \|\beta\|_1\,,$$

where we have used the fact that $\nu^*\nu\,|\beta| = |\beta|$, and thus

$$\mathrm{trace}\,(\nu\,|\beta|\,\nu^*) = \mathrm{trace}\,(\nu^*\nu\,|\beta|) = \mathrm{trace}\,(|\beta|).$$ □

1.3 MATRICES OF OPERATORS

Given an indexed family of Hilbert spaces H_s ($s \in \mathfrak{s}$), the algebraic direct sum $\bigoplus_{s\in\mathfrak{s}} H_s$ is a pre-Hilbert space with the inner product determined by

$$\langle \eta \mid \xi\rangle = \sum_{s\in\mathfrak{s}} \langle \eta_s \mid \xi_s\rangle\,.$$

We also write $H = \bigoplus_{s\in\mathfrak{s}} H_s$ for the Hilbert space completion, which may be identified with $\ell_2(\mathfrak{s}, H_s)$ (see §A.2). Given Hilbert spaces K_s ($s \in \mathfrak{s}$) and operators $b_s : H_s \to K_s$ with $\sup\{\|b_s\|\} < \infty$, if we let K be the Hilbert space $\bigoplus_{s\in\mathfrak{s}} K_s$, then the $\mathfrak{s}$-tuple $(b_s)_{s\in\mathfrak{s}}$ determines a bounded operator from H to K, which is also denoted by $b = (b_s)_{s\in\mathfrak{s}}$. We have

$$\|b\| = \sup\{\|b_s\| : s \in \mathfrak{s}\}\,. \tag{1.3.1}$$

If H and K are Hilbert spaces, then the algebraic tensor product $H \otimes K$ is a pre-Hilbert space with the inner product determined by

$$\langle \eta_1 \otimes \eta_2 \mid \xi_1 \otimes \xi_2\rangle = \langle \eta_1 \mid \xi_1\rangle\langle \eta_2 \mid \xi_2\rangle\,.$$

We also write $H \otimes K$ for the completion of this tensor product. Given Hilbert spaces H_k and K_k and bounded operators $b_k : H_k \to K_k$ where $k = 1, 2$, the corresponding linear mapping of the algebraic tensor products

$$b_1 \otimes b_2 : H_1 \otimes H_2 \to K_1 \otimes K_2$$

satisfies

$$\|b_1 \otimes b_2\| = \|b_1\|\|b_2\|, \tag{1.3.2}$$

and therefore extends to a bounded operator $b_1 \otimes b_2$ of the Hilbert space tensor products with the same property.

Given a Hilbert space H, and $n \in \mathbb{N}$, we have a natural identification

$$\mathbb{M}_n(\mathcal{B}(H)) \cong \mathcal{B}(H^n), \tag{1.3.3}$$

determined by matrix multiplication. To be more explicit, if we write our vectors $\eta \in H^n$ as column matrices, a matrix $b = [b_{i,j}] \in \mathbb{M}_n(\mathcal{B}(H))$ determines an operator on H^n by

$$b\eta = \begin{bmatrix} b_{1,1} & \dots & b_{1,n} \\ \vdots & \ddots & \vdots \\ b_{n,1} & \dots & b_{n,n} \end{bmatrix} \begin{pmatrix} \eta_1 \\ \vdots \\ \eta_n \end{pmatrix} = \begin{pmatrix} \sum b_{1,j}\eta_j \\ \vdots \\ \sum b_{n,j}\eta_j \end{pmatrix}.$$

Equivalently, this corresponds to the action of $\mathbb{M}_n \otimes \mathcal{B}(H)$ on $\mathbb{C}^n \otimes H$ determined by

$$(\alpha \otimes b)(\zeta \otimes \eta) = \alpha(\zeta) \otimes b(\eta).$$

Thus we may regard $\mathbb{M}_n(\mathcal{B}(H))$ as a C^*-algebra, which we henceforth denote by $M_n(\mathcal{B}(H))$. With the above conventions $\mathcal{B}(H)$ is *matrix normed* and *matrix ordered*, in the sense that each matrix space has a corresponding norm and ordering. The norms on these spaces are linked by certain fundamental relations.

Proposition 1.3.1 *Let H be a Hilbert space. Then for all $b \in M_m(\mathcal{B}(H))$, $c \in M_n(\mathcal{B}(H))$, $\alpha \in M_{n,m}$, and $\beta \in M_{m,n}$,*

- **M1** $\|b \oplus c\| = \max\{\|b\|, \|c\|\}$,
- **M2** $\|\alpha b \beta\| \leq \|\alpha\| \, \|b\| \, \|\beta\|$.

Proof **M1** is a special case of (1.3.1). Under the identification

$$M_n(\mathcal{B}(H)) \cong \mathcal{B}(\mathbb{C}^n \otimes H),$$

we have that $\alpha b \beta$ corresponds to $(\alpha \otimes I)b(\beta \otimes I)$, and thus from (1.3.2),

$$\|\alpha b \beta\| = \|(\alpha \otimes I)b(\beta \otimes I)\| \leq \|\alpha\| \, \|b\| \, \|\beta\|.$$

□

It is a simple exercise to verify that the matrix orderings for $\mathcal{B}(H)$ are similarly interrelated. For any $b \in M_m(\mathcal{B}(H))$, $c \in M_n(\mathcal{B}(H))$, and $\alpha \in M_{m,n}$:

- **O1** if $b \geq 0$ and $c \geq 0$, then $b \oplus c \geq 0$,
- **O2** if $b \geq 0$, then $\alpha^* b \alpha \geq 0$.

From elementary spectral theory, the natural ordering on $\mathcal{B}(H)$ determines the norms of self-adjoint elements (see (A.4.2)). By contrast, the matrix orderings for $\mathcal{B}(H)$ determine the norms of *arbitrary* (i.e. not necessarily self-adjoint) matrices of operators.

Proposition 1.3.2 *For any Hilbert space H and operator $b \in M_n(\mathcal{B}(H))$,*

$$\|b\| \leq 1 \text{ if and only if } \begin{bmatrix} I_n & b \\ b^* & I_n \end{bmatrix} \geq 0. \tag{1.3.4}$$

Proof Given $b \in M_n(\mathcal{B}(H))$ with $\|b\| \le 1$, it is evident that

$$\tilde{b} = \begin{bmatrix} 0 & b \\ b^* & 0 \end{bmatrix}$$

is a self-adjoint element in $M_{2n}(\mathcal{B}(H))$ such that

$$\|\tilde{b}\| = \max\{\|b\|, \|b^*\|\} = \|b\| \le 1\,.$$

It follows from (A.4.2) that

$$\begin{bmatrix} I_n & b \\ b^* & I_n \end{bmatrix} = I_{2n} + \tilde{b} \ge 0\,.$$

Conversely, if

$$I_{2n} + \tilde{b} = \begin{bmatrix} I_n & b \\ b^* & I_n \end{bmatrix} \ge 0,$$

then

$$I_{2n} - \tilde{b} = \begin{bmatrix} I_n & -b \\ -b^* & I_n \end{bmatrix} = \begin{bmatrix} I_n & 0 \\ 0 & -I_n \end{bmatrix} \begin{bmatrix} I_n & b \\ b^* & I_n \end{bmatrix} \begin{bmatrix} I_n & 0 \\ 0 & -I_n \end{bmatrix} \ge 0,$$

or equivalently, $-I_{2n} \le \tilde{b} \le I_{2n}$. Therefore, from (A.4.2),

$$\|b\| = \|\tilde{b}\| \le 1\,. \qquad \square$$

Corollary 1.3.3 *For any Hilbert space H and operator $b \in \mathcal{B}(H)$,*

$$\|b\| = \inf\left\{\alpha > 0 : \begin{bmatrix} \alpha I & b \\ b^* & \alpha I \end{bmatrix} \ge 0\right\}. \tag{1.3.5}$$

□

Corollary 1.3.4 *If b is a bounded operator on a Hilbert space H, then*

$$\begin{bmatrix} I & b \\ b^* & 0 \end{bmatrix} \ge 0$$

implies that $b = 0$.

Proof Given $\varepsilon > 0$, we have from the inequality that

$$0 \le \begin{bmatrix} \varepsilon^{1/2} I & 0 \\ 0 & \varepsilon^{-1/2} I \end{bmatrix} \begin{bmatrix} I & b \\ b^* & 0 \end{bmatrix} \begin{bmatrix} \varepsilon^{1/2} I & 0 \\ 0 & \varepsilon^{-1/2} I \end{bmatrix} = \begin{bmatrix} \varepsilon I & b \\ b^* & 0 \end{bmatrix} \le \begin{bmatrix} \varepsilon I & b \\ b^* & \varepsilon I \end{bmatrix},$$

and thus $\|b\| \le \varepsilon$. Since $\varepsilon > 0$ is arbitrary, $b = 0$. □

The following provides additional relations between matrix orderings. We shall not need these until §5.4.

Proposition 1.3.5 *For any Hilbert space H and self-adjoint operators $a, b \in \mathcal{B}(H)$, $-a \le b \le a$ implies that*

$$\begin{bmatrix} a & b \\ b & a \end{bmatrix} \ge 0. \tag{1.3.6}$$

On the other hand, if a_1 *and* a_2 *are positive operators with*

$$\begin{bmatrix} a_1 & b \\ b & a_2 \end{bmatrix} \geq 0, \tag{1.3.7}$$

then $-\tilde{a} \leq b \leq \tilde{a}$, *where* $\tilde{a} = \frac{1}{2}(a_1 + a_2)$.

If b *is an arbitrary operator and* $b = (b_1 - b_2) + \mathrm{i}(b_3 - b_4)$, *where* $b_j \geq 0$, *then*

$$\begin{bmatrix} \sum b_j & b \\ b^* & \sum b_j \end{bmatrix} \geq 0. \tag{1.3.8}$$

Conversely, if one has positive operators a_1 *and* a_2 *with*

$$D = \begin{bmatrix} a_1 & b \\ b^* & a_2 \end{bmatrix} \geq 0,$$

then we can write b *as an explicit combination of positive operators derived from* D*:*

$$b = \frac{1}{4}\sum_{k=0}^{3} \mathrm{i}^k \left[1\ \mathrm{i}^k\right] D \left[1\ \mathrm{i}^k\right]^*. \tag{1.3.9}$$

Proof If $-a \leq b \leq a$, then

$$0 \leq \begin{bmatrix} 1 & -1 \\ 1 & 1 \end{bmatrix} \begin{bmatrix} a+b & 0 \\ 0 & a-b \end{bmatrix} \begin{bmatrix} 1 & 1 \\ -1 & 1 \end{bmatrix} = 2\begin{bmatrix} a & b \\ b & a \end{bmatrix}.$$

On the other hand, if we are given (1.3.7), then

$$0 \leq \left[1\ -1\right] \begin{bmatrix} a_1 & b \\ b & a_2 \end{bmatrix} \begin{bmatrix} 1 \\ -1 \end{bmatrix} = a_1 + a_2 - 2b$$

and

$$0 \leq \left[1\ 1\right] \begin{bmatrix} a & b \\ b & a \end{bmatrix} \begin{bmatrix} 1 \\ 1 \end{bmatrix} = a_1 + a_2 + 2b,$$

and thus $-\tilde{a} \leq b \leq \tilde{a}$.

For any operator b, if $b = (b_1 - b_2) + \mathrm{i}(b_3 - b_4)$, where $b_j \geq 0$ and $a = \sum b_j$, then we may write

$$\begin{bmatrix} a & b \\ b^* & a \end{bmatrix} = \begin{bmatrix} b_1 + b_2 & b_1 - b_2 \\ b_1 - b_2 & b_1 + b_2 \end{bmatrix} + \begin{bmatrix} b_3 + b_4 & \mathrm{i}(b_3 - b_4) \\ -\mathrm{i}(b_3 - b_4) & b_3 + b_4 \end{bmatrix}.$$

The inequalities

$$\begin{bmatrix} b_1 + b_2 & b_1 - b_2 \\ b_1 - b_2 & b_1 + b_2 \end{bmatrix} \geq 0$$

and

$$\begin{bmatrix} b_3 + b_4 & \mathrm{i}(b_3 - b_4) \\ -\mathrm{i}(b_3 - b_4) & b_3 + b_4 \end{bmatrix} = \begin{bmatrix} \mathrm{i} & 0 \\ 0 & 1 \end{bmatrix} \begin{bmatrix} b_3 + b_4 & b_3 - b_4 \\ b_3 - b_4 & b_3 + b_4 \end{bmatrix} \begin{bmatrix} -\mathrm{i} & 0 \\ 0 & 1 \end{bmatrix} \geq 0$$

are evident from (1.3.6), and the last relation (1.3.9) follows from the obvious calculation. □

The Schatten matrix norms have infinite-dimensional generalizations. Let us consider a Hilbert space H with an orthonormal basis $\{e_s\}_{s\in\mathfrak{s}}$. Given an operator $b \in \mathcal{B}(H)^+$, we define

$$\operatorname{trace}(b) = \sum_{s\in\mathfrak{s}} \langle be_s \mid e_s\rangle$$

(this might be infinite). If $b \in \mathcal{B}(H)$, then the *trace class norm* and the *Hilbert–Schmidt norm* are defined by

$$\|b\|_1 = \operatorname{trace}(|b|) \tag{1.3.10}$$

and

$$\|b\|_2 = [\operatorname{trace}(b^*b)]^{1/2}, \tag{1.3.11}$$

respectively. We let $\mathcal{FB}(H)$ denote the finite-rank bounded operators on H, (see §A.2) and we define the *Schatten spaces* $\mathcal{S}_p(H)$ for $p = 1, 2, \infty$ by

$$\mathcal{S}_1(H) = \mathcal{T}(H) = \{b \in \mathcal{B}(H) : \|b\|_1 < \infty\} \tag{1.3.12}$$

with the norm $\|\cdot\|_1$,

$$\mathcal{S}_2(H) = \mathcal{HS}(H) = \{b \in \mathcal{B}(H) : \|b\|_2 < \infty\} \tag{1.3.13}$$

with the norm $\|\cdot\|_2$, and the compact operators

$$\mathcal{S}_\infty(H) = \mathcal{K}(H) = \text{the norm closure of } \mathcal{FB}(H) \tag{1.3.14}$$

(see §A.2) with the operator norm $\|\cdot\|$, and we use the notation $\mathcal{S}_p = \mathcal{S}_p(\ell_2)$ for $p = 1, 2, \infty$. We have the natural isometric dualities

$$\mathcal{S}_\infty(H)^* \cong \mathcal{S}_1(H),\ \ \mathcal{S}_1(H)^* \cong \mathcal{B}(H),\ \ \mathcal{S}_2(H)^* \cong \mathcal{S}_2(H). \tag{1.3.15}$$

Each of these dualities is determined by the absolutely convergent pairing

$$\langle b, c\rangle = \sum_{s,t} b_{s,t}c_{s,t} \tag{1.3.16}$$

(our discussion of (1.2.2) applies in the infinite case as well).

It is occasionally useful to generalize these notions to operators between different Hilbert spaces. Given Hilbert spaces H and K and a bounded operator $b : H \to K$, we have $b^*b \in \mathcal{B}(H)$, and therefore we may use (1.3.10) and (1.3.11) to define $\|b\|_1$ and $\|b_2\|$. We let

$$\mathcal{S}_1(H,K) = \mathcal{T}(H,K) = \{b \in \mathcal{B}(H,K) : \|b\|_1 < \infty\}, \tag{1.3.17}$$

$$\mathcal{S}_2(H,K) = \mathcal{HS}(H,K) = \{b \in \mathcal{B}(H,K) : \|b\|_2 < \infty\}, \tag{1.3.18}$$

$$\mathcal{S}_\infty(H,K) = \mathcal{K}(H,K) = \text{the norm closure of } \mathcal{FB}(H,K), \tag{1.3.19}$$

where $\mathcal{FB}(H,K)$ is the space of all finite-rank bounded operators from H into K.

If we identify H with $\ell_2(\mathfrak{s})$, then we may regard $\mathcal{B}(H)$ and the spaces $\mathcal{S}_p(H)$ $(p = 1, 2, \infty)$ as linear spaces of the matrices $[b_{s,t}]$, where

$$b_{s,t} = \langle be_s \mid e_t \rangle \;\; (s, t \in \mathfrak{s}).$$

With these matrix conventions, we also use the notation $M_{\mathfrak{s}} = \mathcal{B}(H)$, $T_{\mathfrak{s}} = \mathcal{T}(H)$, $K_{\mathfrak{s}} = \mathcal{K}(H)$, and $HS_{\mathfrak{s}} = \mathcal{HS}(H)$. The Hilbert–Schmidt norm on the latter space is given by

$$\|b\|_2 = \left(\sum_{s,t} |b_{s,t}|^2 \right)^{1/2} . \tag{1.3.20}$$

We conclude with the simple remark that the usual identification of vectors in $\ell_2(\mathfrak{s})$ with column matrices on that Hilbert space is isometric since if we are given $\xi = (\xi_s) \in \ell_2(\mathfrak{s})$ and we let b denote the corresponding column matrix, then

$$\|b\|^2 = \|b^* b\| = \sum_{\mathfrak{s}} \overline{\xi}_s \xi_s = \|\xi\|^2 . \tag{1.3.21}$$

A similar remark applies to the representation of vectors with row matrices.

Part I

Examples and Three Basic Theorems

2

The representation theorem

The representation theorem characterizes the operator spaces in terms of their matrix norms. With it one can prove that the category of operator spaces is closed under a wide range of constructions, including quotients, mapping spaces, and tensor products. We present an elementary proof in §2.3 which is based upon a simple minimax result.

When comparing the properties of normed spaces and operator spaces, we shall generally use the letters $E, F, \ldots$ for Banach spaces, and $V, W, \ldots$ for operator spaces.

2.1 CONCRETE AND ABSTRACT OPERATOR SPACES

We define a *concrete function space* on a set $\mathfrak{s}$ to be a linear subspace E of $\ell_\infty(\mathfrak{s})$. All normed spaces arise in this fashion, which is to say that any normed space E is isometric to a function space. This result is a simple consequence of the Hahn–Banach theorem. To see this, let us suppose that we are given an element $x \in E$. From the Hahn–Banach theorem there is a linear functional $f \in E^*$ with $\|f\| = 1$ for which $|f(x)| = \|x\|$. Thus if we let $\mathfrak{s}$ be the closed unit ball of E^*, then we may define an isometry

$$\Phi : E \to \ell_\infty(\mathfrak{s}) \tag{2.1.1}$$

by letting $\Phi(v)(f) = f(v)$ for $f \in \mathfrak{s}$.

From the preceding analysis, we may regard a normed linear space as just an *abstract function space*, and in turn, we may consider a function space to be an *isometric representation* (or '*realization*') of a normed space. The advantage of the abstract (or axiomatic) description is its greater flexibility. If, for example, we are given a function space E on a set $\mathfrak{s}$ and a closed subspace $N \subseteq E$, it is not apparent from the definition that the dual E^* and the quotient E/N are again function spaces. On the other hand, it is immediate that these are normed spaces, and thus they can be represented as function spaces.

The transition to operator spaces is motivated by the following observation. Given a normed space E and $n \in \mathbb{N}$, we recall that $\ell_\infty^n(E)$ consists of the linear space E^n of n-tuples $x = (x_1, \ldots, x_n)$ $(x_j \in E)$ together with the norm

$$\|x\|_\infty = \max\{\|x_j\| : 1 \leq j \leq n\} \tag{2.1.2}$$

(see §A.2). If E is represented as a function space $E \subseteq \ell_\infty(\mathfrak{s})$, then this norm is also determined by the inclusion

$$E^n \subseteq \ell_\infty(\mathfrak{s} \times n), \tag{2.1.3}$$

where we let n stand for the set $\{1, \ldots, n\}$, and thus $\mathfrak{s} \times n$ is a disjoint union of n copies of the set $\mathfrak{s}$.

Replacing functions by operators, and in particular $\ell_\infty(\mathfrak{s})$ by $\mathcal{B}(H)$, we define a (*concrete*) *operator space* V on a Hilbert space H to be a linear subspace of $\mathcal{B}(H)$. Our goal in this section and the next is to axiomatically characterize these spaces. In order to accomplish this, we must first identify the relevant underlying structure. This is done by considering the matrix analogue of (2.1.3). The natural inclusion

$$\mathbb{M}_n(V) \subseteq M_n(\mathcal{B}(H)) = \mathcal{B}(H^n)$$

determines a norm $\|\cdot\|_n$ on $\mathbb{M}_n(V)$, and we let $M_n(V)$ denote the corresponding normed space. In contrast to the situation for vector norms, *there is no analogue of the formula* (2.1.2) relating the norm $\|v\|_n$ of a matrix $v = [v_{i,j}]$ to the norms of its entries $\|v_{i,j}\|$. To put it another way, the matrix norms are not implicit in the norm structure on V, and there is generally a wide variety of essentially distinct operator spaces having the same underlying normed space. Fixing an operator space V, the corresponding matrix norms inherit the properties described in Proposition 1.3.1. This leads us to the obvious axiomatization in terms of the given matrix norms.

We define a *matrix norm* $\|\cdot\|$ on a linear space V to be an assignment of a norm $\|\cdot\|_n$ on the matrix space $\mathbb{M}_n(V)$ for each $n \in \mathbb{N}$. An *abstract operator space* is a linear space V together with a matrix norm $\|\cdot\|$ for which

- **M1** $\|v \oplus w\|_{m+n} = \max\{\|v\|_m, \|w\|_n\}$ and
- **M2** $\|\alpha v \beta\|_n \leq \|\alpha\| \, \|v\|_m \, \|\beta\|$

for all $v \in \mathbb{M}_m(V)$, $w \in \mathbb{M}_n(V)$ and $\alpha \in M_{n,m}$, $\beta \in M_{m,n}$. We let $M_n(V)$ denote $\mathbb{M}_n(V)$ with the given norm $\|\cdot\| = \|\cdot\|_n$ (we usually omit the subscript n). We say that a matrix norm is an *operator space matrix norm* if it satisfies **M1** and **M2**. As in the classical theory, we shall be primarily interested in operator spaces V for which the norm on V is complete (we shall see shortly that this implies that each of the normed spaces $M_n(V)$ is complete). Beginning with Part II, this will be a standing assumption.

Given any abstract operator space V, it follows from condition **M1** that the natural mapping

$$v \mapsto v \oplus 0 = \begin{bmatrix} v & 0 \\ 0 & 0 \end{bmatrix}$$

is an isometry of $M_n(V)$ into $M_{n+1}(V)$, and thus we obtain a norm on $\mathbb{M}_\infty^{\text{fin}}(V)$ (see (1.1.12)) which we again refer to as the *operator space matrix norm* on V. We use the expressions *closed matrix unit ball* and *open matrix unit ball* of V for the sets $\mathbb{M}_\infty^{\text{fin}}(V)_{\|\cdot\|\leq 1}$ and $\mathbb{M}_\infty^{\text{fin}}(V)_{\|\cdot\|<1}$, respectively.

It is evident from the preceding discussion that any concrete operator space is an abstract operator space. Given Hilbert spaces H and K, we may use the identifications

$$\mathbb{M}_n(\mathcal{B}(H,K)) \cong \mathcal{B}(H^n, K^n) \tag{2.1.4}$$

to determine a matrix norm on $\mathcal{B}(H,K)$, and it follows from the argument for Proposition 1.3.1 that we have **M1** and **M2**; hence $\mathcal{B}(H,K)$ is an operator space. Alternatively, we may identify $\mathcal{B}(H,K)$ with the subspace of matrices of the form

$$\begin{bmatrix} 0 & 0 \\ b & 0 \end{bmatrix}$$

in $\mathcal{B}(H \oplus K)$.

Any normed space E is isometric to a linear space of operators. To see this, it suffices to use the embedding (2.1.1) together with the isometry $\ell_\infty(\mathfrak{s}) \hookrightarrow M_\mathfrak{s}$ determined by sending a bounded sequence to the corresponding diagonal matrix. In §3.3 we shall prove that the corresponding abstract operator space, which is denoted by $\min E$, does not depend on the embedding, and that this provides a natural functor from the category of normed spaces into the corresponding category of abstract operator spaces.

If $\mathcal{A}$ is a C^*-algebra, then the norm of $\mathcal{A}$ can be determined by the underlying $*$-algebraic structure on $\mathcal{A}$ (see §A.4). If we fix a faithful $*$-representation of $\mathcal{A}$ on a Hilbert space H, then we may regard $M_n(\mathcal{A})$ as a C^*-algebra of operators on H^n. Since the $*$-algebraic structure of $M_n(\mathcal{A})$ is uniquely determined by that on $\mathcal{A}$, we conclude that the corresponding norm on $M_n(\mathcal{A})$ does not depend on the representation of $\mathcal{A}$. With this matrix norm, $\mathcal{A}$ is an operator space, and we refer to this as the *canonical operator space structure* on $\mathcal{A}$.

Given an abstract operator space V and a matrix $v \in M_n(V)$, we have from **M2** that, if $\mu \in M_n$ is unitary, then

$$\|\mu v\| \leq \|\mu\| \, \|v\| = \|v\| \leq \left\|\mu^{-1}\right\| \|\mu v\| = \|\mu v\| .$$

By symmetry the same applies to right multiplication, and we conclude that

$$\mu \text{ unitary } \Rightarrow \|\mu v\| = \|v \mu\| = \|v\| . \tag{2.1.5}$$

It follows that we may permute rows and columns of v without affecting its norm since such an operation corresponds to multiplication on the left or right by a permutation matrix. Given any finite set $\mathfrak{f}$ with n elements, we may use a bijection of $\mathfrak{f}$ onto $\{1, \ldots, n\}$ to identify the matrix space $\mathbb{M}_\mathfrak{f}(V)$ with $M_n(V)$, and in this manner we obtain a norm on $\mathbb{M}_\mathfrak{f}(V)$. Owing to the unitary invariance (2.1.5), this norm does not depend on the identification of $\mathfrak{f}$ with $\{1,\ldots,n\}$. We let $M_{\mathfrak{f}}(V)$ denote the corresponding normed space.

Proposition 2.1.1 *Let us suppose that V is an abstract operator space. For any matrices $v \in M_n(V)$ and $\alpha \in M_p$,*

$$\|v \otimes \alpha\| = \|\alpha \otimes v\| = \|v\| \, \|\alpha\| . \tag{2.1.6}$$

Proof We may suppose that $\alpha = \mu\,|\alpha|$, where μ is unitary (see Theorem 1.2.1). From the finite-dimensional spectral theorem, there is a unitary matrix $\lambda \in M_p$ and scalars $c_1 \geq \cdots \geq c_p \geq 0$ such that $\|\alpha\| = c_1$ and

$$|\alpha| = \lambda^*(c_1 \oplus \cdots \oplus c_p)\lambda.$$

If we let

$$\tilde{v} = (c_1 \oplus \cdots \oplus c_p) \otimes v = c_1 v \oplus \cdots \oplus c_p v \in M_{pn}(V),$$

then it follows from (2.1.5) and **M1** that

$$\begin{aligned}\|\alpha \otimes v\| &= \|\mu\lambda^*(c_1 \oplus \cdots \oplus c_p)\lambda \otimes v\| \\ &= \|(\mu\lambda^* \otimes I_n)\tilde{v}(\lambda \otimes I_n)\| = \|\tilde{v}\| = \|\alpha\| \, \|v\| .\end{aligned}$$

Since $v \otimes \alpha \cong \alpha \otimes v$, we are done. □

Rectangular matrices over an abstract operator space V also carry a distinguished norm. Given arbitrary $m, n \in \mathbb{N}$, there is a natural norm on $\mathbb{M}_{m,n}(V)$ obtained by regarding $\mathbb{M}_{m,n}(V)$ as a subspace of $M_p(V)$, where $p = \max\{m, n\}$. We use the notation $M_{m,n}(V)$ to denote the linear space $\mathbb{M}_{m,n}(V)$ with the indicated norm. It is easy to verify that **M1** and **M2** hold for rectangular matrices, provided that the matrices in **M2** have multiplicatively compatible dimensions. As above, we obtain a corresponding well-defined normed space $M_{\mathfrak{f},\mathfrak{g}}(V)$ for any finite sets $\mathfrak{f}$ and $\mathfrak{g}$.

For any $v = [v_{i,j}] \in M_n(V)$, we have from (1.1.10) and (1.1.11) that

$$\|v_{i,j}\| = \|E_i v E_j^*\| \leq \|v\| \tag{2.1.7}$$

and

$$\|v\| = \left\|\sum E_i^* v_{i,j} E_j\right\| \leq \sum_{i,j} \|v_{i,j}\| . \tag{2.1.8}$$

Although the inequalities (2.1.7) and (2.1.8) do not determine the matrix norms, they provide important constraints on their properties. In particular, any two such norms must be equivalent on $\mathbb{M}_n(V)$ and a sequence $v(k)$ in $M_n(V)$ ($k \in \mathbb{N}$) converges if and only if the entries $v(k)_{i,j}$ converge. It follows that, if W is a closed subspace of an operator space V, then $M_n(W)$ is closed in $M_n(V)$ for each $n \in \mathbb{N}$. It is also apparent from these inequalities that V is complete if and only if each of the normed spaces $M_n(V)$ is complete.

It is evident that a linear mapping $F = [F_{i,j}] : V \to M_n$ is continuous if and only if each of the functions $F_{i,j} : V \to \mathbb{C}$ is continuous. Thus we may use the pairing

$$\langle\,\cdot\,,\,\cdot\,\rangle : \mathbb{M}_n(V) \times \mathbb{M}_n(V^*) \to \mathbb{C} : \langle v, f\rangle \mapsto \sum f_{i,j}(v_{i,j}) \qquad (2.1.9)$$

to identify the linear space $\mathbb{M}_n(V^*)$ with the Banach dual space $M_n(V)^*$. It must be stressed, however, that this pairing does not determine the natural operator space matrix norm on $\mathbb{M}_n(V^*)$. If, for example, $V = \mathbb{C}$, then the dual norm of $M_n(V) = M_n$ is the trace class norm rather than the usual operator norm on $\mathbb{M}_n$, and therefore $M_n^* = T_n$. We shall return to this topic in §4.1.

2.2 COMPLETELY BOUNDED LINEAR MAPPINGS

Given abstract operator spaces V and W and a linear mapping $\varphi : V \to W$, we recall from §1.1 that for each $n \in \mathbb{N}$, there is a corresponding linear mapping $\varphi_n : M_n(V) \to M_n(W)$ defined by

$$\varphi_n(v) = [\varphi(v_{i,j})]$$

for all $v = [v_{i,j}] \in M_n(V)$. We define the *completely bounded norm* of φ by

$$\|\varphi\|_{cb} = \sup\{\|\varphi_n\| : n \in \mathbb{N}\}$$

(this might be infinite). It is evident from **M1** that the norms $\|\varphi_n\|$ form an increasing sequence

$$\|\varphi\| \leq \|\varphi_2\| \leq \cdots \leq \|\varphi_n\| \leq \cdots \leq \|\varphi\|_{cb}\,. \qquad (2.2.1)$$

We say that φ is *completely bounded* (respectively, *completely contractive*) if $\|\varphi\|_{cb} < \infty$ (respectively, $\|\varphi\|_{cb} \leq 1$). It is a simple matter to verify that $\|\cdot\|_{cb}$ is a norm on the linear space $\mathcal{CB}(V, W)$ of completely bounded linear mappings $\varphi : V \to W$. We define φ to be a *complete isometry* if each mapping $\varphi_n : M_n(V) \to M_n(W)$ is isometric, and a *complete isomorphism* if it is a linear isomorphism with $\|\varphi\|_{cb}\,, \|\varphi^{-1}\|_{cb} < \infty$. Finally, we say that φ is a *complete quotient mapping* if each φ_n is a quotient mapping, and that φ is an *exact* complete quotient mapping if each φ_n is an exact quotient mapping (see §A.2). Given an abstract operator space V and a Hilbert space H, we say that a mapping $\varphi : V \to \mathcal{B}(H)$ is a *realization* of V if it is a completely isometric injection. We shall often use the notation $V \hookrightarrow W$ and $V \twoheadrightarrow W$ to indicate completely isometric injections and complete quotient surjections.

As in the case of normed spaces, we may use either the complete isometries or the completely bounded mappings as the morphisms of a corresponding category of abstract operator spaces.

There are some instances in which the inequalities (2.2.1) stabilize. This depends on the following important observation of Roger Smith, who used it to prove Proposition 2.2.2.

Lemma 2.2.1 *Given $m, n \in \mathbb{N}$ with $m \geq n$ and a vector $\eta \in \mathbb{C}^m \otimes \mathbb{C}^n$, there exists an isometry $\beta : \mathbb{C}^n \hookrightarrow \mathbb{C}^m$ and a vector $\tilde{\eta} \in \mathbb{C}^n \otimes \mathbb{C}^n$ for which $(\beta \otimes I_n)(\tilde{\eta}) = \eta$.*

Proof There exist unique vectors $\eta_j \in \mathbb{C}^m$ $(j = 1, \ldots, n)$ with

$$\eta = \sum_{j=1}^{n} \eta_j \otimes \varepsilon_j^{(n)}.$$

If we let $F \subseteq \mathbb{C}^m$ be the subspace spanned by the vectors η_j, then we have $\dim F \leq n \leq m$. Thus we may find an isometry $\beta : \mathbb{C}^n \hookrightarrow \mathbb{C}^m$ with image containing F. For each j, we have a unique vector $\tilde{\eta}_j \in \mathbb{C}^n$ for which $\beta(\tilde{\eta}_j) = \eta_j$. Thus if $\tilde{\eta} = \sum_{j=1}^{n} \tilde{\eta}_j \otimes \varepsilon_j^{(n)}$, then $\beta \otimes I_n(\tilde{\eta}) = \eta$. □

Proposition 2.2.2 *If V is an abstract operator space and $\varphi : V \to M_n$ is a linear map, then*

$$\|\varphi\|_{cb} = \|\varphi_n\|.$$

Proof It suffices to show that for any integer $m \geq n$, $\|\varphi_m\| \leq \|\varphi_n\|$. Given $\varepsilon > 0$, we choose an element $v \in M_m(V)$ with $\|v\| \leq 1$ for which

$$\|\varphi_m\| - \varepsilon < \|\varphi_m(v)\|,$$

and we then select unit vectors $\eta, \xi \in (\mathbb{C}^n)^m = \mathbb{C}^m \otimes \mathbb{C}^n$ such that

$$\|\varphi_m\| - \varepsilon < |\langle \varphi_m(v)\eta \mid \xi \rangle|. \tag{2.2.2}$$

From Lemma 2.2.1 there exist isometries $\alpha, \beta : \mathbb{C}^n \hookrightarrow \mathbb{C}^m$ and unit vectors $\tilde{\xi}, \tilde{\eta} \in \mathbb{C}^n \otimes \mathbb{C}^n$ for which $\xi = (\alpha \otimes I_n)(\tilde{\xi})$ and $\eta = (\beta \otimes I_n)(\tilde{\eta})$. We thus have from (1.1.9) and (1.1.23),

$$\begin{aligned}\|\varphi_m\| - \varepsilon &< |\langle \varphi_m(v)(\beta \otimes I_n)(\tilde{\eta}) \mid (\alpha \otimes I_n)(\tilde{\xi}) \rangle| \\ &= |\langle (\alpha^* \varphi_m(v)\beta)(\tilde{\eta}) \mid (\tilde{\xi}) \rangle| \\ &= |\langle \varphi_n(\alpha^* v \beta)\tilde{\eta} \mid \tilde{\xi} \rangle| \leq \|\varphi_n\|.\end{aligned}$$

Since $\varepsilon > 0$ is arbitrary, we conclude that $\|\varphi_m\| \leq \|\varphi_n\|$. □

Corollary 2.2.3 *If V is an abstract operator space, then for each linear functional $f : V \to \mathbb{C}$,*

$$\|f\|_{cb} = \|f\|.$$

□

If V is an operator space and $v \in V$ satisfies $\|v\| = 1$, then the mapping

$$\theta_v : \mathbb{C} \to V : \alpha \mapsto \alpha v$$

is completely isometric since from Proposition 2.1.1

$$\|(\theta_v)_n(\alpha)\| = \|\alpha \otimes v\| = \|\alpha\|$$

for all $\alpha \in M_n$. In particular, there is essentially only one operator space of dimension 1.

Corollary 2.2.4 *Given abstract operator spaces V and W with either V or W n-dimensional, any linear mapping $\varphi : V \to W$ satisfies*

$$\|\varphi\|_{cb} \leq n \|\varphi\|.$$

Proof Let us suppose that W has dimension n. We may select an Auerbach basis for W, which by definition is a vector basis $w_1, \ldots, w_n$ with $\|w_j\| = 1$, for which there exist $g_j \in W^*$ with $\|g_j\| = 1$ and $g_j(w_i) = \delta_{i,j}$ (see §A.2). Since

$$id_W = \sum_{j=1}^{n} \theta_{w_j} \circ g_j,$$

we have

$$\varphi = \sum_{j=1}^{n} \theta_{w_j} \circ g_j \circ \varphi,$$

where $\theta_{w_j}(\alpha) = \alpha w_j$ are complete isometries from $\mathbb{C}$ into W, and $g_j \circ \varphi$ are bounded linear functionals on V. It follows from Corollary 2.2.3 that

$$\|\varphi\|_{cb} \leq \sum_{j=1}^{n} \|\theta_{w_j}\|_{cb} \|g_j \circ \varphi\|_{cb} = \sum_{j=1}^{n} \|g_j \circ \varphi\| \leq n \|\varphi\|.$$

Similarly, if V is n-dimensional, then we may replace W by $\varphi(V)$, which has dimension less than or equal to n, and the result follows from the previous argument. □

Corollary 2.2.5 *If V and W are n-dimensional abstract operator spaces, then there exists a linear isomorphism $\varphi : V \to W$ such that*

$$\|\varphi\|_{cb} \|\varphi^{-1}\|_{cb} \leq n^2.$$

Proof We choose Auerbach bases $v_i \in V$ and $w_i \in W$ $(i = 1, \ldots, n)$, together with dual bases $f_i \in V^*$ and $g_i \in W^*$ (see the previous proof). We have that

$$\varphi : V \to W : v \mapsto \sum f_i(v) w_i$$

and

$$\psi : W \to V : w \mapsto \sum g_i(w) v_i$$

are inverse linear mappings. Since

$$\|\varphi\|_{cb} \leq \sum \|f_i\|_{cb} \|\theta_{w_i}\|_{cb} \leq n,$$

and similarly $\|\psi\|_{cb} \leq n$, the result follows. □

For any abstract operator space V and linear functional $\varphi : V \to \mathbb{C}$, it follows from Corollary 2.2.3 that $\|\varphi\|_{cb} = \|\varphi\|$, and we may therefore identify the Banach spaces V^* with $\mathcal{CB}(V, \mathbb{C})$. There are other operator spaces W for which one automatically has that $\mathcal{CB}(V, W) = \mathcal{B}(V, W)$ for all operator spaces V. We shall characterize these spaces in §3.3. Anticipating this discussion, we next show that any commutative C^*-algebra has this property.

Proposition 2.2.6 *Let V be an abstract operator space, and let $\mathcal{A}$ be a commutative C^*-algebra. Then any bounded linear mapping $\varphi : V \to \mathcal{A}$ satisfies*

$$\|\varphi\|_{cb} = \|\varphi\|.$$

Proof We can assume that $\mathcal{A}$ coincides with $C_0(\Omega)$, the complex continuous functions vanishing at ∞ on a locally compact Hausdorff space Ω. We may identify $M_n(C_0(\Omega))$ with $C_0(\Omega, M_n)$, the corresponding matrix valued functions, where given $f = [f_{i,j}] \in M_n(C_0(\Omega))$, we have

$$\|f\| = \sup_{\omega\in\Omega}\{\|[f_{i,j}(\omega)]\|\}.$$

Let us fix an element $v \in M_n(V)$. Taking the supremum over all $\omega \in \Omega$ and $\alpha, \beta \in \mathbb{C}^n$ with $\|\alpha\|_2 = \|\beta\|_2 = 1$, we have

$$\begin{aligned}\|\varphi_n(v)\| &= \|[\varphi(v_{i,j})]\| = \sup_{\omega\in\Omega}\{\|[\varphi(v_{i,j})(\omega)]\|\} \\ &= \sup_{\omega,\alpha,\beta}\{|\langle[\varphi(v_{i,j})(\omega)]\alpha \mid \beta\rangle|\} = \sup_{\omega,\alpha,\beta}\left\{\left|\sum \overline{\beta}_i\varphi(v_{i,j})(\omega)\alpha_j\right|\right\}\end{aligned}$$

and thus letting α and β also stand for column matrices,

$$\|\varphi_n(v)\| = \sup_{\omega,\alpha,\beta}\{|\varphi(\beta^* v\alpha)(\omega)|\} \leq \|\varphi\| \sup_{\alpha,\beta}\{\|\beta^* v\alpha\|\} \leq \|\varphi\|\,\|v\|.$$

This shows that $\|\varphi_n\| \leq \|\varphi\|$ for all $n \in \mathbb{N}$, and thus $\|\varphi\|_{cb} = \|\varphi\|$. □

For any C^*-algebras $\mathcal{A}$ and $\mathcal{B}$, a $*$-homomorphism $\varphi : \mathcal{A} \to \mathcal{B}$ is automatically contractive. Since for each $n \in \mathbb{N}$ the corresponding mapping $\varphi_n : M_n(\mathcal{A}) \to M_n(\mathcal{B})$ is again a $*$-homomorphism, it follows that φ is completely contractive. Similarly, since a $*$-isomorphic injection is necessarily isometric, it is also a complete isometry.

Let us suppose that V is an abstract operator space. Given contractions $\mu \in M_{m,n}$ and $\gamma \in M_{n,m}$, the mapping

$$\varphi : M_n(V) \to M_m(V) : v \mapsto \mu v \gamma$$

is completely contractive since for any $r \in \mathbb{N}$, and $v \in M_{r\times n}(V)$,

$$\varphi_r(v) = (I_r \otimes \mu)v(I_r \otimes \gamma).$$

If $\varphi_i : V \to W_i$ are completely contractive, then that is also the case for the mapping $\varphi(v) = \varphi_1(v) \oplus \cdots \oplus \varphi_n(v)$. The diagonal truncation

$$D_n : M_n(V) \to M_n(V) : \begin{bmatrix} v_{1,1} & \cdots & v_{1,n} \\ \vdots & & \vdots \\ v_{n,1} & \cdots & v_{n,n}\end{bmatrix} \mapsto \begin{bmatrix} v_{1,1} & & 0 \\ & \ddots & \\ 0 & & v_{n,n}\end{bmatrix} \tag{2.2.3}$$

is completely contractive since $D_n(v) = \varepsilon_{1,1}v\varepsilon_{1,1} \oplus \cdots \oplus \varepsilon_{n,n}v\varepsilon_{n,n}$ (see (1.1.1)). We shall prove in §5.3 that any complete contraction $\sigma : M_n \to M_r$ has the form

$$\sigma(\alpha) = \mu(\overbrace{\alpha \oplus \cdots \oplus \alpha}^{nr})\gamma, \tag{2.2.4}$$

where $\mu \in M_{r,rn^2}$ and $\gamma \in M_{rn^2,r}$ are contractions.

Proposition 2.2.7 *For $n < \infty$ the transpose mapping $\mathbf{t} : M_n \to M_n$ is an isometry with $\|\mathbf{t}\|_{cb} = n$. The transpose mapping*

$$\mathbf{t} : K_\infty \to K_\infty$$

is isometric, but not completely bounded.

Proof We showed that $\mathbf{t}$ is isometric in §1.2. Any $\alpha \in M_n$ can be written as a sum of n 'generalized diagonal' matrices

$$\alpha = \begin{bmatrix} \alpha_{1,1} & & \\ & \ddots & \\ & & \alpha_{n,n} \end{bmatrix} + \begin{bmatrix} 0 & \alpha_{1,2} & & \\ & \ddots & \ddots & \\ & & & \alpha_{n-1,n} \\ \alpha_{n,1} & & & 0 \end{bmatrix}$$
$$+ \begin{bmatrix} 0 & 0 & \alpha_{1,3} & \\ & \ddots & \ddots & \\ \alpha_{n-2,n-1} & 0 & & 0 \\ 0 & \alpha_{n-1,n} & & 0 \end{bmatrix} + \cdots .$$

If we let π be the cyclical permutation matrix

$$\pi = \begin{bmatrix} 0\,0 \ldots 0\,1 \\ 1\,0 \ldots 0\,0 \\ \ldots \\ 0\,0 \ldots 1\,0 \end{bmatrix},$$

then multiplication on the left or right results in the corresponding permutation of rows or columns, respectively. A simple calculation shows that we may rewrite the above sum in the form

$$\alpha = \sum_{k=0}^{n-1} D_n(\alpha\pi^k)\pi^{-k}.$$

The unitary π has real entries, and thus

$$\mathbf{t}(\pi^{-k}) = (\pi^{-k})^{tr} = (\pi^{-k})^* = \pi^k .$$

Since $\mathbf{t}$ is a multiplicative anti-isomorphism which leaves diagonal matrices invariant, it follows that

$$\mathbf{t}(\alpha) = \sum_{k=0}^{n-1} \pi^k D_n(\alpha\pi^k).$$

Thus,

$$\|\mathbf{t}\|_{cb} \leq \sum_{k=0}^{n-1} \left\| \alpha \mapsto \pi^k D_n(\alpha \pi^k) \right\|_{cb} \leq n.$$

If we let $\varepsilon = [\varepsilon_{i,j}] \in M_n(M_n)$ be the matrix of matrix units in M_n, then

$$\tilde{\varepsilon} = \mathbf{t}_n(\varepsilon) = \begin{bmatrix} \varepsilon_{11} & \varepsilon_{21} & \cdots \\ \varepsilon_{12} & \varepsilon_{22} & \cdots \\ \vdots & \vdots & \end{bmatrix} = \begin{bmatrix} 1\,0 \ldots 0\,0 \ldots \\ 0\,0 \ldots 1\,0 \ldots \\ \vdots \qquad \vdots \\ 0\,1 \ldots 0\,0 \ldots \\ 0\,0 \ldots 0\,1 \ldots \\ \vdots\;\vdots \qquad \vdots\;\vdots \end{bmatrix} \tag{2.2.5}$$

is a permutation matrix; that is, it has precisely one 1 in each column (and in each row). Thus, $\tilde{\varepsilon}$ is unitary and $\|\tilde{\varepsilon}\| = 1$. On the other hand, if we let ε_i denote the column matrix $\varepsilon_{i,1}^{[n,1]}$ (see §1.1), then we have $\varepsilon_i \varepsilon_j^* = \varepsilon_{i,j}$, and thus

$$\mathbf{t}_n\left(\tilde{\varepsilon}\right) = \varepsilon = \begin{bmatrix} \varepsilon_1 \\ \vdots \\ \varepsilon_n \end{bmatrix} \begin{bmatrix} \varepsilon_1^* \ldots \varepsilon_n^* \end{bmatrix}. \tag{2.2.6}$$

It follows that

$$\left\|\mathbf{t}_n\left(\tilde{\varepsilon}\right)\right\| = \left\| \begin{bmatrix} \varepsilon_1 \\ \vdots \\ \varepsilon_n \end{bmatrix} \right\|^2 = \left\| \begin{bmatrix} \varepsilon_1^* \ldots \varepsilon_n^* \end{bmatrix} \begin{bmatrix} \varepsilon_1 \\ \vdots \\ \varepsilon_n \end{bmatrix} \right\| = n,$$

and therefore

$$\|\mathbf{t}\|_{cb} = n.$$

We have a commutative diagram

$$\begin{array}{ccc} M_n & \xrightarrow{\mathbf{t}} & M_n \\ \downarrow & & \downarrow \\ K_\infty & \xrightarrow{\mathbf{t}} & K_\infty \end{array},$$

where the columns are completely isometric injections, and the rows are isometric. Since the top row has the completely bounded norm n, it is apparent that the bottom row has the completely bounded norm ∞. □

2.3 THE REPRESENTATION THEOREM

As we observed in §2.1, the fact that any normed space may be represented as a function space is a consequence of the Hahn–Banach theorem. In order to prove the corresponding result for abstract operator spaces, we must demonstrate an analogous theorem for *matrix-valued mappings* (see

Lemma 2.3.4). Surprisingly, there does not seem to be an obvious way to derive this from the operator space version of the Hahn–Banach theorem as stated in §4.1. Although the proof has been greatly simplified since its discovery, it is still fairly subtle. Our strategy is to relate matrix-valued functionals on a space V to scalar functionals on the matrix spaces over V (see Lemma 2.3.3). After that it suffices to apply the classical Hahn–Banach theorem.

We begin with a simple minimax lemma. We are indebted to Erik Alfsen for the elegant argument. Let us suppose that K is a convex subset of a real linear space V. A real valued function e on K is said to be *affine* if it preserves convex combinations, i.e.

$$e(\alpha x + (1-\alpha)y) = \alpha e(x) + (1-\alpha)e(y)$$

for $0 \leq \alpha \leq 1$ and $x, y \in K$. We have, for example, that if f is a real linear functional on V and C is a constant, then the restriction of $e = f + C$ to K is an affine function.

Lemma 2.3.1 *Suppose that $\mathcal{E}$ is a cone of real continuous affine functions on a compact convex subset K of a topological linear space E, and that for each $e \in \mathcal{E}$ there is a corresponding point $k_e \in K$ with $e(k_e) \geq 0$. Then there is a point $k_0 \in K$ for which $e(k_0) \geq 0$ for all $e \in \mathcal{E}$.*

Proof We must show that the sets

$$K(e) = \{k \in K : e(k) \geq 0\}$$

have a non-zero intersection. By assumption these sets are non-empty, and they are compact. Thus, it suffices to show that they have the finite intersection property. If this is not the case, then there exist $e_1, \ldots, e_n \in \mathcal{E}$ such that $K(e_1) \cap \cdots \cap K(e_n) = \emptyset$. The mapping $\theta : K \to \mathbb{R}^n$ defined by $\theta(k) = (e_1(k), \ldots, e_n(k))$ is continuous and affine, and thus $\theta(K)$ is a compact convex set in $\mathbb{R}^n$. By assumption,

$$\theta(K) \cap (\mathbb{R}^n)^+ = \emptyset\,.$$

It follows that there is a linear functional f on $\mathbb{R}^n$ such that $f((\mathbb{R}^n)^+) \geq 0$ and $f(\theta(K)) < 0$. This is a consequence of the usual geometric separation lemma, which in finite dimensions is an elementary result. Since

$$f(x_1, \ldots, x_n) = c_1 x_1 + \cdots + c_n x_n$$

for constants $c_j \geq 0$, it follows that $e = f \circ \theta = c_1 e_1 + \cdots + c_n e_n$ is an element of $\mathcal{E}$ for which $K(e) = \emptyset$, contradicting the hypotheses of the lemma. □

Lemma 2.3.2 *If V is an abstract operator space, and $F \in [M_n(V)]^*$ satisfies $\|F\| = 1$, then there exist states p_0 and q_0 on M_n such that*

$$|F(\alpha v \beta)| \leq p_0(\alpha\alpha^*)^{1/2}\|v\| q_0(\beta^*\beta)^{1/2} \tag{2.3.1}$$

for all $\alpha \in M_{n,r}$, $\beta \in M_{r,n}$, *and* $v \in M_r(V)$ ($r \in \mathbb{N}$ *arbitrary*).

Proof We may suppose that $\|v\| = 1$. As in §1.2, we let $\mathbf{S}_n$ denote the state space of M_n. It suffices to find $p_0, q_0 \in \mathbf{S}_n$ with

$$\operatorname{Re} F(\alpha v \beta) \leq p_0(\alpha\alpha^*)^{1/2} q_0(\beta^*\beta)^{1/2} \tag{2.3.2}$$

since we can then replace α by $e^{i\theta}\alpha$ for a suitable $\theta \geq 0$. In turn, it is enough to prove that

$$\operatorname{Re} F(\alpha v \beta) \leq (1/2)[p_0(\alpha\alpha^*) + q_0(\beta^*\beta)].$$

To see this, let us replace α by $t^{1/2}\alpha$ and β by $t^{-1/2}\beta$ for $t > 0$. Then

$$\operatorname{Re} F(\alpha v \beta) \leq (1/2)[t p_0(\alpha\alpha^*) + t^{-1} q_0(\beta^*\beta)].$$

If $p_0(\alpha\alpha^*) \neq 0$ and $q_0(\beta^*\beta) \neq 0$, then we may obtain (2.3.2) by letting $t = p_0(\alpha\alpha^*)^{-1/2} q_0(\beta^*\beta)^{1/2}$. If $p_0(\alpha\alpha^*) = 0$ and $q_0(\beta^*\beta) \neq 0$, and we let $t \to \infty$, then it follows that $\operatorname{Re} F(\alpha v \beta) = 0$, and thus (2.3.2) is trivial. We may use a similar argument if $p_0(\alpha\alpha^*) \neq 0$ and $q_0(\beta^*\beta) = 0$. Finally, if both are zero, then again $\operatorname{Re} F(\alpha v \beta) = 0$, and we have (2.3.2) in all cases.

The Cartesian product $K = \mathbf{S}_n \times \mathbf{S}_n$ is a compact and convex subset of $(M_n \oplus M_n)^*$. We let $A(K)$ denote the linear space of real-valued continuous affine functions on K. Given $\alpha \in M_{n,r}$, $\beta \in M_{r,n}$, and $v \in M_r(V)$ with $\|v\| = 1$, we may define a corresponding function $e_{\alpha,v,\beta} \in A(K)$ by

$$e_{\alpha,v,\beta}(p, q) = p(\alpha\alpha^*) + q(\beta^*\beta) - 2\operatorname{Re} F(\alpha v \beta).$$

We let $\mathcal{E}$ denote the collection of all such functions. We wish to show that there is a point $(p_0, q_0) \in K$ for which $e(p_0, q_0) \geq 0$ for all $e \in \mathcal{E}$. But we have:

(a) Each function $e \in \mathcal{E}$ is non-negative at some point $(p_e, q_e) \in K$. To see this, suppose that $e = e_{\alpha,v,\beta}$ and select states p_e and q_e such that

$$p_e(\alpha\alpha^*) = \|\alpha\alpha^*\| = \|\alpha\|^2$$

and

$$q_e(\beta^*\beta) = \|\beta^*\beta\| = \|\beta\|^2.$$

Then

$$e_{\alpha,v,\beta}(p_e, q_e) = \|\alpha\|^2 + \|\beta\|^2 - 2\operatorname{Re} F(\alpha v \beta) \geq 0$$

since

$$\operatorname{Re} F(\alpha v \beta) \leq |F(\alpha v \beta)| \leq \|\alpha v \beta\| \leq \|\alpha\|\|\beta\| \leq (1/2)[\|\alpha\|^2 + \|\beta\|^2].$$

(b) The collection $\mathcal{E}$ is a cone in $A(K)$, or in other words, $\mathcal{E} + \mathcal{E} \subseteq \mathcal{E}$ and $c\mathcal{E} \subseteq \mathcal{E}$ for $c \geq 0$. The second assertion follows from the relation

$$c e_{\alpha,v,\beta} = e_{c^{1/2}\alpha, v, c^{1/2}\beta}.$$

For the first we note that

$$e_{\alpha,v,\beta} + e_{\alpha',v',\beta'} = e_{\alpha'',v'',\beta''},$$

where $\alpha'' = \begin{bmatrix} \alpha\ \alpha' \end{bmatrix}$, $\beta'' = \begin{bmatrix} \beta \\ \beta' \end{bmatrix}$, and $v'' = v \oplus v'$ satisfies $\|v \oplus v'\| = 1$ by **M1**.

Since the hypotheses of Lemma 2.3.1 are satisfied, we have proved the existence of the desired pair (p_0, q_0). □

Lemma 2.3.3 *Suppose that V is an abstract operator space. Given a linear functional $F \in [M_n(V)]^*$ with $\|F\| = 1$, there exists a complete contraction $\varphi : V \to M_n$ and unit vectors $\xi, \eta \in (\mathbb{C}^n)^n$ such that*

$$F(v) = \langle \varphi_n(v)\eta \mid \xi \rangle$$

for all $v \in M_n(V)$.

Proof Let us fix states p_0 and q_0 on M_n as in Lemma 2.3.2. From the GNS representation theorem (see §A.4), there are corresponding representations π and θ of M_n on finite-dimensional Hilbert spaces H and K, respectively, with cyclic vectors $\xi_0 \in H$ and $\eta_0 \in K$ satisfying

$$p_0(\alpha) = \langle \pi(\alpha)\xi_0 \mid \xi_0 \rangle$$

and

$$q_0(\alpha) = \langle \theta(\alpha)\eta_0 \mid \eta_0 \rangle,$$

respectively, for all $\alpha \in M_n$.

Given a row matrix $\alpha = [\alpha_1, \alpha_2, \ldots, \alpha_n] \in M_{1,n}$, we define $\tilde{\alpha} \in M_n$ by

$$\tilde{\alpha} = \begin{bmatrix} \alpha_1 & \alpha_2 & \cdots \\ 0 & 0 & \cdots \\ & & \ddots \end{bmatrix}.$$

We define $\tilde{M}_{1,n}$ to be the linear space of all such $n \times n$ matrices, and we let $H_0 = \pi(\tilde{M}_{1,n})\xi_0 \subseteq H$ and $K_0 = \theta(\tilde{M}_{1,n})\eta_0 \subseteq K$. If we fix an element $v \in V$, then the sesquilinear form B_v defined on $K_0 \times H_0$ by

$$B_v(\theta(\tilde{\beta})\eta_0, \pi(\tilde{\alpha})\xi_0) = F(\alpha^* v \beta)$$

is bounded (and thus well defined) since from (2.3.1)

$$|F(\alpha^* v \beta)| \le \|\pi(\tilde{\alpha})\xi_0\| \, \|\theta(\tilde{\beta})\eta_0\| \, \|v\|.$$

Thus, there exists a unique bounded operator $\varphi_0(v) : K_0 \to H_0$ for which

$$F(\alpha^* v \beta) = \langle \varphi_0(v)\theta(\tilde{\beta})\eta_0 \mid \pi(\tilde{\alpha})\xi_0 \rangle$$

(see A.3.4). It is a simple matter to verify that the corresponding mapping $\varphi_0 : V \to \mathcal{B}(K_0, H_0)$ is linear. The spaces K_0 and H_0 have dimensions $h, k \le n$, and we may thus identify them with the subspaces $\mathbb{C}^h \oplus 0_{n-h}$ and $\mathbb{C}^k \oplus 0_{n-k}$ of $\mathbb{C}^n$. If we let E be the projection of $\mathbb{C}^n$ onto K_0, and we define

$$\varphi(v) = \varphi_0(v)E : \mathbb{C}^n \to \mathbb{C}^n,$$

then we obtain a mapping $\varphi : V \to M_n$ satisfying

$$F(\alpha^* v \beta) = \langle \varphi(v)\theta(\tilde{\beta})\eta_0 \mid \pi(\tilde{\alpha})\xi_0 \rangle . \tag{2.3.3}$$

Given a matrix $v \in M_n(V)$, we have from (1.1.11)) that

$$F(v) = \sum_{i,j} \langle \varphi(v_{i,j})\theta(\tilde{E}_j)\eta_0 \mid \pi(\tilde{E}_i)\xi_0 \rangle = \langle \varphi_n(v)\eta \mid \xi \rangle,$$

where

$$\xi = \begin{pmatrix} \pi(\tilde{E}_1)\xi_0 \\ \vdots \\ \pi(\tilde{E}_n)\xi_0 \end{pmatrix}, \ \eta = \begin{pmatrix} \theta(\tilde{E}_1)\eta_0 \\ \vdots \\ \theta(\tilde{E}_n)\eta_0 \end{pmatrix} \in (\mathbb{C}^n)^n$$

satisfy

$$\|\xi\|^2 = \sum \|\pi(\tilde{E}_j)\xi_0\|^2 = \sum p_0(E_j^* E_j) = p_0(I) = 1,$$

and similarly, $\|\eta\|^2 = 1$.

To show that φ is completely contractive, it suffices to show that φ_n is contractive (see Proposition 2.2.2). From the definition of φ, it suffices to prove that

$$|\langle (\varphi_0)_n(v)\eta_1 \mid \xi_1 \rangle| \leq \|v\| \, \|\eta_1\| \, \|\xi_1\|$$

for all $\xi_1 \in H_0^n$ and $\eta_1 \in K_0^n$. Given

$$\xi_1 = \begin{pmatrix} \pi(\tilde{\alpha}_1)\xi_0 \\ \vdots \\ \pi(\tilde{\alpha}_n)\xi_0 \end{pmatrix}, \ \eta_1 = \begin{pmatrix} \theta(\tilde{\beta}_1)\eta_0 \\ \vdots \\ \theta(\tilde{\beta}_n)\eta_0 \end{pmatrix},$$

where $\alpha_i, \beta_j \in M_{1,n}$,

$$\|\xi_1\|^2 = \sum \|\pi(\tilde{\alpha}_i)\xi_0\|^2 = \sum p_0(\alpha_i^* \alpha_i) = p_0(\alpha^* \alpha),$$

and similarly $\|\eta_1\|^2 = q_0(\beta^* \beta)$, where

$$\alpha = \begin{bmatrix} \alpha_1 \\ \vdots \\ \alpha_n \end{bmatrix}, \ \beta = \begin{bmatrix} \beta_1 \\ \vdots \\ \beta_n \end{bmatrix} \in M_n .$$

It follows that

$$\begin{aligned} \langle (\varphi_0)_n(v)\eta_1 \mid \xi_1 \rangle &= \sum \langle \varphi_0(v_{i,j})\theta(\tilde{\beta}_j)\eta_0 \mid \pi(\tilde{\alpha}_i)\xi_0 \rangle \\ &= \sum F(\alpha_i^* v_{ij} \beta_j) = F(\alpha^* v \beta), \end{aligned}$$

and thus

$$\begin{aligned} |\langle (\varphi_0)_n(v)\eta_1 \mid \xi_1 \rangle| &\leq |F(\alpha^* v \beta)| \\ &\leq p_0(\alpha^* \alpha)^{1/2} \|v\| q_0(\beta^* \beta)^{1/2} = \|v\| \|\xi_1\| \|\eta_1\| . \end{aligned}$$

□

Lemma 2.3.4 *Suppose that V is an abstract operator space. Given any element $v \in M_n(V)$, there exists a complete contraction $\varphi : V \to M_n$ such that*

$$\|\varphi_n(v)\| = \|v\|.$$

Proof If we are given $v \in M_n(V)$, then we may use the Hahn–Banach theorem to find a linear functional $F \in [M_n(V)]^*$ with $\|F\| = 1$ and $|F(v)| = \|v\|$. From Lemma 2.3.3 there is a corresponding complete contraction $\varphi : V \to \mathbb{M}_n$ for which

$$\|\varphi_n(v)\| \geq |\langle \varphi_n(v)\eta \mid \xi \rangle| = |F(v)| = \|v\|.$$

The reverse inequality is trivial. □

We may now prove the *representation theorem* for operator spaces. Owing to this result, we shall usually not distinguish between abstract and concrete operator spaces.

Theorem 2.3.5 *If V is an abstract operator space, then there is a Hilbert space H, a concrete operator space $W \subseteq \mathcal{B}(H)$, and a complete isometry Φ of V onto W. If V is separable as a normed space, then we can let $H = \ell_2$.*

Proof For each $n \in \mathbb{N}$ we let

$$\mathfrak{s}_n = \mathfrak{s}_n(V) = \mathcal{CB}(V, M_n)_{\|\cdot\|_{cb} \leq 1} \tag{2.3.4}$$

and

$$\mathfrak{s} = \mathfrak{s}(V) = \bigcup_{n \in \mathbb{N}} \mathfrak{s}_n(V). \tag{2.3.5}$$

We define $H = \bigoplus_{\varphi \in \mathfrak{s}} \mathbb{C}^{n(\varphi)}$, where $n(\varphi)$ is the integer n with $\varphi \in \mathfrak{s}_n$ and we let

$$\Phi : V \to \mathcal{B}(H) : v \mapsto (\varphi(v))_{\varphi \in \mathfrak{s}}. \tag{2.3.6}$$

We may identify $\Phi_n : M_n(V) \to \mathcal{B}(H^n)$ with the mapping $v \mapsto (\varphi_n(v))_{\varphi \in \mathfrak{s}}$, and hence it is immediate that Φ is a complete contraction. On the other hand, given a fixed $v \in M_n(V)$, we may use Lemma 2.3.4 to select a $\varphi_0 \in s_n$ with $\|(\varphi_0)_n(v)\| = \|v\|$. This implies that $\|\Phi_n(v)\| \geq \|(\varphi_0)_n(v)\| = \|v\|$, from which the first assertion follows.

If V is separable as a Banach space, then for each $n \in \mathbb{N}$ the space

$$\mathfrak{s}_n = \mathcal{CB}(V, M_n)_{\|\cdot\|_{cb} \leq 1}$$

is compact and metrizable in the point-norm topology. To see this, we note that the mapping

$$\theta : \mathcal{CB}(V, M_n) \to \mathcal{B}(M_n(V), M_n(M_n)) : \varphi \mapsto \varphi_n$$

is a point-norm to weak* homeomorphism of $\mathfrak{s}_n$ onto a weak*-closed subset of

$$\mathcal{B}(M_n(V), M_n(M_n))_{\|\cdot\| \leq 1}$$

(see Proposition 2.2.2). Thus, it suffices to prove that the above set is compact and metrizable in the relative weak* topology. Letting $E = M_n(V)$ and $F = M_n(M_n)$, we may identify $\mathcal{B}(E, F)$ with the Banach dual G^*, where $G = E \otimes^\gamma F^*$ is the usual Banach space projective tensor product (see §7.1). Since E and F are separable Banach spaces, the same is true for $G = E \otimes^\gamma F^*$. Thus, we have reduced our argument to the well-known fact that if G is a separable Banach space, then the closed unit ball of its dual G^* is a compact and metrizable space in the weak* topology.

Since any compact metric space is separable, we can find a countable dense subset $\mathfrak{s}_n^0$ of $\mathfrak{s}_n$. If we replace $\mathfrak{s}$ by $\mathfrak{s}^0 = \bigcup \mathfrak{s}_n^0$, we obtain a complete isometry of V into $\mathcal{B}(\ell_2)$. □

We may use the representation theorem to transfer various results about concrete operator spaces to abstract operator spaces. For example, if we identify an abstract operator space V with a subspace of $\mathcal{B}(H)$ for some Hilbert space H, then for each finite set $\mathfrak{f}$, the norm on $M_{\mathfrak{f}}(V)$ is determined by the inclusion $M_{\mathfrak{f}}(V) \subseteq \mathcal{B}(H^{\mathfrak{f}})$. It follows that we have natural isometries

$$M_{\mathfrak{g}}(M_{\mathfrak{f}}(V)) \cong M_{\mathfrak{g}\times\mathfrak{f}}(V) \cong M_{\mathfrak{f}}(M_{\mathfrak{g}}(V)) .$$

The following alternative version of the representation theorem is often easier to use.

Proposition 2.3.6 *Suppose that V is a linear space, and that we are provided with mappings*

$$\| \cdot \|_n : \mathbb{M}_n(V) \to [0, \infty)$$

for all $n \in \mathbb{N}$, which satisfy

- **M1′** $\|v \oplus w\|_{m+n} \leq \max \{\|v\|_m, \|w\|_n\}$,
- **M2** $\|\alpha v \beta\|_n \leq \|\alpha\| \, \|v\|_m \|\beta\|$,

for all $v \in \mathbb{M}_m(V)$, $w \in \mathbb{M}_n(V)$, and $\alpha \in M_{n,m}$, $\beta \in M_{m,n}$. Then these mappings are seminorms which satisfy **M1** *and* **M2**. *If, in addition, $\| \cdot \|_1$ is a norm, then the same is true for all the given matrix seminorms, and they determine an operator space structure on V.*

Proof Given $v, w \in \mathbb{M}_n(V)$ and $\varepsilon > 0$, we let

$$\alpha = \|v\|_n + \varepsilon \text{ and } \beta = \|w\|_n + \varepsilon.$$

Then $v = \alpha \overline{v}$ and $w = \beta \overline{w}$, where $\|\overline{v}\|, \|\overline{w}\| < 1$. We have

$$v + w = \gamma \begin{bmatrix} \overline{v} & 0 \\ 0 & \overline{w} \end{bmatrix} \gamma^*,$$

where $\gamma = \begin{bmatrix} \alpha^{1/2} I_n & \beta^{1/2} I_n \end{bmatrix}$ satisfies

$$\|\gamma\| \|\gamma^*\| = \|\gamma\gamma^*\| = \alpha + \beta .$$

It follows from **M1**$'$ and **M2** that

$$\|v + w\|_n \le \|\gamma\|\|\gamma^*\| \max\{\|\overline{v}\|_n, \|\overline{w}\|_n\} < \|v\|_n + \|w\|_n + 2\varepsilon,$$

and since $\varepsilon > 0$ is arbitrary, $\|\cdot\|_n$ is subadditive. Homogeneity follows from **M2** and the fact that for $\alpha \in \mathbb{C}$,

$$\alpha v = (\alpha I_n) v\,.$$

Condition **M2** implies that if $v \in M_m(V)$ and $w \in M_n(V)$, then

$$\|v\| = \|\,[I_m\ 0_{m,n}]\,(v \oplus w) \begin{bmatrix} I_m \\ 0_{n,m} \end{bmatrix} \| \le \|v \oplus w\|,$$

and similarly, $\|w\| \le \|v \oplus w\|$. Combining these inequalities with **M1**$'$, we obtain **M1**.

The last assertion is immediate from (2.1.8). □

2.4 NOTES AND REFERENCES

It may be argued that operator space theory began with Arveson's version of the Hahn–Banach theorem for completely bounded operator-valued mappings (Arveson 1969). In the preceding two decades, functional analysis had been successfully applied to the study of function algebras (see Browder 1969). Since these may be regarded as non-self-adjoint subalgebras of commutative C^*-algebras, it was only natural to turn to consider the non-commutative analogues. Kadison and Singer (1960) wrote the first substantial paper in this direction.

A key technique in function algebra theory is to extend linear functionals from a function algebra to functionals (that is to say measures) on the ambient C^*-algebras. Arveson realized that for the non-commutative case it was necessary to replace scalar functionals with operator-valued mappings. The usual Hahn–Banach theorem fails in this context, and this seemed to be a serious obstacle to his programme. Arveson's crucial insight was that one could extend such mappings provided one used the 'matrix orders' and 'completely positive mappings' of Stinespring (1955). He succeeded in proving a powerful extension theorem of this type, and investigated some of its applications (see Arveson 1972). This was given a completely general formulation by Wittstock (1981) (see Theorem 4.1.4).

Arveson's remarkable result alerted others to the fact that linear spaces of operators have an intrinsic 'hidden structure' encoded in their matrix orderings and norms. The ordered theory was more accessible, and a corresponding theory of 'operator systems' was formulated by Choi and Effros (1977*a*). These spaces are the operator analogues of Kadison's 'function systems' (Kadison 1951), a natural category of unital ordered Banach spaces (see §5.1). The resulting theory played a significant role in the theory of nuclear and injective C^*-algebras, and in certain forms of C^*-algebraic K-theory, but this work is not the focus of this monograph.

Perhaps the most important impetus for studying matrix norms was provided by Haagerup's discovery (Haagerup 1979) that if one used completely bounded mappings rather than completely positive mappings, then one could formulate C^*-algebraic invariants that were more inclusive than nuclearity (see §11.2). In light of this result and Wittstock's general form of Arveson's theorem, the time had arrived for an axiomatic approach to operator spaces.

The abstract matrix norm characterization for operator spaces was achieved by Ruan (1988), in which he formulated the abstract axioms **M1** and **M2**. His original proof of the representation theorem (Theorem 2.3.5) was based on Paulsen's 2×2 matrix trick for replacing operator spaces by operator systems, and the Choi–Effros abstract characterization theorem for operator systems (see Paulsen 1986, and Choi and Effros 1977*a*). As we have seen in §2.3, one can use a simple direct argument from Effros and Ruan (1993). An early discussion of operator space theory can be found in Effros (1987*a*).

A better estimate for Corollary 2.2.5 can be found in Pisier (1996*b*), where he proved that there is, in fact, a linear isomorphism $\varphi : V \to W$ with

$$\|\varphi\|_{cb}\|\varphi^{-1}\|_{cb} \leq n.$$

Proposition 2.2.6 was first proved by Loebl (1975), where he also showed that a C^*-algebra $\mathcal{B}$ is commutative if and only if for any C^*-algebra $\mathcal{A}$, every contractive mapping $\varphi : \mathcal{A} \to \mathcal{B}$ is 2-contractive (respectively, n-contractive for all integers n). Proposition 2.2.2 is due to Smith (1983), and Proposition 2.2.7 is due to Tomiyama (1983).

3

Constructions and examples

We apply the representation theorem to show that many Banach space constructions have operator space analogues. Furthermore, we show that Banach spaces generally have elaborate families of distinct quantizations. Among these are the canonically defined 'minimal' and 'maximal' quantizations. In order to illustrate some interesting 'intermediate' quantizations, we examine three distinct quantizations that may be defined for a Hilbert space. These are the column and row Hilbert operator spaces, and Pisier's intriguing self-dual Hilbert operator space.

3.1 SUBSPACES, QUOTIENTS, PRODUCTS, AND CONJUGATES

In this section we let V be a fixed operator space. It is immediate that if W is a subspace of V, then the inclusions $M_n(W) \subseteq M_n(V)$ and the corresponding relative norms determine an operator space matrix norm on W.

For each $p \in \mathbb{N}$ we use the identifications $M_n(M_p(V)) = M_{np}(V)$ to determine matrix norms for $M_p(V)$. These satisfy **M1** and **M2** since, for example, if we are given $\alpha \in M_{n,m}, \beta \in M_{m,n}$, and v is in $M_m(M_p(V))$, which we identify with $M_{mp}(V)$, then

$$\|\alpha v \beta\| = \|(\alpha \otimes I_p) v (\beta \otimes I_p)\| \leq \|\alpha \otimes I_p\| \, \|v\| \, \|\beta \otimes I_p\| = \|\alpha\| \, \|v\| \, \|\beta\|,$$

from which we conclude that $M_p(V)$ is an operator space. If $V \hookrightarrow \mathcal{B}(H)$ is a completely isometric injection, it is immediate that $M_p(V) \hookrightarrow \mathcal{B}(H^p)$ is a corresponding completely isometric injection of $M_p(V)$.

We have already seen that if N is a closed subspace of an operator space V, then $M_n(N)$ is closed in $M_n(V)$ for each $n \in \mathbb{N}$. We may therefore use the identification

$$\mathbb{M}_n(V/N) = M_n(V)/M_n(N) \tag{3.1.1}$$

to define a norm on $\mathbb{M}_n(V/N)$, which we then denote by $M_n(V/N)$. If we let $\pi : V \twoheadrightarrow V/N$ be the quotient mapping, then for each $n \in \mathbb{N}$, $\pi_n : M_n(V) \twoheadrightarrow M_n(V/N)$ is a quotient mapping, and for $\tilde{v} \in M_n(V/N)$,

$$\|\tilde{v}\| = \inf\{\|v\| : v \in M_n(V), \pi_n(v) = \tilde{v}\}.$$

Proposition 3.1.1 *If N is a closed subspace of an operator space V, then V/N is an operator space.*

Proof Given $\alpha \in M_{n,m}$, $\beta \in M_{m,n}$, and $\tilde{v} \in M_m(V/N)$, there exists a $v \in M_m(V)$ such that $\pi_m(v) = \tilde{v}$ and $\|v\| < \|\tilde{v}\| + \varepsilon$. It follows that $\pi_n(\alpha v \beta) = \alpha \tilde{v} \beta$, and thus

$$\|\alpha\tilde{v}\beta\| \leq \|\alpha v \beta\| \leq \|\alpha\| \|v\| \|\beta\| \leq \|\alpha\|(\|\tilde{v}\| + \varepsilon)\|\beta\| .$$

Since $\varepsilon > 0$ is arbitrary, we obtain **M2**.

On the other hand, given $\tilde{w} \in M_n(V/N)$ and an element $w \in M_n(V)$ with $\pi_n(w) = \tilde{w}$ and $\|w\| < \|\tilde{w}\| + \varepsilon$, it follows that $\pi_{m+n}(v \oplus w) = \tilde{v} \oplus \tilde{w}$, and thus

$$\|\tilde{v} \oplus \tilde{w}\| \leq \|v \oplus w\| = \max\{\|v\|, \|w\|\} \leq \max\{\|\tilde{v}\|, \|\tilde{w}\|\} + \varepsilon .$$

Again since $\varepsilon > 0$ is arbitrary, we obtain **M1′**, and from Proposition 2.3.6, V/N is an operator space. □

We call V/N the *quotient operator space* determined by V and N. These spaces have the expected properties. The quotient mapping $\pi : V \to V/N$ is a complete quotient mapping (see §2.2). Each (non-zero) complete contraction $\varphi : V \to W$ with $\ker\varphi = N$ induces a complete contraction $\tilde{\varphi} : V/N \to W$ with $\|\tilde{\varphi}\|_{cb} = \|\varphi\|_{cb}$, and φ is a complete quotient mapping if and only if $\tilde{\varphi}$ is a complete isometry.

Given an indexed family of operator spaces $(V_s)_{s\in\mathfrak{s}}$, we define the *product operator space* $\prod_{s\in\mathfrak{s}} V_s = \ell_\infty(\mathfrak{s};V_s)$ to be the normed space $\ell_\infty(\mathfrak{s}; V_s)$ (see §A.2) together with the matrix norms determined by the identifications

$$M_n\left(\prod_{s\in\mathfrak{s}} V_s\right) = \prod_{s\in\mathfrak{s}} M_n(V_s)$$

for all $n \in \mathbb{N}$. If $\mathfrak{s} = \{1, \ldots, n\}$, then we also use the alternative notation

$$V_1 \oplus \cdots \oplus V_n .$$

Given a completely isometric injection $V_s \hookrightarrow \mathcal{B}(H_s)$ for each $s \in \mathfrak{s}$, we obtain a corresponding completely isometric injection

$$\prod_{s\in\mathfrak{s}} V_s \hookrightarrow \mathcal{B}(H),$$

where $H = \bigoplus_{s\in\mathfrak{s}} H_s$, and thus $\prod_{s\in\mathfrak{s}} V_s$ is an operator space. It is also possible to 'quantize' ℓ_p-direct sums; that is, we can provide them with natural compatible operator space matrix norms, but we shall not pursue this topic.

Given a sequence of operator spaces $(V_n)_{n\in\mathbb{N}}$,

$$\sum_{n\in\mathbb{N}} V_n = \left\{(v_n) \in \prod_{n\in\mathbb{N}} V_n : \lim_{n\to\infty} \|v_n\| = 0\right\}$$

is a closed subspace of the product operator space $\prod_{n\in\mathbb{N}} V_n$. We define the *asymptotic product operator space* to be the quotient operator space $\prod_{n\in\mathbb{N}} V_n / \sum_{n\in\mathbb{N}} V_n$, and we let

$$\pi_\infty : \prod_{n\in\mathbb{N}} V_n \twoheadrightarrow \prod_{n\in\mathbb{N}} V_n \Big/ \sum_{n\in\mathbb{N}} V_n$$

denote the complete quotient mapping. It follows from Proposition A.6.1 that if

$$v = (v_n) \in M_m\left(\prod_{n\in\mathbb{N}} V_n\right),$$

then

$$\|(\pi_\infty)_m((v_n))\| = \limsup_{n\to\infty} \|v_n\|. \tag{3.1.2}$$

We shall discuss ultraproducts in §10.3. In what follows we shall discuss the simplest version of the direct limit notion, and we shall only sketch the standard theory. We suppose that the operator spaces under consideration are all norm complete. If we are given a diagram of operator spaces and completely isometric injections

$$V_1 \overset{\varphi_1}{\hookrightarrow} V_2 \overset{\varphi_2}{\hookrightarrow} \cdots, \tag{3.1.3}$$

then we say that an operator space V together with completely isometric injections $\varphi_{n,\infty} : V_n \hookrightarrow V$ is a *direct limit* of (3.1.3) if for each n,

$$\varphi_{n,\infty} = \varphi_{n+1,\infty} \circ \varphi_n,$$

and $\bigcup \varphi_n(V_n)$ is norm dense in V. If $n < m$, and we define

$$\varphi_{n,m} = \varphi_{m-1} \circ \cdots \circ \varphi_n : V_n \to V_m,$$

then it follows that

$$\varphi_{n,\infty} = \varphi_{m,\infty} \circ \varphi_{n,m}.$$

To illustrate this notion, let us suppose that V is an operator space with closed subspaces

$$V_1 \subseteq V_2 \subseteq \cdots$$

such that $\bigcup V_n$ is norm dense in V. Then V together with the inclusion mappings $V_n \hookrightarrow V$ is a direct limit of the V_n. On the other hand, given any diagram (3.1.3), we can construct a direct limit as follows. For each $n \in \mathbb{N}$, we define a mapping

$$\varphi_{n,\infty} : V_n \to \prod_{n\in\mathbb{N}} V_n \Big/ \sum_{n\in\mathbb{N}} V_n$$

by letting

$$v_n \mapsto \pi_\infty((0, \ldots, 0_{n-1}, v_n, \varphi_{n,n+1}(v), \varphi_{n,n+2}(v), \ldots)).$$

It is not difficult to see that this is a completely isometric injection. It follows that the norm closure

$$V_\infty = \overline{\bigcup_{n\in\mathbb{N}} \varphi_{n,\infty}(V_n)}$$

together with the mappings

$$\varphi_{n,\infty} : V_n \to V_\infty$$

is a direct limit for (3.1.3).

Given a commuting diagram of completely isometric injections

$$\begin{array}{ccccc} V_1 & \xrightarrow{\varphi_1} & V_2 & \xrightarrow{\varphi_2} & \cdots \\ {\scriptstyle\theta_1}\big\downarrow & & {\scriptstyle\theta_2}\big\downarrow & & \\ W_1 & \xrightarrow{\psi_1} & W_2 & \xrightarrow{\psi_2} & \cdots \end{array}$$

and direct limits V_∞ and W_∞ of the rows, the mappings θ_n uniquely determine a completely isometric injection $\theta_\infty : V_\infty \to W_\infty$. One can, in particular, use this to prove that direct limits are essentially unique.

If V is an operator space and we fix a conjugate space $\overline{V}$ (see §A.3), then we may define a matrix norm on $\overline{V}$ by letting

$$\|[\overline{v}_{i,j}]\| = \left\|\overline{[v_{i,j}]}\right\| = \|[v_{i,j}]\|. \tag{3.1.4}$$

It is a simple matter to check that this is an operator space matrix norm on $\overline{V}$, and we refer to $\overline{V}$ with this structure as the *conjugate operator space* of V. Alternatively, let us suppose that H is a Hilbert space with a conjugate space $\overline{H}$. For each $T \in \mathcal{B}(H)$, we define a mapping $\overline{T} : \overline{V} \to \overline{V}$ by letting $\overline{T}(\overline{v}) = \overline{T(v)}$. It is easy to verify that if we are given a completely isometric realization

$$\pi : V \hookrightarrow \mathcal{B}(H),$$

then the preceding operator space structure is also determined by the mapping

$$\overline{\pi} : \overline{V} \hookrightarrow \mathcal{B}(\overline{H}) : \overline{v} \mapsto \overline{\pi(v)}.$$

3.2 DUAL SPACES AND MAPPING SPACES

There is a natural operator space structure on the *mapping space* $\mathcal{CB}(V, W)$ for any operator spaces V and W, which we will describe below. For simplicity we first consider the dual space

$$V^* = \mathcal{B}(V, \mathbb{C}) = \mathcal{CB}(V, \mathbb{C}) \tag{3.2.1}$$

(see Proposition 2.2.2). Our task is to define $M_n(V^*)$ by introducing an appropriate norm on $\mathbb{M}_n(V^*)$. For this purpose, it is instructive to first consider the parallel situation for normed spaces.

If E is a normed space and $n \in \mathbb{N}$, then each $f = (f_1, \ldots, f_n) \in (E^*)^n$ determines a function $f : E \to \mathbb{C}^n$ by

$$f(v) = (f_1(v), \ldots, f_n(v)).$$

This correspondence determines an isometric identification

$$\ell^n_\infty(E^*) = \mathcal{B}(E, \ell^n_\infty).$$

By analogy, each matrix $f = [f_{i,j}] \in M_n(V^*)$ determines a linear mapping $f : V \to M_n$, where

$$f(v) = [f_{i,j}(v)] .$$

This gives us a linear isomorphism $\mathbb{M}_n(V^*) \cong \mathcal{CB}(V, M_n)$, which we use to determine the norm on $\mathbb{M}_n(V^*)$. Thus, if we let $M_n(V^*)$ be the corresponding normed space, we have the isometric identification

$$M_n(V^*) = \mathcal{CB}(V, M_n). \tag{3.2.2}$$

We emphasize that just as $\ell^n_\infty(E^*)$ is not equal to the Banach dual space of $\ell^n_\infty(E)$ (for a Banach space E), $M_n(V^*)$ *is not defined to be the Banach dual of* $M_n(V)$; in other words, we must distinguish $M_n(V^*)$ from $(M_n(V))^*$. Nevertheless, if we use the natural matrix pairing (1.1.27), the norm on $M_n(V)$ *does* determine that on $M_n(V^*)$. For any $f \in M_n(V^*)$, we have from Proposition 2.2.2,

$$\begin{aligned} \|f\| &= \sup\{\|f_n(v)\| : v \in M_n(V), \|v\| \leq 1\} \\ &= \sup\{\|\langle\langle f, v\rangle\rangle\| : v \in M_n(V), \|v\| \leq 1\} . \end{aligned} \tag{3.2.3}$$

Conversely, the norm on $M_n(V^*)$ determines that on $M_n(V)$ since we have from Lemma 2.3.4 that for any $v \in M_n(V)$,

$$\begin{aligned} \|v\| &= \sup\{\|f_n(v)\| : f \in \mathcal{CB}(V, M_n), \|f\|_{cb} \leq 1\} \\ &= \sup\{\|\langle\langle f, v\rangle\rangle\| : f \in \mathcal{CB}(V, M_n), \|f\|_{cb} \leq 1\}. \end{aligned} \tag{3.2.4}$$

The matrix norms on V^* determine an operator space. To see this, let us suppose that we are given $f \in M_m(V^*)$, $\alpha \in M_{n,m}$, and $\beta \in M_{m,n}$. Then

$$\begin{aligned} \|(\alpha f \beta)_r\| &= \|(\alpha \otimes I_r) f_r (\beta \otimes I_r)\| \\ &\leq \|\alpha \otimes I_r\| \, \|f_r\| \, \|\beta \otimes I_r\| \leq \|\alpha\| \, \|f\|_{cb} \, \|\beta\| , \end{aligned}$$

and hence

$$\|\alpha f \beta\|_{cb} \leq \|\alpha\| \, \|f\|_{cb} \, \|\beta\| ,$$

and we have **M2.** On the other hand, given $f \in M_m(V^*)$, $g \in M_n(V^*)$, and $v \in M_r(V)$ with $\|v\| \leq 1$,

$$\begin{aligned} \|(f \oplus g)_r(v)\| &= \|[f(v_{i,j}) \oplus g(v_{i,j})]\| = \|f_r(v) \oplus g_r(v)\| \\ &\leq \max\{\|f_r(v)\| , \|g_r(v)\|\} \leq \max\{\|f\|_{cb} , \|g\|_{cb}\} , \end{aligned}$$

and we have **M1′**:

$$\|f \oplus g\|_{cb} \leq \max\{\|f\|_{cb}, \|g\|_{cb}\}.$$

Since V^* is complete, it follows that this is the case for each matrix space $M_n(V^*)$ (see §2.1).

Given an operator space V, we let $\iota_V : V \hookrightarrow V^{**}$ be the *canonical inclusion* defined by

$$\langle \iota_V(v), f \rangle = \langle f, v \rangle.$$

Proposition 3.2.1 *For any operator space V, the canonical inclusion*

$$\iota_V : V \hookrightarrow V^{**}$$

is completely isometric.

Proof For any $v \in M_n(V)$ and $f \in M_n(V^*)$,

$$((\iota_V)_n(v))_n(f) = [\iota_V(v_{i,j})(f_{k,l})] = [f_{k,l}(v_{i,j})] = \langle\!\langle f, v \rangle\!\rangle. \tag{3.2.5}$$

It follows from (3.2.3) and (3.2.4) that

$$\begin{aligned}\|(\iota_V)_n(v)\|_{cb} &= \sup\{\|(\iota_V)_n(v)_n(f)\| : f \in \mathcal{CB}(V, M_n), \|f\|_{cb} \leq 1\} \\ &= \sup\{\|\langle\!\langle f, v \rangle\!\rangle\| : f \in \mathcal{CB}(V, M_n), \|f\|_{cb} \leq 1\} = \|v\|.\end{aligned}$$

Thus, $(\iota_V)_n$ is isometric for each n, and ι_V is a complete isometry. □

If $\varphi : V \to W$ is a completely bounded mapping of operator spaces, then we let $\varphi^* : W^* \to V^*$ be the dual Banach space mapping (see §A.2). For any $v \in M_n(V)$ and $g \in M_m(W^*)$,

$$\langle\!\langle g, \varphi_n(v) \rangle\!\rangle = [g_{k,l}(\varphi(v_{i,j}))] = [\varphi^*(g_{k,l})(v_{i,j})] = \langle\!\langle (\varphi^*)_m(g), v \rangle\!\rangle. \tag{3.2.6}$$

Proposition 3.2.2 *Given operator spaces V and W, and a completely bounded mapping $\varphi : V \to W$, we have*

$$\|(\varphi^*)_n\| = \|\varphi_n\| \tag{3.2.7}$$

for all $n \in \mathbb{N}$, and

$$\|\varphi^*\|_{cb} = \|\varphi\|_{cb}. \tag{3.2.8}$$

Proof The second relation is immediate from the first. That in turn follows from the calculation

$$\|(\varphi^*)_n\| = \sup\{\|\langle\!\langle (\varphi^*)_n(g), v \rangle\!\rangle\|\} = \sup\{\|\langle\!\langle g, \varphi_n(v) \rangle\!\rangle\|\} = \|\varphi_n\|,$$

where the supremum is taken over all $g \in M_n(V^*)$ and $v \in M_n(V)$ of norm ≤ 1. □

We also note that given a completely bounded mapping $\varphi \in \mathcal{CB}(V, W)$, its second adjoint mapping $\varphi^{**} : V^{**} \to W^{**}$ is in $\mathcal{CB}(V^{**}, W^{**})$ with $\|\varphi^{**}\|_{cb} = \|\varphi\|_{cb}$ and φ^{**} restricted to V is equal to φ.

Given an operator space V, we say that a net

$$f_\lambda = [f^\lambda_{k,l}] \in M_m(V^*) = \mathcal{CB}(V, M_m),$$

with λ in a directed index set Λ, *converges in the weak* topology* to

$$f = [f_{k,l}] \in M_m(V^*)$$

if for any $v \in M_n(V)$ ($n \in \mathbb{N}$ arbitrary) the net of matrices

$$\langle\langle f_\lambda, v\rangle\rangle = [f^\lambda_{k,l}(v_{i,j})] \quad (\lambda \in \Lambda)$$

converges in the norm topology to

$$\langle\langle f, v\rangle\rangle = [f_{k,l}(v_{i,j})].$$

This will be the case if and only if each net $(f^\lambda_{k,l})_{\lambda\in\Lambda}$ converges weakly* to $f_{k,l}$. Thus, in contrast to the norm on V^*, the weak* topology on V^* determines the weak* topology on each $M_n(V^*)$. In particular, if W is a weak* closed subspace of V^*, then each space $M_n(W)$ is weak* closed in $M_n(V^*)$.

It is evident that a mapping

$$f = [f_{k,l}] : V \to M_n$$

is continuous if and only if the linear functionals $f_{k,l} : V \to \mathbb{C}$ are continuous. Similarly, a mapping

$$F = [F_{i,j}] : V^* \to M_n$$

is continuous in the weak* topology if and only if each $F_{i,j}$ is weak* continuous. The latter implies that $F_{i,j} = \iota_V(v_{i,j})$ for some $v_{i,j} \in V$, and thus $F = (\iota_V)_n(v)$. If we let $\mathcal{CB}^\sigma(V^*, W^*)$ denote the weak* continuous completely bounded mappings $\varphi : V^* \to W^*$, and $\iota_V : V \hookrightarrow V^{**}$ be the canonical injection, then

$$\mathcal{CB}^\sigma(V^*, M_n) = (\iota_V)_n(M_n(V)). \tag{3.2.9}$$

In order to illustrate the notion of duality for operator spaces, let us show that the first two Banach space identifications in (1.3.15) are in fact operator space dualities. Given a Hilbert space H, we consider the compact operators $\mathcal{K}(H)$ as an operator subspace of $\mathcal{B}(H)$. On the other hand, we use the scalar pairing (1.3.16) to identify the trace class operators $\mathcal{T}(H)$ with the subspace of weak* continuous functionals in the operator space dual $\mathcal{B}(H)^*$.

Theorem 3.2.3 *Given a Hilbert space H, we have the operator space dualities*

$$\mathcal{B}(H) \cong \mathcal{T}(H)^*, \quad \mathcal{T}(H) \cong \mathcal{K}(H)^*.$$

Proof With the identification

$$M_n(\mathcal{B}(H)^*) \cong \mathcal{CB}(\mathcal{B}(H), M_n),$$

it is clear that $M_n(\mathcal{B}(H)_*)$ consists of the weak* continuous linear functionals $\omega : \mathcal{B}(H) \to M_n$. To show that $\mathcal{B}(H)$ is the operator space dual of $\mathcal{T}(H)$, it suffices to prove that the isometry $\mathcal{B}(H) \to (\mathcal{T}(H))^*$ is a complete isometry. This is equivalent to proving that if $b_0 \in M_n(\mathcal{B}(H))$, then

$$\|b_0\| = \sup\{\|\langle\langle b_0, \omega\rangle\rangle\| : \|\omega\| = 1, \omega \in M_n(\mathcal{T}(H))\}.$$

For any contractive $\omega \in M_n(\mathcal{T}(H))$,

$$\|\langle\langle b_0, \omega\rangle\rangle\| = \|\omega_n(b_0)\| \leq \|\omega\|_{cb}\|b_0\| \leq \|b_0\|.$$

On the other hand, given $\varepsilon > 0$, we may find unit vectors $\eta = (\eta_j)$, $\xi = (\xi_j) \in H^n$ for which

$$|\langle b_0\eta \mid \xi\rangle| \geq \|b_0\| - \varepsilon.$$

We let H_1 (respectively, H_2) be the linear span of the vectors $\eta_j \in H$ (respectively, $\xi_j \in H$), and we fix isometries s_k of H_k into $\mathbb{C}^n$ $(k = 1, 2)$. If we let $r_k = s_k e_k$, where e_k is the projection of H onto H_k, then

$$\omega : \mathcal{B}(H) \to M_n : b \mapsto r_2 b r_1^*$$

is a weak* continuous complete contraction, and for any $b \in M_n(\mathcal{B}(H))$, we have

$$\omega_n(b) = r_2^{(n)} b r_1^{(n)*},$$

where $r_k^{(n)} = r_k \oplus \cdots \oplus r_k$. Since $\eta \in H_1^n$ and $\xi \in H_2^n$,

$$\begin{aligned}\|\langle\langle b_0, \omega\rangle\rangle\| = \|\omega_n(b_0)\| &= \|r_1^{(n)} b_0 r_0^{(n)*}\| \\ &\geq |\langle b_0\eta \mid \xi\rangle| \geq \|b_0\| - \varepsilon,\end{aligned}$$

and thus $\mathcal{B}(H)$ is the operator dual of $\mathcal{T}(H)$.

Turning to the second duality, the isometric identification

$$\mathcal{B}(H^n) \cong \mathcal{K}(H^n)^{**}$$

implies that the unit ball of $M_n(\mathcal{K}(H)) = \mathcal{K}(H^n)$ is weak* dense in the unit ball of $M_n(\mathcal{B}(H)) = \mathcal{B}(H^n)$. Thus, if $\omega \in M_n(\mathcal{T}(H))$, then it follows from the preceding result that

$$\begin{aligned}\|\omega\| &= \sup\{\|\langle\langle b, \omega\rangle\rangle\| : \|b\| = 1, b \in M_n(\mathcal{B}(H))\} \\ &= \sup\{\|\langle\langle k, \omega\rangle\rangle\| : \|k\| = 1, k \in M_n(\mathcal{K}(H))\},\end{aligned}$$

and we conclude that the usual isometry $\mathcal{T}(H) \to \mathcal{K}(H)^*$ is a complete isometry. □

If W is a weak* closed subspace of $\mathcal{B}(H)$ for some Hilbert space H, then we have from Banach space theory the natural weak* homeomorphic isometry

$$W = (\mathcal{T}(H)/W_\perp)^*,$$

where $W_\perp$ is the preannihilator of W in $\mathcal{T}(H)$. We shall see in Proposition 4.2.2 that this is a *complete* isometry, i.e. W is the operator space dual of the complete operator space $V = \mathcal{T}(H)/W_\perp$. In the next result we show that all dual operator spaces arise in this fashion. Given an operator space W which is the dual of a complete operator space V, and a Hilbert space H, we say that a mapping $\varphi : W \to \mathcal{B}(H)$ is a *dual realization* of W on H if it is a weak* homeomorphic completely isometric injection.

Proposition 3.2.4 *If V is a complete operator space, then V^* has a dual realization on a Hilbert space H.*

Proof We let $\mathfrak{s}_n^\sigma = M_n(V)_{\|\cdot\|\leq 1}$. We have from Lemma 2.2.2 that if $f \in M_n(V^*) = \mathcal{CB}(V, M_n)$, then

$$\|f\| = \sup\{\|\langle\langle\varphi, f\rangle\rangle\| : \varphi \in \mathfrak{s}_n^\sigma\}.$$

We define $\mathfrak{s}^\sigma = \bigcup_{n\in\mathbb{N}} \mathfrak{s}_n^\sigma$ and we let $H = \bigoplus_{\varphi\in\mathfrak{s}^\sigma} \mathbb{C}^{n(\varphi)}$, where $n = n(\varphi)$ is the integer with $\varphi \in \mathfrak{s}_n^\sigma$. The argument in the proof of Theorem 2.3.5 shows that the mapping

$$\Phi : V^* \to \mathcal{B}(H) : f \mapsto (\varphi(f))_{\varphi\in\mathfrak{s}^\sigma} \tag{3.2.10}$$

is a complete isometry. It is obvious that the mapping Φ is continuous in the weak* topology, and, since $V^*_{\|\cdot\|\leq 1}$ is weak* compact, that is also the case for $\Phi(V^*_{\|\cdot\|\leq 1})$. But the latter is just the closed unit ball of $\Phi(V^*)$ and, since V is complete, $\Phi(V^*)$ is a weak* closed subspace of $\mathcal{B}(\mathcal{H})$ (see the appendix). Finally, Φ is one-to-one and weak* continuous on $V^*_{\|\cdot\|\leq 1}$, and thus it is a weak* homeomorphism on that set. Since V is complete, Φ maps V^* weak* homeomorphically onto its image (see §A.2). □

As in the classical theory of normed spaces, pairings may be used to define *dual operator space structures*. Let us suppose that we are given a linear space V, an operator space V', and a pairing $\langle\cdot,\cdot\rangle : V \times V' \to \mathbb{C}$ for which the functionals

$$V' \to \mathbb{C} : w \mapsto \langle v, w\rangle$$

are bounded for all $v \in V$. We use the injection $V \hookrightarrow (V')^*$ to determine the *dual operator structure* on V. Given a matrix $v \in M_n(V)$, we have from (3.2.3) that the *dual operator space matrix norm* $\|\cdot\|'$ is determined by

$$\|v\|' = \sup\{\|\langle\langle v, w\rangle\rangle\| : w \in M_n(V'), \|w\| \leq 1\}. \tag{3.2.11}$$

Turning from dual spaces to general mapping spaces, let us suppose that we are given operator spaces V and W. Each $\varphi = [\varphi_{i,j}] \in M_m(\mathcal{CB}(V, W))$ determines a mapping $\varphi : V \to M_m(W)$ by the convention $\varphi(v) = [\varphi_{i,j}(v)]$ (see (1.1.31)). We use the resulting linear identification

$$\mathbb{M}_m(\mathcal{CB}(V, W)) \cong \mathcal{CB}(V, M_m(W))$$

and the completely bounded norm on the second space to define a norm on $\mathbb{M}_m(\mathcal{CB}(V, W))$, and we let $M_m(\mathcal{CB}(V, W))$ denote the corresponding normed space. Thus, by definition we have the natural isometric identification

$$M_m(\mathcal{CB}(V,W)) = \mathcal{CB}(V, M_m(W)) . \tag{3.2.12}$$

Exactly the same argument used for dual spaces may be used to show that $\mathcal{CB}(V,W)$ is an operator space. It is useful, however, to consider another approach, which avoids the representation theorem. For each integer n we let

$$\mathfrak{d}_n = M_n(V)_{\|\cdot\|\leq 1},$$

and for each $v \in \mathfrak{d}_n$ we let $W_v = W$. We have a natural mapping

$$\mathcal{CB}(V,W) \hookrightarrow \prod_{n\in\mathbb{N}} \prod_{v\in\mathfrak{d}_n} M_n(W_v) : \varphi \mapsto ((\varphi_n(v))),$$

where for each $\varphi \in \mathcal{CB}(V,W)$, $(\varphi_n(v))_{v\in\mathfrak{d}_n}$ is bounded since

$$\|\varphi_n(v)\| \leq \|\varphi\|_{cb} \|v\| \leq \|\varphi\|_{cb} .$$

It is immediate that this determines a complete isometry of $\mathcal{CB}(V,W)$ into the product operator space, and thus $\mathcal{CB}(V,W)$ is an operator space.

Proposition 3.2.5 *If W is complete, then so is $\mathcal{CB}(V,W)$.*

Proof Let us suppose that W is complete. It suffices to show that $\mathcal{CB}(V,W)$ is a closed subspace of $\mathcal{B}(V,W)$. Given any Cauchy sequence $\{\varphi_n\} \in \mathcal{CB}(V,W)$, it is clear that $\{\varphi_n\}$ is a Cauchy sequence in the space of all bounded linear mappings $\mathcal{B}(V,W)$. From classical Banach space theory, $\mathcal{B}(V,W)$ is complete, and thus there is a bounded linear mapping $\varphi : V \to W$ such that φ_n converges to φ in the norm topology, i.e. $\|\varphi_n - \varphi\| \to 0$.

Since $\{\varphi_n\}$ is Cauchy in $\mathcal{CB}(V,W)$, for any $\varepsilon > 0$ there exists a sufficiently large integer $N(\varepsilon) > 0$ such that whenever $n, m > N(\varepsilon)$, we have

$$\|\varphi_n - \varphi_m\|_{cb} < \varepsilon .$$

Given any $v = [v_{i,j}] \in M_p(V)$ and $p \in \mathbb{N}$, we have

$$\|(\varphi_n - \varphi_m)_p(v)\| \leq \|\varphi_n - \varphi_m\|_{cb}\|v\| < \varepsilon\|v\| .$$

Since $\varphi_m(v_{i,j})$ converges to $\varphi(v_{i,j})$ in W,

$$\|(\varphi_n - \varphi)_p(v)\| \leq \varepsilon\|v\|,$$

and thus $\|\varphi_n - \varphi\|_{cb} \leq \varepsilon$. It follows that $\varphi \in \mathcal{CB}(V,W)$ and φ_n converges to φ in $\mathcal{CB}(V,W)$. □

Let us suppose that we are instead given a *normed* space E and an operator space W. Then we may also regard $\mathcal{B}(E,W)$ as an operator space. We again identify an $n \times n$ matrix $\varphi = [\varphi_{i,j}]$ of bounded linear mappings $\varphi_{i,j} : E \to W$ with a bounded linear mapping $\varphi : E \to M_n(W)$.

The corresponding isomorphism

$$\mathbb{M}_n(\mathcal{B}(E,W)) \cong \mathcal{B}(E, M_n(W)) \tag{3.2.13}$$

determines a norm on $\mathbb{M}_n(\mathcal{B}(E,W))$. If we let $\mathfrak{d}_1 = E_{\|\cdot\|\leq 1}$ and $W_v = W$, then for each $v \in \mathfrak{d}_1$ we have a natural completely isometric inclusion

$$\mathcal{B}(E,W) \hookrightarrow \prod_{v\in\mathfrak{d}_1} W_v : \varphi \mapsto (\varphi(v)),$$

and thus $\mathcal{B}(E,W)$ is an operator space with these matrix norms. We shall use this construction in our discussion of the 'minimal quantization' in the next section.

3.3 THE MIN AND MAX QUANTIZATIONS

In §2.1 we observed that every normed space is isometric to a linear space of operators. In this section we begin a systematic investigation of 'quantization' procedures.

We let $\mathfrak{N}$ denote the category of normed spaces, in which the objects are the normed spaces and the morphisms are the bounded linear mappings. Similarly, we let $\mathfrak{O}$ be the category of operator spaces with the morphisms being the completely bounded mappings. We have a natural 'forgetful' functor $N : \mathfrak{O} \to \mathfrak{N}$ which maps an operator space into its underlying normed space. We say that a functor $Q : \mathfrak{N} \to \mathfrak{O}$ is a *strict quantization* if for each normed space E,

$$N \circ Q(E) = E,$$

and for each bounded linear mapping of normed spaces $\varphi : E \to F$, the corresponding mapping $Q(\varphi) : Q(E) \to Q(F)$ satisfies

$$\|Q(\varphi)\|_{cb} = \|\varphi\|.$$

For any Banach space E, we let

$$\mathfrak{b}_r(E) = \mathcal{B}(E, M_r)_{\|\cdot\|\leq 1}$$

and

$$\mathfrak{b}(E) = \bigcup_{r\in\mathbb{N}} \mathfrak{b}_r(E),$$

and we define the matrix norms $\|x\|_{\min}$ and $\|x\|_{\max}$ for $x \in M_n(E)$ by

$$\|x\|_{\min} = \sup\{\|f_n(x)\| : f \in \mathfrak{b}_1(E)\} \tag{3.3.1}$$

and

$$\|x\|_{\max} = \sup\{\|f_n(x)\| : f \in \mathfrak{b}(E)\}. \tag{3.3.2}$$

To see that these are indeed operator space matrix norms, it suffices to consider the linear injections

$$E \hookrightarrow \ell_\infty(\mathfrak{b}_1(E)) : x \mapsto (f(x)) \tag{3.3.3}$$

and

$$E \hookrightarrow \prod_{r\in\mathbb{N}} \ell_\infty(\mathfrak{b}_r(E), M_r) : x \mapsto ((f(x))_{f\in\mathfrak{b}_r(E)})_{r\in\mathbb{N}}, \tag{3.3.4}$$

respectively. We have the natural operator space identifications

$$\ell_\infty(\mathfrak{b}_1(E)) = \prod_{\mathfrak{b}_1(E)} \mathbb{C}$$

and

$$\prod_{r\in\mathbb{N}} \ell_\infty(\mathfrak{b}_r(E), M_r) = \prod_{r\in\mathbb{N}} \prod_{\mathfrak{b}_r(E)} M_r.$$

Since the relative matrix norms on E are given by (3.3.1) and (3.3.2), it is evident that these determine operator spaces, which we denote by $\min E$ and $\max E$, respectively. We refer to these operator spaces as the *minimal* and the *maximal quantization* of E.

If V is an operator space and $v \in M_n(V)$, then it follows from Lemma 2.3.4 that

$$\|v\| = \sup\{\|f_n(v)\| : f \in \mathfrak{s}_n(V)\}. \tag{3.3.5}$$

Since

$$\mathfrak{b}_1(V) \subseteq \mathfrak{s}_n(V) \subseteq \mathfrak{b}(V),$$

we conclude that

$$\|v\|_{\min} \leq \|v\| \leq \|v\|_{\max} \tag{3.3.6}$$

for any $v \in M_n(V)$.

For any operator space V, we have the isometric identification

$$\mathcal{CB}(V, \min E) = \mathcal{B}(V, E), \tag{3.3.7}$$

or equivalently, for any linear mapping $\varphi : V \to E$,

$$\|\varphi : V \to \min E\|_{cb} = \|\varphi : V \to E\|. \tag{3.3.8}$$

To see this, let us suppose that $v \in M_n(V)$ and that $\|v\| \leq 1$. Then

$$\|\varphi_n(v)\|_{\min} = \sup\{\|f_n \circ \varphi_n(v)\| : \|f : E \to \mathbb{C}\| \leq 1\}.$$

But $\|f\| \leq 1$ implies that

$$\|(f\circ\varphi)_n\| \leq \|f\circ\varphi\|_{cb} = \|f\circ\varphi\| \leq \|f\|\,\|\varphi\| \leq \|\varphi\|,$$

and thus $\|\varphi_n(v)\| \leq \|\varphi\|$ for all n.

If $\varphi : E \to F$ is a contraction, then since $\varphi : \min E \to F$ is a contraction,

$$\min \varphi = \varphi : \min E \to \min F$$

is completely contractive. We conclude that min is a strict quantization functor. If $\varphi : E \to F$ is an isometric injection, then it follows from (3.3.1) that $\min \varphi$ is completely isometric since we may extend any $f \in \mathfrak{b}_1(E)$ to a functional $\tilde{f} \in \mathfrak{b}_1(F)$.

For any normed space E and operator space W,

$$\mathcal{CB}(\max E, W) = \mathcal{B}(E, W), \tag{3.3.9}$$

i.e. for any linear mapping $\varphi : E \to W$,

$$\|\varphi : \max E \to W\|_{cb} = \|\varphi : E \to W\|. \tag{3.3.10}$$

To prove this, it suffices to show that if $\|\varphi\| \leq 1$, then $\|\varphi\|_{cb} \leq 1$. If $v \in M_n(\max E)$, then

$$\begin{aligned}\|\varphi_n(v)\| &= \sup\{\|f_n \circ \varphi_n(v)\| : \|f : W \to M_r\|_{cb} \leq 1, r \in \mathbb{N}\} \\ &\leq \sup\{\|(f \circ \varphi)_n(v)\| : \|f : W \to M_r\| \leq 1, r \in \mathbb{N}\} \\ &\leq \sup\{\|g_n(v)\| : \|g : V \to M_r\| \leq 1 : r \in \mathbb{N}\} = \|v\|_{\max}.\end{aligned}$$

In particular, if we are given normed spaces E and F and a contraction $\varphi : E \to F$, then since $\varphi : E \to \max F$ is a contraction,

$$\max \varphi : \max E \to \max F$$

is a complete contraction. Thus max is a strict quantization.

If $\varphi : E \to F$ is an isometric injection, then $\max \varphi$ need not be completely isometric. However, if there is, in addition, a contraction $\psi : F \to E$ such that $\psi \circ \varphi = id_E$, then $\max \varphi$ is completely isometric since

$$\max \psi \circ \max \varphi = id_{\max E}.$$

This is also the case for the canonical injection $E \hookrightarrow E^{**}$ since any contraction $f : E \to M_n$ automatically extends to the contraction $f^{**} : E^{**} \to M_n$.

Given an operator space V, let us consider the diagram of identity mappings $V \to V \to V$. If we regard the first and third spaces V as normed spaces, then we obtain the diagram of complete contractions

$$\max V \to V \to \min V \tag{3.3.11}$$

and thus we have re-proved (3.3.6). We say that an operator space V is *minimal* or *abelian* if $V = \min V$ and *maximal* if $V = \max V$.

If D is a subset of $V^*_{\|\cdot\|\leq 1}$ for which the absolutely convex hull $|\mathrm{co}|(D)$ (see §A.2) is weak* dense in $V^*_{\|\cdot\|\leq 1}$, then for any $v \in M_n(V)$,

$$\|v\|_{\min} = \sup\{\|f_n(v)\| : f \in D\}.$$

To see this, let us suppose that $\|f_n(v)\| \leq 1$ for all $f \in D$. If $g = \sum_k t_k f_k$, where $f_k \in D$ and $\sum |t_k| \leq 1$, then

$$\|g_n(v)\| = \left\| \sum_k t_k (f_k)_n(v) \right\| \leq \sum_k |t_k| \leq 1.$$

Given an arbitrary element $g \in V^*_{\|\cdot\|\leq 1}$, we may find a net $g_\beta \in |\mathrm{co}|(D)$ converging to g in the weak* topology. Then $g_\beta(v_{i,j})$ converges to $g(v_{i,j})$

for each i, j, and thus the matrices $(g_\alpha)_n(v)$ converge to $g_n(v)$ in the norm topology. It follows that $\|g_n(v)\| \leq 1$, and thus $\|v\|_{\min} \leq 1$.

If Ω is a locally compact Hausdorff space and $\mathcal{Z} = C_0(\Omega)$ is the corresponding commutative C^*-algebra, then we have a natural mapping

$$\delta : \Omega \to \mathcal{Z}^* : \delta(x)(v) = v(x).$$

It is a simple consequence of the bipolar theorem that $|\mathrm{co}|\,(\delta(\Omega))$ is weak* dense in $\mathcal{Z}^*_{\|\cdot\|\leq 1}$, and from our preceding observation, if $a = [a_{i,j}]$ is an element of $\in M_n(\mathcal{Z})$, then

$$\|a\|_{\min} = \sup\left\{\|[f(a_{i,j})]\| : f \in \delta(\Omega)\right\} = \sup\left\{\|[a_{i,j}(x)]\| : x \in \Omega\right\},$$

which is the canonical norm on the C^*-algebra

$$M_n(\mathcal{Z}) \cong C_0(\Omega, M_n).$$

We conclude that as an operator space, $\mathcal{Z}$ is just the minimal quantization of its underlying Banach space, i.e. $\mathcal{Z} = \min \mathcal{Z}$. This result, together with (3.3.7), provides another proof of Proposition 2.2.6. Similarly, if E is a normed space, then each isometric injection $E \hookrightarrow C_0(\Omega)$ determines a completely isometric injection $\min E \hookrightarrow C_0(\Omega)$.

For any $v \in \mathbb{M}_n(V)$, the linear mappings

$$V^* \to M_n : f \mapsto [f(v_{i,j})]$$

are just the weak* linear mappings from V^* into M_n, and thus from (3.3.1) we have the isometric identification

$$M_n(\min E) = \mathcal{B}^\sigma(E^*, M_n). \tag{3.3.12}$$

Given a normed space E, $n \in \mathbb{N}$ and a linear mapping $f : E \to M_n$, the second adjoint f^{**} provides an extension of f to a weak* continuous mapping $E^{**} \to M_n$. This provides us with a natural identification

$$\mathcal{B}(E, M_n) = \mathcal{B}^\sigma(E^{**}, M_n).$$

Applying (3.2.1), we have the isometries

$$\begin{aligned} M_n((\max E)^*) &= \mathcal{CB}(\max E, M_n) = \mathcal{B}(E, M_n) \\ &= \mathcal{B}^\sigma(E^{**}, M_n) = M_n(\min E^*), \end{aligned}$$

and thus the identification

$$(\max E)^* = \min E^*. \tag{3.3.13}$$

Given a normed space E, and an isometric injection $E \hookrightarrow \mathcal{Z}$ where $\mathcal{Z}$ is a commutative C^*-algebra, we have a corresponding commutative diagram

$$\begin{array}{ccc} E & \longrightarrow & \mathcal{Z} \\ \downarrow & & \downarrow \\ E^{**} & \longrightarrow & \mathcal{Z}^{**} \end{array},$$

where the first column is an isometry, the second column is a complete isometry, and both rows are isometric. Since $\mathcal{Z}^{**}$ is again a commutative C^*-algebra, it determines the minimal operator space structure on E^{**}; hence

$$(\min E)^{**} = \min E^{**}. \tag{3.3.14}$$

Thus, we have the complete isometries

$$(\max E^*)^* = \min E^{**} = (\min E)^{**},$$

and since these identifications are compatible with the dualities, we have the complete isometry

$$\max E^* = (\min E)^* . \tag{3.3.15}$$

Proposition 3.3.1 *If V is an operator space, then V is minimal if and only if it is completely isometric to a subspace of a commutative C^*-algebra.*

Proof If V is a minimal operator space and we let $V \hookrightarrow \mathcal{Z}$ be an isometric inclusion into a commutative C^*-algebra, then $V = \min V \hookrightarrow \min \mathcal{Z} = \mathcal{Z}$ is completely isometric. Conversely, if we have a complete isometry $V \hookrightarrow \mathcal{Z}$, then since

$$\min V \hookrightarrow \min \mathcal{Z} = \mathcal{Z}$$

is also completely isometric, we must have $\min V = V$. □

We next consider the dual result. Strictly speaking, this is out of order since we use a result from the next chapter.

Proposition 3.3.2 *A complete operator space V is maximal if and only if there is a set $\mathfrak{s}$ and a complete quotient mapping $\psi : \max \ell_1(\mathfrak{s}) \twoheadrightarrow V$.*

Proof If V is a maximal operator space, then from (3.3.13), V^* is minimal. From (3.3.1), for each $f \in M_n(V^*)$,

$$\|f\| = \sup\{\|[F(f_{i,j})]\| : F \in V^{**}, \|F\| \leq 1\}.$$

Given a contraction $F \in V^{**}$, we may choose a net of contractions $v_\nu \in V$ which converges weakly* to F. It follows that the finite matrices $[f_{i,j}(v_\nu)]$ converge in the norm topology to $[F(f_{i,j})]$, and we conclude that

$$\|f\| = \sup\{[f_{i,j}(x)] : x \in V, \|x\| \leq 1\}.$$

If we let $\mathfrak{s} = \mathfrak{d}_1$ denote the closed unit ball of V, the correspondence $f \mapsto (f(x))_{x\in\mathfrak{s}}$ determines a weak* homeomorphic completely isometric mapping φ of V^* onto a weak* closed subspace E of $\ell_\infty(\mathfrak{s})$ (see the proof of Proposition 3.2.4). Since

$$\min \ell_\infty(\mathfrak{s}) = (\max \ell_1(\mathfrak{s}))^*,$$

it follows from Corollary 4.1.9 that φ is the adjoint of a complete quotient mapping $\psi : \max \ell_1(\mathfrak{s}) \twoheadrightarrow V$.

Conversely, given such a mapping ψ, for any contraction $\varphi : V \to M_n$, the composition $\varphi \circ \psi : \ell_1(\mathfrak{s}) \to M_n$ is a contraction, and thus the composition $\varphi \circ \psi : \max \ell_1(\mathfrak{s})) \to M_n$ is completely contractive. Since ψ is a complete quotient mapping, it follows that $\varphi : V \to M_n$ is completely contractive. For any matrix $v \in M_n(V)$, we conclude from (3.3.5) and (3.3.2) that $\|v\| = \|v\|_{\max}$. □

Finally, we conclude this section with a concrete description of the operator spaces $\min \ell_1^n$ and $\max \ell_1^n$.

Lemma 3.3.3 *Let z_j and u_j ($j = 1, \ldots, n$) be the canonical generators of $C^*(\mathbb{Z}^n)$ and $C^*(\mathbb{F}_n)$, respectively. Consider the operator spaces*

$$Z_n = \mathbb{C}z_1 + \cdots + \mathbb{C}z_n \subseteq C^*(\mathbb{Z}^n)$$

and

$$E_n = \mathbb{C}u_1 + \cdots + \mathbb{C}u_n \subseteq C^*(\mathbb{F}_n).$$

The linear mappings determined by $e_j \mapsto z_j$ and $e_j \mapsto u_j$ provide complete isometries

$$\min \ell_1^n \cong Z_n$$

and

$$\max \ell_1^n \cong E_n.$$

Proof We define

$$\theta : \ell_1^n \to Z_n : \alpha \mapsto \sum_{j=1}^{n} \alpha_j z_j.$$

The commutative group C^*-algebra $C^*(\mathbb{Z}^n)$ is $*$-isomorphic to $C(\mathbb{T}^n)$, the space of all continuous functions on the n-torus $\mathbb{T}^n$. In this case, each z_j ($j = 1, \ldots, n$) corresponds to the canonical jth variable function on the $\mathbb{T}^n$. It follows that

$$\begin{aligned}\left\|\sum \alpha_j z_j\right\|_{C^*(\mathbb{Z}^n)} &= \left\|\sum \alpha_j z_j\right\|_{C(\mathbb{T}^n)} \\ &= \sup\left\{\left|\sum \alpha_j z_j\right| : |z_j| \le 1, z_j \in \mathbb{C}\right\} = \sum |\alpha_j| = \|\alpha\|_1 .\end{aligned}$$

Therefore, θ is an isometry, from which it follows that

$$\theta : \min \ell_1^n \to \min Z_n = Z_n$$

is a complete isometry.

It is evident that the mapping

$$\psi : \ell_1^n \to E_n : \alpha \mapsto \sum \alpha_j u_j$$

is contractive, and thus the mapping

$$\psi : \max \ell_1^n \to E_n$$

is completely contractive. Let us fix a complete isometry

$$\varphi : \max \ell_1^n \to \mathcal{B}(H)$$

for some Hilbert space H. Then for each j, $b_j = \varphi(e_j)$ is a contractive operator on H, and we may let

$$v_j = \begin{bmatrix} b_j & \sqrt{1 - b_j b_j^*} \\ \sqrt{1 - b_j^* b_j} & -b_j^* \end{bmatrix} \in \mathcal{B}(H^2) \tag{3.3.16}$$

be the 'unitary dilation' of b_j. From the universal property of $\mathbb{F}_n$, the mapping $u_j \mapsto v_j$ extends uniquely to a $*$-homomorphism

$$\pi : C^*(\mathbb{F}_n) \to C^*(v_1, \ldots, v_n),$$

where we let $C^*(v_1, \ldots, v_n)$ denote the C^*-subalgebra of $\mathcal{B}(H^2)$ generated by $v_1, \ldots, v_n$. If we let V_n be the operator subspace of $\mathcal{B}(H^2)$ generated by $\{v_1, \ldots, v_n\}$, then the restriction $\pi_0 = \pi_{|E_n} : E_n \to V_n$ is completely contractive. If we define

$$P : \mathcal{B}(H^2) \to \mathcal{B}(H) : [a_{i,j}] \mapsto a_{11},$$

and we let $P_0 = P_{|V_n}$, then we obtain the commutative diagram of complete contractions

$$\begin{array}{ccccc}
\max \ell_1^n & \xrightarrow{\psi} & E_n & \subseteq & C^*(\mathbb{F}_n) \\
{\scriptstyle \varphi^{-1}} \uparrow & & {\scriptstyle \pi_0} \downarrow & & \downarrow {\scriptstyle \pi} \\
\varphi(\max \ell_1^n) & \xleftarrow{P_0} & V_n & \subseteq & C^*(v_1, \ldots, v_n) \\
\cap\!| & & \cap\!| & & \cap\!| \\
\mathcal{B}(H) & \xleftarrow{P} & \mathcal{B}(H^2) & = & \mathcal{B}(H^2)
\end{array}.$$

It is clear that the composition $\varphi^{-1} \circ P_0 \circ \pi_0$ is the inverse of ψ and is completely contractive. $\square$

These results can be refined. If we let $z_0 = 1$ in $C^*(\mathbb{Z}^{n-1})$ and $u_0 = I$ in $C^*(\mathbb{F}_{n-1})$, then the mappings $e_j \mapsto z_{j-1}$ and $e_j \mapsto u_{j-1}$ determine complete isometries

$$\theta' : \min \ell_1^n \to \min Z_{n-1} \subseteq C^*(\mathbb{Z}^{n-1})$$

and

$$\varphi' : \max \ell_1^n \to E_{n-1} \subseteq C^*(\mathbb{F}_{n-1}).$$

This can be proved by using the same argument as above. In particular, since $\mathbb{F}_{2-1} = \mathbb{Z}$, we see that

$$\max \ell_1^2 = \min \ell_1^2,$$

and taking the dual operator spaces (see (3.3.13) and (3.3.15))

$$\min \ell_\infty^2 = \max \ell_\infty^2.$$

We conclude that the two-dimensional Banach spaces ℓ_1^2 and ℓ_∞^2 can be quantized in only one way. Pisier (1996*b*) has shown that the two-dimensional Hilbert space ℓ_2^2 has a continuum of quantizations, three of which will be considered in the next section. Paulsen (1992, Theorem 2.14) proved that any Banach space of dimension greater than or equal to 5 has more than one quantization.

3.4 COLUMN AND ROW HILBERT OPERATOR SPACES

As we have already indicated, a normed space has a multiplicity of operator space matrix norms. These structures are particularly interesting in the case of Hilbert spaces. In this section we introduce the row and column Hilbert operator spaces, which play an important role in the theory of the Haagerup tensor product.

We recall that if we are given Hilbert spaces H and K, then the identification (2.1.4) determines an operator space structure on $\mathcal{B}(H, K)$. Thus, we may use the *column* isometry

$$C : H \cong \mathcal{B}(\mathbb{C}, H), \tag{3.4.1}$$

where $C(\xi)(a) = a\xi$ for $a \in \mathbb{C}$, to define an operator space matrix norm on H. To be more specific, if $\xi \in M_n(H)$, then we have a corresponding mapping

$$C_n(\xi) : \mathbb{C}^n \to H^n,$$

and we define a *column matrix norm* by

$$\|\xi\|_c = \|C_n(\xi)\|.$$

We let H_c denote H with this operator space structure, and we refer to it as the *column Hilbert operator space* or simply the *column Hilbert space* determined by H. We note that this formula may also be used to calculate the norms of rectangular matrices $\xi \in M_{m,n}(H_c)$.

The adjoint operator of $C(\xi)$ is given by

$$C(\xi)^* : H \to \mathbb{C} : \zeta \mapsto \langle \zeta \mid \xi \rangle$$

since for any $c \in \mathbb{C}$, $d = C(\xi)^*(\zeta)$ satisfies

$$c\bar{d} = \langle c \mid C(\xi)^*\zeta \rangle = \langle c\xi \mid \zeta \rangle = c\overline{\langle \zeta \mid \xi \rangle}.$$

It follows that for any $\xi, \eta \in H$,

$$C(\xi)^* C(\eta) = \langle \eta \mid \xi \rangle.$$

The norm of a matrix $\xi = [\xi_{i,j}] \in M_{m,n}(H_c)$ is given by

$$\begin{aligned} \|\xi\|_c &= \|C_{m,n}(\xi)\| = \|C_{m,n}(\xi)^* C_{m,n}(\xi)\|^{1/2} \\ &= \left\| \left[\sum_{k=1}^m C(\xi_{k,i})^* C(\xi_{k,j}) \right] \right\|^{1/2} = \left\| \left[\sum_{k=1}^m \langle \xi_{k,j} \mid \xi_{k,i} \rangle \right] \right\|^{1/2}. \end{aligned} \tag{3.4.2}$$

From the definition, we have the natural isometry

$$\varphi : M_{m,n}(H_c) \cong \mathcal{B}(\mathbb{C}^n, H^m)$$

for any $m, n \in \mathbb{N}$. However, both sides of this relation are operator spaces, and with these structures this is a complete isometry since for any $p \in \mathbb{N}$, we have the commutative diagram

$$\begin{array}{ccc} M_p(M_{m,n}(H_c)) & = & M_{pm,pn}(H_c) \\ {\scriptstyle \varphi_p}\downarrow & & \downarrow{\scriptstyle \varphi} \\ M_p(\mathcal{B}(\mathbb{C}^n, H^m)) & = & \mathcal{B}(\mathbb{C}^{pn}, H^{pm}) \end{array}$$

with φ isometric. In particular, we have the complete isometry

$$M_{m,1}(H_c) \cong H_c^m \tag{3.4.3}$$

since we have the complete isometry

$$M_{m,1}(H_c) \cong M_{m,1}(\mathcal{B}(\mathbb{C}, H)) = \mathcal{B}(\mathbb{C}, H^m) \cong H_c^m.$$

There is an alternative formula for this operator space matrix norm that is quite convenient. If we are given $\alpha^{(h)} \in M_n$ and orthonormal vectors $e_h \in H$, where $1 \leq h \leq p$, then we have from (3.4.2),

$$\begin{aligned} \left\| \sum_h \alpha^{(h)} \otimes e_h \right\|_c &= \left\| \left[\sum_h \alpha_{i,j}^{(h)} e_h \right] \right\|_c \\ &= \left\| \left[\sum_{g,h,k} \langle \alpha_{k,j}^{(h)} e_h \mid \alpha_{k,i}^{(g)} e_g \rangle \right] \right\|^{1/2} = \left\| \left[\sum_{k,h} \alpha_{k,j}^{(h)} \overline{\alpha}_{k,i}^{(h)} \right] \right\|^{1/2} \\ &= \left\| \sum_h \alpha^{(h)*} \alpha^{(h)} \right\|^{1/2} = \left\| \begin{bmatrix} \alpha^{(1)} \\ \vdots \\ \alpha^{(p)} \end{bmatrix} \right\|. \end{aligned} \tag{3.4.4}$$

The same formula may be used for rectangular matrices.

In order to define the row Hilbert operator space structure on H, we use the identification $\theta : \overline{H} \to H^*$ (see §A.3). The natural isometry

$$R : H \to H^{**} = \mathcal{B}(H^*, \mathbb{C}) = \mathcal{B}(\overline{H}, \mathbb{C}) \tag{3.4.5}$$

is given by

$$R(\zeta)(\overline{\xi}) = \theta(\overline{\xi})(\zeta) = \langle \zeta \mid \xi \rangle.$$

$\mathcal{B}(\overline{H}, \mathbb{C})$ is an operator space, and it thus determines an operator space matrix norm on H. We call this the *row Hilbert operator space* or simply the *row Hilbert space* structure, and we let H_r denote the corresponding operator space. By definition, each $\xi \in M_n(H)$ determines a corresponding operator

$$R_n(\xi) : \overline{H}^n \to \mathbb{C}^n,$$

and the corresponding *row matrix norm* is given by

$$\|\xi\|_r = \|R_n(\xi)\|.$$

For any $\xi \in H$, the adjoint mapping $R(\xi)^* : \mathbb{C} \to \overline{H}$ is given by $R(\xi)^*(c) = c\overline{\xi}$ since for any $\zeta \in H$,

$$\langle R(\xi)^*(c) \mid \overline{\zeta}\rangle = c\,\overline{(R(\xi)\overline{\zeta})} = c\,\overline{\langle \xi \mid \zeta\rangle} = \langle \zeta \mid \overline{c}\xi\rangle = \langle c\overline{\xi} \mid \overline{\zeta}\rangle.$$

It follows that for any $\xi, \eta \in H$,

$$R(\eta)R(\xi)^* = \langle \eta \mid \xi\rangle.$$

From the definition, if $\xi = [\xi_{i,j}] \in M_{m,n}(H_r)$, then

$$\begin{aligned}\|\xi\|_r &= \|R_{m,n}(\xi)\| = \|R_{m,n}(\xi)R_{m,n}(\xi)^*\|^{1/2} \\ &= \left\|\left[\sum_{k=1}^{n} R(\xi_{i,k})R(\xi_{j,k})^*\right]\right\|^{1/2} = \left\|\left[\sum_{k=1}^{n}\langle \xi_{i,k} \mid \xi_{j,k}\rangle\right]\right\|^{1/2}.\end{aligned}$$

By essentially the same argument we used above, we have the complete isometry

$$M_{1,n}(H_r) \cong H_r^n. \tag{3.4.6}$$

Furthermore, if e_h $(1 \leq h \leq p)$ are orthonormal in H, then

$$\begin{aligned}\left\|\sum_{h=1}^{p} \alpha^{(h)} \otimes e_h\right\|_r &= \left\|\sum_{h=1}^{p} \alpha^{(h)}\alpha^{(h)*}\right\|^{1/2} \\ &= \left\|\left[\alpha^{(1)} \ldots \alpha^{(p)}\right]\right\|.\end{aligned} \tag{3.4.7}$$

In order to illustrate the difference between these operator space matrix norms on H, let us suppose that $e_1, \ldots, e_p$ are orthonormal vectors in H, and then compare the column and row norms of the row matrix $[e_1 \ldots e_p]$. We have from (3.4.4),

$$\|[e_1 \ldots e_p]\|_c = \left\|\sum_{j=1}^{p} \varepsilon_{1,j} \otimes e_j\right\| = \left\|\sum_{j=1}^{p} \varepsilon_{j,1}\varepsilon_{1,j}\right\|^{1/2} = \left\|I_p\right\|^{1/2} = 1, \tag{3.4.8}$$

whereas from (3.4.7),

$$\|[e_1 \ldots e_p]\|_r = \left\|\sum_{h=1}^{p} 1\right\|^{1/2} = \sqrt{p}.$$

If $n = 1$, i.e. if $H = \mathbb{C}$, then for any $\xi \in \mathbb{C}$, $R(\xi) = C(\xi) = \xi$, and thus if $\xi = [\xi_{i,j}] \in M_n(\mathbb{C})$, then

$$\|\xi\|_c = \|\xi\|_r = \|[\xi_{i,j}]\|.$$

Theorem 3.4.1 *For any Hilbert spaces H and K, there is a natural completely isometric identification*

$$\mathcal{B}(H,K) \cong \mathcal{CB}(H_c, K_c).$$

Proof For any matrix

$$T = [T_{k,l}] \in M_n(\mathcal{B}(H,K)) = \mathcal{B}(H^n, K^n),$$

the corresponding mapping

$$\tilde{T} \in M_n(\mathcal{CB}(H_c, K_c)) = \mathcal{CB}(H_c, M_n(K_c))$$

is defined by $\tilde{T}(\xi) = [T_{k,l}(\xi)]$ for all $\xi \in H_c$. We wish to show that

$$\|\tilde{T}\|_{cb} = \|T\|.$$

If we fix a vector $\xi \in M_p(H_c)$, we can assume that

$$\xi = \sum_{j=1}^{r} \alpha_j \otimes f_j,$$

where $\alpha_j \in M_p$ and $f_j \in H$ $(1 \leq j \leq r)$ are orthonormal vectors. We let H_0 be the linear span of the f_j, and we fix an orthonormal basis g_i $(1 \leq i \leq q)$ for the linear span $K_0 = \sum_{k,l} T_{k,l}(H_0)$. We define $T_{k,l}(i,j)$ by the relations

$$T_{k,l}(f_j) = \sum_{i=1}^{q} T_{k,l}(i,j) g_i.$$

If we let $T_0(i,j) = [T_{k,l}(i,j)] \in M_{q,r}$, then the matrix

$$T_0 = [T_0(i,j)] \in M_{nq,nr}$$

satisfies $\|T_0\| \leq \|T\|$ since it is the matrix of the restriction of T to a mapping from H_0^n into K_0^n.

The mapping

$$\tilde{T}_p = id \otimes \tilde{T} : M_p \otimes H_c \to M_p \otimes M_n \otimes K_c$$

satisfies

$$\begin{aligned}
\tilde{T}_p(\xi) &= \sum_{j=1}^{r} \alpha_j \otimes \tilde{T}(f_j) = \sum_{j} \alpha_j \otimes \sum_{k,l=1}^{n} e_{k,l} \otimes T_{k,l}(f_j) \\
&= \sum_{j,k,l} \alpha_j \otimes e_{k,l} \otimes \sum_{i=1}^{q} T_{k,l}(i,j) g_i = \sum_{i,j} \alpha_j \otimes \sum_{k,l} T_{k,l}(i,j) e_{k,l} \otimes g_i \\
&= \sum_{i,j} \alpha_j \otimes T_0(i,j) \otimes g_i,
\end{aligned}$$

and thus

$$\left\|\tilde{T}_p(\xi)\right\|_{M_{pn}(H_c)} = \left\|\begin{bmatrix}\sum_j \alpha_j \otimes T_0(1,j)\\ \vdots \\ \sum_j \alpha_j \otimes T_0(q,j)\end{bmatrix}\right\|$$

$$= \left\|\begin{bmatrix} I_p \otimes T_0(1,1) \dots I_p \otimes T_0(1,r) \\ \vdots \qquad\qquad \vdots \\ I_p \otimes T_0(q,1) \dots I_p \otimes T_0(q,r)\end{bmatrix}\begin{bmatrix}\alpha_1 \otimes I_n \\ \vdots \\ \alpha_r \otimes I_n\end{bmatrix}\right\|$$

$$\leq \|I_p \otimes T_0\| \left\|\begin{bmatrix}\alpha_1 \otimes I_n \\ \vdots \\ \alpha_r \otimes I_n\end{bmatrix}\right\| \leq \|T\| \, \|\xi\|_{M_p(H_c)}.$$

It follows that

$$\|\tilde{T} : H_c \to M_n(H_c)\|_{cb} \leq \|T\|.$$

On the other hand, if $\xi = (\xi_l) \in H^n$, and we let e_j $(1 \leq j \leq p)$ be an orthonormal basis for the linear span of the ξ_l, then

$$\xi_l = \sum_{l,j} c_{(l,j)} e_j,$$

where the constants $c_{(l,j)} = \langle \xi_l \mid e_j \rangle$ satisfy

$$\sum \left|c_{(l,j)}\right|^2 = \|\xi\|^2.$$

If we use the isometric identification (3.4.3) of K^n and $M_{n,1}(K_c)$, then $T(\xi) \in K^n$ is given by

$$T(\xi) = \left[\sum_l T_{k,l}(\xi_l)\right]_{k=1,\dots,n} = \overbrace{\left[\sum_{l,j} T_{k,l}(e_j) c_{(l,j)}\right]}^{\in M_{n\times p,1}}$$

$$= \overbrace{[T_{k,l}(e_1) \dots T_{k,l}(e_p)]}^{\in M_{n,n\times p}(K_c)} \begin{bmatrix} c_{(1,1)} \\ \vdots \\ c_{(l,j)} \\ \vdots \\ c_{(n,p)}\end{bmatrix}$$

$$= \tilde{T}_{1,p}\left([e_1 \dots e_p]\right)\left[\, c_{(l,j)} \,\right],$$

and thus

$$\|T(\xi)\| \leq \left\|\tilde{T}_{1,p}\left([e_1 \dots e_p]\right)\right\| \left\|\left[\, c_{(l,j)} \,\right]\right\| \leq \left\|\tilde{T}\right\|_{cb} \|[e_1 \dots e_p]\|_c \, \|\xi\|.$$

From (3.4.8),

$$\|[e_1 \ldots e_p]\|_c = 1.$$

Hence,

$$\|T(\xi)\| \leq \|\tilde{T}\|_{cb}\ \|\xi\|,$$

and $\|T\| \leq \|\tilde{T}\|_{cb}$. □

We have from Theorem 3.4.1 the complete isometries

$$(H_c)^* = \mathcal{CB}(H_c, \mathbb{C}) = \mathcal{B}(H, \mathbb{C}) = \mathcal{B}(H^{**}, \mathbb{C}) = (H^*)_r. \tag{3.4.9}$$

Dually, if we let $K = H^*$, then

$$(K_r)^* = (H_c)^{**} = H_c = (K^*)_c.$$

Proposition 3.4.2 *For any Hilbert spaces H and K, we have the complete isometry*

$$\mathcal{B}(K^*, H^*) \cong \mathcal{CB}(H_r, K_r).$$

Proof The result follows from the following complete isometries,:

$$\begin{aligned} \mathcal{CB}(H_r, K_r) &\cong \mathcal{CB}((K_r)^*, (H_r)^*) \\ &\cong \mathcal{CB}((K^*)_c, (H^*)_c) \\ &\cong \mathcal{B}(K^*, H^*) \end{aligned}$$

where we use the adjoint to define the first and fourth isometries. □

Finally, it should be noted that the canonical isometry $\theta : \overline{H} \to H^*$ does not define a complete isometry of the conjugate and dual operator spaces of H_c, and in this sense the column Hilbert operator space is not 'self-dual'. The correct relationship is

$$(H_c)^* = (H^*)_r = (\overline{H})_r = \overline{H_r},$$

where we have used (3.4.9) and the following result.

Proposition 3.4.3 *For any Hilbert space H, there is a natural completely isometric identification*

$$\overline{H_c} = (\overline{H})_c \text{ and } \overline{H_r} = (\overline{H})_r. \tag{3.4.10}$$

Proof We shall just prove the first equality. Let us suppose that we are given orthonormal $e_h \in H$ and matrices $\alpha^{(h)} \in M_n$, and a corresponding element $u = \sum_h \alpha^h \otimes e_h \in M_n(H_c)$. From the definition of the conjugate operator space structure (see (3.1.4)),

$$\begin{aligned} \|\overline{u}\|_{M_n(\overline{H_c})} &= \left\| \sum \overline{\alpha^{(h)}} \otimes \overline{e_h} \right\|_{M_n(\overline{H_c})} = \left\| \left[\sum \overline{\alpha}_{i,j}^{(h)} \overline{e_h} \right] \right\|_{M_n(\overline{H_c})} \\ &= \left\| \left[\sum \alpha_{i,j}^{(h)} e_h \right] \right\|_{M_n(H_c)} = \left\| \sum \alpha^{(h)*} \alpha^{(h)} \right\|^{1/2}. \end{aligned}$$

On the other hand, the vectors $\overline{e_h}$ are orthonormal since

$$\langle \overline{e_h}, \overline{e_k} \rangle = \langle e_k, e_h \rangle,$$

and thus by the computation for the column matrix norm (3.4.4),

$$\|\overline{u}\|_{M_n((\overline{H})_c)} = \left\| \sum \overline{\alpha}^{(h)*} \overline{\alpha}^{(h)} \right\|^{1/2} = \left\| \sum \alpha^{(h)*} \alpha^{(h)} \right\|^{1/2},$$

and we have proved (3.4.10). □

3.5 PISIER'S SELF-DUAL HILBERT OPERATOR SPACES

As we have remarked above, if H is a Hilbert space and there is an operator space structure on H, then there are corresponding operator structures on both the conjugate $\overline{H}$ and the dual H^*. In this section we shall prove Pisier's remarkable result that there is a *unique* operator structure on H for which the canonical isometry $\theta : \overline{H} \to H^*$ is a complete isometry. We are indebted to Soren Winkler for showing us the following simple argument.

Given Hilbert spaces H and K, we have a natural Hilbert space isometry

$$V : H \otimes \overline{K} \to \mathcal{HS}(K, H)$$

determined by letting

$$V(\eta \otimes \overline{\xi}) = x_{\eta \otimes \overline{\xi}},$$

where for all $\zeta \in K$, we define

$$x_{\eta \otimes \overline{\xi}}(\zeta) = \langle \zeta \mid \xi \rangle \eta.$$

We thus have a unitary equivalence

$$\sigma : \mathcal{B}(H \otimes \overline{K}) \to \mathcal{B}(\mathcal{HS}(K, H)) : u \mapsto VuV^{-1}.$$

If $b \in \mathcal{B}(H)$ and $a \in \mathcal{B}(K)$, then we have for any $x \in \mathcal{HS}(H, K)$ that

$$\sigma(b \otimes \overline{a})(x) = bxa^*.$$

To see this, let us suppose that $x = x_{\eta \otimes \overline{\xi}}$. We have $V^{-1}(x_{\eta \otimes \overline{\xi}}) = \eta \otimes \overline{\xi}$, and thus

$$\begin{aligned} \sigma(b \otimes \overline{a})(x_{\eta \otimes \overline{\xi}}) &= V(b \otimes \overline{a})(\eta \otimes \overline{\xi}) = V(b\eta \otimes \overline{a\xi}) \\ &= x_{b\eta \otimes \overline{a\xi}} = bx_{\eta \otimes \overline{\xi}} a^* \end{aligned}$$

since if $\zeta \in H$, then

$$x_{b\eta \otimes \overline{a\xi}}(\zeta) = \langle \zeta \mid a\xi \rangle b\eta = b(\langle a^* \zeta \mid \xi \rangle \eta) = b(x_{\eta \otimes \overline{\xi}}(a^* \zeta)).$$

It follows that for any element $u = \sum b_i \otimes \overline{a}_i \in \mathcal{B}(H) \otimes \mathcal{B}(\overline{K})$,

$$\|u\| = \sup \left\{ \left\| \sum b_i x a_i^* \right\|_2 : \|x\|_2 \leq 1 \right\}. \tag{3.5.1}$$

Given a Hilbert space H, the sesquilinear form $\langle \cdot \mid \cdot \rangle$ determines a *matrix sesquilinear form*

$$M_m(H) \otimes M_n(H) \to M_m \otimes M_n : (\eta, \xi) \mapsto \langle\langle \eta \mid \xi \rangle\rangle = [\langle \eta_{k,l} \mid \xi_{i,j} \rangle].$$

Given $\xi, \eta \in M_n(H) = M_n \otimes H$, let us assume that

$$\xi = \sum_{h=1}^{p} \alpha^{(h)} \otimes e_h$$

and

$$\eta = \sum_{h=1}^{p} \beta^{(h)} \otimes e_h,$$

where $e_h \in H$ are orthonormal and $\alpha^{(h)}, \beta^{(h)} \in M_n$. We have

$$\langle\langle \eta \mid \xi \rangle\rangle = \left[\sum_{h=1}^{p} \beta_{k,l}^{(h)} \overline{\alpha_{i,j}^{(h)}} \right] = \sum_{h=1}^{p} \beta^{(h)} \otimes \overline{\alpha^{(h)}} \in M_n \otimes \overline{M_n}.$$

The following remarkable analogue of the Schwarz inequality is due to Haagerup.

Theorem 3.5.1 *For any Hilbert space H, $n \in \mathbb{N}$ and $\xi, \eta \in M_n(H)$,*

$$\| \langle\langle \eta \mid \xi \rangle\rangle \| \leq \| \langle\langle \eta \mid \eta \rangle\rangle \|^{1/2} \| \langle\langle \xi \mid \xi \rangle\rangle \|^{1/2} .$$

Proof If we use the notation defined above, we have from (3.5.1) (where we let $K = H = \mathbb{C}^n$),

$$\begin{aligned}
\| \langle\langle \eta \mid \xi \rangle\rangle \| &= \left\| \sum \beta^{(h)} \otimes \overline{\alpha^{(h)}} \right\| \\
&= \sup \left\{ \left| \left\langle \sum \beta^{(h)} x \alpha^{(h)*} \,\middle|\, y \right\rangle \right| : \|x\|_2, \|y\|_2 \leq 1 \right\} \\
&= \sup \left\{ \left| \text{trace} \left(\sum \beta^{(h)} x \alpha^{(h)*} y^* \right) \right| : \|x\|_2 , \|y\|_2 \leq 1 \right\}.
\end{aligned}$$

Let us fix x and y with $\|x\|_2 , \|y\|_2 \leq 1$, and let $x = v\,|x|$ and $y = w\,|y|$ be the corresponding polar decompositions. We have the factorizations $x = x_1 x_2$ and $y = y_1 y_2$, where $x_1 = v\,|x|^{1/2}$, $x_2 = |x|^{1/2}$, $y_1 = w\,|y|^{1/2}$, and $y_2 = |y|^{1/2}$. Furthermore,

$$\begin{aligned}
x_1 x_1^* &= |x^*| , \quad & y_1 y_1^* &= |y^*| , \\
x_2^* x_2 &= |x| , \quad & y_2^* y_2 &= |y| ,
\end{aligned}$$

where $\||x|\|_2 = \|x\|_2$ and $\||y|\|_2 = \|y\|_2$ (see Theorem 1.2.1). Applying the matrix and then the classical Schwarz inequalities (see (1.2.3)),

$$\begin{aligned}
&\left| \text{trace} \left(\sum \beta^{(h)} x \alpha^{(h)*} y^* \right) \right| \\
&\leq \sum \left| \text{trace} \left(\beta^{(h)} x_1 x_2 \alpha^{(h)*} y_2^* y_1^* \right) \right| \\
&\leq \sum \left[\text{trace} \left((y_1^* \beta^{(h)} x_1)(y_1^* \beta^{(h)} x_1)^* \right) \right]^{1/2} \\
&\qquad\qquad \left[\text{trace} \left((x_2 \alpha^{(h)*} y_2^*)^* (x_2 \alpha^{(h)*} y_2^*) \right) \right]^{1/2}
\end{aligned}$$

$$\leq \left[\sum \operatorname{trace}\left(y_1^* \beta^{(h)} x_1 x_1^* \beta^{(h)*} y_1\right)\right]^{1/2}$$
$$\left[\sum \operatorname{trace}\left(y_2 \alpha^{(h)} x_2^* x_2 \alpha^{(h)*} y_2^*\right)\right]^{1/2}$$
$$= \left|\sum \operatorname{trace}\left(\beta^{(h)} \left|x^*\right| \beta^{(h)*} \left|y^*\right|\right)\right|^{1/2}$$
$$\left|\sum \operatorname{trace}\left(\alpha^{(h)} \left|x\right| \alpha^{(h)*} \left|y\right|\right)\right|^{1/2}$$
$$= \left|\left\langle \sigma\left(\sum \beta^{(h)} \otimes \overline{\beta^{(h)}}\right)(|x^*|) \;\middle|\; |y^*|\right\rangle\right|^{1/2}$$
$$\left|\left\langle \sigma\left(\sum \alpha^{(h)} \otimes \overline{\alpha^{(h)}}\right)(|x|) \;\middle|\; |y|\right\rangle\right|^{1/2}$$
$$\leq \left\|\sum \beta^{(h)} \otimes \overline{\beta^{(h)}}\right\|^{1/2} \left\|\sum \alpha^{(h)} \otimes \overline{\alpha^{(h)}}\right\|^{1/2},$$

from which the desired result follows. □

Given a Hilbert space H, we define the OH *matrix norm* on H by letting

$$\|\xi\|_o = \|\langle\langle \xi \mid \xi \rangle\rangle\|^{1/2}$$

for any matrix $\xi \in M_n(H)$.

Proposition 3.5.2 *For any Hilbert space H, the OH matrix norm on H is an operator space matrix norm. If we let $\overline{H}$ and H^* have the induced conjugate and dual operator space matrix norms, then $\|\cdot\|_o$ is the unique operator space matrix norm for which the corresponding mapping*

$$\psi : \overline{H} \to H^*$$

is completely isometric.

Proof We have

$$\begin{aligned}
\|\xi \oplus \eta\|_o^2 &= \|\langle\langle \xi \oplus \eta \mid \xi \oplus \eta \rangle\rangle\| \\
&= \|\langle\langle \xi \mid \xi \rangle\rangle \oplus \langle\langle \eta \mid \eta \rangle\rangle \oplus \langle\langle \xi \mid \eta \rangle\rangle \oplus \langle\langle \eta \mid \xi \rangle\rangle\| \\
&= \max\left\{\|\langle\langle \xi \mid \xi \rangle\rangle\|, \|\langle\langle \eta \mid \eta \rangle\rangle\|, \|\langle\langle \xi \mid \eta \rangle\rangle\|, \|\langle\langle \eta \mid \xi \rangle\rangle\|\right\} \\
&\leq \max\left\{\|\xi\|_o^2, \|\eta\|_o^2, \|\xi\|_o \|\eta\|_o\right\} = \max\left\{\|\xi\|_o^2, \|\eta\|_o^2\right\},
\end{aligned}$$

and we conclude that

$$\|\xi \oplus \eta\|_o \leq \max\left\{\|\xi\|_o, \|\eta\|_o\right\}.$$

A simple calculation shows that

$$\langle\langle \alpha\xi\beta \mid \alpha\xi\beta \rangle\rangle = (\alpha \otimes \overline{\alpha})\langle\langle \xi \mid \xi \rangle\rangle(\beta \otimes \overline{\beta}),$$

and since

$$\|\alpha \otimes \overline{\alpha}\| = \|\alpha\| \, \|\overline{\alpha}\| = \|\alpha\|^2,$$

and

$$\left\|\beta \otimes \overline{\beta}\right\| = \|\beta\| \left\|\overline{\beta}\right\| = \|\beta\|^2 ,$$

it follows that

$$\|\alpha\xi\beta\|_o \leq \|\alpha\| \, \|\xi\|_o \, \|\beta\| \, .$$

Turning to duality, let us suppose that $\overline{\xi} \in M_n(\overline{H}_o)$, where $(\overline{\xi})_{k,l} = \overline{\xi_{k,l}}$ (see (3.1.4)). We can determine the completely bounded norm of

$$\varphi = \psi_n(\overline{\xi}) \in M_n(H_o^*) = \mathcal{CB}(H_o, M_n)$$

by choosing $\eta \in M_m(H)$, and considering the norm of

$$\begin{aligned}\varphi_m(\eta) &= [\varphi(\eta_{i,j})] = \left[\psi_n(\overline{\xi})(\eta_{i,j})\right] \\ &= \left[\psi(\overline{\xi_{k,l}})(\eta_{i,j})\right] = [\langle \eta_{i,j} \mid \xi_{k,l} \rangle] = \langle\langle \eta \mid \xi \rangle\rangle.\end{aligned}$$

From Theorem 3.5.1,

$$\|\varphi_m(\eta)\| \leq \|\xi\|_o \, \|\eta\|_o \, ,$$

and thus

$$\left\|\psi_n(\overline{\xi})\right\|_{cb} \leq \|\xi\|_o \, .$$

If we let $\eta = \xi / \|\xi\|_o$, then

$$\left\|\psi_n(\overline{\xi})(\eta)\right\| = \|\langle\langle \xi \mid \xi \rangle\rangle / \|\xi\|_o\| = \|\xi\|_o \, ,$$

and we conclude that ψ_n is isometric, and ψ is completely isometric.

Let us suppose that there is another operator space matrix norm $\|\cdot\|'$ on H for which ψ is a complete isometry. Repeating the previous calculations,

$$\|\xi\|' = \sup\left\{\|\langle\langle \eta \mid \xi \rangle\rangle\| : \|\eta\|' \leq 1\right\}.$$

If we let $\eta = \xi / \|\xi\|'$, then

$$\|\xi\|' \geq \|\langle\langle \xi \mid \xi \rangle\rangle\| \, / \, \|\xi\|'$$

and thus

$$\|\xi\|' \geq \|\xi\|_o \, .$$

Conversely, this inequality holds for any vector η, and thus from Theorem 3.5.1,

$$\begin{aligned}\|\xi\|' &= \sup\left\{\|\langle\langle \eta \mid \xi \rangle\rangle\| : \|\eta\|' \leq 1\right\} \\ &\leq \sup\left\{\|\langle\langle \eta \mid \xi \rangle\rangle\| : \|\eta\|_o \leq 1\right\} = \|\langle\langle \xi \mid \xi \rangle\rangle\|^{1/2} = \|\xi\|_o \, .\end{aligned}$$

□

3.6 NOTES AND REFERENCES

The fact that the quotients of operator spaces are again operator spaces was first proved by Ruan (1988) as an application of the representation theorem. The operator space matrix norms on the mapping spaces $\mathcal{CB}(V, W)$ and $\mathcal{B}(E, W)$ were first studied in Effros and Ruan (1988*b*).

There are many additional constructions for operator spaces which parallel those used in Banach space theory, but space does not allow us to present all of them. Of particular interest are the interpolated operator spaces of Pisier (see Pisier 1996*b*), and a discussion of various alternative matrix norms that can be placed on a direct sum of operator spaces.

The realization that dual spaces of operator spaces may be regarded as operator spaces required an important change of perspective. If $\mathcal{A}$ is a C^*-algebra, then it is customary to identify $M_n(\mathcal{A}^*)$ with the Banach space dual $M_n(\mathcal{A})^*$ of the C^*-algebra $M_n(\mathcal{A})$. The difficulty with this approach is that these matrix norms on $\mathcal{A}^*$ do not satisfy axiom **M1** (see details in Ruan 1988, and Effros and Ruan 1988*a*). This constituted a fundamental obstacle to generalizing Banach space theory. The notion that one could reintepret $M_n(\mathcal{A}^*)$ as $\mathcal{CB}(\mathcal{A}, M_n)$ was independently discovered by both Blecher and Paulsen (1991*a*) and Effros and Ruan (1991*a*). The 'old' duality theory is nonetheless still relevant in some contexts, and it underlies the discussion of the Wittstock–Arveson–Hahn–Banach theorem in §4.1. Proposition 3.2.4 is due to Blecher (1992*a*).

The minimal operator space quantization was discussed in Effros and Ruan (1988*a*) and both the minimal and maximal operator space matrix norms were systematically investigated by Blecher and Paulsen (1991*a*). The min and max dualities (3.3.13) and (3.3.15) are proven in Blecher (1992*a*). The concrete description of the maximal operator space matrix norm of ℓ_1^n in Lemma 3.3.3 is due to Paulsen (1996). For $n > 2$, ℓ_1^n is an important example of a non-exact finite-dimensional operator space (see Theorem 14.5.4).

The finite-dimensional column and row Hilbert operator spaces were first considered by Blecher and Paulsen (1991*a*). This was extended to the infinite-dimensional case by Effros and Ruan (1991*b*) and Blecher (1992*b*). Pisier (1996*b*) introduced the self-dual Hilbert operator spaces. They played a particularly important role in his interpolation theory for operator spaces. We are indebted to Soren Winkler, who explained several aspects of Pisier's theory to us.

4
The extension theorem

Arveson's matrix-valued analogue of the classical Hahn–Banach theorem chronologically preceded the representation theorem by nearly twenty years. We have included a simplified proof of Wittstock's version of this result in §4.1. In §4.2 we use the theorem to prove various duality results for operator spaces.

4.1 THE ARVESON–WITTSTOCK THEOREM AND INJECTIVITY

We begin with an obvious vector generalization of the Hahn–Banach theorem. Given normed spaces $E \subseteq F$ and an integer $n \in \mathbb{N}$, any bounded linear mapping $f = (f_1, \ldots, f_n) : E \to \ell^n_\infty$ has an *isometric extension* $\tilde{f} : F \to \ell^n_\infty$. This is trivial since it suffices to let $\tilde{f} = (\tilde{f}_1, \ldots, \tilde{f}_n)$, where $\tilde{f}_j$ is an isometric extension of f_j. However, there is another approach to this result, which has the advantage that it may be applied to operator spaces. It is immediate that

$$\ell^n_1(E)^* = \ell^n_\infty(E^*) = \mathcal{B}(E, \ell^n_\infty), \tag{4.1.1}$$

and thus we have a commutative diagram

$$\begin{array}{ccc} \mathcal{B}(F, \ell^n_\infty) & \xrightarrow{\rho} & \mathcal{B}(E, \ell^n_\infty) \\ \| & & \| \\ \ell^n_1(F)^* & \xrightarrow{\rho} & \ell^n_1(E)^* \end{array}, \tag{4.1.2}$$

where ρ is the restriction mapping. From the formula

$$\|(x_1, \ldots, x_n)\|_1 = \sum \|x_i\|, \tag{4.1.3}$$

it is also apparent that the inclusion $\ell^n_1(E) \hookrightarrow \ell^n_1(F)$ is isometric, and thus its adjoint ρ is an exact quotient mapping (see §A.2).

A special case of the Arveson–Wittstock theorem states that if $V \subseteq W$ are operator spaces, and $\varphi : V \to M_n$ is a complete contraction, then there exists a completely contractive extension $\tilde{\varphi} : W \to M_n$, or equivalently, that the restriction

$$\rho : \mathcal{CB}(W, M_n) \to \mathcal{CB}(V, M_n)$$

is an exact quotient mapping. In this situation it no longer suffices to isometrically extend each of the entries $\varphi_{i,j}$ of φ to a functional $\tilde{\varphi}_{i,j}$ on F since the resulting mapping $\tilde{\varphi}$ need not be completely contractive. We must instead find an analogue of the relation (4.1.1), or more specifically, we must find a norm $\|\cdot\|_1$ on $\mathbb{M}_n(V)$ which plays the role of $\|\cdot\|_1$ on E^n for normed spaces E. We shall subsequently show that the resulting normed space $T_n(V)$ is an operator space with respect to a natural matrix norm (see (7.1.20)); this additional structure is not needed at this point.

In order to motivate the definition, let us find an explicit expression for the norm of a matrix

$$f \in M_n(V^*) = \mathcal{CB}(V, M_n).$$

We have

$$\begin{aligned}\|f\| &= \sup\{\|f_r(\tilde{v})\| : \|\tilde{v}\| \le 1, \tilde{v} \in M_r(V), r \in \mathbb{N}\}\\ &= \sup\{\|\langle\langle f, \tilde{v}\rangle\rangle\| : \|\tilde{v}\| \le 1, \tilde{v} \in M_r(V),\, r \in \mathbb{N}\},\end{aligned}$$

where $\langle\langle f, \tilde{v}\rangle\rangle \in M_{n\times r}$. It follows that if we let $D_{r\times n}$ be the closed unit ball of $\ell_2^{r\times n}$ (this is just $\mathbb{C}^{r\times n}$ with the norm $\|\cdot\|_2$), then

$$\begin{aligned}\|f\| &= \sup\{|\langle\langle\langle f, \tilde{v}\rangle\rangle\eta \mid \xi\rangle| : \eta, \xi \in D_{r\times n}\}\\ &= \sup\left\{\left|\sum_{i,j,k,l} f_{k,l}(\tilde{v}_{i,j})\eta_{(j,l)}\overline{\xi}_{(i,k)}\right| : \eta, \xi \in D_{r\times n}\right\}\\ &= \sup\left\{\left|\sum_{k,l}\left\langle f_{k,l}, \sum_{i,j}\overline{\xi}_{(i,k)}\tilde{v}_{i,j}\eta_{(j,l)}\right\rangle\right| : \eta, \xi \in D_{r\times n}\right\},\end{aligned}$$

where each supremum is taken over all $\tilde{v} \in M_r(V)$ with $\|\tilde{v}\| \le 1$ and $r \in \mathbb{N}$. Given vectors $\eta, \xi \in D_{r\times n}$, we let $\alpha_{k,i} = \overline{\xi}_{(i,k)}$ and $\beta_{j,l} = \eta_{(j,l)}$. The matrices

$$\alpha = [\alpha_{k,i}] \in HS_{n,r}, \qquad \beta = [\beta_{j,l}] \in HS_{r,n}$$

satisfy $\|\alpha\|_2 = \|\xi\|$ and $\|\beta\|_2 = \|\eta\|$, and we have

$$\begin{aligned}\|f\| &= \sup\left\{\left|\sum\langle f_{k,l}, (\alpha\tilde{v}\beta)_{k,l}\rangle\right| : \|\tilde{v}\|, \|\alpha\|_2, \|\beta\|_2 \le 1\right\}\\ &= \sup\{|\langle f, \alpha\tilde{v}\beta\rangle| : \|\tilde{v}\|, \|\alpha\|_2, \|\beta\|_2 \le 1\}\\ &= \sup\{|\langle f, v\rangle| : v = \alpha\tilde{v}\beta, \|\tilde{v}\|, \|\alpha\|_2, \|\beta\|_2 \le 1\},\end{aligned}$$

where each of these suprema is taken over all $\tilde{v} \in M_r(V)$ with $\|v\| \le 1$ and $r \in \mathbb{N}$. Thus, if we define $\|\cdot\|_1 : \mathbb{M}_n(V) \to [0, \infty)$ by

$$\|v\|_1 = \inf\{\|\alpha\|_2 \|\tilde{v}\| \|\beta\|_2 : v = \alpha\tilde{v}\beta\}, \tag{4.1.4}$$

where $\alpha \in HS_{n,r}$, $\beta \in HS_{r,n}$, and $\tilde{v} \in M_r(V)$ with r arbitrary, then we conclude that

$$\|f\| = \sup\{|\langle f, v\rangle| : \|v\|_1 \le 1\}. \tag{4.1.5}$$

We next show that $\|\cdot\|_1$ is a norm on $\mathbb{M}_n(V)$, after which we shall let $T_n(V)$ denote the corresponding normed space. Before turning to this result, we note that for any $v \in \mathbb{M}_n(V)$,

$$\|v\| \leq \|v\|_1 \leq n\,\|v\|. \tag{4.1.6}$$

To see this let us suppose that $v \in \mathbb{M}_n(V)$ satisfies $\|v\|_1 < 1$. Then we may assume that

$$v = \alpha\tilde{v}\beta$$

with $\|\alpha\|_2, \|\tilde{v}\|, \|\beta\|_2 < 1$. It follows that

$$\|v\| \leq \|\alpha\|\,\|\tilde{v}\|\,\|\beta\| \leq \|\alpha\|_2\,\|\tilde{v}\|\,\|\beta\|_2 < 1.$$

On the other hand, if $\|v\| \leq 1$, then since $v = IvI$, where $I \in M_n$ is the identity matrix, we have

$$\|v\|_1 \leq \|I\|_2\,\|v\|\,\|I\|_2 = n\,\|v\|.$$

Lemma 4.1.1 *Suppose that V is an operator space and $n \in \mathbb{N}$. Then $\|\cdot\|_1$ is a norm on $\mathbb{M}_n(V)$. The scalar pairing* (1.1.24) (*with* $V' = V^*$) *determines the isometric identifications*

$$T_n(V)^* \cong M_n(V^*), \tag{4.1.7}$$

and

$$M_n(V)^* \cong T_n(V^*). \tag{4.1.8}$$

Proof Given $v_1, v_2 \in \mathbb{M}_n(V)$ and $\varepsilon > 0$, let us suppose that $v_i = \alpha_i\tilde{v}_i\beta_i$, where $\|\tilde{v}_i\| \leq 1$ and $\|\alpha_i\|_2 = \|\beta_i\|_2 < (\|v_i\|_1 + \varepsilon)^{1/2}$. If we let $\alpha = \begin{bmatrix}\alpha_1 & \alpha_2\end{bmatrix}$ and $\beta = \begin{bmatrix}\beta_1 & \beta_2\end{bmatrix}^{tr}$, and $\tilde{v} = \tilde{v}_1 \oplus \tilde{v}_2$, then

$$\|\alpha\|_2^2 = \|\alpha_1\|_2^2 + \|\alpha_2\|_2^2,$$
$$\|\beta\|_2^2 = \|\beta_1\|_2^2 + \|\beta_2\|_2^2,$$

and $\|\tilde{v}\| \leq 1$. Since

$$v_1 + v_2 = \alpha\tilde{v}\beta,$$

it follows that

$$\begin{aligned}\|v_1 + v_2\|_1 &\leq \|\alpha\|_2\,\|\beta\|_2 \leq \frac{1}{2}(\|\alpha\|_2^2 + \|\beta\|_2^2)\\ &= \frac{1}{2}(\|\alpha_1\|_2^2 + \|\beta_1\|_2^2 + \|\alpha_2\|_2^2 + \|\beta_2\|_2^2)\\ &< \|v\|_1 + \|v_2\|_1 + 2\varepsilon.\end{aligned}$$

Since $\varepsilon > 0$ is arbitrary, we have subadditivity. For any $c \in \mathbb{C}$, we have $cv_1 = \alpha(c\tilde{v}_1)\beta$, and hence $\|cv\|_1 \leq |c|\,\|v\|_1$. If we replace c by c^{-1} for $c \neq 0$, then we see that $\|cv\|_1 = |c|\,\|v\|_1$. It follows from (4.1.6) that if $\|v\|_1 = 0$, then $v = 0$, and thus $\|\cdot\|_1$ is a norm. The duality (4.1.7) is a consequence of (4.1.5).

Given $f \in M_n(V)^*$ with $\|f\| < 1$, we have from Lemma 2.3.3 that there exists a mapping

$$\tilde{f} \in M_n(V^*) = \mathcal{CB}(V, M_n)$$

with $\|\tilde{f}\|_{cb} < 1$ and vectors $\xi, \eta \in \mathbb{C}^{n^2}$ for which

$$\begin{aligned} f(v) &= \langle \tilde{f}_n(v)\eta \mid \xi \rangle \\ &= \sum \tilde{f}_{k,l}(v_{i,j})\eta_{(j,l)}\overline{\xi}_{(i,k)} \\ &= \sum_{i,j} \left[\sum_{k,l} \overline{\xi}_{(i,k)} \tilde{f}_{k,l} \eta_{(j,l)} \right] (v_{i,j}) \,. \end{aligned}$$

It follows that $f = \alpha \tilde{f} \beta$, where $\alpha_{i,k} = \overline{\xi}_{(i,k)}$ and $\beta_{l,j} = \eta_{(j,l)}$ determine the $n \times n$ matrices α and β with $\|\alpha\|_2, \|\beta\|_2 < 1$; hence $\|f\|_1 < 1$.

Conversely, given such a decomposition $f = \alpha \tilde{f} \beta$ with $\|\tilde{f}\|_{cb} < 1$ and $\|\alpha\|_2, \|\beta\|_2 < 1$, we may use the previous relations to find contractive vectors $\xi, \eta \in \mathbb{C}^{n^2}$ for which

$$|\langle f, v \rangle| = \left| \langle \langle\langle \tilde{f}, v \rangle\rangle \eta \mid \xi \rangle \right| \le \left\| \langle\langle \tilde{f}, v \rangle\rangle \right\| \le \|v\| \,,$$

and therefore $\|f\|_{M_n(V)^*} \le 1$. □

It is apparent from (3.2.3) that we can restrict to matrices

$$\tilde{v} \in M_n(V)_{\|\cdot\| \le 1}$$

in the definition (4.1.4). The following result provides an important refinement of this observation.

Lemma 4.1.2 *Suppose that V is an operator space and that $v \in \mathbb{M}_n(V)$. Then*

$$\|v\|_1 < 1 \text{ if and only if } v = \alpha \tilde{v} \beta, \tag{4.1.9}$$

where $\tilde{v} \in M_n(V)$, $\alpha \in HS_n$, and $\beta \in HS_n$ satisfy $\|\tilde{v}\| < 1, \|\alpha\|_2 < 1$, and $\|\beta\|_2 < 1$, and furthermore we may suppose that α and β are invertible matrices.

Proof Let us suppose that we are given a decomposition

$$v = \alpha w \beta,$$

where $w \in M_p(V), \alpha \in HS_{n,p}$, and $\beta \in HS_{p,n}$ satisfy $\|w\| < 1, \|\alpha\|_2 < 1$, and $\|\beta\|_2 < 1$. We may regard β as a linear mapping $\mathbb{C}^n \to \mathbb{C}^p$ and let

$$\beta = \nu |\beta|$$

be the corresponding polar decomposition (see Theorem 1.2.1), where

$$\| \, |\beta| \, \|_2 = \|\beta\|_2 < 1.$$

If we let P be the projection of $\mathbb{C}^n$ onto the range of $|\beta|$, then $\nu(I-P)=0$, and for $\varepsilon > 0$, $\beta_1 = |\beta| + \varepsilon(I-P)$ is an invertible $n \times n$ matrix with

$$\beta = \nu\beta_1 .$$

If ε is sufficiently small, then we may assume that $\|\beta_1\|_2 < 1$. Similarly, if we can take the adjoint of the polar decomposition of α^*, we find that

$$\alpha = \alpha_1 \rho,$$

where ρ is a partial isometry and α_1 is an invertible $n \times n$ matrix with $\|\alpha_1\|_2 < 1$. It follows that if we let $\tilde{v} = \rho w \nu$, then

$$v = \alpha w \beta = \alpha_1 \tilde{v} \beta_1$$

is the desired decomposition of v. □

Corollary 4.1.3 *If V is a subspace of an operator space W, the inclusion mapping $T_n(V) \hookrightarrow T_n(W)$ is isometric for each $n \in \mathbb{N}$.*

Proof If $v \in \mathbb{M}_n(V)$, then

$$\|v\|_{T_n(W)} \leq \|v\|_{T_n(V)}$$

since there are more decompositions for v in W. On the other hand, given $\|v\|_{T_n(W)} < 1$, let us select a decomposition $v = \alpha\tilde{w}\beta$ as in Lemma 4.1.2, with $\tilde{w} \in M_n(W)$. Since α and β are invertible, $\tilde{w} = \alpha^{-1} v \beta^{-1} \in M_n(V)$, and thus $\|v\|_{T_n(V)} < 1$. □

Corollary 4.1.4 *Given a subspace V of an operator space W, any completely bounded mapping $\varphi : V \to M_n$ has an extension $\tilde{\varphi} : W \to M_n$ satisfying $\|\tilde{\varphi}\|_{cb} = \|\varphi\|_{cb}$.*

Proof We have a commutative diagram

$$\begin{array}{ccc} M_n(W^*) & \longrightarrow & M_n(V^*) \\ \| & & \| \\ \mathcal{CB}(W, M_n) & \longrightarrow & \mathcal{CB}(V, M_n) \end{array},$$

where the top row is the adjoint of the inclusion mapping $T_n(V) \to T_n(W)$, the bottom row is the restriction mapping, and the columns are isometric identifications. Since the inclusion mapping is isometric, the top row, and thus the bottom row, are exact quotient mappings (see §A.2). □

We are now ready to prove the Arveson–Wittstock–Hahn–Banach theorem in the general form due to Wittstock.

Theorem 4.1.5 *If V is a subspace of an operator space W, and H is a Hilbert space, then any complete contraction $\varphi : V \to \mathcal{B}(H)$ has a completely contractive extension $\Phi : W \to \mathcal{B}(H)$.*

Proof We have proved this for $H = \mathbb{C}^n$ in Corollary 4.1.4. Given an arbitrary Hilbert space H, we let $\mathcal{F}$ be all finite-rank orthogonal projections $F \in \mathcal{B}(H)$. If $F(H)$ has dimension n, then we may identify $\mathcal{B}(F(H))$ with M_n. It follows that there exists a completely contractive extension

$$\psi_F : W \to \mathcal{B}(H)$$

of the mapping

$$F\varphi F : V \to \mathcal{B}(F(H)) \subseteq \mathcal{B}(H).$$

If we order $\mathcal{F}$ in the usual manner, then we may regard $\{\psi_F\}_{F\in\mathcal{F}}$ as a net of contractions in $\mathcal{CB}(W, \mathcal{B}(H))$. Owing to the fact that the unit ball of $\mathcal{B}(H)$ is compact in the weak operator topology, the unit ball of $\mathcal{CB}(V, \mathcal{B}(H))$ is also compact in the point-weak operator topology. We let ψ be an arbitrary limit point of the net $\{\psi_F\}_{F\in\mathcal{F}}$ in this topology. For each $v \in V$ and $\xi \in H$ we let F_0 be the projection onto $\mathbb{C}\ \xi + \mathbb{C}\varphi(v)\xi$. If $F \geq F_0$, then

$$\psi_F(v)\xi = F\varphi(v)F\xi = \varphi(v)\xi.$$

We conclude that $\psi(v) = \varphi(v)$, and ψ is the desired extension of φ. □

A Banach space V is said to be *injective* if for any inclusion of Banach spaces $W_0 \subseteq W$, every bounded linear mapping $\varphi_0 : W_0 \to V$ has a linear extension $\varphi : W \to V$ with $\|\varphi\| = \|\varphi_0\|$. It is often helpful to picture the situation using the commutative diagram

$$\begin{array}{ccc} W & & \\ \cup\!| & \searrow^{\varphi} & \\ W_0 & \xrightarrow{\varphi_0} & V \end{array} \quad . \tag{4.1.10}$$

From the classical Hahn–Banach theorem, $\mathbb{C}$ is an injective Banach space, from which it easily follows that $\ell_\infty(\mathfrak{s})$ is injective for any set $\mathfrak{s}$ (see the discussion at the beginning of this chapter).

More generally, it can be shown that a dual Banach space is injective if and only if it is isometric to $L^\infty(X, \mathcal{S}, \mu)$ for some measure space $(X, \mathcal{S}, \mu)$. The general injective complex Banach spaces V were characterized by Hasumi, who showed that a Banach space V is injective if and only if it is isometric to a commutative C^*-algebra $C(X)$, where X is a compact Hausdorff space such that the closure of any open set is again an open set (the real case was proved five years earlier by Kelley, Nachbin and Goodner). We return to this result in §6.1.

In the same manner, we say that an operator space V is *injective* if for any operator spaces $W_0 \subseteq W$, every completely bounded linear mapping $\varphi_0 : W_0 \to V$ has a linear extension $\varphi : W \to V$ satisfying $\|\varphi\|_{cb} = \|\varphi_0\|_{cb}$. We may again use the commutative diagram (4.1.10) in this context.

From Theorem 4.1.5, $\mathcal{B}(H)$ is an injective operator space. As in the case of Banach space theory, the injective operator spaces have a simple 'categorical' interpretation. If we are given a linear space B, then we say that a linear mapping $\Phi : B \to B$ is a *projection* if $\Phi^2 = \Phi$.

Proposition 4.1.6 *If B is an injective operator space and $\Phi : B \to B$ is a completely contractive projection, then $V = \Phi(B)$ is again injective. Conversely, if V is an injective operator space and $V \subseteq \mathcal{B}(H)$, then there is a completely contractive projection of $\mathcal{B}(H)$ onto V.*

Proof For the first result, we let W_0, W, and φ_0 be as before. Then the result is apparent from the commutative diagram

$$\begin{array}{ccc} W & \xrightarrow{\varphi} & B \\ \cup\!| & & \cup\!| \downarrow \Phi \,. \\ W_0 & \xrightarrow{\varphi_0} & V \end{array} \tag{4.1.11}$$

The second assertion is immediate from the diagram

$$\begin{array}{ccc} \mathcal{B}(H) & & \\ \cup\!| & \overset{\Phi}{\searrow} & , \\ V & = & V \end{array}$$

where we let Φ be a completely contractive extension of the identity mapping $V \to V \subseteq \mathcal{B}(H)$. □

In contrast to the commutative situation, von Neumann algebras need not be injective operator spaces. The classification of the injective von Neumann algebras, arguably one of the most remarkable accomplishments of modern analysis, is outside the domain of this book. On the other hand, the injective C^*-algebras remain unclassified. In Chapter 6, we shall show how one may reduce the theory of injective operator spaces to the theory of injective C^*-algebras.

The following dual characterization of injectivity will be useful in §7.1.

Lemma 4.1.7 *An operator space V is injective if and only if for any inclusion of operator spaces $W_0 \subseteq W$, the restriction mapping*

$$\rho : \mathcal{CB}(W, V) \to \mathcal{CB}(W_0, V)$$

is an exact complete quotient mapping.

Proof The restriction mapping ρ is an exact complete quotient mapping if and only if for all $n \in \mathbb{N}$, the corresponding restriction mapping

$$\rho_n : \mathcal{CB}(W, M_n(V)) \to \mathcal{CB}(W_0, M_n(V))$$

is an exact quotient mapping. By definition, the latter will be the case if and only if $M_n(V)$ is injective for all $n \in \mathbb{N}$. Thus, it suffices to show that if V is injective, then $M_n(V)$ is injective for all $n \in \mathbb{N}$. If we are given a completely isometric representation $V \hookrightarrow \mathcal{B}(H)$, then the corresponding mapping $M_n(V) \hookrightarrow M_n(\mathcal{B}(H))$ is completely isometric for each n. We may choose a completely contractive surjective projection $P : \mathcal{B}(H) \to V$, and it is evident that $P_n : M_n(\mathcal{B}(H)) \to M_n(V)$ is a completely contractive surjective projection. Thus, since $\mathcal{B}(H^n)$ is injective, the same is true for $M_n(V)$. □

There is a useful refinement of the technique that was used in the proof of Theorem 4.1.5. Given a linear mapping of operator spaces $\varphi : V \to W$, we can relate the metric properties of the mappings

$$\varphi_n = id \otimes \varphi : M_n(V) \to M_n(W)$$

and

$$T_n(\varphi) = id \otimes \varphi : T_n(V) \to T_n(W)$$

for a fixed integer n. We note that

$$T_n(\varphi)^* = (id \otimes \varphi)^* = id \otimes \varphi^* = (\varphi^*)_n, \tag{4.1.12}$$

and similarly,

$$(\varphi_n)^* = T_n(\varphi^*). \tag{4.1.13}$$

Theorem 4.1.8 *Given operator spaces V and W and a linear mapping $\varphi : V \to W$, then*

(i) $\|T_n(\varphi)\| = \|\varphi_n\|$;

(ii) *the following are equivalent:*

- (a) *φ_n is an isometric injection,*
- (b) *$T_n(\varphi)$ is an isometric injection,*
- (c) *$(\varphi^*)_n$ is a quotient mapping,*
- (d) *$T_n(\varphi^*)$ is a quotient mapping;*

(iii) *if V is complete, then the following are equivalent:*

- (a) *φ_n is a quotient mapping,*
- (b) *$T_n(\varphi)$ is a quotient mapping,*
- (c) *$(\varphi^*)_n$ is an isometric injection,*
- (d) *$T_n(\varphi^*)$ is an isometric injection.*

Furthermore, if φ_n is an isometric injection, then both $(\varphi^*)_n$ and $T_n(\varphi^*)$ are exact quotient mappings.

Proof We claim that it suffices to prove the following results:

(α) $\|T_n(\varphi)\| \leq \|\varphi_n\|$,

(β) if φ_n is a quotient mapping, then so is $T_n(\varphi)$,

(γ) if φ_n is an isometric injection, then so is $T_n(\varphi)$.

If we assume these results, then from Proposition 3.2.2 and (4.1.12),

$$\|T_n(\varphi)\| \leq \|\varphi_n\| = \|(\varphi^*)_n\| = \|T_n(\varphi)^*\| = \|T_n(\varphi)\|,$$

and we have (i). From the coimplications (A.2.1) and (A.2.3),

$$\varphi_n \text{ is an isometric injection} \Rightarrow T_n(\varphi) \text{ is an isometric injection}$$

$$\Rightarrow (\varphi^*)_n \text{ is a quotient mapping}$$
$$\Rightarrow T_n(\varphi^*) \text{ is a quotient mapping}$$
$$\Rightarrow \varphi_n \text{ is an isometric injection}\,.$$

If V is complete, then each matrix space $M_n(V)$ is complete (see the discussion following (2.1.8)), and

$$\begin{aligned}\varphi_n \text{ is a quotient mapping} &\Rightarrow T_n(\varphi) \text{ is a quotient mapping}\\ &\Rightarrow (\varphi^*)_n \text{ is an isometric injection}\\ &\Rightarrow T_n(\varphi^*) \text{ is an isometric injection}\\ &\Rightarrow \varphi_n \text{ is a quotient mapping}\,.\end{aligned}$$

If φ_n is isometric, then from (A.2.1), $T_n(\varphi^*) = (\varphi_n)^*$ is an exact quotient mapping. Similarly, if φ_n is isometric, then so is $T_n(\varphi)$. It follows that $(\varphi^*)_n = T_n(\varphi)^*$ is an exact quotient mapping.

We turn to the proofs of (α)–(γ).

(α) It suffices to show that if φ_n is a contraction, then so is $T_n(\varphi)$. Given $v \in T_n(V)$ with $\|v\|_1 < 1$, we may assume that $v = \alpha\tilde{v}\beta$, where $\tilde{v} \in M_n(V)$ and $\alpha, \beta \in \mathbb{M}_n$ satisfy $\|\tilde{v}\|, \|\alpha\|_2, \|\beta\|_2 < 1$. It follows that

$$\begin{aligned}T_n(\varphi)(v) &= \left[\varphi\left(\sum_{k,l} \alpha_{i,k}\tilde{v}_{k,l}\beta_{l,j}\right)\right]\\ &= \left[\sum_{k,l} \alpha_{i,k}\varphi(\tilde{v}_{k,l})\beta_{l,j}\right] = \alpha\varphi_n(\tilde{v})\beta.\end{aligned}$$

By hypothesis, φ_n is contractive, and thus $\|T_n(\varphi)(v)\|_1 < 1$.

(β) Let us suppose that φ_n is a quotient mapping. For any $w \in T_n(W)$ with $\|w\|_1 < 1$, $w = \alpha\tilde{w}\beta$, where $\tilde{w} \in M_n(W)$ and $\alpha, \beta \in \mathbb{M}_n$ satisfy $\|\tilde{w}\|, \|\alpha\|_2, \|\beta\|_2 < 1$. By hypothesis, we may choose an element $\tilde{v} \in M_n(V)$ with $\|\tilde{v}\| < 1$, for which $\varphi_n(\tilde{v}) = \tilde{w}$. If we let $v = \alpha\tilde{v}\beta$, then it follows that $\|v\|_1 < 1$ and $T_n(\varphi)(v) = w$.

(γ) We have proved this in Corollary 4.1.3. □

Corollary 4.1.9 *Let us suppose that we are given operator spaces V and W, and a linear mapping $\varphi : V \to W$. Then φ is a complete isometry if and only if $\varphi^* : W^* \to V^*$ is an exact complete quotient mapping.*

If V and W are complete, then $\varphi : V \to W$ is a complete quotient mapping if and only if φ^* is a complete isometry. In the latter case, $\varphi^*(W^*)$ is weak closed, and φ^* is a weak* homeomorphism in the topologies defined by V and W, respectively.*

Proof All the assertions follow from the preceding results, as well as the usual properties of dual Banach spaces (see §A.2). □

4.2 DUALITY FOR SUBSPACES AND QUOTIENTS

As in the case of normed spaces, there is a natural duality between subspaces and quotient spaces.

Proposition 4.2.1 *If N is a closed subspace of an operator space V, then we have the complete isometries*

$$(V/N)^* \cong N^{\perp} \quad \textit{and} \quad N^* \cong V^*/N^{\perp},$$

where

$$N^{\perp} = \{f \in V^* : \ f(v) = 0 \ \textit{ for all } v \in N\}.$$

Proof Since $\pi : V \to V/N$ is a complete quotient mapping, we have from the proof of Corollary 4.1.9 that $\pi^* : (V/N)^* \to V^*$ is a complete isometry with range $N^{\perp}$. On the other hand, the adjoint of the inclusion mapping $N \hookrightarrow V$ is the restriction mapping $V^* \to N^*$, and from Corollary 4.1.9 this is a complete quotient mapping. This implies the second assertion. □

Proposition 4.2.2 *Let us suppose that V is a complete operator space and that N is a weak* closed subspace of V^*. If we let*

$$N_{\perp} = \{v \in V : f(v) = 0 \textit{ for all } f \in N\},$$

then the natural weak homeomorphic isometry*

$$\theta : (V/N_{\perp})^* \to N$$

is a complete isometry.

Proof By definition, θ is the Banach space adjoint of the quotient mapping $\theta_* : V \to V/N_{\perp}$. From the definition of the matrix norms on $V/N_{\perp}$, θ_* is a complete quotient mapping, and thus from Corollary 4.1.9, θ is a complete isometry. □

From Proposition 4.2.2, weak* closed subspaces of $\mathcal{B}(H)$, and in particular von Neumann algebras, are dual operator spaces. Conversely, from Proposition 3.2.4, every dual operator space can be identified with a weak* closed subspace of $\mathcal{B}(H)$ for some Hilbert space H. This has the following dual interpretation.

Proposition 4.2.3 *If V is a complete operator space, then there is a Hilbert space H and a complete quotient mapping $\varphi : \mathcal{T}(H) \to V$.*

Proof From Proposition 3.2.4 there is a Hilbert space and we have a completely isometric weak* homeomorphic injection

$$\Phi : V^* \hookrightarrow \mathcal{B}(H).$$

Since Φ is weak* continuous, $\Phi = \varphi^*$ for a bounded linear mapping

$$\varphi : \mathcal{T}(H) \to V.$$

From Corollary 4.1.9, φ is a complete quotient map. □

From Lemma 4.1.1 we have the natural isometries

$$M_n(V^{**}) \cong T_n(V^*)^* \cong M_n(V)^{**}. \tag{4.2.1}$$

It is instructive to put the identification $M_n(V^{**}) \cong M_n(V)^{**}$ in an explicit form. In either $M_n(V^{**})$ or $M_n(V)^{**}$, elements have the form $F = [F_{i,j}]$, $F_{i,j} \in V^{**}$. As an element of $M_n(V)^{**}$, the norm of F is determined by the scalar pairing

$$\begin{aligned} \|F\| &= \sup\{|\langle F, f\rangle| : f \in M_n(V)^*, \|f\| < 1\} \\ &= \sup\left\{\left|\sum F_{i,j}(f_{i,j})\right| : f \in M_n(V)^*, \|f\| < 1\right\}, \end{aligned}$$

and this is equal to its norm as an element of $M_n(V^{**}) = \mathcal{CB}(V^*, M_n)$, which is given by the matrix pairing

$$\begin{aligned} \|F\|_{cb} &= \sup\{\|\langle\langle F, g\rangle\rangle\| : g \in M_n(V^*) = \mathcal{CB}(V, M_n), \|g\|_{cb} < 1\} \\ &= \sup\{\|[F_{i,j}(g_{k,l})]\| : g \in M_n(V^*) = \mathcal{CB}(V, M_n), \|g\|_{cb} < 1\}. \end{aligned}$$

The weak topologies on matrices that arise in the above contexts generally coincide. Given dual operator spaces V and W, it is easy to see that the scalar duality

$$\mathbb{M}_n(V) \times \mathbb{M}_n(W) \to \mathbb{C} : (v, f) \mapsto \langle v, f\rangle = \sum f_{i,j}(v_{i,j}) \tag{4.2.2}$$

and the matrix duality

$$\mathbb{M}_m(V) \times \mathbb{M}_n(W) \to \mathbb{M}_{m\times n} : (v, f) \mapsto \langle\langle v, f\rangle\rangle = [f_{i,j}(v_{k,l})] = f_m(v)$$

for any m, determine the same weak topology on $\mathbb{M}_n(W)$. In particular, since each of the isomorphisms in (4.2.1) is determined by the pairing (4.2.2), we see that these mappings are weak* homeomorphisms, and that, moreover, a net $f^\gamma \in M_n(V^*)$ converges in the weak* topology if and only if $(f^\gamma)_n(v)$ converges to $f_n(v)$ in $\mathbb{M}_{m\times n}$.

Lemma 4.2.4 *With the above definitions, the identification*

$$M_n(V^{**}) \cong M_n(V)^{**}$$

is a weak homeomorphism.*

Proof A net $F_\gamma = [F^\gamma_{i,j}] \in M_n(V)^{**}$ converges in the weak* topology to $F = [F_{i,j}]$ if and only if for each i and j, $F^\gamma_{i,j}$ converges weakly* to $F_{i,j}$. It follows that the scalar matrices $[F^\gamma_{i,j}(g)]$ converge in the norm topology to $[F_{ij}(g)]$ for $g \in V^*$, and thus it is equivalent to assume that the net $F_\gamma : V^* \to M_n$ converges in the weak* topology to F. □

The usual density result for a normed space regarded as a subspace of its second dual also has a matrix analogue.

Proposition 4.2.5 *Given any complete contraction $F : V^* \to M_n$, there is a net of weak* continuous complete contractions $F_\gamma : V^* \to M_n$ which* *converges to F in the point-norm topology. For each γ there is a unique element $v_\gamma \in M_n(V)$ for which*

$$F_\gamma(f) = f_n(v_\gamma)$$

for all $f \in V^*$.

Proof This is apparent from the commutative diagram

$$\begin{array}{ccc} M_n(V) & \longrightarrow & M_n(V)^{**} \\ \downarrow & & \downarrow \\ \mathcal{CB}^\sigma(V^*, M_n) & \longrightarrow & \mathcal{CB}(V^*, M_n), \end{array}$$

in which the top row is the canonical embedding of a Banach space into its second dual. □

4.3 NOTES AND REFERENCES

We have already discussed the history and importance of the Arveson–Wittstock–Hahn–Banach theorem in §2.4. The Hahn–Banach theorem is the key initial ingredient of classical functional analysis, and the same is true in the non-commutative context.

There are several ways to prove the Arveson–Wittstock–Hahn–Banach theorem. Wittstock's initial proof (Wittstock 1981, §2.3.1) was quite difficult. Although it was formulated there for self-adjoint mappings, one need only replace a complete contraction φ by the self-adjoint complete contraction

$$\tilde{\varphi} = \begin{bmatrix} 0 & \varphi \\ \varphi^* & 0 \end{bmatrix}$$

to get the general result. The first elementary proof, which was based upon the theory of operator systems, was given by Paulsen (1984). The approach used in §4.1 is a refinement of that described in Effros and Ruan (1988*a*).

The duality for subspaces and quotients in §4.2 is mainly due to Blecher (1992*a*).

5

Operator systems and decompositions

The *decomposition theorem* for completely bounded operator-valued mappings is an important generalization of the Gelfand–Naimark–Segal representation for states on a C^*-algebra. The first such principle was discovered by Stinespring in his pioneering investigation of the completely positive mappings. We follow Paulsen's approach, which uses the theory of matrix ordered spaces, or more precisely of operator systems, to reduce the general theorem to Stinespring's result.

For our purposes, it suffices to consider only the 'concrete operator systems', which are described in §5.1 and §5.2. We use these results, together with Paulsen's '2×2 matrix trick', to prove the decomposition theorem in §5.3. The decomposability theorem cannot be generalized to mappings into arbitrary von Neumann algebras. In fact, Haagerup proved that a von Neumann algebra will have this property if and only if it is injective. In his deep investigation of this problem, he introduced the notion of the decomposability norm, which provides a quantitative measure of the extent to which a mapping can be decomposed. We have included a discussion of Haagerup's theory in §5.4 since it provides a very effective technique for some of our later proofs. In §5.5 we briefly indicate how one can use the decomposition theorem to introduce the notion of 'matrix convexity'.

5.1 OPERATOR SYSTEMS AND COMPLETE POSITIVITY

A (concrete) *operator system* V on a Hilbert space H is, by definition, a norm closed linear subspace $V \subseteq \mathcal{B}(H)$ which is self-adjoint ($v \in V$ implies $v^* \in V$) and unital ($I \in V$). We let $M_n(V)$ have the canonical $*$-operation, together with the relative ordering determined by the cone

$$M_n(V)^+ = M_n(V) \cap \mathcal{B}(H^n)^+ . \tag{5.1.1}$$

It follows that **O1** and **O2** of §1.3 as well as Proposition 1.3.2 and Corollaries 1.3.3 and 1.3.4 hold for matrices over V. In particular, for any $v \in M_n(V)_{sa}$,

$$v = (v + \|v\| I) - \|v\| I,$$

where $v + \|v\| I \geq 0$ and $\|v\| I \geq 0$, and thus

$$M_n(V)_{sa} = M_n(V)^+ - M_n(V)^+ . \tag{5.1.2}$$

As one might expect, an abstract characterization for the operator systems incorporating these conditions **O1** and **O2** can be given. Even though we shall not explicitly use this approach, it is best to think of an operator system as a *$*$-linear space with distinguished matrix norms, matrix orderings, and an order unit*, which can be concretely represented in the above manner. It will usually not be necessary to indicate the particular Hilbert space on which the operator system is situated. As in the case of the matrix norm, if $\mathcal{A}$ is a unital C^*-algebra, then each matrix space $M_n(\mathcal{A})$ has an intrinsic ordering determined by the C^*-algebraic structure on that space, and $\mathcal{A}$ is an operator system.

Given operator systems V and W, a linear mapping $\varphi : V \to W$ is called *completely positive* if $\varphi_n \geq 0$ for all $n \in \mathbb{N}$, and we then write $\varphi \geq_{cp} 0$.

Lemma 5.1.1 *If $\varphi : V \to W$ is a completely positive linear mapping of operator systems, then φ is completely bounded with*

$$\|\varphi\|_{cb} = \|\varphi\| = \|\varphi(I)\| .$$

Proof Let us assume that V and W are operator systems on Hilbert spaces H and K. We let I indicate the identity operator on both H and K (which is being used will be clear from the context), and I_n be the corresponding identity operators on H^n and K^n. Given any contractive element $v = [v_{ij}] \in M_n(V)$, we have from (1.3.4) that

$$\begin{bmatrix} I_n & v \\ v^* & I_n \end{bmatrix} \geq 0 .$$

It follows from the complete positivity of φ that φ_n is self-adjoint, and

$$\varphi_{2n}\left(\begin{bmatrix} I_n & v \\ v^* & I_n \end{bmatrix}\right) = \begin{bmatrix} \varphi_n(I_n) & \varphi_n(v) \\ \varphi_n(v)^* & \varphi_n(I_n) \end{bmatrix} \geq 0.$$

Since $\varphi_n(I_n) \geq 0$, we have

$$0 \leq \varphi_n(I_n) \leq \|\varphi(I)\|\, I_n.$$

If we let $\alpha = \|\varphi(I)\|$, then

$$\begin{bmatrix} \alpha I_n & \varphi_n(v) \\ \varphi_n(v^*) & \alpha I_n \end{bmatrix} = \begin{bmatrix} \alpha I_n - \varphi_n(I_n) & 0 \\ 0 & \alpha I_n - \varphi_n(I_n) \end{bmatrix} + \begin{bmatrix} \varphi_n(I_n) & \varphi_n(v) \\ \varphi_n(v)^* & \varphi_n(I_n) \end{bmatrix} \geq 0,$$

and thus $\|\varphi_n(v)\| \leq \|\varphi(I)\|$. This implies that

$$\|\varphi(I)\| \leq \|\varphi\| \leq \|\varphi\|_{cb} \leq \|\varphi(I)\|,$$

and hence

$$\|\varphi\|_{cb} = \|\varphi\| = \|\varphi(I)\|.$$

□

Corollary 5.1.2 *If $\varphi : V \to W$ is a linear mapping of operator systems such that $\varphi(I) = I$, then φ is completely positive if and only if φ is a complete contraction.*

Proof If φ is completely positive and unital, we have from Lemma 5.1.1 that $\|\varphi\|_{cb} = \|\varphi(I)\| = 1$. Conversely, suppose that φ is completely contractive and unital. If $-I \leq v \leq I$, then we have from Lemma A.4.2 that

$$\|v - itI\| \leq \sqrt{1 + t^2}$$

for all $t \in \mathbb{R}$. It follows that

$$\|\varphi(v) - itI\| = \|\varphi(v - itI)\| \leq \sqrt{1 + t^2}$$

for all $t \in \mathbb{R}$ and thus $-I \leq \varphi(v) \leq I$. If we are given $0 \leq v' \leq I$, and we let $v = 2v' - I$, we see that $-I \leq \varphi(v) \leq I$ and $0 \leq \varphi(v') \leq I$. It follows that φ is positive. Since the same argument may be applied to the mapping φ_n, φ is completely positive. □

Lemma 5.1.3 *If V is an operator system and $\varphi : V \to M_n$ is a linear mapping, then φ is completely positive if and only if $\varphi_n \geq 0$.*

Proof Let us suppose that $\varphi_n \geq 0$. We must show that if $m \geq n$ and $v \in M_m(V)^+$, then $\varphi_m(v) \geq 0$. Given a vector $\eta \in (\mathbb{C}^n)^m$, we have from Lemma 2.2.1 that there exists an isometry $\beta : \mathbb{C}^n \hookrightarrow \mathbb{C}^m$ and a vector $\tilde{\eta} \in \mathbb{C}^n \otimes \mathbb{C}^n$ for which $(\beta \otimes I_n)(\tilde{\eta}) = \eta$. It follows from **O2** that

$$\langle \varphi_m(v)\eta \mid \eta \rangle = \langle \varphi_n(\beta^* v \beta)\tilde{\eta} \mid \tilde{\eta} \rangle \geq 0,$$

and thus $\varphi_m \geq 0$. □

In particular, if f is a positive linear functional on an operator system V, then it is completely positive. We refer to the positive (and thus completely positive) linear functionals $p : V \to \mathbb{C}$ with $p(I) = 1$ as *states* on V, and the completely positive mappings $\varphi : V \to M_n$ with $\varphi(I) = I_n$ as *matrix states*. More generally, we say that a mapping between operator systems $\varphi : V \to W$ is a *morphism* if it is completely positive and $\varphi(I) = I$.

The following is analogous to Proposition 2.2.6.

Lemma 5.1.4 *If $\mathcal{A}$ is a commutative unital C^*-algebra and V is an arbitrary operator system, then any positive mapping $\varphi : V \to \mathcal{A}$ is completely positive.*

Proof Without loss of generality, we may assume that $\mathcal{A} = C(\Omega)$ for some compact Hausdorff space Ω. To prove that $\varphi : V \to C(\Omega)$ is completely positive, it suffices to show that if $v = [v_{i,j}] \in M_n(V)$ is positive, then $[\varphi_n(v_{i,j})]$ is a positive element in $M_n(C(\Omega)) \cong C(\Omega, M_n)$. Thus, it is enough to show that for each $\omega \in \Omega$, $[\varphi(v_{i,j})(\omega)] \in M_n^+$. Given any $\alpha \in \mathbb{C}^n$, we have

$$\langle [\varphi(v_{i,j})(\omega)]\alpha \mid \alpha\rangle = \sum \overline{\alpha}_i \varphi(v_{i,j})(\omega)\alpha_j = \varphi\Big(\sum \overline{\alpha}_i v_{i,j}\alpha_j\Big)(\omega) \geq 0.$$

Therefore, $\varphi : V \to C(\Omega)$ is completely positive. □

In contrast to completely bounded mappings, positivity and complete positivity also coincide for commutative domains.

Theorem 5.1.5 *If $\mathcal{A}$ is a commutative unital C^*-algebra and V is any operator system, then any positive mapping $\varphi : \mathcal{A} \to V$ is completely positive.*

Proof We may assume that $\mathcal{A} = C(\Omega)$ for some compact Hausdorff space Ω. Given an operator system $V \subseteq \mathcal{B}(H)$ and a positive linear mapping $\varphi : C(\Omega) \to V$, we must prove that if $[f_{i,j}] \in M_n(C(\Omega))^+$, then

$$\varphi_n([f_{i,j}]) \in M_n(V)^+ \subseteq M_n(\mathcal{B}(H))^+.$$

It suffices to show that for any n-tuple of vectors $(\xi_1, \dots, \xi_n) \in H^n$,

$$\sum_{i,j} \langle \varphi(f_{i,j})\xi_j \,|\, \xi_i\rangle \geq 0.$$

Since φ is positive,

$$\|\varphi\| \leq 2\,\|\varphi(1)\|\,.$$

To see this we note that $\varphi(\mathcal{A}_{sa}) \subseteq V_{sa}$ and

$$\left\|\varphi_{|\mathcal{A}_{sa}}\right\| \leq \|\varphi(1)\|$$

since $-1 \leq f \leq 1$ implies that

$$-\|\varphi(1)\|\, I \leq -\varphi(1) \leq \varphi(f) \leq \varphi(1) \leq \|\varphi(1)\|\, I.$$

Thus, for arbitrary $f \in C(\Omega)$,

$$\|\varphi(f)\| = \|\varphi(\mathrm{Re} f)\| + \|\varphi(\mathrm{Im} f)\| \leq 2\,\|\varphi(1)\|\,\|f\|\,.$$

Each of the linear functionals

$$f \in C(\Omega) \to \langle \varphi(f)\xi_j \,|\, \xi_i\rangle \in \mathbb{C}$$

is bounded on $C(\Omega)$ since

$$|\langle \varphi(f)\xi_i \,|\, \xi_j\rangle| \leq 2\,\|\varphi\|\,\|f\|\,.$$

From the Riesz representation theorem, there exists a regular Borel measure $\mu_{i,j}$ on Ω such that

$$\langle \varphi(f)\xi_j \,|\, \xi_i\rangle = \int_\Omega f\,\mathrm{d}\mu_{i,j}$$

for all $f \in C(\Omega)$. We let $\mu = \sum_{i,j} |\mu_{i,j}|$. For any i and j,

$$|\mu_{i,j}(f)| \leq |\mu_{i,j}|\,(|f|) \leq \mu(|f|) = \|f\|_1\,,$$

where we are using the norm on $L_1(\Omega, \mu)$. Since $C(\Omega)$ is dense in $L_1(\Omega, \mu)$, $\mu_{i,j}$ extends to a contractive linear functional on the latter Banach space. It follows that there is a Borel function $h_{i,j} \in L_\infty(\Omega, \mu)$ such that

$$\langle \varphi(f)\xi_j \,|\, \xi_i \rangle = \int_\Omega f h_{i,j} \,\mathrm{d}\mu$$

for all $f \in L_1(\Omega, \mu)$, and in particular for all $f \in C(\Omega)$.

If $\alpha = (\alpha_1, \ldots, \alpha_n) \in \mathbb{C}^n$, then we have that

$$\begin{aligned}\int_\Omega f \sum h_{i,j}\overline{\alpha}_i\alpha_j \,\mathrm{d}\mu &= \sum \langle \varphi(f)\alpha_j\xi_j \,|\, \alpha_i\xi_i \rangle \\ &= \Big\langle \varphi(f)\Big(\sum \alpha_j\xi_j\Big) \,\Big|\, \Big(\sum \alpha_i\xi_i\Big) \Big\rangle \geq 0\end{aligned}$$

for all $f \in C(\Omega)^+$, which implies that

$$\sum h_{i,j}(\omega)\overline{\alpha}_i\alpha_j \geq 0$$

for $\mu-$almost all $\omega \in \Omega$. If we apply this to a countable dense set of $\alpha \in \mathbb{C}^n$, then we see that $[h_{i,j}(\omega)] \geq 0$ for $\omega \notin N$, where N is a Borel set with $\mu(N) = 0$. Modifiying $h_{i,j}$ to be zero on N, we may assume that $[h_{i,j}(\omega)] \geq 0$ for all $\omega \in \Omega$. Since $[f_{i,j}(\omega)] \geq 0$ for all ω, it follows that the Schur product $[f_{i,j}(\omega)h_{i,j}(\omega)] \geq 0$ for all ω, and thus

$$\sum f_{i,j}(\omega)h_{i,j}(\omega) = [1 \ldots 1][f_{i,j}(\omega)h_{i,j}(\omega)][1 \ldots 1]^* \geq 0.$$

We conclude that

$$\sum \langle \varphi(f_{i,j})\xi_j \,|\, \xi_i \rangle = \int_\Omega \sum f_{i,j}h_{i,j} \,\mathrm{d}\mu \geq 0.$$

□

We can use the example presented in §2.2 to show that positive linear mappings are generally not completely positive. The transpose mapping $\mathbf{t} : M_2 \to M_2$ is obviously self-adjoint and unital. Since $\mathbf{t}$ is isometric, it follows from (A.4.2) that it is also an order isomorphism. From (2.2.6),

$$\varepsilon = \begin{bmatrix} \varepsilon_{1,1} & \varepsilon_{1,2} \\ \varepsilon_{2,1} & \varepsilon_{2,2} \end{bmatrix}$$

is a positive matrix, but

$$\mathbf{t}_2(\varepsilon) = \begin{bmatrix} \varepsilon_{1,1} & \varepsilon_{2,1} \\ \varepsilon_{1,2} & \varepsilon_{2,2} \end{bmatrix} = \begin{bmatrix} 1 & 0 & 0 & 0 \\ 0 & 0 & 1 & 0 \\ 0 & 1 & 0 & 0 \\ 0 & 0 & 0 & 1 \end{bmatrix}$$

is not, since the middle submatrix $\begin{bmatrix} 0 & 1 \\ 1 & 0 \end{bmatrix}$ is not positive (consider its determinant).

The following result is useful for analysing completely positive mappings that are not unital.

Lemma 5.1.6 *Suppose that H and K are Hilbert spaces, $V \subseteq \mathcal{B}(H)$ is an operator system, and that $\varphi : V \to \mathcal{B}(K)$ is a completely positive mapping with $b = \varphi(I)$. Then there is a morphism $\psi : V \to \mathcal{B}(K)$ such that*

$$\varphi(v) = b^{1/2}\psi(v)b^{1/2} \tag{5.1.3}$$

for all $v \in V$.

Proof Since φ is positive, $b = \varphi(I)$ is a positive element in $\mathcal{B}(K)$. If we let $f = \chi_{\mathrm{sp}\,(b)\setminus\{0\}}$ denote the characteristic function of the set $\mathrm{sp}\,(b)\setminus\{0\}$, then it follows from elementary spectral theory that $p = f(b)$ is the *support projection* of b, in other words it is the minimal projection with $bp = b$. The sequence of functions

$$f_n : [0,\infty) \to [0,1)$$

given by

$$f_n(x) = x^{1/2}(x + 1/n)^{-1/2}$$

is increasing and converges pointwise to $f(x)$ on $\mathrm{sp}\,(b)$. It follows from spectral theory that the sequence

$$f_n(b) = b^{1/2}(b + I/n)^{-1/2} \in \mathcal{B}(K)^+$$

is increasing and converges to p in the strong operator topology.

We let $\psi^n : V \to \mathcal{B}(K)$ be the completely positive mapping defined by

$$\psi^n(v) = (b + I/n)^{-1/2}\varphi(v)(b + I/n)^{-1/2} + \langle v\xi \mid \xi\rangle(I - p),$$

where ξ is a fixed unit vector in H. If v is an element in V such that $0 \leq v \leq I$, we have $0 \leq \varphi(v) \leq b$. A simple argument from operator algebra theory (see Dixmier 1981, I.1.6, Lemma 2) shows that there is an operator $S_v \in \mathcal{B}(K)$ such that

$$\varphi(v)^{1/2} = S_v b^{1/2} = b^{1/2}S_v^*.$$

It follows that

$$\varphi(v)^{1/2}(b + I/n)^{-1/2} = S_v b^{1/2}(b + I/n)^{-1/2} \to S_v p$$

and

$$(b + I/n)^{-1/2}\varphi(v)^{1/2} = (b + I/n)^{-1/2}b^{1/2}S_v^* \to pS_v^*$$

in the strong operator topology. Multiplication is jointly continuous on bounded sets in the strong operator topology. The inequality

$$\|\varphi(v)^{1/2}(b + I/n)^{-1/2}\| = \|(b + I/n)^{-1/2}\varphi(v)^{1/2}\| \leq \|S_v\|,$$

shows that the sequence is bounded, and thus

$$(b + I/n)^{-1/2}\varphi(v)(b + I/n)^{-1/2} \to pS_v^*S_v p \tag{5.1.4}$$

strongly. Since

$$V = (V^+ - V^+) + \mathrm{i}\,(V^+ - V^+),$$

it follows that for any $v \in V$ the sequence

$$\psi^n(v) = (b + I/n)^{-1/2}\varphi(v)(b + I/n)^{-1/2} + \langle v\xi \mid \xi\rangle(I - p)$$

converges in the strong topology. We denote the limit by $\psi(v)$. It is easy to see that $\psi : V \to \mathcal{B}(K)$ is a completely positive linear map, and it is unital because

$$\psi(I) = \lim_n \psi^n(I) = \lim_n b(b + I/n)^{-1} + (I - p) = I\,.$$

Moreover, $\varphi = b^{1/2}\psi b^{1/2}$ since for every $v \in V$,

$$\varphi(v) = \lim_n (b + I/n)^{1/2}\psi^n(v)(b + I/n)^{1/2} = b^{1/2}\psi(v)b^{1/2}\,.$$ □

We can now prove the completely positive version of the Arveson–Hahn–Banach theorem.

Theorem 5.1.7 *Given Hilbert spaces H and K, and operator systems $W \subseteq V$ on H, any completely positive mapping $\varphi : W \to \mathcal{B}(K)$ has a completely positive extension $\Phi : V \to \mathcal{B}(K)$.*

Proof Let us first assume that φ is a morphism, i.e. $\varphi(I) = I$. Then φ is completely contractive by Lemma 5.1.2 and it has a completely contractive extension $\Phi : V \to \mathcal{B}(K)$. From Lemma 5.1.2, Φ is completely positive.

In general, if $\varphi(I) = b$, then from Lemma 5.1.6 there exists a morphism $\psi : W \to \mathcal{B}(K)$ such that

$$\varphi(v) = b^{1/2}\psi(v)b^{1/2}$$

for all $v \in V$. From the preceding argument, ψ has a unital completely positive extension $\Psi : V \to \mathcal{B}(K)$, and

$$\Phi = b^{1/2}\Psi b^{1/2}$$

is a completely positive extension of φ. □

5.2 THE STINESPRING THEOREM AND ITS CONSEQUENCES

The following theorem of Stinespring is a natural generalization of the Gelfand–Naimark–Segal theorem to operator valued mappings. The proof is closely patterned on the usual argument of that earlier result.

Theorem 5.2.1 *If $\mathcal{A}$ is a unital C^*-algebra, H is a Hilbert space, and $\varphi : \mathcal{A} \to \mathcal{B}(H)$ is a completely contractive and completely positive mapping, then there exists a Hilbert space K, a contraction $T : H \to K$, and a unital $*$-representation $\pi : \mathcal{A} \to \mathcal{B}(K)$ such that*

$$\varphi(a) = T^*\pi(a)T\,. \tag{5.2.1}$$

If φ is a morphism, then T is an isometry.

Proof We let $\langle \cdot \mid \cdot \rangle$ be the sesquilinear form on $\mathcal{A} \otimes H$ defined by

$$\left\langle \sum b_j \otimes \eta_j \,\middle|\, \sum a_i \otimes \xi_i \right\rangle = \sum \langle \varphi(a_i^* b_j)\eta_j \mid \xi_i \rangle$$

for $a_i, b_j \in \mathcal{A}$ and $\xi_i, \eta_j \in H$ $(i = 1, \ldots, r; j = 1, \ldots, s)$. In particular, if $\eta = (\eta_1, \ldots, \eta_s)$, then

$$\begin{aligned}\left\langle \sum b_j \otimes \eta_j \middle| \sum b_i \otimes \eta_i \right\rangle &= \sum \langle \varphi(b_i^* b_j)\eta_j \mid \eta_i \rangle \\ &= \langle \varphi_n \left([b_i^* b_j] \right) \eta \mid \eta \rangle \geq 0\end{aligned}$$

since

$$[b_i^* b_j] = \begin{bmatrix} b_1^* \, 0 \ldots \\ b_2^* \, 0 \ldots \\ \vdots \;\; \vdots \end{bmatrix} \begin{bmatrix} b_1 \; b_2 \ldots \\ 0 \;\; 0 \ldots \\ \vdots \;\; \vdots \end{bmatrix} \geq 0$$

and $\varphi_n \geq 0$. Thus, we have a positive semidefinite sesquilinear form, and if we let

$$N = \{ u \in \mathcal{A} \otimes H : \langle u \mid u \rangle = 0 \},$$

then $\langle \cdot \mid \cdot \rangle$ induces a Hilbert space inner product on $(\mathcal{A} \otimes H)/N$. We let K denote the completion of the pre-Hilbert space $(\mathcal{A} \otimes H)/N$.

For each $a \in \mathcal{A}$, we let $\pi(a)$ be the linear mapping on $\mathcal{A} \otimes H$ defined by

$$\pi(a)\Big(\sum a_j \otimes \xi_j \Big) = \sum a\, a_j \otimes \xi_j \,.$$

For any $\sum a_j \otimes \xi_j \in \mathcal{A} \otimes H$,

$$\begin{aligned}\left\langle \pi(a)\Big(\sum a_j \otimes \xi_j \Big) \middle| \pi(a)\Big(\sum a_i \otimes \xi_i \Big) \right\rangle &= \sum \langle \varphi(a_i^* a^* a a_j)\xi_j \mid \xi_i \rangle \\ &= \langle \varphi_n([a_i^* a^* a a_j])[\xi_j] \mid [\xi_i] \rangle \\ &\leq \|a\|^2 \langle \varphi_n([a_i^* a_j])[\xi_j] \mid [\xi_i] \rangle \\ &= \|a\|^2 \left\langle \sum a_j \otimes \xi_j \middle| \sum a_i \otimes \xi_i \right\rangle,\end{aligned}$$

and in particular $\pi(a)$ maps N into N. It determines a bounded linear mapping on the inner product space $(\mathcal{A} \otimes H)/N$, which we also denote by $\pi(a)$, and it is evident that $\|\pi(a)\| \leq \|a\|$. Thus, $\pi(a)$ extends to a bounded linear operator on K, which we again denote by $\pi(a)$. It is a simple exercise to verify that $\pi : \mathcal{A} \to \mathcal{B}(K)$ is a unital $*$-homomorphism.

We define $T : H \to K$ by

$$T(\xi) = 1 \otimes \xi + N \,.$$

T is a contraction since for any $\xi \in H$,

$$\|T(\xi)\|^2 = \langle T(\xi) \mid T(\xi) \rangle = \langle 1 \otimes \xi \mid 1 \otimes \xi \rangle = \langle \varphi(1)\xi \mid \xi \rangle \leq \|\xi\|^2 \,.$$

If φ is a morphism, then T is an isometry since $\varphi(1) = I$ and thus

$$\|T(\xi)\|^2 = \langle \varphi(1)\xi \mid \xi \rangle = \|\xi\|^2 \,.$$

Finally, for any $a \in \mathcal{A}$ and $\xi, \eta \in H$, we have

$$\begin{aligned}\langle T^*\pi(a)T\xi \mid \eta\rangle &= \langle \pi(a)(1 \otimes \xi) \mid 1 \otimes \eta\rangle \\ &= \langle a \otimes \xi \mid 1 \otimes \eta\rangle = \langle \varphi(a)\xi \mid \eta\rangle .\end{aligned}$$

It follows that $\varphi = T^*\pi T$. □

If p is a state on a C^*-algebra $\mathcal{A}$ and (π, ξ) is the corresponding cyclic representation, then applying the Schwarz inequality,

$$|p(a)|^2 = |\langle \pi(a)\xi \mid \xi\rangle|^2 \leq \|\pi(a)\xi\|^2 \|\xi\|^2 = p(a^*a).$$

For this reason, the first part of the following result is often referred to as the *Schwarz inequality for contractive completely positive mappings.*

Corollary 5.2.2 *Suppose that $\varphi : \mathcal{A} \to \mathcal{B}(H)$ is a completely contractive and completely positive mapping from a unital C^*-algebra $\mathcal{A}$ into $\mathcal{B}(H)$. Then for any $a \in \mathcal{A}$,*

$$\varphi(a)^*\varphi(a) \leq \varphi(a^*a). \tag{5.2.2}$$

Furthermore, if $\varphi(a)^*\varphi(a) = \varphi(a^*a)$, then

$$\varphi(ba) = \varphi(b)\varphi(a)$$

for all $b \in A$.

Proof We have from the above Stinespring decomposition theorem that there is a Hilbert space K, a unital $*$-representation $\pi : \mathcal{A} \to \mathcal{B}(K)$, and a contraction $T : H \to K$ such that

$$\varphi(a) = T^*\pi(a)T .$$

It follows that

$$\varphi(a)^*\varphi(a) = T^*\pi(a^*)TT^*\pi(a)T \leq T^*\pi(a^*a)T = \varphi(a^*a) .$$

On the other hand, let us suppose that $\varphi(a)^*\varphi(a) = \varphi(a^*a)$. Then applying (5.2.2) to the mapping φ_2, we have for any $b \in \mathcal{A}$,

$$\varphi_2\left(\begin{bmatrix} b^* & a \\ a^* & 0 \end{bmatrix}\right)^* \varphi_2\left(\begin{bmatrix} b^* & a \\ a^* & 0 \end{bmatrix}\right) \leq \varphi_2\left(\begin{bmatrix} b & a \\ a^* & 0 \end{bmatrix}\begin{bmatrix} b^* & a \\ a^* & 0 \end{bmatrix}\right),$$

and thus

$$\begin{bmatrix} \varphi(b)\varphi(b^*) + \varphi(a)\varphi(a^*) & \varphi(b)\varphi(a) \\ \varphi(a^*)\varphi(b^*) & \varphi(a^*)\varphi(a) \end{bmatrix} \leq \begin{bmatrix} \varphi(bb^*) + \varphi(aa^*) & \varphi(ba) \\ \varphi(a^*b^*) & \varphi(a^*a) \end{bmatrix}.$$

Subtracting, it follows that

$$\begin{bmatrix} * & \varphi(ba) - \varphi(b)\varphi(a) \\ * & 0 \end{bmatrix} \geq 0,$$

where we have not bothered to compute the first column. From Corollary 1.3.4,

$$\varphi(ba) = \varphi(b)\varphi(a).$$

□

In the next result we see that the underlying operator system completely determines the algebraic structure of a unital C^*-algebra.

Corollary 5.2.3 *Let $\mathcal{A}$ and $\mathcal{B}$ be two unital C^*-algebras. Each unital complete order isomorphism $\varphi : \mathcal{A} \to \mathcal{B}$ is $*$-isomorphism.*

Proof If we apply (5.2.2) to φ and to φ^{-1}, then for any $a \in \mathcal{A}$,

$$\varphi(a)^*\varphi(a) \leq \varphi(a^*a)$$

and

$$a^*a = \varphi^{-1}(\varphi(a)^*)\varphi^{-1}(\varphi(a)) \leq \varphi^{-1}(\varphi(a)^*\varphi(a)) \leq \varphi^{-1}(\varphi(a^*a)) = a^*a\,.$$

Therefore, we must have

$$a^*a = \varphi^{-1}(\varphi(a)^*\varphi(a)),$$

or equivalently,

$$\varphi(a^*a) = \varphi(a^*)\varphi(a)$$

for all $a \in \mathcal{A}$. From Corollary 5.2.2, φ is a $*$-isomorphism. □

5.3 DECOMPOSITIONS OF COMPLETE CONTRACTIONS

The Gelfand–Naimark–Segal theorem has a natural extension to arbitrary linear functionals on a C^*-algebra $\mathcal{A}$. Given a contractive functional f in $\mathcal{A}^*$, there exists a Hilbert space L, a $*$-representation $\pi : \mathcal{A} \to \mathcal{B}(L)$, and unit vectors $\xi, \eta \in L$ such that

$$f(a) = \langle \pi(a)\eta \mid \xi \rangle.$$

We wish to extend this to operator-valued linear mappings on $\mathcal{A}$. For this purpose we shall use Paulsen's powerful 'off-diagonal trick' for embedding an operator space in an operator system.

Let us suppose that $\mathcal{A}$ is a C^*-algebra, H is a Hilbert space, and that $\varphi : \mathcal{A} \to \mathcal{B}(H)$ is a complete contraction. We may assume that $\mathcal{A}$ is faithfully represented on a Hilbert space K. We let I stand for the identity operators on both H and K. The linear space

$$\mathcal{L} = \left\{ \begin{bmatrix} \alpha I & b \\ c & \delta I \end{bmatrix} : b, c \in \mathcal{A}, \alpha, \delta \in \mathbb{C} \right\} \tag{5.3.1}$$

is an operator system on K^2. We define a linear mapping

$$\tilde{\varphi} : \mathcal{L} \to M_2(\mathcal{B}(H)) = \mathcal{B}(H^2)$$

by

$$\tilde{\varphi}\left(\begin{bmatrix} \alpha I & b \\ c & \delta I \end{bmatrix} \right) = \begin{bmatrix} \alpha I & \varphi(b) \\ \varphi^*(c) & \delta I \end{bmatrix}. \tag{5.3.2}$$

Lemma 5.3.1 *If $\varphi : \mathcal{A} \to \mathcal{B}(H)$ is a completely contractive linear mapping, then $\tilde{\varphi} : \mathcal{L} \to \mathcal{B}(H^2)$ is a morphism.*

Proof For each $n \in \mathbb{N}$, $M_n(\mathcal{L})$ is an operator system in $M_n(M_2(\mathcal{B}(K)))$ and

$$\tilde{\varphi}_n : M_n(\mathcal{L}) \to M_n(M_2(\mathcal{B}(H)))$$

is a unital mapping. A typical matrix element in $M_n(\mathcal{L})$ has the form

$$u = \left[\begin{bmatrix} \alpha_{i,j} I & b_{i,j} \\ c_{i,j} & \delta_{i,j} I \end{bmatrix}\right]_{i,j=1,\ldots,n},$$

where $b_{i,j}, c_{i,j} \in \mathcal{A}$ and $\alpha_{i,j}, \delta_{i,j} \in \mathbb{C}$. Under the identification

$$M_n(M_2(\mathcal{B}(K))) \cong M_2(M_n(\mathcal{B}(K))),$$

u corresponds to the 2×2 matrix

$$\begin{bmatrix} \alpha I_n & [b_{i,j}] \\ [c_{i,j}] & \delta I_n \end{bmatrix},$$

and under the parallel identification $M_n(M_2(\mathcal{B}(H)) \cong M_2(M_n(\mathcal{B}(H))$,

$$\tilde{\varphi}_{2n}(u) = \left[\begin{bmatrix} \alpha_{i,j} I & \varphi(b_{i,j}) \\ \varphi^*(c_{i,j}) & \delta_{i,j} I \end{bmatrix}\right]$$

corresponds to

$$\begin{bmatrix} \alpha I_n & [\varphi(b_{i,j})] \\ [\varphi^*(c_{i,j})] & \delta I_n \end{bmatrix} = \begin{bmatrix} \alpha I_n & \varphi_n(b) \\ (\varphi_n)^*(c) & \delta I_n \end{bmatrix},$$

where we have used the relation $(\varphi^*)_n = (\varphi_n)^*$.

If u is a positive element in $M_n(\mathcal{L})$, then we can write

$$u = \begin{bmatrix} \alpha I_n & b \\ b^* & \delta I_n \end{bmatrix}$$

with $\alpha, \delta \in M_n^+$. For any $\varepsilon > 0$, $\alpha + \varepsilon I$ and $\delta + \varepsilon I$ are positive invertible matrices, and we have

$$u + \varepsilon I_{2n} = \begin{bmatrix} (\alpha + \varepsilon I) I_n & b \\ b^* & (\delta + \varepsilon I) I_n \end{bmatrix} \geq 0$$

in $M_n(\mathcal{L})$. If we let

$$\beta_\varepsilon = \begin{bmatrix} \alpha + \varepsilon I & 0 \\ 0 & \delta + \varepsilon I \end{bmatrix},$$

then we have

$$\begin{bmatrix} I_n & b_\varepsilon \\ b_\varepsilon^* & I_n \end{bmatrix} = \beta_\varepsilon^{-1/2}(u + \varepsilon I_{2n})\beta_\varepsilon^{-1/2} \geq 0,$$

where $b_\varepsilon = (\alpha + \varepsilon I)^{-1/2} b (\delta + \varepsilon I)^{-1/2}$. It follows from Proposition 1.3.2 that $\|b_\varepsilon\| \leq 1$, and thus $\|\varphi_n(b_\varepsilon)\| \leq 1$ since φ is completely contractive. Therefore, from Proposition 1.3.2,

$$\begin{bmatrix} (\alpha + \varepsilon I) I_n & \varphi_n(b) \\ \varphi_n(b)^* & (\delta + \varepsilon I) I_n \end{bmatrix} = \beta_\varepsilon^{1/2} \begin{bmatrix} I_n & \varphi_n(b_\varepsilon) \\ \varphi_n(b_\varepsilon)^* & I_n \end{bmatrix} \beta_\varepsilon^{1/2} \geq 0 .$$

Since $\varepsilon > 0$ is arbitrary, we conclude that $\tilde{\varphi}_{2n}(u) \geq 0$, and that

$$\tilde{\varphi} : \mathcal{L} \to \mathcal{B}(H^2)$$

is a unital completely positive map. □

Theorem 5.3.2 *Let $\mathcal{A}$ be a unital C^*-algebra and let $\varphi : \mathcal{A} \to \mathcal{B}(H)$ be a complete contraction. Then there exist morphisms ψ_1 and ψ_2 from $\mathcal{A}$ into $\mathcal{B}(H)$ such that*

$$\Phi = \begin{bmatrix} \psi_1 & \varphi \\ \varphi^* & \psi_2 \end{bmatrix} : M_2(\mathcal{A}) \to M_2(\mathcal{B}(H)) : \begin{bmatrix} a & b \\ c & d \end{bmatrix} \mapsto \begin{bmatrix} \psi_1(a) & \varphi(b) \\ \varphi^*(c) & \psi_2(d) \end{bmatrix} \quad (5.3.3)$$

is a morphism.

Proof We let $\mathcal{L} \subseteq M_2(\mathcal{B}(K^2))$ be the operator system, and let

$$\tilde{\varphi} : \mathcal{L} \to \mathcal{B}(H^2)$$

be the unital completely positive mapping defined in (5.3.1) and (5.3.2), respectively. The set $\mathcal{L}$ is a subspace of $M_2(\mathcal{A})$ since, by hypothesis, $\mathcal{A}$ is unital. From the Arveson–Wittstock–Hahn–Banach theorem, $\tilde{\varphi}$ has a unital completely positive extension

$$\Phi : M_2(\mathcal{A}) \to M_2(\mathcal{B}(H)) .$$

If we let e_1 and e_2 denote the projections $I \oplus 0$ and $0 \oplus I$ in $M_2(\mathcal{A})$, and let $\tilde{e}_1 = I \oplus 0$ and $\tilde{e}_2 = 0 \oplus I$ be the corresponding projections in $M_2(\mathcal{B}(H))$, then we have

$$\Phi(e_i) = \tilde{\varphi}(e_i) = \tilde{e}_i,$$

and thus

$$\Phi(e_i^* e_i) = \Phi(e_i^*)\Phi(e_i)$$

$(i = 1, 2)$. From Corollary 5.2.2, $\Phi(ue_i) = \Phi(u)\tilde{e}_i$ and $\Phi(e_i u) = \tilde{e}_i \Phi(u)$ for all $u \in M_2(\mathcal{A})$. We have $e_i = v_i^* v_i$, where $v_1 = [I\, 0]$ and $v_2 = [0\, I]$, and for any

$$a = [a_{i,j}] \in M_2(\mathcal{A}),$$

we have $a_{i,j} = v_i a v_j^*$. Similarly, $\tilde{e}_i = \tilde{v}_i^* \tilde{v}_i$, where $\tilde{v}_1 = [I\, 0]$ and $\tilde{v}_2 = [0\, I]$, and we define $\Phi_{i,j} : \mathcal{A} \to \mathcal{B}(H)$ by

$$\Phi_{i,j}(a_0) = \tilde{v}_i \Phi(v_i^* a_0 v_j) \tilde{v}_j^* .$$

With these definitions we have

$$\begin{aligned} \Phi\,(a) &= \sum \Phi(e_i a e_j) = \sum \tilde{e}_i \Phi(e_i a e_j) \tilde{e}_j \\ &= \sum \tilde{v}_i^* \tilde{v}_i \Phi(v_i^* v_i a v_j^* v_j) \tilde{v}_j^* \tilde{v}_j = \sum \tilde{v}_i^* \Phi_{i,j}(a_{i,j}) \tilde{v}_j \\ &= \begin{bmatrix} \Phi_{1,1}(a_{1,1}) & \Phi_{1,2}(a_{1,2}) \\ \Phi_{2,1}(a_{2,1}) & \Phi_{2,2}(a_{2,2}) \end{bmatrix}. \end{aligned}$$

In particular,

$$\Phi\left(\begin{bmatrix} 0 & b \\ c & 0 \end{bmatrix}\right) = \tilde{\varphi}\left(\begin{bmatrix} 0 & b \\ c & 0 \end{bmatrix}\right) = \begin{bmatrix} 0 & \varphi(b) \\ \varphi^*(c) & 0 \end{bmatrix}$$

and thus $\Phi_{1,2} = \varphi$ and $\Phi_{2,1} = \varphi^*$. On the other hand,

$$\Phi\left(\begin{bmatrix} I & 0 \\ 0 & I \end{bmatrix}\right) = \tilde{\varphi}\left(\begin{bmatrix} I & 0 \\ 0 & I \end{bmatrix}\right) = \begin{bmatrix} I & 0 \\ 0 & I \end{bmatrix};$$

hence the completely positive mappings $\psi_i = \Phi_{i,i}$ $(i = 1, 2)$ are unital and we have the desired representation (5.3.3). □

Theorem 5.3.3 *If $\mathcal{A}$ is a C^*-algebra and $\varphi : \mathcal{A} \to \mathcal{B}(H)$ is a completely contractive mapping, then there exists a $*$-representation $\pi : \mathcal{A} \to \mathcal{B}(K)$ and a diagram of contractions*

$$H \xrightarrow{T} K \xrightarrow{S} H$$

such that

$$\varphi(a) = S\pi(a)T\,.$$

Proof Without loss of generality we may suppose that $\mathcal{A}$ is a unital C^*-algebra. After selecting mappings ψ_1 and ψ_2 as in Theorem 5.3.2, and applying Stinespring's theorem to the morphism $\Phi : M_2(\mathcal{A}) \to \mathcal{B}(H^2)$, we obtain

$$\begin{bmatrix} \psi_1(a) & \varphi(b) \\ \varphi^*(c) & \psi_2(d) \end{bmatrix} = V^*\pi\left(\begin{bmatrix} a & b \\ c & d \end{bmatrix}\right)V,$$

where $\pi : M_2(\mathcal{A}) \to \mathcal{B}(K)$ is a unital $*$-representation and $V : H^2 \to K$ is an isometry. It follows that

$$\pi_0 : \mathcal{A} \to \mathcal{B}(K) : a \mapsto \pi\left(\begin{bmatrix} a & 0 \\ 0 & a \end{bmatrix}\right)$$

is a unital $*$-representation. Then we obtain the desired decomposition

$$\varphi(a) = [I \;\; 0]V^*\pi\left(\begin{bmatrix} a & 0 \\ 0 & a \end{bmatrix}\begin{bmatrix} 0 & I \\ 0 & 0 \end{bmatrix}\right)V\begin{bmatrix} 0 \\ I \end{bmatrix} = S\pi_0(a)T,$$

where

$$S = [I \;\; 0]V^* \quad \text{and} \quad T = \pi\left(\begin{bmatrix} 0 & I \\ 0 & 0 \end{bmatrix}\right)V\begin{bmatrix} 0 \\ I \end{bmatrix}$$

are contractions. □

Theorem 5.3.4 *Suppose that V is an operator space and that $1 \leq r < \infty$.*

(i) *If $\varphi : M_n(V) \to M_r$ is a complete contraction, then there exists a complete contraction $g : V \to M_{rn}$ and contractive matrices $\alpha \in M_{r,rn^2}$ and $\beta \in M_{rn^2,r}$, such that for all $v \in M_n(V)$,*

$$\varphi(v) = \alpha g_n(v)\beta. \qquad (5.3.4)$$

(ii) *If* $\varphi : K_\infty(V) \to M_r$ *is a complete contraction, then there exists a complete contraction* $g : V \to M_{r\times\infty}$ *and matrices* $\alpha \in M_{r,r\times\infty^2}$ *and* $\beta \in M_{r\times\infty^2,r}$, *such that for all* $v \in K_\infty(V)$,

$$\varphi(v) = \alpha g_\infty(v)\beta. \tag{5.3.5}$$

Proof (i) We can suppose that V is an operator subspace of a unital C^*-algebra $\mathcal{A}$, and thus that $M_n(V)$ is an operator subspace of the unital C^*-algebra $M_n(\mathcal{A})$. From the Arveson–Wittstock–Hahn–Banach extension theorem (Theorem 4.1.5), we may extend φ to a complete contraction

$$\psi : M_n(\mathcal{A}) \to M_r.$$

Thus, from the decomposition theorem (Theorem 5.3.3),

$$\psi(a) = \alpha\pi(a)\beta,$$

where π is a unital $*$-representation of $M_n(\mathcal{A})$ on a Hilbert space L, and

$$\mathbb{C}^r \xrightarrow{\beta} L \xrightarrow{\alpha} \mathbb{C}^r$$

is a diagram of contractions.

We define a unital $*$-representation π_{M_n} of M_n on L by letting

$$\pi_{M_n}(\gamma) = \pi(\gamma \otimes I_{\mathcal{A}}).$$

The $*$-representations of M_n have a very simple structure: they are unitarily equivalent to a multiple of the canonical $*$-representation of M_n on ℓ_2^n. If we let $H = \pi_{M_n}(\varepsilon_{1,1})L$, then we can find a unitary $u : L \to \ell_2^n \otimes H$ such that

$$u\pi_{M_n}(\gamma)u^* = \gamma \otimes I_H.$$

If we replace π by $u\pi u^*$ and α and β by αu^* and $u\beta$, respectively, then we can suppose that $L = \ell_2^n \otimes H$ and that $\pi(\gamma) = \gamma \otimes I_H$. If $a \in \mathcal{A}$, then

$$\pi(I_n \otimes a) \in (M_n \otimes \mathbb{C}I_H)' \subseteq \mathbb{C}I_n \otimes \mathcal{B}(H).$$

Thus, we have a $*$-representation $\pi_{\mathcal{A}}$ of $\mathcal{A}$ on H such that

$$\pi(I_n \otimes a) = I_n \otimes \pi_{\mathcal{A}}(a).$$

For any $\gamma \in M_n$ and $a \in \mathcal{A}$,

$$\pi(\gamma \otimes a) = \gamma \otimes \pi_{\mathcal{A}}(a),$$

and thus

$$\psi(\gamma \otimes a) = \alpha(\gamma \otimes \pi_{\mathcal{A}}(a))\beta.$$

If we let E_2 be the projection on the range of β, and E_1 be the projection on $(\ker\alpha)^\perp$, then $\alpha E_1 = \alpha$ and $E_2\beta = \beta$ with $\dim E_i(L) \leq r$ for $i = 1, 2$. The subspaces

$$L_i = (M_n \otimes \mathbb{C}I)(E_i L)$$

are invariant for $\pi_{|M_n \otimes \mathbb{C}I}$, and dim $L_i \leq rn^2$. If we let Q_i be the projection of L onto L_i, then it follows that

$$Q_i \in (M_n \otimes I_H)' = \mathbb{C}I_n \otimes \mathcal{B}(H),$$

and therefore $Q_i = I_n \otimes P_i$ for some projection P_i on H. It follows that $L_i = \mathbb{C}^n \otimes H_i$, where $H_i = P_i(H)$ and

$$\dim H_i = (\dim L_i) / n \leq rn.$$

Since $E_i \leq I_n \otimes P_i$, we have $\alpha(I_n \otimes P_1) = \alpha$ and $(I_n \otimes P_2)\beta = \beta$, and thus

$$\psi(\gamma \otimes a) = \alpha(\gamma \otimes P_1 \pi_{\mathcal{A}}(a) P_2)\beta.$$

We fix isometries $u_i : H_i \hookrightarrow \mathbb{C}^{rn}$ and we let $\tilde{P}_1 = u_1 P_1$ and $\tilde{P}_2 = P_2 u_2^*$. We then have

$$\begin{aligned}\psi(\gamma \otimes a) &= \alpha(I_n \otimes u_1^*)(\gamma \otimes \tilde{P}_1 \pi_{\mathcal{A}}(a) \tilde{P}_2)(I_n \otimes u_2)\beta \\ &= \tilde{\alpha}(\gamma \otimes g(a))\tilde{\beta},\end{aligned}$$

where

$$\tilde{\alpha} = \alpha(I \otimes u_1^*) \in \mathcal{B}(\mathbb{C}^{rn^2}, \mathbb{C}^n)$$

and

$$\tilde{\beta} = (I \otimes u_2)\beta \in \mathcal{B}(\mathbb{C}^n, \mathbb{C}^{rn^2})$$

are contractions, and

$$g(a) = \tilde{P}_1 \pi_{\mathcal{A}}(a) \tilde{P}_2$$

is a complete contraction of $\mathcal{A}$ into M_{rn}. We conclude that if

$$v = [v_{i,j}] \in M_n(V) \subseteq M_n(\mathcal{A}),$$

then

$$\varphi(v) = \sum \varphi(\varepsilon_{i,j} \otimes v_{i,j}) = \sum \tilde{\alpha}(\varepsilon_{i,j} \otimes g(v_{i,j}))\tilde{\beta} = \tilde{\alpha} g_n(v) \tilde{\beta}.$$

(ii) The argument is essentially the same, although we have to be more careful with the representation theory of the non-unital C^*-algebra K_∞. As before,

$$K_\infty(V) \subseteq K_\infty(\mathcal{A}) = K_\infty \otimes \mathcal{A},$$

where $\mathcal{A}$ is a unital C^*-algebra, and we let $\otimes$ denote the usual spatial C^*-algebraic closure of the algebraic tensor product (see §8.1 for the corresponding operator space notion). By the previous arguments, we may extend φ to a complete contraction

$$\psi : K_\infty \otimes \mathcal{A} \to M_r,$$

and find a decomposition $\psi(a) = \alpha \pi(a) \beta$ with π a $*$-representation of the (non-unital) algebra $\mathcal{A}$.

We may assume that the $*$-representation π is non-degenerate in the sense that

$$\overline{\pi(K_\infty \otimes \mathcal{A})L} = L.$$

To see this we let P be the projection of L onto $\tilde{L} = \overline{\pi(K_\infty \otimes \mathcal{A})L}$. Then

$$\tilde{\pi}(a) = \pi(a)_{|\tilde{L}} : \tilde{L} \to \tilde{L}$$

is a non-degenerate $*$-representation of $K_\infty \otimes \mathcal{A}$ with $\pi(a) = P\tilde{\pi}(a)P$. It follows that

$$\psi(a) = (\alpha P)\tilde{\pi}(a)(P\beta)$$

is a decomposition with $\tilde{\pi}$ non-degenerate.

We can extend π to the unital extension

$$K_\infty(\mathcal{A})^\dagger = K_\infty(\mathcal{A}) \oplus (I_{K_\infty} \otimes I_{\mathcal{A}})$$

by letting $\pi(I_{K_\infty} \oplus I_{\mathcal{A}}) = I_L$. It follows that the representation π_0 of K_∞ defined by $\pi_{K_\infty}(\gamma) = \pi(\gamma \otimes I_{\mathcal{A}})$ is non-degenerate since if $\pi_{K_\infty}(\gamma)\xi = 0$ for all $\gamma \in K_\infty$, then

$$\pi(\gamma \otimes a_0)\xi = \pi(1 \otimes a)\pi_{K_\infty}(\gamma)\xi = 0.$$

The non-degenerate $*$-representations of K_∞ are unitarily equivalent to multiples of the obvious $*$-representation on ℓ_2. Thus, we may assume that

$$L = \ell_2 \otimes H,$$

where $\pi_{K_\infty}(\alpha) = \alpha \otimes I_H$. The commutant argument again applies, and we have a unital $*$-representation $\pi_{\mathcal{A}}$ of $\mathcal{A}$ with

$$\pi(\alpha \otimes a) = \alpha \otimes \pi_{\mathcal{A}}(a).$$

The remainder of the argument is identical to that for (i). □

Part (i) of the following corollary completes the discussion of (2.2.4). It is a special case of the above theorem. The other parts are proved in precisely the same manner.

Corollary 5.3.5 *Suppose that $\sigma : M_n \to M_r$ is a linear mapping. Then*

(i) *σ is a complete contraction if and only if there exist contractive matrices $\mu \in M_{r,rn^2}$ and $\gamma \in M_{rn^2,r}$ such that*

$$\sigma(\alpha) = \mu(\overbrace{\alpha \oplus \cdots \oplus \alpha}^{nr})\gamma. \tag{5.3.6}$$

(ii) *σ is completely positive if and only if there exists a matrix $\gamma \in M_{rn^2,r}$ such that*

$$\sigma(\alpha) = \gamma^*(\overbrace{\alpha \oplus \cdots \oplus \alpha}^{nr})\gamma. \tag{5.3.7}$$

(iii) *σ is a morphism if and only if there exists and a matrix $\gamma \in M_{rn^2,r}$ such that $\gamma^*\gamma = I_r$ and*

$$\sigma(\alpha) = \gamma^*(\overbrace{\alpha \oplus \cdots \oplus \alpha}^{nr})\gamma. \tag{5.3.8}$$

□

It is often convenient to express (5.3.6) in a more compact form. If we let

$$\mu = \begin{bmatrix} \mu_1 \ \dots \ \mu_{nr} \end{bmatrix} \quad \text{and} \quad \gamma = \begin{bmatrix} \gamma_1 \\ \vdots \\ \gamma_{nr} \end{bmatrix},$$

where $\mu_i \in M_{r,n}$ and $\gamma_i \in M_{n,r}$, then

$$\sigma(\alpha) = \sum_{i=1}^{nr} \mu_i \alpha \gamma_i \,,$$

where

$$\left\| \left[\mu_1 \ \dots \ \mu_{nr} \right] \right\| = \left\| \sum_{i=1}^{nr} \mu_i \mu_i^* \right\|^{1/2} \leq 1$$

and similarly, $\sum_{i=1}^{nr} \gamma_i^* \gamma_i \leq I_r$. In the same manner, we may rewrite (5.3.8) as

$$\sigma(\alpha) = \sum_{i=1}^{nr} \gamma_i^* \alpha \gamma_i,$$

where $\sum_{i=1}^{nr} \gamma_i^* \gamma_i = I_r$.

5.4 DECOMPOSABILITY

In general, one cannot replace $\mathcal{B}(H)$ by an arbitrary C^*-algebra $\mathcal{B}$ in Theorem 5.3.2. This phenomenon was studied in detail by Haagerup, who developed a tool for measuring the 'decomposability' of a mapping. We shall present only a small portion of his theory, in order to provide a key example of a completely contractive mapping $\varphi : \mathcal{A} \to \mathcal{B}$ for which one cannot find a morphism

$$\Phi : M_2(\mathcal{A}) \to M_2(\mathcal{B})$$

satisfying (5.3.3). This will play an important role in our discussion of exactness (see the proof of Lemma 14.5.3).

Given two C^*-algebras $\mathcal{A}$ and $\mathcal{B}$, the set of completely positive mappings $\mathcal{CP}(\mathcal{A}, \mathcal{B})$ is a cone in the operator space $\mathcal{CB}(\mathcal{A}, \mathcal{B})$. Haagerup carefully analysed the subtle interactions between the corresponding partial ordering and its related matrix orderings with the completely bounded matrix norms. In particular, he showed that the partial orderings determine an alternative norm for mappings, rather analogous to the order unit norms one encounters in classical functional analysis.

A mapping of C^*-algebras $\varphi : \mathcal{A} \to \mathcal{B}$ is *decomposable* if it is in the linear span of the completely positive mappings from $\mathcal{A}$ into $\mathcal{B}$. The relevance of the following result to our earlier discussion will become apparent in Proposition 5.4.2.

Proposition 5.4.1 *Given C^*-algebras $\mathcal{A}$ and $\mathcal{B}$, a mapping $\varphi : \mathcal{A} \to \mathcal{B}$ is decomposable if and only if there exist two completely positive mappings $\psi_1, \psi_2 : \mathcal{A} \to \mathcal{B}$ for which the mapping*

$$\Phi : \mathcal{A} \to M_2(\mathcal{B}) : a \mapsto \begin{bmatrix} \psi_1(a) & \varphi(a) \\ \varphi^*(a) & \psi_2(a) \end{bmatrix} \tag{5.4.1}$$

is completely positive.

Proof If φ is decomposable, then $\varphi = \varphi_1 - \varphi_2 + \mathrm{i}(\varphi_3 - \varphi_4)$, where φ_k are completely positive. It follows that if we let $\psi = \sum \varphi_j$, then

$$\Phi = \begin{bmatrix} \psi & \varphi \\ \varphi^* & \psi \end{bmatrix}$$

has the desired properties. To see this, we note that if $a \in \mathcal{A}^+$ and we let $c_j = \varphi_j(a)$, then

$$\begin{aligned} \begin{bmatrix} \psi(a) & \varphi(a) \\ \varphi^*(a) & \psi(a) \end{bmatrix} &= \begin{bmatrix} c_1 + c_2 + c_3 + c_4 & (c_1 - c_2) + \mathrm{i}(c_3 - c_4) \\ (c_1 - c_2) - \mathrm{i}(c_3 - c_4) & c_1 + c_2 + c_3 + c_4 \end{bmatrix} \\ &= \begin{bmatrix} c_1 + c_2 & c_1 - c_2 \\ c_1 - c_2 & c_1 + c_2 \end{bmatrix} + \begin{bmatrix} c_3 + c_4 & \mathrm{i}(c_3 - c_4) \\ -\mathrm{i}(c_3 - c_4) & c_3 + c_4 \end{bmatrix} \end{aligned}$$

is positive by (1.3.8). Since a similar argument applies to matrices a in $M_n(\mathcal{A})^+$, we conclude that

$$\Phi = \begin{bmatrix} \psi & \varphi \\ \varphi^* & \psi \end{bmatrix} \geq_{cp} 0.$$

Conversely, if we have this relation, then

$$\varphi = \frac{1}{4}(R_1 - R_2 + \mathrm{i}(R_3 - R_4)),$$

where

$$\begin{aligned} R_1 &= [1\ 1]\, \Phi\, [1\ 1]^*, \\ R_2 &= [1\ {-1}]\, \Phi\, [1\ {-1}]^*, \\ R_3 &= [1\ \mathrm{i}]\, \Phi\, [1\ \mathrm{i}]^*, \end{aligned}$$

and

$$R_4 = [1\ {-\mathrm{i}}]\, \Phi\, [1\ {-\mathrm{i}}]^*$$

are completely positive mappings. □

The following result of Haagerup shows that the above notion coincides with that used in (5.3.3).

Proposition 5.4.2 *Suppose that $\varphi : \mathcal{A} \to \mathcal{B}$ is a complete contraction. Then given completely positive mappings ψ_1 and ψ_2 from $\mathcal{A}$ into $\mathcal{B}$, the mapping*

$$\Phi = \begin{bmatrix} \psi_1 & \varphi \\ \varphi^* & \psi_2 \end{bmatrix} : \mathcal{A} \to M_2(\mathcal{B}) : a \mapsto \begin{bmatrix} \psi_1(a) & \varphi(a) \\ \varphi^*(a) & \psi_2(a) \end{bmatrix}$$

is completely positive if and only if

$$\Psi = \begin{bmatrix} \psi_1 & \varphi \\ \varphi^* & \psi_2 \end{bmatrix} : M_2(\mathcal{A}) \to M_2(\mathcal{B}) : \begin{bmatrix} a & b \\ c & d \end{bmatrix} \mapsto \begin{bmatrix} \psi_1(a) & \varphi(b) \\ \varphi^*(c) & \psi_2(d) \end{bmatrix}$$

is completely positive.

Proof We have $\Phi = \Psi \circ P$, where P is the completely positive mapping

$$P : \mathcal{A} \to M_2(\mathcal{A}) : a \mapsto \begin{bmatrix} a & a \\ a & a \end{bmatrix},$$

and thus $\Psi \geq_{cp} 0$ implies that $\Phi \geq_{cp} 0$.

Conversely, if Φ is completely positive, then the same is true for Ψ, since

$$\Psi\left(\begin{bmatrix} a & b \\ c & d \end{bmatrix}\right) = \begin{bmatrix} \psi_1(a) & \varphi(b) \\ \varphi^*(c) & \psi_2(d) \end{bmatrix} = \begin{bmatrix} 1\,0\,0\,0 \\ 0\,0\,0\,1 \end{bmatrix} \Phi_2\left(\begin{bmatrix} a & b \\ c & d \end{bmatrix}\right) \begin{bmatrix} 1\,0 \\ 0\,0 \\ 0\,0 \\ 0\,1 \end{bmatrix},$$

where Φ_2 is also completely positive. □

Given a decomposable mapping $\varphi : \mathcal{A} \to \mathcal{B}$, we define the *decomposable norm* $\|\varphi\|_{dec}$ by

$$\|\varphi\|_{dec} = \inf\left\{\max\{\|\psi_1\|, \|\psi_2\|\} : \begin{bmatrix} \psi_1 & \varphi \\ \varphi^* & \psi_2 \end{bmatrix} \geq_{cp} 0\right\},$$

where $\psi_j : \mathcal{A} \to \mathcal{B}$ are completely positive mappings. From Proposition 5.4.2, we can interpret $\begin{bmatrix} \psi_1 & \varphi \\ \varphi^* & \psi_2 \end{bmatrix}$ as a mapping from $\mathcal{A}$ into $M_2(\mathcal{B})$.

In order to obtain an equivalent formulation, let us suppose that we are given mappings φ, ψ_1, and ψ_2 from $\mathcal{A}$ into $\mathcal{B}$ with $\psi_1, \psi_2 \geq_{cp} 0$. Then

$$\begin{bmatrix} \psi_1 & \varphi \\ \varphi^* & \psi_2 \end{bmatrix} = \Psi + \tilde{\varphi},$$

where $\tilde{\varphi}$ is the self-adjoint mapping

$$\tilde{\varphi}(a) = \begin{bmatrix} 0 & \varphi(a) \\ \varphi^*(a) & 0 \end{bmatrix}$$

and Ψ is the completely positive mapping

$$\Psi(a) = \begin{bmatrix} \psi_1(a) & 0 \\ 0 & \psi_2(a) \end{bmatrix}.$$

Furthermore,
$$\|\Psi\|_{cb} = \|\Psi\| = \max\{\|\psi_1\|, \|\psi_2\|\}.$$
If $\mathcal{A}$ is unital, then this is immediate from the relation
$$\|\Psi\| = \|\Psi(1)\| = \max\{\|\psi_1(1)\|, \|\psi_2(1)\|\},$$
and in the non-unital case it suffices to consider approximate identities. Finally,
$$\|\tilde{\varphi}\|_{cb} = \left\|(\varphi \oplus \varphi^*)\begin{bmatrix} 0 & 1 \\ 1 & 0 \end{bmatrix}\right\|_{cb} = \|\varphi \oplus \varphi^*\|_{cb} = \|\varphi\|_{cb}.$$

We have
$$\Psi - \tilde{\varphi} = \begin{bmatrix} \psi_1 & -\varphi \\ -\varphi^* & \psi_2 \end{bmatrix} = \alpha^* \begin{bmatrix} \psi_1 & \varphi \\ \varphi^* & \psi_2 \end{bmatrix} \alpha,$$
where $\alpha = \begin{bmatrix} 1 & 0 \\ 0 & -1 \end{bmatrix} \in M_2$. It follows that
$$\begin{bmatrix} \psi_1 & \varphi \\ \varphi^* & \psi_2 \end{bmatrix} \geq_{cp} 0 \Leftrightarrow \Psi \pm \tilde{\varphi} \geq_{cp} 0$$
$$\Leftrightarrow -\Psi \leq_{cp} \tilde{\varphi} \leq_{cp} \Psi$$
and
$$\|\varphi\|_{dec} = \inf\{\|\Psi\| : -\Psi \leq_{cp} \tilde{\varphi} \leq_{cp} \Psi\}.$$

It is evident that $\|\cdot\|_{dec}$ is a norm on the linear space $\mathcal{D}(\mathcal{A}, \mathcal{B})$ of decomposable mappings $\varphi : \mathcal{A} \to \mathcal{B}$. If φ is not decomposable, then we let $\|\varphi\|_{dec} = \infty$.

Lemma 5.4.3 *Given C^*-algebras $\mathcal{A}$ and $\mathcal{B}$ and a mapping $\varphi : \mathcal{A} \to \mathcal{B}$, we have $\|\varphi\|_{cb} \leq \|\varphi\|_{dec}$. If $\mathcal{B}$ is injective, then $\|\varphi\|_{dec} = \|\varphi\|_{cb}$.*

Proof If $\|\varphi\|_{dec} < 1$, then we may find completely positve mappings ψ_1 and ψ_2 with
$$0 \leq_{cp} \Psi = \begin{bmatrix} \psi_1 & \varphi \\ \varphi^* & \psi_2 \end{bmatrix} : M_2(\mathcal{A}) \to M_2(\mathcal{B}),$$
and $\max\{\|\psi_j\|\} < 1$. If $b \in M_n(\mathcal{A})$ and $\|b\| \leq 1$, then
$$\begin{bmatrix} 1_n & b \\ b^* & 1_n \end{bmatrix} \geq 0$$
implies that
$$0 \leq \begin{bmatrix} (\psi_1)_n(1_n) & \varphi_n(b) \\ \varphi_n(b)^* & (\psi_2)_n(1_n) \end{bmatrix} \leq \begin{bmatrix} 1_n & \varphi_n(b) \\ \varphi_n(b)^* & 1_n \end{bmatrix},$$
and thus $\|\varphi_n(b)\| \leq 1$. The first assertion follows.

Let us suppose that $\mathcal{B}$ is injective and that $\|\varphi\|_{cb} \leq 1$. We can suppose that $\mathcal{B}$ is a C^*-subalgebra of $\mathcal{B}(H)$ for some Hilbert space H. Since $\mathcal{B}$ is injective, there is a completely contractive projection P of $\mathcal{B}(H)$ onto $\mathcal{B}$. From Corollary 5.1.2 P is completely positive. From Theorem 5.3.2 there exist completely positive contractions $\psi_1, \psi_2 \in \mathcal{CP}(\mathcal{A}, \mathcal{B}(H))$ such that

$$\Phi = \begin{bmatrix} \psi_1 & \varphi \\ \varphi^* & \psi_2 \end{bmatrix} \geq_{cp} 0.$$

It follows that

$$\begin{bmatrix} P \circ \psi_1 & \varphi \\ \varphi^* & P \circ \psi_2 \end{bmatrix} = P_2 \circ \Phi \geq_{cp} 0,$$

and thus

$$\|\varphi : \mathcal{A} \to \mathcal{B}\|_{dec} \leq \max\{\|P \circ \psi_1\|, \|P \circ \psi_2\|\} \leq 1. \qquad \square$$

In the remainder of this section we construct an important example due to Haagerup (1985*b*) of a mapping φ for which $\|\varphi\|_{dec} > \|\varphi\|_{cb}$.

Theorem 5.4.4 *Suppose that $\mathcal{R}$ is a finite factor and that $u_1, \ldots, u_n \in \mathcal{R}$ are arbitrary unitaries. If we define $\varphi : \ell^n_\infty \to \mathcal{R}$ by letting $\varphi(e_k) = u_k$, then*

$$\|\varphi\|_{dec} = n.$$

Proof $p(\sum c_k e_k) = \sum c_k$ is a positive linear functional on ℓ^n_∞, and thus

$$\psi : \ell^n_\infty \to \mathcal{R} : \sum c_k e_k \mapsto \left(\sum c_k\right)1$$

is a completely positive mapping with $\|\psi\| = n$. We let $\Psi = \psi \oplus \psi$ and

$$\tilde{\varphi} = \begin{bmatrix} 0 & \varphi \\ \varphi^* & 0 \end{bmatrix}.$$

We have

$$\Psi(e_k) = \begin{bmatrix} 1 & 0 \\ 0 & 1 \end{bmatrix}$$

and

$$\tilde{\varphi}(e_k) = \begin{bmatrix} 0 & \varphi(e_k) \\ \varphi^*(e_k) & 0 \end{bmatrix} = \begin{bmatrix} 0 & u_k \\ u_k^* & 0 \end{bmatrix}$$

since $e_k^* = e_k$. The latter 2×2 matrix is a self-adjoint unitary, and thus

$$-\Psi(e_k) = -1_2 \leq \tilde{\varphi}(e_k) \leq 1_2 = \Psi(e_k).$$

We conclude that $-\Psi \leq \tilde{\varphi} \leq \Psi$, and since ℓ^n_∞ is abelian,

$$-\Psi \leq_{cp} \tilde{\varphi} \leq_{cp} \Psi$$

(see Theorem 5.1.5). Thus, $\|\varphi\|_{dec} \leq n$.

On the other hand, if $\varepsilon > 0$, then we may find completely positive mappings $\psi_j : \ell^n_\infty \to \mathcal{R}$ for which the mapping $\Psi = \psi_1 \oplus \psi_2$ satisfies $-\Psi \leq_{cp} \tilde{\varphi} \leq_{cp} \Psi$ and

$$\|\Psi\| \leq \|\varphi\|_{dec} + \varepsilon.$$

If we let $x_k = \Psi(e_k)$ and $\tilde{u}_k = \tilde{\varphi}(e_k)$, then $\tilde{u}_k$ is a self-adjoint unitary with

$$-x_k \leq \tilde{u}_k \leq x_k.$$

Thus,

$$x_k = \frac{1}{2}(x_k - \tilde{u}_k) + \frac{1}{2}(x_k + \tilde{u}_k),$$

where $x_k \pm \tilde{u}_k \geq 0$. If we let τ be the unique trace on $M_2(\mathcal{R})$ with $\tau(I_2) = 1$, then it follows that

$$\begin{aligned}\tau(x_k) &= \frac{1}{2}\|x_k - \tilde{u}_k\|_1 + \frac{1}{2}\|x_k + \tilde{u}_k\|_1 \\ &\geq \frac{1}{2}\|(x_k - \tilde{u}_k) - (x_k + \tilde{u}_k)\|_1 = \|\tilde{u}_k\|_1,\end{aligned}$$

where

$$\|\tilde{u}_k\|_1 = \tau((\tilde{u}_k^* \tilde{u}_k)^{1/2}) = \tau(1_2) = 1.$$

Thus,

$$\begin{aligned}\|\Psi\| &= \|\Psi(1)\| = \|\Psi(e_1) + \cdots + \Psi(e_n)\| \\ &= \|x_1 + \cdots + x_n\| \geq |\tau(x_1 + \cdots + x_n)| \geq n,\end{aligned}$$

and we conclude that $\|\varphi\|_{dec} \geq n$. □

Lemma 5.4.5 *Let $s_1, \ldots, s_n$ be the generators of $\mathbb{F}_n$, where $n \geq 2$, and suppose that $\beta_j > 0$ $(1 \leq j \leq n)$. Then*

$$\left\| \sum \beta_j \lambda(s_j) \right\| \leq \inf_{\alpha > 0} \left\{ 2\alpha + \sum \left(\sqrt{\alpha^2 + \beta_j^2} - \alpha \right) \right\}.$$

Proof In the following argument we shall use the simple algebraic identity that for any real numbers $\alpha, \beta > 0$,

$$\left[\frac{\sqrt{\alpha^2 + \beta^2} - \alpha}{\beta} \right]^{-1} = \frac{\sqrt{\alpha^2 + \beta^2} + \alpha}{\beta}.$$

Given an element $t \in \mathbb{F}_n$, we define

$$p_j(t, \alpha) = \begin{cases} \dfrac{\sqrt{\alpha^2 + \beta_j^2} - \alpha}{\beta_j} & \text{if } t \text{ does not begin with } s_j, \\ \dfrac{\sqrt{\alpha^2 + \beta_j^2} + \alpha}{\beta_j} & \text{if } t \text{ begins with } s_j. \end{cases}$$

If we let

$$c(\alpha) = 2\alpha + \sum_j \left(\sqrt{\alpha^2 + \beta_j^2} - \alpha \right),$$

then $\sum_j \beta_j p_j(t,\alpha) \leq c(\alpha)$. This follows since if t begins with s_i, then

$$\begin{aligned}\beta_i p_i(t,\alpha) + \sum_{j\neq i} \beta_j p_j(t,\alpha) &= \left(\sqrt{\alpha^2+\beta_i^2}+\alpha\right) + \sum_{j\neq i}\left(\sqrt{\alpha^2+\beta_j^2}-\alpha\right) \\ &= c(\alpha),\end{aligned}$$

whereas if t does not begin with s_i for any i, then

$$\sum_j \beta_j p_j(t,\alpha) = \sum_j \left(\sqrt{\alpha^2+\beta_j^2}-\alpha\right) \leq c(\alpha).$$

If $b = \sum \beta_j \lambda(s_j)$, then we have for any $\xi \in \ell_2(\mathbb{F}_2)$ with finite support,

$$\begin{aligned}|(b(\xi))(t)|^2 &= \left|\sum_j \beta_j^{1/2} p_j(t,\alpha)^{1/2} \beta_j^{1/2} p_j(t,\alpha)^{-1/2}\xi(s_j^{-1}t)\right|^2 \\ &\leq \left(\sum_j \beta_j p_j(t,\alpha)\right)\left(\sum_j \beta_j p_j^{-1}(t,\alpha)\left|\xi(s_j^{-1}t)\right|^2\right) \\ &\leq c(\alpha)\sum_j \beta_j p_j^{-1}(t,\alpha)\left|\xi(s_j^{-1}t)\right|^2,\end{aligned}$$

and thus

$$\begin{aligned}\|b(\xi)\|^2 &\leq c(\alpha)\sum_{j,t}\beta_j p_j^{-1}(t,\alpha)\left|\xi(s_j^{-1}t)\right|^2 \\ &= c(\alpha)\sum_{j,t}\beta_j p_j^{-1}(s_j t,\alpha)\left|\xi(t)\right|^2 \\ &= c(\alpha)\sum_t\left(\sum_j \beta_j p_j^{-1}(s_j t,\alpha)\right)|\xi(t)|^2.\end{aligned}$$

If we fix t, then

$$p_j^{-1}(s_j t,\alpha) = \begin{cases} \dfrac{\sqrt{\alpha^2+\beta_j^2}+\alpha}{\beta_j} & \text{if } s_j t \text{ does not begin with } s_j, \\[2ex] \dfrac{\sqrt{\alpha^2+\beta_j^2}-\alpha}{\beta_j} & \text{if } s_j t \text{ begins with } s_j. \end{cases}$$

For a given i, $s_i t$ does not begin with s_i precisely when the reduced form of t begins with s_i^{-1}. If that occurs, then for $j \neq i$, the reduced form of $s_j t$ begins with s_j. It follows that if t begins with s_i^{-1}, for some i, then

$$\begin{aligned}\sum_j \beta_j p_j^{-1}(s_j t,\alpha)\left|\xi(t)\right|^2 &= \left(\sqrt{\alpha^2+\beta_i^2}+\alpha\right)+\sum_{j\neq i}\left(\sqrt{\alpha^2+\beta_j^2}-\alpha\right) \\ &= c(\alpha),\end{aligned}$$

whereas if t does not begin with s_i^{-1} for any i,

$$\sum_j \beta_j p_j^{-1}(s_j t, \alpha) = \sum_j \left(\sqrt{\alpha^2 + \beta_j^2} - \alpha\right) \leq c(\alpha).$$

We conclude that

$$\|b(\xi)\|^2 \leq c(\alpha)^2 \|\xi\|^2,$$

and the lemma follows. □

Corollary 5.4.6 *If $s_1, \ldots, s_n$ are the generators of $\mathbb{F}_n$, then*

$$\left\|\sum \lambda(s_j)\right\| \leq 2\sqrt{n-1}.$$

Proof From the above result,

$$\left\|\sum \lambda(s_j)\right\| \leq \inf_{\alpha>0} \left\{2\alpha + n\left(\sqrt{\alpha^2+1} - \alpha\right)\right\}$$

and indicated infimum is a consequence of elementary calculus. □

If $n > 1$, the von Neumann algebra $L(\mathbb{F}_n)$ generated by $\lambda(\mathbb{F}_n)$, where $\lambda : \mathbb{F}_n \to \mathcal{B}(\ell_2(\mathbb{F}_n))$ is the left regular representation, is a finite factor.

Theorem 5.4.7 *If $n > 1$, then the mapping*

$$\varphi : \ell_\infty^n \to L(\mathbb{F}_n) : \alpha \mapsto \sum \alpha_j \lambda(s_j)$$

satisfies $\|\varphi\|_{cb} \leq 2\sqrt{n-1}$ and $\|\varphi\|_{dec} = n$.

Proof The equality follows from Theorem 5.4.4. Since any contraction in the C^*-algebra $M_r \otimes \ell_\infty^n$ is a convex combination of unitaries, it suffices to prove that for each unitary operator $U \in M_r \otimes \ell_\infty^n$,

$$\|\varphi_r(U)\| \leq \left\|\sum \lambda(s_j)\right\|.$$

We begin by noting that any unitary operator U in the C^*-algebra $M_r \otimes \ell_\infty^n$ has the form $\sum U_j \otimes e_j$, where each U_j is a unitary in M_r. It follows that

$$\varphi_r(U) = \sum U_j \otimes \lambda(s_j).$$

The mapping $s_j \mapsto U_j$ determines a representation

$$\pi : \mathbb{F}_n \to M_r$$

with $\pi(s_j) = U_j$, and thus we have

$$\varphi_r(U) = \sum \pi(s_j) \otimes \lambda(s_j) = \sum (\pi \times \lambda)(s_j).$$

Owing to the well-known tensor product absorption property of the regular representation (see Fell 1962 and Dixmier 1964, §13.11.3), there is a unitary

$$v : \mathbb{C}^r \otimes H \to \mathbb{C}^r \otimes H$$

such that $\pi \times \lambda = v^*(\iota_r \times \lambda)v$, where ι_r is the trivial representation,

$$\iota_r : \mathbb{F}_n \to M_r : s_k \mapsto I_r.$$

We have

$$(\iota_r \times \lambda)(s_j) = I_r \otimes \lambda(s_j),$$

and thus

$$\begin{aligned}\|\varphi_r(U)\| &= \left\|\sum(\pi \times \lambda)(s_j)\right\| \\ &= \left\|v^*(I_r \otimes \sum \lambda(s_j))v\right\| \\ &= \left\|\sum \lambda(s_j)\right\|.\end{aligned}$$

□

5.5 MATRIX CONVEXITY AND THE TRACE CLASS OPERATORS

Underlying the ideas of norms and orderings in linear spaces is the classical notion of convexity. This again has a 'matrix' version, which is implicit in many operator algebraic arguments. Although we have chosen not to base our treatment on this notion, it is occasionally useful for motivational purposes. We begin by reformulating the usual scalar definition.

Let us suppose that E is an arbitrary linear space and $n \in \mathbb{N}$. If $\sigma \in (\ell^n_\infty)^* = \ell^n_1$, and $\tilde{x} = (\tilde{x}_1, \ldots, \tilde{x}_n) \in E^n$, then

$$x = \sigma \otimes id(\tilde{x}) = \sigma_1\tilde{x}_1 + \cdots + \sigma_n\tilde{x}_n$$

is just another way of stating that x is a linear combination of the $\tilde{x}_i$. In particular, if σ is a probability measure, i.e. $\sigma_i \geq 0$ and $\sum \sigma_i = 1$, then x is a convex combination of the $\tilde{x}_j$, whereas if $\sum |\sigma_i| \leq 1$, then we obtain an absolutely convex combination (see §A.2). It follows that a set $K \subseteq V$ is convex (respectively, absolutely convex) if and only if $\sigma \otimes id(K^n) \subseteq K$ for each probability measure σ (respectively, each contractive measure σ) on $\{1, \ldots, n\}$ where $n \in \mathbb{N}$ is arbitrary. This has a natural generalization to morphisms and contractions $\sigma : \ell^n_\infty \to \ell^m_\infty$. If K is convex or absolutely convex, then $\sigma \otimes id(K^n) \subseteq K^m$ for the corresponding mappings σ.

By analogy, let us suppose that V is a linear space and $n \in \mathbb{N}$. If

$$\sigma = [\sigma_{k,l}] \in M_r(T_n) = \mathcal{CB}(M_n, M_r)$$

and $\tilde{v} = [\tilde{v}_{i,j}] \in \mathbb{M}_n(V)$, then we say that $v = \sigma \otimes id(\tilde{v})$ is a *matrix combination* of the entries of $\tilde{v}$. If $\sigma_{k,l} = [\sigma_{k,l}(i,j)] \in T_n$, then using the scalar pairing, $\sigma_{k,l}(i,j) = \sigma_{k,l}(\varepsilon^{[n]}_{i,j})$, and we have the more explicit expression

$$v = \sigma \otimes id(\tilde{v}) = \sum_{i,j} \sigma(\varepsilon^{[n]}_{i,j}) \otimes \tilde{v}_{i,j}$$

$$= \sum_{i,j} \sigma_{k,l}(i,j)\varepsilon_{k,l}^{[r]} \otimes \tilde{v}_{i,j} = \left[\sum_{i,j} \sigma_{k,l}(i,j) v_{i,j} \right].$$

We say that v is a *matrix convex combination* if σ is a morphism, and an *matrix absolutely convex combination* if $\|\sigma\| \leq 1$.

Given a linear space V and a collection of sets (K_n) with $K_n \subseteq M_n(V)$, we say that (K_n) is *matrix convex* (respectively, *matrix absolutely convex*) if for all $m, n \in \mathbb{N}$,

MC1 $K_1^n = K_1 \oplus \cdots \oplus K_1 \subseteq K_n$,

MC2 for each morphism (respectively, complete contraction)

$$\sigma : M_n \to M_m,$$

we have

$$\sigma \otimes id(K_n) \subseteq K_m.$$

It is evident from Corollary 5.3.5 and axioms **M1** and **M2** that the matrix unit balls of an operator space constitute an 'matrix absolutely convex set'. From this example we see that K_n is generally larger than K^n, and therefore we cannot expect the set K_1 to 'determine' the sets K_n.

Proposition 5.5.1 *Suppose that V is a linear space, and that we are given a collection of sets $K_n \subseteq M_n(V)$ with $K^n \subseteq K_n$. Then*

(i) *(K_n) is matrix convex if and only if given $\tilde{v} \in K_n$ and matrices γ_i in $M_{n,m}$ $(1 \leq i \leq p)$ satisfying*

$$\sum \gamma_i^* \gamma_i = I_m,$$

we have

$$\gamma_1^* \tilde{v} \gamma_1 + \cdots + \gamma_p^* \tilde{v} \gamma_p \in K_m;$$

(ii) *(K_n) is absolutely matrix convex if and only if given $\tilde{v} \in K_n$ and matrices $\gamma_i \in M_{n,m}$ and $\mu_i \in M_{m,n}$ $(1 \leq i \leq p)$ satisfying*

$$\sum \gamma_i^* \gamma_i \leq I_m \text{ and } \sum \mu_i \mu_i^* \leq I_m,$$

respectively, we have

$$\mu_1 \tilde{v} \gamma_1 + \cdots + \mu_p \tilde{v} \gamma_p \in K_m .$$ □

To illustrate these notions, any element f of $(\ell_1^n)_{\|\cdot\| \leq 1}$ is an absolute convex combination of the functions ε_i $(1 \leq i \leq n)$ since

$$f = \sum f(i)\varepsilon_i.$$

For the analogous matrix calculation, let us suppose that

$$t = [t_{k,l}] \in M_r(T_n)_{\|\cdot\| \leq 1} = \mathcal{CB}(M_n, M_r)_{\|\cdot\|_{cb} \leq 1},$$

where

$$t_{k,l} = [t_{k,l}(i,j)] \in T_n.$$

With the identification $T_n = M_n^*$, we have $t_{k,l}(i,j) = t_{k,l}(\varepsilon_{i,j}^{[n]})$, and therefore

$$t(\varepsilon_{i,j}^{[n]}) = [t_{k,l}(\varepsilon_{i,j}^{[n]})] = \sum_{k,l} t_{k,l}(i,j)\varepsilon_{k,l}^{[r]}.$$

It follows that

$$\begin{aligned} t &= \sum_{k,l} \varepsilon_{k,l}^{[r]} \otimes t_{k,l} = \sum_{i,j,k,l} \varepsilon_{k,l}^{[r]} \otimes t_{k,l}(i,j)\varepsilon_{i,j}^{[n]} \\ &= \sum_{i,j,k,l} t_{k,l}(i,j)\varepsilon_{k,l}^{[r]} \otimes \varepsilon_{i,j}^{[n]} = \sum_{i,j} t(\varepsilon_{i,j}^{[n]}) \otimes \varepsilon_{i,j}^{[n]} = t \otimes id_{T_n}(\varepsilon^{[n]}) \end{aligned}$$

and t is an absolute matrix convex combination of the entries of $\varepsilon^{[n]}$ in $M_n(T_n)$. If $\alpha \in M_n$, then

$$\varepsilon^{[n]}(\alpha) = [\varepsilon_{i,j}^{[n]}(\alpha)] = [\alpha_{i,j}] = \alpha;$$

hence $\varepsilon^{[n]} = id : M_n \to M_n$, and

$$\left\|\varepsilon^{[n]}\right\|_{M_n(T_n)} = \|id\|_{cb} = 1.$$

If we are given an operator space V and an element $v \in M_n(V)$ with $\|v\| \leq 1$, then we may define a mapping $\varphi : T_n \to V$ by letting

$$\varphi(\varepsilon_{i,j}^{[n]}) = v_{i,j}.$$

It follows from the above formalism that φ is completely contractive since it maps matrix absolutely convex combinations of the entries of the matrix $\varepsilon^{[n]}$ to matrix absolutely convex combinations of the entries of v. In detail, let us suppose that

$$t \in M_r(T_n)_{\|\cdot\| \leq 1} = \mathcal{CB}(M_n, M_r)_{\|\cdot\|_{cb} \leq 1}.$$

From the calculation of (5.5.1),

$$\varphi_r(t) = \sum_{i,j} t(\varepsilon_{i,j}^{[n]}) \otimes \varphi(\varepsilon_{i,j}^{[n]}) = \sum_{i,j} t(\varepsilon_{i,j}^{[n]}) \otimes v_{i,j} = (t \otimes id_V)(v),$$

and thus

$$\|\varphi_r(t)\| = \|(t \otimes id_V)(v)\| \leq \|v\| \leq 1.$$

This result has a natural converse, and we shall prove by using tensor products that we have a natural isometric identification

$$M_n(V) \cong \mathcal{CB}(T_n, V)$$

(see Proposition 8.1.2 and Corollary 8.1.3).

5.6 NOTES AND REFERENCES

Theorem 5.1.5 was proved by Stinespring (1955). His purpose was to find a general C^*-algebraic version of a theorem of Naimark (1943). In the latter paper, Naimark proved that, under suitable conditions, a non-negative operator-valued measure is a compression of a projection-valued measure. The remaining results in §5.1 are more difficult to track down, but some versions of them may be found in Arveson (1969,1972) and Choi and Effros (1977a).

Theorem 5.2.1 is due to Stinespring (1955). The second statement in Corollary 5.2.2 is due to Choi (1974), in which he also proved that the Schwarz inequality (5.2.2) actually holds for 2-positive mappings. The proof of Theorem 5.3.3 was first explicitly given in Paulsen (1984). A different approach to the same result can be found in the unpublished notes of Haagerup (1980). We have followed Paulsen's 2×2 'off-diagonalization technique' for Lemma 5.3.1, Theorem 5.3.2, and Theorem 5.3.3. Corollary 5.3.5 (ii) and (iii) were first proved by Choi (1975).

The material in §5.4 is due to Haagerup (1985*b*). Lemma 5.4.5 was first proved by Akemann and Ostrand (1976). We have given the elegant proof due to Picardello and Pytlik (1988). The discussion of matrix convexity in §5.5 appeared in Effros and Winkler (1997).

Over the years, Banach spaces have attracted much more interest than the corresponding theory of ordered Banach spaces. This stems, in part, from the difficulties that arise when one takes quotients and mapping spaces of ordered spaces. A similar situation holds for operator systems. Although they continue to have significant applications in operator algebra theory (see, for example, Wassermann 1994), the general theory of operator systems has not received much attention.

Despite these remarks, one can make the case that ordered and matrix ordered spaces are more relevant to applications. As pointed out by Kadison (1951), the correspondence between function systems and their state spaces provides a natural contravariant isomorphism between the category of function systems and compact convex subsets of locally convex Hausdoff spaces. Alfsen and Schultz have used function systems in their characterization of the state spaces of C^*-algebras (see Alfsen and Schultz 1998). Their axioms have physical interpretations, and the theory has provided the most convincing explanation to date of why one must model the observables of quantum mechanics with Hilbert space operators.

On the other hand, there are compelling arguments for regarding the completely positive mappings and the associated matrix orderings as more natural than the positive mappings in quantum mechanics. The reader can find a discussion of this point of view in K. Kraus (1983).

6

Injectivity

6.1 THE INJECTIVE OPERATOR SPACES

As in the previous chapter, operator systems provide a very convenient tool for studying extensions. We say that an operator system V is *injective* if for any operator systems $W_0 \subseteq W \subseteq \mathcal{B}(K)$, every morphism $\varphi_0 : W_0 \to V$ may be extended to a morphism $\varphi : W \to V$. Thus, we have the commutative diagram (4.1.10) with morphisms rather than completely bounded mappings.

Lemma 6.1.1 *If $V \subseteq \mathcal{B}(H)$ is an operator system, then V is injective as an operator space if and only if it is injective as an operator system.*

Proof If V is an injective operator space, then there exists a completely contractive projection Φ of $\mathcal{B}(H)$ onto V. Since $\Phi(I) = I$, it follows from Corollary 5.1.2 that Φ is completely positive and thus a morphism. As above (see (4.1.11)), it follows that V is an injective operator system. Conversely, if V is an injective operator system, then repeating the above argument, there is a projection morphism Φ of $\mathcal{B}(H)$ onto V, and since morphisms are necessarily completely contractive, V is an injective operator space. □

Lemma 6.1.2 *Suppose that $\Phi : \mathcal{B}(H) \to \mathcal{B}(H)$ is a completely contractive and completely positive projection. Then for every $a, b \in \mathcal{B}(H)$, we have*

$$\Phi(\Phi(a)b) = \Phi(\Phi(a)\Phi(b)) = \Phi(a\Phi(b)). \tag{6.1.1}$$

Proof By linearity, it suffices to consider the case in which a and b are self-adjoint. Then

$$d = \begin{bmatrix} 0 & \Phi(a) \\ \Phi(a) & b \end{bmatrix} \quad \text{and} \quad \Phi_2(d) = \begin{bmatrix} 0 & \Phi(a) \\ \Phi(a) & \Phi(b) \end{bmatrix}$$

are self-adjoint elements of $M_2(\mathcal{B}(H))$. Since $\Phi_2 : M_2(\mathcal{B}(H)) \to M_2(\mathcal{B}(H))$ is also completely contractive and completely positive, we have from Lemma 5.2.2 that

$$\Phi_2(d)^2 \leq \Phi_2(d^2),$$

and hence

$$\begin{bmatrix} \Phi(a)^2 & \Phi(a)\Phi(b) \\ \Phi(b)\Phi(a) & \Phi(a)^2 + \Phi(b)^2 \end{bmatrix} \leq \begin{bmatrix} \Phi(\Phi(a)^2) & \Phi(\Phi(a)b) \\ \Phi(b\Phi(a)) & \Phi(\Phi(a)^2 + b^2) \end{bmatrix}.$$

Applying Φ_2 on both sides,

$$\begin{bmatrix} \Phi(\Phi(a)^2) & \Phi(\Phi(a)\Phi(b)) \\ \Phi(\Phi(b)\Phi(a)) & \Phi(\Phi(a)^2 + \Phi(b)^2) \end{bmatrix} \leq \begin{bmatrix} \Phi(\Phi(a)^2) & \Phi(\Phi(a)b) \\ \Phi(b\Phi(a)) & \Phi(\Phi(a)^2 + b^2) \end{bmatrix},$$

and thus if we let '$*$' indicate an entry we do not need to compute, then

$$\begin{bmatrix} 0 & \Phi(\Phi(a)b) - \Phi(\Phi(a)\Phi(b)) \\ \Phi(b\Phi(a)) - \Phi(\Phi(b)\Phi(a)) & * \end{bmatrix} \geq 0.$$

Therefore from Corollary 1.3.4,

$$\Phi(\Phi(a)b) = \Phi(\Phi(a)\Phi(b)) \quad \text{and} \quad \Phi(\Phi(b)\Phi(a)) = \Phi(b\Phi(a)). \tag{6.1.2}$$

Since a and b are arbitrary self-adjoint elements, we obtain (6.1.1). □

A C^*-algebra $\mathcal{A}$ is said to be (conditionally) *monotonically complete* if any bounded increasing net $a_\nu \in \mathcal{A}_{sa}$ has a least upper bound. It is not difficult to show that a monotonically complete C^*-algebra must be unital. Furthermore, given a compact Hausdorff space Ω, the commutative C^* -algebra $C(\Omega)$ is monotonically complete if and only if Ω is Stonean (see §A.6). In the context of Banach spaces, this explains why an injective commutative C^* -algebra must have the form $C(\Omega)$ with Ω Stonean. For the converse, one may imitate the proof of the ordered version of the classical Hahn–Banach theorem, using the fact that $C(\Omega)_{sa}$ is a (conditionally) monotonically complete lattice. It is proved in operator algebra theory that $\mathcal{B}(H)$ is monotonically complete, and we shall see in the next result that this is true for any injective operator system.

Theorem 6.1.3 *If $V \subseteq \mathcal{B}(H)$ is an injective operator system, then there is a unique multiplication*

$$\cdot : V \times V \to V$$

for which V, together with its given $*$-operation and norm, is a C^*-algebra with multiplicative identity I. The C^*-algebra is monotonically complete, and its canonical matrix ordering coincides with the given matrix ordering on V.

Proof Let us suppose that V is an injective operator system. The uniqueness of the multiplication follows from Corollary 5.2.3. As in the proof of Lemma 6.1.1, we may fix a projection morphism Φ of $\mathcal{B}(H)$ onto V. If $a \in V$, then $a = \Phi(b)$ for some $b \in \mathcal{B}(H)$, and thus

$$\Phi(a) = \Phi(\Phi(b)) = \Phi(b) = a. \tag{6.1.3}$$

Given $a, b \in V$, we define an operation $\cdot$ on V by

$$a \cdot b = \Phi(ab). \tag{6.1.4}$$

This is an associative multiplication on V since we have from Lemma 6.1.2 and (6.1.3) that

$$a \cdot (b \cdot c) = \Phi(a\Phi(bc)) = \Phi(\Phi(a)bc) = \Phi(abc), \tag{6.1.5}$$

and similarly,

$$(a \cdot b) \cdot c = \Phi(\Phi(ab)c) = \Phi(ab\Phi(c)) = \Phi(abc)$$

for all $a, b, c \in V$. The identity operator I is a unital element in V since for every $a \in V$,

$$I \cdot a = \Phi(Ia) = \Phi(a) = a$$

and

$$a \cdot I = \Phi(aI) = \Phi(a) = a\,.$$

Since Φ is positive, it is self-adjoint and

$$(a \cdot b)^* = \Phi(ab)^* = \Phi(b^*a^*) = b^* \cdot a^*$$

for all $a, b \in V$; hence V is a unital $*$-algebra. For any $a \in V$,

$$a^*a = \Phi(a)^*\Phi(a) \leq \Phi(a^*a) \leq \|a\|^2 I\,.$$

With the norm on V (inherited from $\mathcal{B}(H)$), we have

$$\|a^* \cdot a\| = \|\Phi(a^*a)\| = \|a\|^2\,.$$

We let $\mathcal{A}_\Phi$ denote V with this C^*-algebraic structure.

The morphism Φ determines a morphism

$$\Phi_n : M_n(\mathcal{B}(H)) \to M_n(\mathcal{B}(H)),$$

and thus a corresponding C^*-algebraic structure on the range $\mathcal{A}_{\Phi_n}$. We have a natural $*$-algebraic isomorphism

$$M_n(\mathcal{A}_\Phi) \cong \mathcal{A}_{\Phi_n}$$

since if we are given $a, b \in M_n(\mathcal{A}_\Phi)$, the canonical product is given by

$$\begin{aligned} a \cdot b &= \left[\sum_k a_{i,k} \cdot b_{k,j}\right] = \left[\sum_k \Phi(a_{i,k}b_{k,j})\right] \\ &= \Phi_n\left(\left[\sum_k a_{i,k}b_{k,j}\right]\right) = \Phi_n(ab)\,. \end{aligned}$$

Since a $*$-isomorphism of C^*-algebras is necessarily isometric, we conclude that the norm on $M_n(V)$ (regarded as a subspace of $\mathcal{B}(H^n)$) and the canonical norm on $M_n(\mathcal{A}_\Phi)$ coincide. But the norm and the identity operator I_n determine the order on both of these spaces (see (A.4.2)), and thus V and $\mathcal{A}_\Phi$ coincide as operator systems.

Finally, let us suppose that $a_\nu \in (\mathcal{A}_\Phi)_{sa}$ is an increasing net such that $\|a_\nu\| \leq k$ for some constant k. Since $\mathcal{B}(H)$ is conditionally order complete (see §A.4), the collection $S = \{a_\nu\}$ has a least upper bound $b \in \mathcal{B}(H)$. We claim that $a = \Phi(b)$ is a least upper bound for S in $\mathcal{A}_\Phi$. We have

$$a_\nu = \Phi(a_\nu) \leq \Phi(b) = a$$

for all ν. On the other hand, if $a' \in (\mathcal{A}_\Phi)_{sa}$ is such that $a_\nu \leq a'$ for all ν, then $b \leq a'$ implies that

$$\Phi(b) \leq \Phi(a') = a' .$$

□

The C^*-algebra $\mathcal{A}_\Phi$ constructed in the above proof is generally not a C^*-subalgebra of $\mathcal{B}(H)$. If, for example, we let τ be the normalized trace on M_2, then the operator space

$$\{a \oplus \tau(a) : a \in M_2\} \subseteq M_3$$

is completely order isomorphic to M_2, but it is not a subalgebra of M_3.

In contrast to the characterization of injective Banach spaces, an injective operator space is not necessarily completely isometric to a C^*-algebra. To see this we note that the mapping

$$\Phi : M_2 \to M_2 : \alpha \to \varepsilon_{1,1}\alpha$$

is a completely contractive projection of M_2 onto the operator space

$$V = \varepsilon_{1,1}M_2 = \left\{ \begin{bmatrix} \alpha_1 & \alpha_2 \\ 0 & 0 \end{bmatrix} : \alpha_i \in \mathbb{C} \right\}, \tag{6.1.6}$$

and thus V is injective. On the other hand, let us suppose that V is completely isometric to a C^*-algebra $\mathcal{A}$. Any finite-dimensional C^*-algebra is isomorphic to a direct sum of matrix algebras, and hence $\dim V = 2$ implies that $\mathcal{A} = \ell^2_\infty$. It follows that every contractive mapping into V is automatically completely contractive (see Theorem 2.2.6). To see that this is not the case we note that since the transpose mapping $\mathbf{t} : M_2 \to M_2$ is isometric,

$$\varphi = \Phi \circ \mathbf{t} : M_2 \to V$$

is a contraction. From (2.2.5) and (2.2.6),

$$\tilde{\varepsilon} = [\mathbf{t}(\varepsilon_{i,j})] \in M_2(M_2)$$

is of norm 1, but

$$\varphi_2(\tilde{\varepsilon}) = \Phi_2(\mathbf{t}_2(\tilde{\varepsilon})) = \begin{bmatrix} \Phi(\varepsilon_{11}) & \Phi(\varepsilon_{12}) \\ \Phi(\varepsilon_{21}) & \Phi(\varepsilon_{22}) \end{bmatrix} = \begin{bmatrix} 1 & 0 & 0 & 1 \\ 0 & 0 & 0 & 0 \\ 0 & 0 & 0 & 0 \\ 0 & 0 & 0 & 0 \end{bmatrix}$$

has norm $\sqrt{2}$.

We define a *partial projection* φ on $\mathcal{B}(H)$ to be a linear mapping

$$\varphi : W \to \mathcal{B}(H),$$

where W is a subspace of $\mathcal{B}(H)$, such that $\varphi(W) \subseteq W$ and $\varphi^2 = \varphi$. It is immediate from the Arveson–Wittstock–Hahn–Banach theorem that any completely contractive partial projection may be extended to a completely contractive mapping $\psi : \mathcal{B}(H) \to \mathcal{B}(H)$. We shall next show that we can suppose that ψ is a projection.

Given a reflexive and transitive relation $\preceq$ on a set X, we shall write $x \sim y$ if $x \preceq y$ and $y \preceq x$. We say that an element $x_0 \in X$ is *minimal* if for any $x \in X$, $x \preceq x_0$ implies that $x \sim x_0$.

Lemma 6.1.4 *Suppose that $\preceq$ is a reflexive and transitive relation on a compact Hausdorff space X such that for any $x \in X$, the set*

$$F_x = \{y \in X : y \preceq x\}$$

is closed. Then X has a minimal element.

Proof We say that a subset $L \subseteq X$ is *linear* if for all $y, z \in L$, either $y \preceq z$ or $z \preceq y$. There exists a maximal linear set $L_0 \subseteq X$. To see this, let $\mathcal{F}$ be the collection of all linear subsets of X, ordered by inclusion. From Zorn's lemma there exists a maximal subfamily $\mathcal{F}_0$ which is totally ordered by inclusion. Since it is evident that a union of linear sets is linear, it easily follows that $L_0 = \bigcup \mathcal{F}_0$ is a maximal linear set.

The family of sets $\{F_x : x \in L_0\}$ satisfies the finite intersection property since if we are given $x_1, \ldots, x_n \in L$, we may choose a k with $x_k \preceq x_i$ for all i. By assumption these are closed sets, and thus they have non-empty intersection. We may let x_0 be any element in this intersection. For any $x \in L_0$, we have $x_0 \preceq x$. Thus, if $y \preceq x_0$, then $y \preceq x$ for all $x \in L_0$, and $\{y\} \cup L_0$ is a linear set. Since L_0 is maximal, $y \in L_0$ and $x_0 \preceq y$. □

Lemma 6.1.5 *Suppose that φ_0 is a completely contractive partial projection on $\mathcal{B}(H)$. Then it may be extended to a completely contractive projection $\Phi : \mathcal{B}(H) \to \mathcal{B}(H)$.*

Proof We let $\mathcal{E}_{\varphi_0}$ be the set of all complete contractions $\varphi : \mathcal{B}(H) \to \mathcal{B}(H)$ such that $\varphi(w) = \varphi_0(w)$ for all $w \in W$. Owing to the fact that the unit ball of $\mathcal{B}(H)$ is compact in the weak operator topology, it is evident that $\mathcal{E}_{\varphi_0}$ is compact in the topology of point-weak operator convergence. We define a relation $\preceq$ on $\mathcal{E}_{\varphi_0}$ by letting $\varphi \preceq \psi$ if $\|\varphi(b)\| \leq \|\psi(b)\|$ for all $b \in \mathcal{B}(H)$. This is a transitive and reflexive relation. We write $\psi \sim \varphi$ if $\varphi \preceq \psi$ and $\psi \preceq \varphi$. For each $\varphi \in \mathcal{E}_{\varphi_0}$, the set

$$F_\varphi = \{\psi \in \mathcal{E}_\varphi : \psi \preceq \varphi\}$$

is closed since if $\psi_\nu \in F_\varphi$ is a net converging to a mapping $\psi \in \mathcal{E}_{\varphi_0}$, then the semicontinuity of the norm in the weak operator topology on $\mathcal{B}(H)$ implies that

$$\|\psi(b)\| \leq \limsup \{\|\psi_\nu(b)\|\} \leq \|\varphi(b)\| .$$

We conclude from Lemma 6.1.4 that there is a minimal element Φ in $\mathcal{E}_{\varphi_0}$. Let us fix $k \in \mathbb{N}$, and define

$$\Psi(b) = \frac{1}{k}(\Phi + \cdots + \Phi^k)(b)\,.$$

Then Ψ is an element of $\mathcal{E}_{\varphi_0}$ which satisfies $\Psi \preceq \Phi$, since if $b \in \mathcal{B}(H)$, then

$$\|\Psi(b)\| \leq \|\Phi(b)\|\,.$$

It follows that $\Psi \sim \Phi$. Applying this to the element $b - \Phi(b)$, we have

$$\begin{aligned}\|\Phi(b) - \Phi^2(b)\| &= \|\Phi(b - \Phi(b))\| = \|\Psi(b - \Phi(b))\| \\ &= \frac{1}{k}\left\|\Phi(b - \Phi(b)) + \cdots + \Phi^k(b - \Phi(b))\right\| \\ &= \frac{1}{k}\left\|\Phi(b) - \Phi^{k+1}(b)\right\| \leq \frac{2}{k}\,\|b\|\,,\end{aligned}$$

and since k is arbitrary, we conclude that $\Phi^2 = \Phi$. □

Theorem 6.1.6 *Let W be an injective operator space. Then there exists an injective C^*-algebra $\mathcal{A}$ and a projection $e \in \mathcal{A}$ such that W is completely isometric to $e\mathcal{A}(1 - e)$.*

Proof Let W be an injective operator subspace of $\mathcal{B}(H)$. Then there is a completely contractive projection φ from $\mathcal{B}(H)$ onto W, which extends the identity mapping on W. We let $\mathcal{L}$ denote the operator system on H^2 given by

$$\mathcal{L} = \left\{ \begin{bmatrix} \alpha I & b \\ c^* & \delta I \end{bmatrix} : \alpha, \delta \in \mathbb{C}, b, c \in \mathcal{B}(H) \right\},$$

and we let $\tilde{\varphi} : \mathcal{L} \to M_2(\mathcal{B}(H))$ be the morphism defined by

$$\tilde{\varphi}\left(\begin{bmatrix} \alpha I & b \\ c^* & \delta I \end{bmatrix}\right) = \begin{bmatrix} \alpha I & \varphi(b) \\ \varphi(c)^* & \delta I \end{bmatrix}.$$

It is evident that $\tilde{\varphi}(\mathcal{L}) \subseteq \mathcal{L}$ and $\tilde{\varphi}^2 = \tilde{\varphi}$.

From Lemma 6.1.5 there is a minimal completely contractive projection

$$\Phi : M_2(\mathcal{B}(H)) \to M_2(\mathcal{B}(H))$$

such that $\Phi|_{\mathcal{L}} = \tilde{\varphi}$. If we let $\mathcal{A} = \Phi(M_2(\mathcal{B}(H)))$ have the relative norm and $*$-operation, and the multiplication

$$a \cdot b = \Phi(ab),$$

then we have from Theorem 6.1.3 that $\mathcal{A}$ is an injective C^*-algebra with unit $1 = \begin{bmatrix} I & 0 \\ 0 & I \end{bmatrix}$. Let $e = \begin{bmatrix} I & 0 \\ 0 & 0 \end{bmatrix}$. Then e is a projection in $\mathcal{A}$ since $e^* = e$ and

$$e \cdot e = \Phi(e^2) = \Phi(e) = \tilde{\varphi}(e) = e.$$

Moreover, since

$$\Phi(e)^* \cdot \Phi(e) = e^* \cdot e = \Phi(e^*e),$$

it follows from Corollary 5.2.2 that for any $x \in M_2(\mathcal{B}(H))$,

$$\Phi(ex) = \Phi(e) \cdot \Phi(x) = e \cdot \Phi(x).$$

Similarly, $1 - e$ is a projection in $\mathcal{A}$ such that

$$\Phi(x(1-e)) = \Phi(x) \cdot \Phi(1-e) = \Phi(x) \cdot (1-e).$$

The mapping

$$\theta : W \to M_2(\mathcal{B}(H)) : w \mapsto \begin{bmatrix} 0 & w \\ 0 & 0 \end{bmatrix}$$

is a complete isometry onto its image since it coincides with the composition

$$w \mapsto \begin{bmatrix} w & 0 \\ 0 & 0 \end{bmatrix} \mapsto \begin{bmatrix} w & 0 \\ 0 & 0 \end{bmatrix} \begin{bmatrix} 0 & I \\ I & 0 \end{bmatrix},$$

where the last matrix is unitary. Thus, we have

$$\begin{aligned} W &\cong \left\{ \begin{bmatrix} 0 & w \\ 0 & 0 \end{bmatrix} : w \in W \right\} \\ &= \left\{ \tilde{\varphi}\left(\begin{bmatrix} 0 & b \\ 0 & 0 \end{bmatrix} \right) : b \in \mathcal{B}(H) \right\} \\ &= \tilde{\varphi}(eM_2(\mathcal{B}(H))(1-e)) \\ &= \Phi(eM_2(\mathcal{B}(H))(1-e)) \\ &= e \cdot \Phi(M_2(\mathcal{B}(H))) \cdot (1-e) \\ &= e \cdot \mathcal{A} \cdot (1-e), \end{aligned}$$

which is the desired result. □

If e and f are projections in a C^*-algebra $\mathcal{A}$ with unit 1, then the space $V = e\mathcal{A}f$ can be regarded as a left $e\mathcal{A}e$ and right $f\mathcal{A}f$-bimodule. However, this algebraic structure can be degenerate since, for example, if e is central and $f = 1 - e$, then $V = 0$. This situation can be remedied by the following result of Smith.

Given an algebra $\mathcal{A}$, we say that a left $\mathcal{A}$-module V is *faithful* if $av = 0$ for all v implies that $a = 0$, and we similarly define the notion of a faithful right $\mathcal{A}$-module.

Lemma 6.1.7 *Suppose that an operator space V is completely isometric to $e\mathcal{A}f$, where e is a projection in a unital C^*-algebra $\mathcal{A}$ and $f = 1 - e$. Then V is completely isometric to $e_0\mathcal{B}f_0$, where $\mathcal{B}$ is a C^*-algebraic quotient of $\mathcal{A}$, e_0 and f_0 are the images of e and f, and in addition, V is a faithful left $e_0\mathcal{B}e_0$ and right $f_0\mathcal{B}f_0$-bimodule.*

Proof Let us assume that $\mathcal{A}$ is represented on a Hilbert space H. We may identify H with $H_1 \oplus H_2$, where $H_1 = eH$ and $H_2 = fH$. We can represent the elements of $\mathcal{A}$ as matrices

$$\begin{bmatrix} x & v \\ w^* & y \end{bmatrix},$$

where $x \in e\mathcal{A}e, y \in f\mathcal{A}f$, and $v, w \in V = e\mathcal{A}f$. The linear spaces

$$J = \{j \in e\mathcal{A}e : jV = 0\}$$

and

$$K = \{k \in f\mathcal{A}f : Vk = 0\}$$

are closed two-sided ideals in $e\mathcal{A}e$ and $f\mathcal{A}f$, respectively, and thus they are self-adjoint. Furthermore,

$$\mathcal{I} = J \oplus K$$

is a closed two-sided ideal in $\mathcal{A}$ since if $j \in J$ and $k \in K$

$$\begin{bmatrix} x & v \\ w^* & y \end{bmatrix} \begin{bmatrix} j & 0 \\ 0 & k \end{bmatrix} = \begin{bmatrix} xj & vk \\ (j^*w)^* & yk \end{bmatrix} = \begin{bmatrix} xj & 0 \\ 0 & yk \end{bmatrix},$$

and on the other hand,

$$\begin{bmatrix} j & 0 \\ 0 & k \end{bmatrix} \begin{bmatrix} x & v \\ w^* & y \end{bmatrix} = \begin{bmatrix} jx & jv \\ (wk^*)^* & ky \end{bmatrix} = \begin{bmatrix} jx & 0 \\ 0 & ky \end{bmatrix}.$$

If we let $\mathcal{B} = \mathcal{A}/\mathcal{I}, e_0 = e + \mathcal{I}$, and $f_0 = f + \mathcal{I}$, then the mapping

$$\theta : V = e\mathcal{A}f \to e_0\mathcal{B}f_0$$

is a complete quotient. Since

$$v(j \oplus k) = \begin{bmatrix} 0 & v \\ 0 & 0 \end{bmatrix} \begin{bmatrix} j & 0 \\ 0 & k \end{bmatrix} = \begin{bmatrix} 0 & vk \\ 0 & 0 \end{bmatrix} = 0,$$

and similarly, $(j \oplus k)v = 0$,

$$\begin{aligned} \|(eaf - j \oplus k)\|^2 &= \|(eaf - j \oplus k)(eaf - j \oplus k)^*\| \\ &= \|(eaf)(eaf)^* + (j \oplus k)(j \oplus k)^*\| \geq \|eaf\|^2. \end{aligned}$$

This shows that θ is an isometry. If we use matrices over V, then the same calculation shows that θ is a complete isometry. If we identify V with $\theta(V)$, we see that V is a faithful left $e_0\mathcal{B}e_0$ and right $f_0\mathcal{B}f_0$-bimodule. □

Corollary 6.1.8 *If V is a finite-dimensional injective operator space, then there exist $m_k, n_k \in \mathbb{N}$ $(1 \leq k \leq p)$ such that V is completely isometric to*

$$M_{m_1,n_1} \oplus \cdots \oplus M_{m_p,n_p}. \tag{6.1.7}$$

Proof We may assume that $V = e\mathcal{A}(1 - e)$, where e is a projection in an injective C^*-algebra $\mathcal{A}$, and V is a faithful left $e\mathcal{A}e$ and right $f\mathcal{A}f$-bimodule. Since left multiplication determines a linear isomorphism of $e\mathcal{A}e$ into the finite-dimensional space $\mathbb{L}(V)$, $e\mathcal{A}e$ is finite-dimensional. If we use right multiplication, then we see that $f\mathcal{A}f$ is also finite-dimensional. By hypothesis, $e\mathcal{A}(1-e)$ is finite-dimensional, and taking adjoints, $(1-e)\mathcal{A}e$ is finite-dimensional. For any $a \in \mathcal{A}$,

$$a = eae + ea(1-e) + (1-e)ae + (1-e)a(1-e);$$

hence

$$\mathcal{A} \subseteq e\mathcal{A}e + e\mathcal{A}(1-e) + (1-e)\mathcal{A}(1-e) + (1-e)\mathcal{A}(1-e),$$

and $\mathcal{A}$ is finite-dimensional. We may assume that

$$\mathcal{A} = M_{r_1} \oplus \cdots \oplus M_{r_p}.$$

If we let $p = C(e)C(1-e)$, where $C(e)$ and $C(1-e)$ denote the central covers of e and $1-e$, respectively, we have $V = e(p\mathcal{A})(1-e)$, and thus we may initially assume that $C(e) = C(1-e) = 1$. If we let $e = e_1 + \cdots + e_p$ with $e_k \in M_{r_k}$, then it follows that e_k and $1_k - e_k$ are non-zero projections in M_{r_k}. It is easy to see that $e_k M_{r_k}(1-e_k)$ is completely isometric to M_{m_k,n_k}, where m_k and n_k are the ranks of e_k and $1-e_k$, and in turn that V is completely isometric to (6.1.7). □

More generally, there is a natural characterization for the dual injective operator spaces in terms of injective von Neumann algebras.

Theorem 6.1.9 *If V is a dual operator space, then the following are equivalent:*

(i) *V is injective;*
(ii) *there is a weak* homeomorphic complete isometry from V onto an operator space of the form $eR(1-e)$, where R is an injective von Neumann algebra, and e is a projection in R.*

We conclude this section with the simple observation that if the projections e and f in the C^*-algebra $\mathcal{A}$ are equivalent in the usual von Neumann algebraic sense, then $e\mathcal{A}f$ is completely isometric to the C^*-algebras $e\mathcal{A}e$ and $f\mathcal{A}f$. We leave the details to the reader.

6.2 INJECTIVE ENVELOPES

We say that an inclusion of operator spaces $V \subseteq W$ is *rigid* if for each complete contraction $\varphi : W \to W$, we have that $\varphi_{|V} = id_V$ implies that $\varphi = id_W$. We say that $V \subseteq W$ is *essential* if given any operator space Z and a complete contraction $\varphi : W \to Z$, we have that if $\varphi_{|V}$ is a complete isometry, then $\varphi : W \to Z$ is a completely isometric injection.

Given an operator space $V \subseteq \mathcal{B}(H)$, we define $\mathcal{E}_V$ to be the set of all completely contractive mappings $\varphi : \mathcal{B}(H) \to \mathcal{B}(H)$ such that $\varphi_{|V} = id_V$. In the notation of the previous section, $\mathcal{E}_V = \mathcal{E}_\iota$, where $\iota : V \hookrightarrow \mathcal{B}(H)$ is the inclusion mapping, and we have a corresponding partial ordering $\preceq$ on this set.

Theorem 6.2.1 *Given operator spaces $V \subseteq W \subseteq \mathcal{B}(H)$ with W injective, the following are equivalent:*

(i) *$V \subseteq W$ is rigid;*
(ii) *$V \subseteq W$ is essential;*
(iii) *for any injective operator space W_0 with $V \subseteq W_0 \subseteq W$, we have that $W_0 = W$;*
(iv) *there exists a completely contractive projection Φ of $\mathcal{B}(H)$ onto W which is minimal in $\mathcal{E}_V$.*

Proof (i)⇒(ii). Let us suppose that $\varphi : W \to Z$ is completely contractive and that $\varphi_{|V}$ is completely isometric. We have a commutative diagram

$$\begin{array}{ccccc} W & \xrightarrow{\varphi} & Z & \xrightarrow{\psi} & W \\ \cup| & & \cup| & & \cup|, \\ V & \xrightarrow{\varphi_{|V}} & \varphi(V) & \xrightarrow{(\varphi_{|V})^{-1}} & V \end{array}$$

where we have used the injectivity of W to find a completely contractive extension ψ of $(\varphi_{|V})^{-1}$. Since $(\psi \circ \varphi)_{|V} = id_V$, we have from (i) that $\psi \circ \varphi = id_W$, and in particular, φ is completely isometric.

(ii)⇒(iii). Given $V \subseteq W_0 \subseteq W$ with W_0 injective, there exists a completely contractive projection P of W onto W_0. Since $P_{|V} = id_V$ is a complete isometry, we have from (ii) that is also the case for P. It follows that $\ker P$ is zero; hence, $id_W - P = 0$ and $W = W_0$.

(iii)⇒(iv). Let Ψ be an arbitrary completely contractive projection of $B(H)$ onto W. If Φ is a minimal projection in $\mathcal{E}_V$, then $\Psi \circ \Phi \in \mathcal{E}_V$ and $\Psi \circ \Phi \preceq \Phi$, and therefore $\Psi \circ \Phi$ is minimal in $\mathcal{E}_V$. It follows that $\Psi \circ \Phi$ is a completely contractive projection. The range space $W_0 = (\Psi \circ \Phi)(\mathcal{B}(H))$ is an injective operator subspace of W which contains V. From (iii) we have $W_0 = W$, and thus $\Psi \circ \Phi$ is the desired projection.

(iv)⇒(i). Let us suppose that there is a completely contractive projection Φ of $\mathcal{B}(H)$ onto W which is minimal in $\mathcal{E}_V$. If there is a complete contraction $\varphi : W \to W$ with $\varphi_{|V} = id_V$, then we let $\Psi = \varphi \circ \Phi$. Since $\Psi \preceq \Phi$, it follows that Ψ is minimal in $\mathcal{E}_V$; hence, Ψ is a projection with $\Psi \sim \Phi$. If $b \in \mathcal{B}(H)$, then $\Psi(b) \in W$ implies that $\Phi \circ \Psi = \Psi$. It follows that

$$\|\Phi(b) - \Psi(b)\| = \|\Phi(b - \Psi(b))\| = \|\Psi(b - \Psi(b))\| = 0,$$

and thus, $\Psi = \Phi$ and $\varphi = id_W$. □

We define an *injective envelope* of an operator space V to be a pair (W, ι), where $\iota : V \to W$ is a completely isometric injection, W is injective, and the inclusion $\iota(V) \subseteq W$ satisfies any of the equivalent conditions in Theorem 6.2.1. It is often convenient to identify V with its image $\iota(V)$, and write $\mathcal{I}(V)$ for W. The latter notation is justified by the following result, which shows that the injective envelopes are essentially unique.

Corollary 6.2.2 *If an operator space V has injective envelopes (W_1, ι_1) and (W_2, ι_2), then there exists a unique complete isometry $\varphi : W_1 \to W_2$ such that $\varphi \circ \iota_1 = \iota_2$.*

Proof We have a commutative diagram

$$\begin{array}{ccccc} W_1 & \xrightarrow{\varphi} & W_2 & \xrightarrow{\psi} & W_1 \\ \cup\!| & & \cup\!| & & \cup\!| \, , \\ V & \xrightarrow{id_V} & V & \xrightarrow{id_V} & V \end{array}$$

where we use the injectivity of W_1 and W_2 to find completely contractive extensions of the identity mapping $id_V : V \to V$. Since $(\psi \circ \varphi)_{|V} = id_V$, it follows that $\psi \circ \varphi = id_{W_1}$, and similarly, $\varphi \circ \psi = id_{W_2}$. Thus, φ has the desired properties. If $\varphi' : W_1 \to W_2$ is another complete isometry with $\varphi' \circ \iota_1 = \iota_2$, then the same argument shows that $(\varphi')^{-1} \circ \varphi = id_{W_1}$ and thus, $\varphi' = \varphi$. □

Let us suppose that V is an operator subspace of $\mathcal{B}(H)$ on some Hilbert space H which is *unital*, i.e. it contains the identity I. We may suppose that $V \subseteq \mathcal{I}(V) \subseteq \mathcal{B}(H)$ and we may fix a completely contractive projection Φ on $\mathcal{B}(H)$ with range $\mathcal{I}(V)$. Since $\Phi(I) = I$, Φ must be completely positive. It follows that $\mathcal{I}(V)$ is an injective operator system in $\mathcal{B}(H)$. From Theorem 6.1.3 $\mathcal{I}(V)$ has a C^*-algebra multiplication $\cdot$ defined by

$$a \cdot b = \Phi(ab)$$

for all $a, b \in \mathcal{I}(V)$. Since any completely isometric unital mapping of unital C^*-algebras is a C^*-algebraic isomorphism (see Corollary 5.2.3) the mapping φ in Lemma 6.2.2 must be a complete C^*-algebraic isomorphism, and in this sense the C^*-algebraic structure on $\mathcal{I}(V)$ is *canonical.*

If $V = \mathcal{A}$ is a closed unital subalgebra of $\mathcal{B}(H)$, then the mapping $\mathcal{A} \to \mathcal{I}(\mathcal{A})$ is a multiplicative homomorphism since for any a and $b \in \mathcal{A}$, $ab \in \mathcal{A}$ implies that $a \cdot b = ab$. If $\mathcal{A}$ is a unital C^*-subalgebra of $\mathcal{B}(H)$, then this mapping is also self-adjoint, and we see that $\mathcal{A} \to \mathcal{I}(\mathcal{A})$ is a $*$-homomorphism. We conclude this section by showing that this result also holds for non-unital C^*-algebras.

Lemma 6.2.3 *Suppose that $\varphi : \mathcal{A} \to \mathcal{B}$ is a contractive linear mapping of unital C^*-algebras. If there is a projection $e \in \mathcal{A}$ such that $\varphi(e) = 1$, then $\varphi(1) = 1$.*

Proof For any $z \in \mathbb{C}$ with $|z| = 1$, we have

$$\|1 - z\varphi(1-e)\| = \|\varphi(e - z(1-e))\| \leq \|e - z(1-e)\| \leq 1,$$

where we have used the fact that e and $1 - e$ are orthogonal projections in $\mathcal{A}$. If we let $u = \varphi(1-e)$ and $z = \pm 1$, then

$$\|1 \pm \mathrm{Re}\, u\| = \|\mathrm{Re}\,(1 \pm u)\| \leq \|1 \pm u\| \leq 1$$

and from spectral theory $\operatorname{Re} u = 0$. A similar argument shows that $\operatorname{Im} u = 0$, and thus $u = 0$, and $\varphi(1 - e) = 0$. □

Theorem 6.2.4 *Let $\mathcal{A}$ be a C^*-algebra. Then $\mathcal{I}(\mathcal{A})$ has a canonical C^*-algebraic structure, and the mapping $\iota : \mathcal{A} \to \mathcal{I}(\mathcal{A})$ is a C^*-algebraic isomorphism onto its image.*

Proof Let us suppose that $\mathcal{A} \subseteq \mathcal{B}(H)$ is a non-degenerate representation of $\mathcal{A}$, i.e. $\mathcal{A}H$ is norm dense in H. From the above we may assume that $\mathcal{I}(\mathcal{A}) = \Phi(\mathcal{B}(H))$, where $\Phi : \mathcal{B}(H) \to \mathcal{B}(H)$ is a minimal element of $\mathcal{E}_\iota$.

If we let

$$\Psi = \Phi^*_{\mathcal{T}(\mathcal{H})} : \mathcal{T}(H) \to \mathcal{B}(H)^*,$$

then

$$\tilde{\Phi} = \Psi^* : \mathcal{B}(H)^{**} \to \mathcal{B}(H)$$

is a weak* continuous complete contraction which agrees with Φ on $\mathcal{B}(H)$. The canonical mapping of $\mathcal{B}(H)$ into $\mathcal{B}(H)^{**}$ is unital, and thus if we make the usual identification, then I_H is also the identity for $\mathcal{B}(H)^{**}$.

We may choose an increasing approximate identity $\{a_\gamma\}_{\gamma \in \Gamma}$ $\mathcal{A}$ (see §A.4). This net converges to I_H in the weak* topology on $\mathcal{B}(H)$ since it is bounded and for any ξ and η in H and $b \in \mathcal{A}$,

$$\langle \tilde{\Phi}(e)b\xi \mid \eta \rangle = \lim \langle \tilde{\Phi}(a_\gamma)b\xi \mid \eta \rangle = \lim \langle (a_\gamma b)\xi \mid \eta \rangle = \langle b\xi \mid \eta \rangle.$$

On the other hand, we may regard $\mathcal{A}^{**}$ as a von Neumann subalgebra of $\mathcal{B}(H)^{**}$. The net a_γ converges to its least upper bound e in the weak* topology on $\mathcal{B}(H)^{**}$ (see §A.4). It is easy to see that e is the unit for $\mathcal{A}^{**}$, and thus it is a projection in $\mathcal{B}(H)^{**}$. Since $\tilde{\Phi}$ is continuous with respect to the weak* topologies on $\mathcal{B}(H)^{**}$ and $\mathcal{B}(H)$, respectively, we conclude that

$$\tilde{\Phi}(e) = \lim \tilde{\Phi}(a_\gamma) = \lim a_\gamma = I_H.$$

It follows from Lemma 6.2.3 that $\Phi(I_H) = \tilde{\Phi}(I_H) = I_H$, and thus from Corollary 5.1.2, Φ is a completely positive projection. From our earlier theory, $\mathcal{I}(\mathcal{A}) = \Phi(\mathcal{B}(H))$ is a C^*-algebra with the product $a \cdot b = \Phi(ab)$. In particular, if a and b are in $\mathcal{A}$, then so is ab; hence $a \cdot b = ab$, and the inclusion $\mathcal{A} \to \mathcal{I}(\mathcal{A})$ is a C^*-algebraic isomorphism onto its image. □

6.3 NOTES AND REFERENCES

Lemma 6.1.2 and Theorem 6.1.3 are due to Choi and Effros (1977*a*). Theorem 6.1.6 was proved by Ruan (1989). Lemma 6.1.7 and Corollary 6.1.8 may be found in Smith (1999). Theorem 6.1.9 is proved in Effros *et al.* (1999). The argument used some important results of Zettl (1983).

Motivated by Hamana's results on the injective envelope of operator systems (see Hamana 1979), Ruan proved the existence and uniqueness of the injective envelopes for operator spaces in Ruan (1989). Lemma 6.2.3 is due to Kirchberg (1993). Theorem 6.2.1 and Theorem 6.2.4 (the latter in the context of completely positive projections) are due to Hamana (1979). Hamana has also investigated the canonical ternary algebraic structure on the injective envelope of a general operator space (see Hamana 1989, 1992).

Part II

Tensor Products

Tensor product arguments are ubiquitous in operator space theory, and in fact it is difficult to find a paper on the subject in which they are not used. Since this is not the case for Banach spaces, we shall briefly sketch the reasons for this phenomenon.

Schatten was the first person to systematically study various norms that can be defined on tensor products of Banach spaces (see Schatten 1943, 1950). In his fundamental paper on topological linear spaces, Grothendieck emphasized the importance of relating mapping spaces to tensor products. Despite the success of his approach, many Banach space analysts have preferred a less 'algebraic' framework. For this purpose they have used Pietsch's notion of 'operator ideals' to incorporate Grothendieck's ideas into the subject without explicitly using tensor products.

Although the 'categorical' flavour of Grothendieck's theory may not have appealed to Banach space theorists, it is just that feature of his approach that made it well-suited to quantization. The basic Banach space tensor products have simple analogues in operator space theory, which can be found by considering their categorical properties. To illustrate this principle, the notion of 'uniform convergence on compact sets' has played an important role in Banach space theory. The appropriate replacement in operator space theory initially seemed rather perplexing. Fortunately, there is a simple tensor product approach to this form of convergence (see below). Since the tensor product theory for operator spaces had already been explored, this enabled researchers to find the correct analogue. On the other hand, the successful quantization for various Banach space properties, such as 'smoothness' and 'uniform convexity', remain a fascinating challenge for operator space researchers. These difficulties might again be overcome if we discovered more abstract approaches to the Banach space techniques.

We would be remiss not to mention another aspect of this history. Owing in large part to the work of Pisier, the perception of tensor products by Banach specialists has changed in the last decade. Pisier returned to the tensor product theory, and proceeded to settle a number of problems proposed by Grothendieck. After that work, Pisier was uniquely qualified to make fundamental contributions to operator space theory. A perusal of this monograph or of the literature will make evident the remarkable extent to which he has done that.

In classical tensor product theory, one generally restricts one's attention to complete spaces. The reason for this is that the usual tensor products are defined to be completions of algebraic tensor products. This is the case, for example, with the projective tensor product $\otimes^\gamma$ (see below), and it is desirable to have such relations as $V \otimes^\gamma \mathbb{C} = V$. We shall adopt the same convention for operator spaces. *Unless otherwise indicated, all operator spaces in the remainder of the monograph are assumed complete.*

Although we shall not try to parallel the systematic theory of Banach space tensor products, it will be convenient to use some of the terminology.

When introducing a tensor product $\otimes^\alpha$ of operator spaces, we shall require that complete contractions $\varphi_i : V_i \to W_i$ $(i = 1, 2)$ determine complete contractions $\varphi_1 \otimes \varphi_2 : V_1 \otimes^\alpha W_1 \to V_2 \otimes^\alpha W_2$. We say that $\otimes^\alpha$ is *injective* if completely isometric mappings $\varphi_i : V_i \hookrightarrow W_i$ determine a completely isometric mapping $\varphi_1 \otimes \varphi_2$. We say that $\otimes^\alpha$ is *projective* if complete quotient mappings $\varphi_i : V_i \twoheadrightarrow W_i$ determine a complete quotient mapping $\varphi_1 \otimes \varphi_2$. In a somewhat unfortunate choice of terminology, these terms are also used to identify particular tensor products with those properties, namely the *projective* and *injective tensor products.* The third, and perhaps most interesting property is the *Haagerup* tensor product, which happens to be both projective and injective.

7

The projective tensor product

7.1 DEFINITION AND ELEMENTARY PROPERTIES

Given Banach spaces E and F, a norm $\|\cdot\|_\mu$ on $E \otimes F$ is said to be a *subcross-norm* (respectively, a *cross-norm*) if $\|x \otimes y\|_\mu \leq \|x\| \, \|y\|$ (respectively, $\|x \otimes y\|_\mu = \|x\| \, \|y\|$) for all $x \in E$ and $y \in F$. Given a subcross-norm $\|\cdot\|_\mu$ and a linear combination

$$u = \sum x_i \otimes y_i \in E \otimes F, \tag{7.1.1}$$

we have

$$\|u\|_\mu \leq \sum \|x_i\| \, \|y_i\| \, .$$

Thus, if we define

$$\|u\|_\gamma = \inf\left\{\sum \|x_i\| \, \|y_i\| : u = \sum x_i \otimes y_i\right\}, \tag{7.1.2}$$

then it follows that $\|u\|_\mu \leq \|u\|_\gamma$. It is a simple matter to verify that $\|\cdot\|_\gamma$ is in fact a cross-norm, which is called the *projective tensor product norm* on $E \otimes F$. We let

$$E \otimes_\gamma F = (E \otimes F, \|\cdot\|_\gamma)$$

and we define the *Banach space projective tensor product* $E \otimes^\gamma F$ to be the completion of this space.

If E, F, and G are normed spaces and $\varphi : E \times F \to G$ is a bilinear mapping, then we define

$$\|\varphi\| = \sup \{\|\varphi(v, w)\| : \|v\|, \|w\| \leq 1\}. \tag{7.1.3}$$

We let $\mathcal{B}(E \times F, G)$ denote the normed space of all such mappings φ with $\|\varphi\| < \infty$, and the norm $\|\cdot\|$. The linear isomorphisms (A.1.3) determine the isometries

$$\mathcal{B}(E \otimes^\gamma F, G) \cong \mathcal{B}(E \times F, G) \cong \mathcal{B}(E, \mathcal{B}(F, G)). \tag{7.1.4}$$

For any $n \in \mathbb{N}$ we have a natural isometry

$$\ell_1^n \otimes^\gamma E \cong \ell_1^n(E). \tag{7.1.5}$$

More generally, if $X = (X, \mathcal{S}, \mu)$ is a measure space, then for any Banach space E,

$$L_1(X) \otimes^\gamma E \cong L_1(X, E), \tag{7.1.6}$$

where $L_1(X, E)$ is the Banach space of integrable functions $f : X \to E$, and thus for any other measure space $Y = (Y, \mathcal{T}, \nu)$,

$$L_1(X) \otimes^\gamma L_1(Y) \cong L_1(X \times Y). \tag{7.1.7}$$

For any operator spaces V and W we say that an operator space matrix norm $\|\cdot\|_\mu$ on $V \otimes W$ is a *subcross matrix norm* if

$$\|v \otimes w\|_\mu \leq \|v\| \, \|w\|$$

for all $v \in M_p(V)$ and $w \in M_q(W)$. If, in addition,

$$\|v \otimes w\|_\mu = \|v\| \, \|w\| ,$$

then we say that $\|\cdot\|_\mu$ is a *cross matrix norm* on $V \otimes W$. Given an element u in $M_n(V \otimes W)$, we define

$$\|u\|_\wedge = \inf \{\|\alpha\| \, \|v\| \, \|w\| \, \|\beta\| : u = \alpha(v \otimes w)\beta\} , \tag{7.1.8}$$

where the infimum is taken over arbitrary decompositions with $v \in M_p(V)$, $w \in M_q(W), \alpha \in M_{n,p\times q}$, and $\beta \in M_{p\times q,n}$, with $p, q \in \mathbb{N}$ arbitrary. We next show that this is indeed an operator space matrix norm. We let

$$V \otimes_\wedge W = (V \otimes W, \|\cdot\|_\wedge),$$

and we define the *operator space projective tensor product* (or simply, the *projective tensor product*) $V \widehat{\otimes} W$ to be the completion of this space.

Theorem 7.1.1 *For any operator spaces V and W, $\|\cdot\|_\wedge$ is the largest operator space subcross matrix norm on $V \otimes W$.*

Proof We verify the criteria of Proposition 2.3.6.

Given $u_1 \in M_m(V \otimes W)$, $u_2 \in M_n(V \otimes W)$, and $\varepsilon > 0$, we may find decompositions

$$u_i = \alpha_i(v_i \otimes w_i)\beta_i$$

with $\|v_i\| = \|w_i\| = 1$ and $\|\alpha_i\| = \|\beta_i\| \leq (\|u_i\|_\wedge + \varepsilon)^{1/2}$. If $v = v_1 \oplus v_2$ and $w = w_1 \oplus w_2$, then we have the natural identification

$$v \otimes w = \begin{bmatrix} v_1 & 0 \\ 0 & v_2 \end{bmatrix} \otimes \begin{bmatrix} w_1 & 0 \\ 0 & w_2 \end{bmatrix} = \begin{bmatrix} v_1 \otimes w_1 & 0 & 0 & 0 \\ 0 & v_1 \otimes w_2 & 0 & 0 \\ 0 & 0 & v_2 \otimes w_1 & 0 \\ 0 & 0 & 0 & v_2 \otimes w_2 \end{bmatrix},$$

and thus

$$u_1 \oplus u_2 = \begin{bmatrix} \alpha_1 & 0 & 0 & 0 \\ 0 & 0 & 0 & \alpha_2 \end{bmatrix} (v \otimes w) \begin{bmatrix} \beta_1 & 0 \\ 0 & 0 \\ 0 & 0 \\ 0 & \beta_2 \end{bmatrix},$$

where $\|v\|, \|w\| = 1$. If we let α and β be the indicated scalar matrices, then

$$\begin{aligned}\|u_1 \oplus u_2\|_\wedge &\leq \|\alpha\| \|\beta\| = \|\alpha\alpha^*\|^{1/2} \|\beta^*\beta\|^{1/2} \\ &= \|\alpha_1\alpha_1^* \oplus \alpha_2\alpha_2^*\|^{1/2} \|\beta_1^*\beta_1 \oplus \beta_2^*\beta_2^*\|^{1/2} \\ &\leq (\max\{\|\alpha_i\|^2\})^{1/2} (\max\{\|\beta_j\|^2\})^{1/2} \leq \max\{\|u_i\|\} + \varepsilon,\end{aligned}$$

and since $\varepsilon > 0$ is arbitrary, we have **M1$'$**. On the other hand, if $\gamma \in M_{p,m}$ and $\delta \in M_{m,p}$, then

$$\gamma u_1 \delta = (\gamma\alpha_1)(v_1 \otimes w_1)(\beta_1\delta),$$

and thus

$$\|\gamma u_1 \delta\|_\wedge \leq \|\gamma\alpha_1\| \|\beta_1\delta\| \leq \|\gamma\| \|\delta\| (\|u_1\|_\wedge + \varepsilon).$$

Since $\varepsilon > 0$ is arbitrary, we have established **M2**.

If

$$0 \neq u = \sum v_i \otimes w_i \in V \otimes W,$$

then we can suppose that the w_i are linearly independent, and that $v_i \neq 0$. We select elements $g_i \in W^*_{\|\cdot\| \leq 1}$ with $g_i(w_i) \neq 0$ and $g_i(w_j) = 0$ for $j \neq i$. If we choose $f \in V^*_{\|\cdot\| \leq 1}$ with $f(v_1) \neq 0$ and let $g = g_1$, then it follows that $f \otimes g(u) \neq 0$. Now let us suppose that

$$u = \alpha(v \otimes w)\beta,$$

with $v \in M_m(V)$ and $w \in M_n(W)$. Then since $\|f_m\| = \|f\| \leq 1$ and $\|g_n\| = \|g\| \leq 1$,

$$|(f \otimes g)(u)| = |\alpha(f_m(v) \otimes g_n(w))\beta| \leq \|\alpha\| \|v\| \|w\| \|\beta\|$$

and thus,

$$0 \neq |(f \otimes g)(u)| \leq \|u\|_\wedge .$$

If $v \in M_m(V)$ and $w \in M_n(W)$, then

$$v \otimes w = I_{m\times n}(v \otimes w)I_{m\times n},$$

and thus

$$\|v \otimes w\|_\wedge \leq \|v\| \|w\|,$$

i.e. $\|\cdot\|_\wedge$ is a subcross matrix norm. If $\|\cdot\|_\mu$ is an arbitrary subcross matrix norm on $V \otimes W$, then for any $u \in M_n(V \otimes W)$ and decomposition as in (7.1.8),

$$\|u\|_\mu \leq \|\alpha\| \|v\| \|w\| \|\beta\|,$$

and thus $\|u\|_\mu \leq \|u\|_\wedge$. □

We shall see in (8.1.13) that $\|\cdot\|_\wedge$ is a cross matrix norm. Given operator spaces V, W, and X, and $p, q \in \mathbb{N}$, each bilinear mapping $\varphi : V \times W \to X$ determines a bilinear mapping

$$\varphi_{p,q;r,s} : M_{p,q}(V) \times M_{r,s}(W) \to M_{p\times r, q\times s}(X), \tag{7.1.9}$$

where

$$\varphi_{p,q;r,s}(v, w) = [\varphi(v_{i,j}, w_{k,l})],$$

or in terms of elementary tensors,

$$\varphi_{p,q;r,s}(\alpha \otimes v_0, \beta \otimes w_0) = \alpha \otimes \beta \otimes \varphi(v_0, w_0).$$

If $q = p$ and $r = s$, we let $\varphi_{p;r} = \varphi_{p,q;r,s}$. We define

$$\begin{aligned}\|\varphi\|_{cb} &= \sup\{\|\varphi_{p;q}\| : p, q \in \mathbb{N}\}\\ &= \sup\{\|\varphi_{p;p}\| : p \in \mathbb{N}\}\end{aligned} \tag{7.1.10}$$

(the equality is trivial), and we say that φ is *completely bounded* (respectively, *completely contractive*) if $\|\varphi\|_{cb} < \infty$ (respectively, $\|\varphi\|_{cb} \le 1$). As in the one variable case, it is easy to check that the rectangular versions of φ will have the same bound. It is immediate that $\|\cdot\|_{cb}$ is a norm on the linear space $\mathcal{CB}(V \times W, X)$ of all such completely bounded bilinear mappings. We may define a matrix norm on $\mathcal{CB}(V \times W, X)$ by using the identification

$$M_n(\mathcal{CB}(V \times W, X)) = \mathcal{CB}(V \times W, M_n(X)).$$

Proposition 7.1.2 *If $V, W,$ and X are operator spaces, then there are natural completely isometric identifications*

$$\mathcal{CB}(V \hat{\otimes} W, X) \cong \mathcal{CB}(V \times W, X) \cong \mathcal{CB}(V, \mathcal{CB}(W, X)). \tag{7.1.11}$$

Proof Given a bilinear mapping $\varphi : V \times W \to X$, we have a corresponding linear mapping $\overline{\varphi} : V \otimes W \to X$. If

$$u = \alpha(v \otimes w)\beta \in M_n(V \otimes W),$$

where $v \in M_p(V)$ and $w \in M_q(W)$, then

$$\overline{\varphi}_n(u) = \alpha(\varphi_{p;q}(v, w))\beta$$

and thus

$$\|\overline{\varphi}_n(u)\| \le \|\alpha\| \, \|\varphi_{p;q}\| \, \|v\| \, \|w\| \, \|\beta\| \le \|\varphi\|_{cb} \, \|\alpha\| \, \|v\| \, \|w\| \, \|\beta\| \, .$$

It follows that

$$\|\overline{\varphi}_n(u)\| \le \|\varphi\|_{cb} \, \|u\|_\wedge \, ,$$

from which we conclude that $\|\overline{\varphi}\|_{cb} \le \|\varphi\|_{cb}$. Conversely, if $\varepsilon > 0$, then we may find $v \in M_p(V)_{\|\cdot\| \le 1}$ and $w \in M_q(W)_{\|\cdot\| \le 1}$ for which

$$\|\varphi\|_{cb} \le \|\varphi_{p;q}(v, w)\| + \varepsilon.$$

Since $u = v \otimes w$ satisfies

$$\|v \otimes w\|_\wedge \leq \|v\| \, \|w\| \leq 1,$$

it follows that

$$\|\varphi\|_{cb} \leq \left\|\overline{\varphi}_{p\times q}(v \otimes w)\right\| + \varepsilon \leq \|\overline{\varphi}\|_{cb} + \varepsilon,$$

and the first identification is completely isometric.

Any bounded bilinear mapping $\varphi : V \times W \to X$ determines a linear mapping $\tilde{\varphi} : V \to \mathcal{B}(W, X)$ by

$$\tilde{\varphi}(v)(w) = \varphi(v, w).$$

If $\varphi \in \mathcal{CB}(V \times W, X)$, then for $v \in V$ and $w \in M_q(W)$,

$$\|\tilde{\varphi}(v)_q(w)\| = \|\varphi_{1;q}(v, w)\| \leq \|\varphi\|_{cb}\|v\|\|w\|,$$

and thus $\|\tilde{\varphi}(v)\|_{cb} \leq \|\varphi\|_{cb}\|v\|$. It follows that $\tilde{\varphi}(V) \subseteq \mathcal{CB}(W, X)$. If $v \in M_p(V)$ and $w \in M_q(W)$, then

$$(\tilde{\varphi}_p(v))_q(w) = [(\tilde{\varphi}(v_{i,j}))(w_{k,l})] = [\varphi(v_{i,j}, w_{k,l})] = \varphi_{p;q}(v, w),$$

from which it is evident that $\|\tilde{\varphi}\|_{cb} = \|\varphi\|_{cb}$, and the corresponding mapping

$$\theta : \mathcal{CB}(V \times W, X) \to \mathcal{CB}(V, \mathcal{CB}(W, X)) : \varphi \mapsto \tilde{\varphi}$$

is an isometry. To see that it is surjective, suppose that we are given a completely bounded mapping $\psi \in \mathcal{CB}(V, \mathcal{CB}(W, X))$. We may define a bilinear mapping $\varphi : V \times W \to X$ by letting $\varphi(v, w) = \psi(v)(w)$. Replacing $\tilde{\varphi}$ by ψ in the above calculation, it is immediate that φ is completely bounded, and that $\tilde{\varphi} = \psi$.

The fact that θ_r is an isometry for each $r \in \mathbb{N}$, and thus that θ is a complete isometry, is apparent from the commutative diagram

$$\begin{array}{ccc} M_r(\mathcal{CB}(V \times W, X)) & \xrightarrow{\theta_r} & M_r(\mathcal{CB}(V, \mathcal{CB}(W, X))) \\ \downarrow & & \downarrow \\ \mathcal{CB}(V \times W, M_r(X)) & \xrightarrow{\theta} & \mathcal{CB}(V, \mathcal{CB}(W, M_r(X)) \end{array},$$

and this completes the proof. □

To illustrate this result, we recall that if V is an operator space, then for each $v \in M_m(V)$ and $f \in M_n(V^*)$,

$$\|\langle\langle v, f \rangle\rangle\| \leq \|v\| \, \|f\|,$$

and thus the bilinear mapping

$$V \times V^* \to \mathbb{C} : (v, f) \to f(v)$$

extends to a complete contraction

$$\text{trace} : V \hat{\otimes} V^* \to \mathbb{C} \qquad (7.1.12)$$

(see (A.2.4)).

Corollary 7.1.3 *Suppose that V, V_1, W, and W_1 are operator spaces. Given complete contractions $\varphi : V \to V_1$ and $\psi : W \to W_1$, the corresponding mapping*

$$\varphi \otimes \psi : V \otimes W \to V_1 \otimes W_1$$

extends to a complete contraction

$$\varphi \otimes \psi : V \hat{\otimes} W \to V_1 \hat{\otimes} W_1 .$$

Proof The complete contractions φ and ψ determine a bilinear mapping

$$\varphi \times \psi : V \times W \to V_1 \hat{\otimes} W_1 : (v, w) \mapsto \varphi(v) \otimes \psi(w),$$

which is a complete contraction since for any matrices $v \in M_p(V)$ and $w \in M_q(W)$,

$$\|(\varphi \times \psi)_{p;q}(v, w)\|_\wedge = \|\varphi_p(v) \otimes \psi_q(w)\|_\wedge \leq \|\varphi_p(v)\|\|\psi_q(w)\| \leq \|v\|\|w\| .$$

It follows from (7.1.11) that the linear extension $\varphi\otimes\psi$ of $\varphi\times\psi$ is a complete contraction. Since the image space is complete, it extends to the completed tensor product, and we are done. □

It is an easy exercise from the definition that the associativity and commutativity relations for the tensor products of linear spaces readily extend to the projective tensor product. We leave the details to the reader.

Proposition 7.1.4 *Given operator spaces V, W, and X, we have the completely isometric isomorphisms*

$$V \hat{\otimes} W \cong W \hat{\otimes} V \tag{7.1.13}$$

and

$$(V \hat{\otimes} W) \hat{\otimes} X \cong V \hat{\otimes} (W \hat{\otimes} X). \tag{7.1.14}$$

□

The following special case of Proposition 7.1.2 and Proposition 7.1.4 is of particular importance.

Corollary 7.1.5 *If V and W are operator spaces, then we have natural completely isometries*

$$\lambda : (V \hat{\otimes} W)^* \cong \mathcal{CB}(V, W^*)$$

and

$$\rho : (V \hat{\otimes} W)^* \cong \mathcal{CB}(W, V^*),$$

where for each $u \in (V \hat{\otimes} W)^*$, the mappings

$$\lambda(u) : V \to W^*$$

and

$$\rho(u) : W \to V^*$$

are defined by

$$(\lambda(u)(v))(w) = \langle u, v \otimes w \rangle = (\rho(u)(w))(v). \qquad \square$$

From Corollary 7.1.5 we have the complete isometries

$$(V \mathbin{\widehat{\otimes}} W)^* \cong \mathcal{CB}(V, W^*) \cong \mathcal{CB}(W, V^*). \tag{7.1.15}$$

Since $T_n^* = M_n$, we have

$$(T_n \mathbin{\widehat{\otimes}} V)^* \cong \mathcal{CB}(V, M_n) = M_n(V^*). \tag{7.1.16}$$

If $V = T_r$, then

$$(T_n \mathbin{\widehat{\otimes}} T_r)^* \cong M_n(M_r) = M_{n\times r} = T_{n\times r}^*, \tag{7.1.17}$$

and thus

$$T_n \mathbin{\widehat{\otimes}} T_r \cong T_{n\times r}\,. \tag{7.1.18}$$

We recall that to prove the Arveson–Wittstock–Hahn–Banach theorem, we began by introducing a Banach space

$$T_n(V) = (\mathbb{M}_n(V), \|\cdot\|_1)$$

for which

$$T_n(V)^* \cong M_n(V^*) \tag{7.1.19}$$

(see (4.1.7)). It follows from (7.1.16) that we have a natural isometry

$$T_n(V) \cong T_n \mathbin{\widehat{\otimes}} V. \tag{7.1.20}$$

In more detail, $\|\cdot\|_\wedge$ and $\|\cdot\|_1$ are natural norms on $\mathbb{M}_n \otimes V$ (see §A.2), in the sense that if we use the basis $\varepsilon_{i,j}$ for $\mathbb{M}_n$, a sequence $v(k) \in \mathbb{M}_n \otimes V$ converges to $v \in \mathbb{M}_n \otimes V$ if and only if its entries $v(k)_{i,j}$ converge to $v_{i,j}$ in V. Thus, each determines a natural dual norm on $\mathbb{M}_n \otimes V^*$. But the relations (7.1.16) and (7.1.19) imply that these dual norms coincide with the completely bounded norm on $\mathbb{M}_n \otimes V^*$. It follows that if $u \in \mathbb{M}_n \otimes V$, then

$$\|u\|_{T_n \mathbin{\widehat{\otimes}} V} = \sup\{|\langle u, f\rangle| : f \in M_n(V^*), \|f\|_{cb} \le 1\} = \|u\|_{T_n(V)}\,.$$

We may use (7.1.20) to define an operator space matrix norm on $T_n(V)$. With this convention, (4.1.7) is a complete isometry since we have the complete isometries

$$T_n(V)^* \cong (T_n \mathbin{\widehat{\otimes}} V)^* \cong \mathcal{CB}(V, M_n) = M_n(V^*).$$

From (4.1.8), we also have a natural isometry

$$T_n \mathbin{\widehat{\otimes}} V^* \to M_n(V)^*. \tag{7.1.21}$$

To prove that this is a complete isometry, it suffices to show that for each $r \in \mathbb{N}$, the corresponding mapping

$$T_r \mathbin{\widehat{\otimes}} T_n \mathbin{\widehat{\otimes}} V^* \to T_r \mathbin{\widehat{\otimes}} M_n(V)^*$$

is isometric (see Theorem 4.1.8). This is apparent from the commutative diagram

$$\begin{array}{ccc} T_r \widehat{\otimes} T_n \widehat{\otimes} V^* & \longrightarrow & T_r \widehat{\otimes} (M_n(V))^* \\ \downarrow & & \downarrow \\ T_{r\times n} \widehat{\otimes} V^* & \longrightarrow & (M_{r\times n}(V))^* = (M_r(M_n(V)))^* \end{array}$$

since (7.1.21) implies that the bottom and vertical mappings are isometric. Summarizing, we have the following result.

Proposition 7.1.6 *Let us suppose that V is an operator space and that $n \in \mathbb{N}$. Then we have the following completely isometric identifications:*

$$T_n(V)^* \cong M_n(V^*), \tag{7.1.22}$$

$$M_n(V)^* \cong T_n(V^*), \tag{7.1.23}$$

$$M_n(V^{**}) \cong M_n(V)^{**}. \tag{7.1.24}$$

□

Let us suppose that we are given operator spaces $V \subseteq W$ and X. If we use duality arguments, it is evident that the corresponding mapping $V \widehat{\otimes} X \to W \widehat{\otimes} X$ is completely isometric if and only if the restriction mapping

$$\mathcal{CB}(W, X^*) \to \mathcal{CB}(V, X^*)$$

is an exact complete quotient mapping. For a fixed operator space X, this will be the case for all completely isometric inclusions $V \hookrightarrow W$ if and only if X^* is an injective operator space (see Proposition 4.1.7). In particular, for any index set $\mathfrak{s}$ and completely isometric inclusion $V \hookrightarrow W$, we have the completely isometric injection

$$V \widehat{\otimes} T_{\mathfrak{s}} \hookrightarrow W \widehat{\otimes} T_{\mathfrak{s}}. \tag{7.1.25}$$

From this discussion it is evident that, in general, a complete isometry $V \to W$ does not induce a complete isometry

$$V \widehat{\otimes} X \to W \widehat{\otimes} X\,. \tag{7.1.26}$$

Nonetheless, there are certain cases when this is the case. If V is an operator subspace of W and there is a completely contractive projection P of W onto V, then the composition of complete contractions

$$V \widehat{\otimes} X \longrightarrow W \widehat{\otimes} X \xrightarrow{P \otimes id} V \widehat{\otimes} X \tag{7.1.27}$$

is just the identity mapping on $V \widehat{\otimes} X$, and thus the first mapping is completely isometric.

The usual inclusion mapping $V \hookrightarrow V^{**}$ induces the completely isometric injection

$$V \widehat{\otimes} W \hookrightarrow V^{**} \widehat{\otimes} W\,. \tag{7.1.28}$$

To see this we note that the inclusion mapping $\iota_V : V \to V^{**}$ determines a completely contractive retraction

$$P = (\iota_V)^* : V^{***} \to V^*.$$

The mapping $\varphi \to P \circ \varphi$ provides us with a quotient mapping in the top row of the diagram

$$\begin{array}{ccc} \mathcal{CB}(W, V^{***}) & \longrightarrow & \mathcal{CB}(W, V^*) \\ \Big\| & & \Big\| \\ (V^{**} \widehat{\otimes} W)^* & \xrightarrow[(\iota_V \otimes id)^*]{} & (V \widehat{\otimes} W)^* \end{array},$$

since if we are given a complete contraction $\psi \in \mathcal{CB}(W, V^*)$, then $\iota_{V^*} \circ \psi$ is the desired completely contractive preimage. The diagram commutes since if $\varphi \in \mathcal{CB}(W, V^{***})$ and

$$u = \sum_{i=1}^{n} v_i \otimes w_i \in V \otimes_\wedge W,$$

then

$$\langle P \circ \varphi, u \rangle = \sum_{i=1}^{n} P \circ \varphi(w_i)(v_i) = \sum_{i=1}^{n} \varphi(w_i)(\iota(v_i)) = \langle \varphi, (\iota \otimes id)(u) \rangle .$$

Thus, the bottom row is also a complete quotient mapping, and thus $\iota \otimes id$ is a complete isometry (Corollary 4.1.9).

By contrast, the projective tensor product behaves very well with respect to complete quotient mappings. Recall that we are assuming that our operator spaces are complete.

Proposition 7.1.7 *Suppose that $V, V_1, W,$ and W_1 are operator spaces. Given the complete quotient mappings $\varphi : V \twoheadrightarrow V_1$ and $\psi : W \twoheadrightarrow W_1$, the corresponding mapping*

$$\varphi \otimes \psi : V \otimes W \to V_1 \otimes W_1$$

extends to a complete quotient mapping

$$\varphi \otimes \psi : V \widehat{\otimes} W \twoheadrightarrow V_1 \widehat{\otimes} W_1 .$$

Furthermore,

$$\ker(\varphi \otimes \psi) = [(\ker \varphi) \otimes W + V \otimes (\ker \psi)]^- . \tag{7.1.29}$$

Proof Let us suppose that φ and ψ are complete quotient mappings. Since each matrix space $M_n(V \widehat{\otimes} W)$ is complete, it suffices to show that $(\varphi \otimes \psi)_n$ maps the open unit ball of $M_n(V \otimes_\wedge W)$ onto the open unit ball of $M_n(V_1 \otimes_\wedge W_1)$. Given $\tilde{u} \in M_n(V_1 \otimes_\wedge W_1)$ with $\|\tilde{u}\| < 1$, we may suppose that $\tilde{u} = \alpha(\tilde{v} \otimes \tilde{w})\beta$, where $\tilde{v} \in M_p(V_1)$, $\tilde{w} \in M_q(W_1)$, and $\|\alpha\|, \|\tilde{v}\|,$

$\|\tilde{w}\|, \|\beta\| < 1$. But then we may choose corresponding $v \in M_p(V)$ and $w \in M_q(W)$ with $\varphi_p(v) = \tilde{v}$ and $\psi_q(w) = \tilde{w}$, where $\|v\| < 1$ and $\|w\| < 1$. If $u = \alpha(v \otimes w)\beta$, then we see that $\varphi_n(u) = \bar{u}$ and $\|u\|_\wedge < 1$.

It follows from the bipolar theorem that to prove (7.1.29), it suffices to show that

$$(\ker \varphi \otimes \psi)^\perp = [(\ker \varphi) \otimes W + V \otimes (\ker \psi)]^\perp .$$

Given F in the right-hand annihilator, we may identify it with a bilinear mapping on $V \times W$. We define a bilinear mapping

$$F_1 : V_1 \times W_1 \to \mathbb{C}$$

by letting

$$F_1(v_1, w_1) = F(v, w),$$

where $\varphi(v) = v_1$ and $\psi(w) = w_1$. Owing to the annihilator condition this is well-defined. The complete boundedness of F_1 follows immediately from the assumption that φ and ψ are complete quotient mappings. Thus, F_1 determines a completely bounded mapping

$$F_1 : V_1 \hat{\otimes} W_1 \to \mathbb{C} .$$

We have

$$F(v \otimes w) = F(v, w) = F_1(\varphi(v), \psi(w)) = (F_1 \circ (\varphi \otimes \psi))(v \otimes w) .$$

Since $V \otimes W$ is dense in $V \hat{\otimes} W$, we conclude that $F = F_1 \circ (\varphi \otimes \psi)$, and thus it is immediate that $F \in (\ker \varphi \otimes \psi)^\perp$. Since the converse is trivial, this completes the proof. □

7.2 TRACE CLASS OPERATORS AND A FUBINI THEOREM

In this section we consider the operator space analogue of (7.1.7). This is useful in a variety of contexts, (see, for example, Theorem 15.1.1).

Let us suppose that we have Hilbert spaces $H_0 \subseteq H$, and that e is the orthogonal projection of H onto H_0. We use the completely isometric injection

$$j : \mathcal{K}(H_0) \to \mathcal{K}(H) : x \mapsto xe$$

to identify $\mathcal{K}(H_0)$ with a subspace of $\mathcal{K}(H)$. It is evident that

$$p : \mathcal{K}(H) \to \mathcal{K}(H_0) : x \mapsto ex|_{H_0}$$

is a completely contractive mapping with range $\mathcal{K}(H_0)$.

If we use the usual pairing between $\mathcal{K}(H)$ and $\mathcal{T}(H)$, then it follows that

$$J = p^* : \mathcal{T}(H_0) \to \mathcal{T}(H) : t \mapsto te$$

is completely isometric, and we can use it to identify $\mathcal{T}(H_0)$ with an operator subspace of $\mathcal{T}(H)$. On the other hand,

$$P = j^* : \mathcal{T}(H) \to \mathcal{T}(H_0) : t \mapsto et|_{H_0}$$

is a complete quotient mapping and

$$\pi = J \circ P : \mathcal{T}(H) \to \mathcal{T}(H) : t \mapsto ete \tag{7.2.1}$$

is a completely contractive projection of $\mathcal{T}(H)$ onto $\mathcal{T}(H_0)$.

Proposition 7.2.1 *For any Hilbert spaces $H$and K, we have a natural complete isometry*

$$\mathcal{T}(H) \hat{\otimes} \mathcal{T}(K) \cong \mathcal{T}(H \otimes K).$$

Proof Given finite-dimensional subspaces $H_0 \subseteq H$ and $K_0 \subseteq K$, we have a commutative diagram

$$\begin{array}{ccc} \mathcal{T}(H_0) \hat{\otimes} \mathcal{T}(K_0) & \longrightarrow & \mathcal{T}(H_0 \otimes K_0) \\ \big\downarrow & & \big\downarrow \\ \mathcal{T}(H) \hat{\otimes} \mathcal{T}(K) & \longrightarrow & \mathcal{T}(H \otimes K) \end{array}.$$

It follows from (7.1.18) that the top row is a complete isometry. Since the mappings $\mathcal{T}(H_0) \to \mathcal{T}(H)$ and $\mathcal{T}(K_0) \to \mathcal{T}(K)$ are completely isometric, the same is true for the left column(see (7.2.1)), and the right column is also completely isometric. Since the union of the spaces $\mathcal{T}(H_0)$ is dense in $\mathcal{T}(H)$, and the same is true for the union of the $\mathcal{T}(K_0)$ in $\mathcal{T}(K)$, the union of the spaces $\mathcal{T}(H_0) \otimes \mathcal{T}(K_0)$ is norm dense in $\mathcal{T}(H) \hat{\otimes} \mathcal{T}(K)$. Similarly, the union of the spaces $\mathcal{T}(H_0 \otimes K_0)$ is norm dense in $\mathcal{T}(H \otimes K)$. The mappings in the top row are coherent, and thus by continuity they determine a complete isometry in the bottom row. □

We next consider the *slice mappings* of Tomiyama (1970). Given a functional $\omega_1 \in \mathcal{B}(H)_*$, we can define a linear mapping

$$\omega_1 \otimes id : \mathcal{B}(H) \otimes \mathcal{B}(K) \to \mathcal{B}(K)$$

by

$$(\omega_1 \otimes id)(a \otimes b) = \omega_1(a)b.$$

If $u = a \otimes b$, then

$$\omega_2((\omega_1 \otimes id)(u)) = \langle a \otimes b, \omega_1 \otimes \omega_2 \rangle.$$

Lemma 7.2.2 *For any $\omega_1 \in \mathcal{B}(H)_*$, $\omega_1 \otimes id$ has a unique weak* continuous linear extension*

$$R(\omega_1) : \mathcal{B}(H \otimes K) \to \mathcal{B}(K)$$

and $\|R(\omega_1)\|_{cb} \leq \|\omega_1\|$.

Proof If $u \in \mathcal{B}(H \otimes K)$, then we define $R(\omega_1) \in \mathcal{B}(K)$ by letting

$$(R(\omega_1)(u))(\omega_2) = \langle u, \omega_1 \otimes \omega_2 \rangle \tag{7.2.2}$$

for $\omega_2 \in \mathcal{B}(H)_*$. This is a linear extension of $\omega_1 \otimes id$. From Theorem 7.2.1 and Corollary 7.1.5 we have the identifications

$$\mathcal{B}(H \otimes K) \cong (\mathcal{B}(H)_* \widehat{\otimes} \mathcal{B}(K)_*)^*$$

and

$$\lambda : \mathcal{B}(H \otimes K) \cong \mathcal{CB}(\mathcal{B}(H)_*, \mathcal{B}(K)).$$

If $u \in \mathcal{B}(H \otimes K)$, then

$$(R(\omega_1)(u))(\omega_2) = \langle u, \omega_1 \otimes \omega_2 \rangle = (\lambda(u)(\omega_1))(\omega_2),$$

and thus

$$R(\omega_1)(u) = \lambda(u)(\omega_1).$$

If $u \in M_n(\mathcal{B}(H \otimes K))$, then

$$R(\omega_1)_n(u) = [R(\omega_1)(u_{i,j})] = [\lambda(u_{i,,j})(\omega_1)] = \lambda_n(u)(\omega_1);$$

hence

$$\|R(\omega_1)_n(u)\| \leq \|\lambda_n(u)\| \, \|\omega_1\| = \|u\| \, \|\omega_1\|$$

and

$$\|R(\omega_1)\|_{cb} \leq \|\omega_1\|.$$

Since $\omega_1 \otimes \omega_2$ is weak* continuous on $\mathcal{B}(H \otimes K)$, it is evident from (7.2.2) that $R(\omega_1)$ is continuous in the weak* topologies, and since $\mathcal{B}(H) \otimes \mathcal{B}(K)$ is weak* dense in $\mathcal{B}(H \otimes K)$, the extension is unique. □

We call $R(\omega_1)$ the *right slice mapping* determined by ω_1, and we also write $\omega_1 \otimes id$ for $R(\omega_1)$. Each $\omega_2 \in \mathcal{B}(K)_*$ similarly determines a *left slice mapping*

$$id \otimes \omega_2 = L(\omega_2) : \mathcal{B}(H \otimes K) \to \mathcal{B}(H).$$

Given dual operator spaces V^* and W^*, we may find dual realizations $\pi_1 : V^* \hookrightarrow \mathcal{B}(H)$ and $\pi_2 : W^* \hookrightarrow \mathcal{B}(K)$ (see Proposition 3.2.4). We have two 'dual' tensor products associated with these representations. On the one hand, we define the *normal spatial tensor product* $V^* \overline{\otimes} W^*$ to be the closure of $V^* \otimes W^*$ in $\mathcal{B}(H \otimes K)$, in the *weak** topology defined by

$$\mathcal{B}(H \otimes K)_* = \mathcal{B}(H)_* \widehat{\otimes} \mathcal{B}(K)_*.$$

If, for example, $\mathcal{R}$ and $\mathcal{S}$ are von Neumann algebras on Hilbert spaces H and K, respectively, and thus $\mathcal{R} \subseteq \mathcal{B}(H)$ and $\mathcal{S} \subseteq \mathcal{B}(K)$ are dual realizations (see the discussion preceding Proposition 3.2.4), then $\mathcal{R} \overline{\otimes} \mathcal{S}$ is the usual (spatial) von Neumann algebraic tensor product. On the other hand, we define the *normal Fubini tensor product* $V^* \overline{\otimes}_{\mathcal{F}} W^*$ by

$$V^* \overline{\otimes}_{\mathcal{F}} W^* = \{u \in B(H \otimes K) : (\omega_1 \otimes id)(u) \in W^*, (id \otimes \omega_2)(u) \in V^* \text{ for all } \omega_1 \in \mathcal{B}(H)_*, \, \omega_2 \in \mathcal{B}(K)_*\}.$$

Since $\omega_1 \otimes id$ and $id \otimes \omega_2$ are continuous in the weak* topologies, it is immediate that

$$V^* \overline{\otimes} W^* \subseteq V^* \overline{\otimes}_{\mathcal{F}} W^*.$$

These dual tensor products may be characterized abstractly. To see this, we let $\overline{V^* \otimes W^*}$ denote the weak* closure of $V^* \otimes W^*$ in $(V \mathbin{\widehat{\otimes}} W)^*$. Given dual realizations π_1 and π_2 as above, we have from Corollary 4.1.9 that π_1 and π_2 are the adjoints of complete quotient maps $\pi_{1*} : \mathcal{B}(H)_* \twoheadrightarrow V$ and $\pi_{2*} : \mathcal{B}(K)_* \twoheadrightarrow W$. It follows that

$$\pi_* = \pi_{1*} \otimes \pi_{2*} : \mathcal{B}(H)_* \mathbin{\widehat{\otimes}} \mathcal{B}(K)_* = \mathcal{B}(H \otimes K)_* \twoheadrightarrow V \mathbin{\widehat{\otimes}} W$$

is a complete quotient mapping, and in turn,

$$\pi = (\pi_{1*} \otimes \pi_{2*})^* : (V \mathbin{\widehat{\otimes}} W)^* \hookrightarrow \mathcal{B}(H \otimes K)$$

is a dual realization of $(V \mathbin{\widehat{\otimes}} W)^*$.

Theorem 7.2.3 *Suppose that V^* and W^* are operator space duals with dual realizations π_i as above, and that π is defined as above. Then π is a weak* homeomorphic completely isometric mapping of $(V \mathbin{\widehat{\otimes}} W)^*$ onto $V^* \overline{\otimes}_{\mathcal{F}} W^*$, and it carries $\overline{V^* \otimes W^*}$ onto $V^* \overline{\otimes} W^*$.*

Proof Since $\pi((V \mathbin{\widehat{\otimes}} W)^*)$ is weak* closed,

$$\pi((V \mathbin{\widehat{\otimes}} W)^*) = (\ker \pi_*)^{\perp},$$

and thus from Proposition 7.1.7,

$$\pi((V \mathbin{\widehat{\otimes}} W)^*) = (\mathcal{B}(H)_* \otimes \ker \pi_{2*} + \ker \pi_{1*} \otimes \mathcal{B}(K)_*)^{\perp}.$$

It follows that if $u \in \mathcal{B}(H \otimes K)$, then $u \in \pi((V \widehat{\otimes} W)^*)$ if and only if it is annihilated by the functionals $\omega_1 \otimes \omega_2$ for which either $\pi_{1*}(\omega_1) = 0$ or $\pi_{2*}(\omega_2) = 0$. Thus, u is in the image of π if and only if

$$\omega_2((\omega_1 \otimes id)(u)) = \omega_1 \otimes \omega_2(u) = 0$$

for all $\omega_2 \in \ker \pi_{2*}$, or equivalently,

$$(\omega_1 \otimes id)(u) \in (\ker \pi_{2*})^{\perp} = W^*,$$

and similarly,

$$(id \otimes \omega_2)(u) \in V^*$$

for all $\omega_2 \in \mathcal{B}(K)_*$, if and only if $u \in V^* \overline{\otimes}_{\mathcal{F}} W^*$. Since π is a weak* homeomorphism, the second statement is obvious. □

We shall generally use the notation $V^* \overline{\otimes} W^*$ for the 'abstract' normal spatial tensor product $\overline{V^* \otimes W^*}$. We conclude with the result we promised at the beginning of this section. We shall return to the general theory in §11.2.

Theorem 7.2.4 *Given von Neumann algebras $\mathcal{R} \subseteq \mathcal{B}(H)$ and $\mathcal{S} \subseteq \mathcal{B}(K)$, we have a natural complete isometry*

$$(\mathcal{R} \overline{\otimes} \mathcal{S})_* = \mathcal{R}_* \hat{\otimes} \mathcal{S}_*.$$

Proof We must show that $\mathcal{R} \overline{\otimes}_{\mathcal{F}} \mathcal{S} \subseteq \mathcal{R} \overline{\otimes} \mathcal{S}$, or since $\mathcal{R} \overline{\otimes} \mathcal{S}$ is a von Neumann algebra, that each $u \in \mathcal{R} \overline{\otimes}_{\mathcal{F}} \mathcal{S}$ commutes with all operators in the commutant $(\mathcal{R} \overline{\otimes} \mathcal{S})'$. Owing to a result of Tomita and Takesaki (see Theorem A.4.5), the latter is just $\mathcal{R}' \overline{\otimes} \mathcal{S}'$, and thus it suffices to show that $u(r' \otimes I) = (r' \otimes I)u$ for $r' \in \mathcal{R}'$ and $u(I \otimes s') = (I \otimes s')u$ for $s' \in \mathcal{S}'$. Turning to the latter, the identification $\mathcal{B}(H \otimes K) \cong \mathcal{CB}(\mathcal{B}(H)_*, \mathcal{B}(K))$ is determined by associating to each $x \in \mathcal{B}(H \otimes K)$ the function $\omega_1 \mapsto (\omega_1 \otimes id)(x)$. Thus, it suffices to show that

$$(\omega_1 \otimes id)(u(I \otimes s')) = (\omega_1 \otimes id)((I \otimes s')u),$$

for all $\omega \in R_*$, or equivalently,

$$((\omega_1 \otimes id)(u))s' = s'((\omega_1 \otimes id)(u)).$$

Since $u \in \mathcal{R} \overline{\otimes}_{\mathcal{F}} \mathcal{S}$ implies that $(\omega \otimes id)(u) \in \mathcal{S}$, we are done. The first relation follows in the same manner. □

7.3 NOTES AND REFERENCES

We refer the reader to Grothendieck (1955), Pisier (1986), Tomczak-Jaegermann (1989), and Defant and Floret (1993) for the elementary theory of Banach space projective and injective tensor products.

The theory of the operator space projective tensor product provided one of the most convincing justifications of the duality theory used in operator space theory (see the discussion in §3.6). It was discovered independently by Blecher and Paulsen, and Effros and Ruan in the late 1980s. The major results in §7.1 can be found in Blecher and Paulsen (1991*a*), and in Effros and Ruan (1991*a*).

The slice mappings were introduced by Tomiyama (1970). Kraus (1983, 1991) investigated the slice mapping problems associated with the approximation properties for the weak* closed subspaces of von Neumann algebras. The normal Fubini product for dual operator spaces was first considered by Ruan (1992), and further studied by Effros *et al.* (1993). Theorem 7.2.4 is due to Effros and Ruan (1990, Theorem 3.2).

8

The injective tensor product

Although the theory outlined in this chapter closely parallels the Banach theory, we shall subsequently find that some of the most novel aspects of operator space theory arise when one considers its behaviour under duality (see §14.2).

8.1 DEFINITION AND ELEMENTARY PROPERTIES

Given Banach spaces E and F, the *Banach space injective tensor product norm* of an element $u \in E \otimes F$ is defined by

$$\|u\|_\lambda = \sup\{|(f \otimes g)(u)| : f \in E^*, g \in F^*, \|f\|, \|g\| \leq 1\}. \tag{8.1.1}$$

We let

$$E \otimes_\lambda F = (E \otimes F, \|\cdot\|_\lambda).$$

The completion $E \otimes^\lambda F$ is called the *Banach space injective tensor product.* This space is also determined by the natural injection $E \otimes F \hookrightarrow \mathcal{B}(E^*, F)$. More precisely, we have the natural identification

$$E \otimes^\lambda F \cong \mathcal{L}^\sigma(E^*, F), \tag{8.1.2}$$

where $\mathcal{L}^\sigma(E^*, F)$ denotes the uniform closure of the finite-rank weak* continuous linear mappings $\varphi : E^* \to F$.

In contrast to the Banach space projective tensor product, it is a simple matter to realize the injective tensor product of two Banach spaces as a function space. If we are given isometries $E \hookrightarrow \ell_\infty(\mathfrak{s})$ and $F \hookrightarrow \ell_\infty(\mathfrak{t})$, then the corresponding mapping $E \otimes F \hookrightarrow \ell_\infty(\mathfrak{s} \times \mathfrak{t})$ extends to an isometry $E \otimes^\lambda F \hookrightarrow \ell_\infty(\mathfrak{s} \times \mathfrak{t})$. It follows that if we are given an isometry $E \hookrightarrow E_1$, then for any Banach space F, the corresponding mapping

$$E \otimes^\lambda F \hookrightarrow E_1 \otimes^\lambda F \tag{8.1.3}$$

is isometric since we may choose an isometry $E_1 \hookrightarrow \ell_\infty(\mathfrak{s}_1)$, and then use the corresponding composition

$$E \hookrightarrow E_1 \hookrightarrow \ell_\infty(\mathfrak{s}_1).$$

Given a locally compact space Ω and a Banach space E,

$$C_0(\Omega) \otimes^\lambda E \cong C_0(\Omega, E), \tag{8.1.4}$$

where $C_0(\Omega, E)$ is the Banach space of continuous functions $f : \Omega \to E$ for which $x \mapsto \|f(x)\|$ vanishes at ∞. In particular, if $\mathcal{Z} = C_0(\Omega)$, then for any C^*-algebra $\mathcal{A}$, $\mathcal{Z} \otimes^\lambda \mathcal{A} \cong C_0(\Omega, \mathcal{A})$ is again a C^*-algebra. If $\mathcal{A} = M_n$, then $M_n \otimes^\lambda \mathcal{Z} \cong \mathcal{Z} \otimes^\lambda M_n$ is $*$-isomorphic to the C^*-algebra $M_n(\mathcal{Z})$. Since a $*$-isomorphism must be isometric, we have the natural isometric identification

$$M_n(\mathcal{Z}) \cong M_n \otimes^\lambda \mathcal{Z}. \tag{8.1.5}$$

Since the bilinear mapping

$$E \times F \to E \otimes^\lambda F : (v, w) \mapsto v \otimes w$$

is norm-decreasing, it extends to what we shall call the *canonical* mapping

$$\Phi : E \otimes^\gamma F \to E \otimes^\lambda F. \tag{8.1.6}$$

As we shall see, the study of both the kernel and image of Φ are of central interest in Banach space theory.

Turning to operator spaces V and W, we define the *injective matrix norm* $\|\cdot\|_\vee$ on $V \otimes W$ by letting

$$\|u\|_\vee = \sup\{\|(f \otimes g)_n(u)\| : f \in M_p(V^*), \\ g \in M_q(W^*), \|f\|, \|g\| \leq 1\} \tag{8.1.7}$$

for each matrix $u \in M_n(V \otimes W)$. It follows from the following result that this is an operator space matrix norm.

Proposition 8.1.1 *Suppose that V and W are operator spaces. Then the injective matrix norm* (8.1.7) *is determined by the natural embedding*

$$\theta : V \otimes W \hookrightarrow \mathcal{CB}(V^*, W).$$

Proof The embedding θ is determined by

$$\theta(v \otimes w)(f) = f(v)w.$$

It follows that if $g \in W^*$, then

$$g(\theta(v \otimes w)(f)) = (f \otimes g)(v \otimes w),$$

and by linearity,

$$g(\theta(u)(f)) = (f \otimes g)(u)$$

for all $u \in V \otimes W$.

For any $u \in M_n(V \otimes W)$,

$$\theta_n(u) \in M_n(\mathcal{CB}(V^*, W)) = \mathcal{CB}(V^*, M_n(W))$$

and thus for $f \in M_p(V^*)$,

$$\theta_n(u)_p(f) \in M_p(\mathcal{CB}(V^*, M_n(W))) = \mathcal{CB}(V^*, M_{p \times n}(W)).$$

If $g \in M_q(W^*)$, then

$$g_{p\times n}(\theta_n(u)_p(f)) = g_{p\times n}([\theta(u_{s,t})(f_{i,j})]) = [g(\theta(u_{s,t}(f_{i,j}))]$$
$$= [g_{k,l}(\theta(u_{s,t})(f_{i,j}))] = [f_{i,j} \otimes g_{k,l}(u_{s,t})] = (f \otimes g)_n(u)\,.$$

From Lemma 2.3.4,

$$\|\theta_n(u)_p(f)\| = \sup\{\|g_{p\times n}(\theta_n(u)_p(f))\| : g \in M_q(W^*),$$
$$\|g\| \leq 1, q \in \mathbb{N}\},$$

and thus

$$\|\theta_n(u)\|_{cb} = \sup\{\|g_{p\times n}(\theta_n(u)_p(f))\| : f \in M_p(V^*),$$
$$g \in M_q(W^*), \|f\|, \|g\| \leq 1, p, q \in \mathbb{N}\}$$
$$= \sup\{\|(f \otimes g)_n(u)]\| : f \in M_p(V^*),$$
$$g \in M_q(W^*), \|f\|, \|g\| \leq 1, p, q \in \mathbb{N}\} = \|u\|_\vee\,.$$

□

Given operator spaces V and W, we let

$$V \otimes_\vee W = (V \otimes W,\ \|\cdot\|_\vee)$$

and we define the *operator space injective tensor product* (or simply, the *injective tensor product*) $V \check{\otimes} W$ to be the completion of this operator space. The following is a useful variation of Proposition 8.1.1.

Proposition 8.1.2 *Given operator spaces V and W, the natural embedding*

$$\theta : V^* \check{\otimes} W \hookrightarrow \mathcal{CB}(V, W) \tag{8.1.8}$$

is completely isometric.

Proof The embedding θ is determined by

$$\theta(f \otimes w)(v) = f(v)w\,.$$

If we regard V as a subspace of V^{**}, it is apparent from the proof of Proposition 8.1.1 that if $g \in W^*$, then for $u \in V^* \otimes W$,

$$g(\theta(u)(v)) = (v \otimes g)(u)\,.$$

Following that argument, we see that if $u \in M_n(V^* \otimes W)$, $v \in M_p(V)$, and $g \in M_q(W^*)$, then

$$g_{p\times n}(\theta_n(u)_p(v)) = (v \otimes g)_n(u)\,.$$

Given $F \in M_p(V^{**}) = \mathcal{CB}(V^*, M_p)$ with $\|F\|_{cb} \leq 1$, we may use Proposition 4.2.5 to find a net $v_\gamma \in M_n(V) = \mathcal{CB}^\sigma(V^*, M_n)$ with $\|v_\gamma\|_{cb} \leq 1$, which approximates F in the point-norm topology. Then if we consider elements $v \in M_p(V), g \in M_q(W^*)$, and $F \in M_p(V^{**})$, we have

$$\|\theta_n(u)\|_{cb} = \sup\{\|g_{p\times n}(\theta_n(u)_p(v))\| : \|v\|, \|g\| \leq 1\}$$

$$= \sup\{\|(v \otimes g)_n(u)\| : \|v\|, \|g\| \leq 1\}$$
$$= \sup\{\|(F \otimes g)_n(u)\| : \|F\|, \|g\| \leq 1\} = \|u\|_\vee .$$

□

Corollary 8.1.3 *For any operator space V and $n \in \mathbb{N}$, we have a natural complete isometry*

$$M_n(V) \cong M_n \check{\otimes} V .$$

Proof This is apparent from the diagram

$$M_n \check{\otimes} V \hookrightarrow \mathcal{CB}^\sigma(V^*, M_n) \cong M_n(V),$$

where from Proposition 8.1.1, the first mapping is completely isometric, and the second identification was given in (3.2.9). □

Corollary 8.1.4 *For any operator spaces V and W, the identity mapping on $V \otimes W$ induces a contractive injective mapping $V \check{\otimes} W \to V \otimes^\lambda W$.*

Proof We have a commutative diagram

$$\begin{array}{ccc} V \check{\otimes} W & \longrightarrow & \mathcal{CB}(V^*, W) \\ \downarrow & & \downarrow \\ V \otimes^\lambda W & \longrightarrow & \mathcal{B}(V^*, W) \end{array}$$

in which the rows are isometric, and the right column is a contractive injection. It follows that the same is true for the left column. □

If $\varphi : V \to W$ is a complete quotient mapping of operator spaces, then we say that a completely bounded mapping $\varphi' : W \to V$ is a *lifting* for φ if $\varphi \circ \varphi' = id_V$, i.e. φ' is a 'right inverse' for the quotient mapping.

Proposition 8.1.5 *Suppose that V, V_1, W, and W_1 are operator spaces. Given the complete contractions $\varphi : V \to V_1$ and $\psi : W \to W_1$, the corresponding mapping*

$$\varphi \otimes_0 \psi : V \otimes W \to V_1 \otimes W_1$$

extends to a complete contraction

$$\varphi \otimes \psi : V \check{\otimes} W \to V_1 \check{\otimes} W_1 .$$

If φ and ψ are completely isometric injections (respectively, complete quotient mappings with completely contractive liftings), then $\varphi \otimes \psi$ is a complete isometry (respectively, a complete quotient mapping).

Proof If $f_1 \in M_p(V_1^*) = \mathcal{CB}(V_1, M_p)$ and $g_1 \in M_q(W_1^*) = \mathcal{CB}(W_1, M_q)$ are complete contractions, then that is also the case for

$$f_1 \circ \varphi \in M_p(V^*) = \mathcal{CB}(V, M_p)$$

and

$$g_1 \circ \psi \in M_q(W^*) = \mathcal{CB}(W, M_q) .$$

Thus, if $u \in M_n(V \otimes_\vee W)$ and we take the suprema over all such complete contractions, then

$$\begin{aligned}\|(\varphi \otimes \psi)_n(u)\|_\vee &= \sup\{\|(f_1 \otimes g_1)_n \circ (\varphi \otimes \psi)_n(u)\|\} \\ &= \sup\{\|((f_1 \circ \varphi) \otimes (g_1 \circ \psi))_n(u)\|\} \leq \|u\|_\vee.\end{aligned}$$

If φ and ψ are completely isometric injections, then we may identify V and W with operator subspaces of V_1 and W_1, respectively. Given $\varepsilon > 0$, we choose complete contractions

$$f \in M_p(V^*) = \mathcal{CB}(V, M_p) \text{ and } g \in M_q(W^*) = \mathcal{CB}(W, M_q)$$

with $\|(f \otimes g)_n(u)\| > \|u\|_\vee - \varepsilon$. We may use the Arveson–Wittstock–Hahn–Banach theorem to extend these mappings to complete contractions $f_1 \in M_p(V_1^*) = \mathcal{CB}(V_1, M_p)$ and $g_1 \in M_q(W_1^*) = \mathcal{CB}(W_1, M_q)$. Thus, we have

$$\|(\varphi \otimes \psi)_n(u)\|_\vee \geq \|(f_1 \otimes g_1)_n(u)\| > \|u\|_\vee - \varepsilon .$$

The last assertion is immediate if one utilizes the tensor products of the liftings. □

Proposition 8.1.6 *Given operator subspaces $V \subseteq \mathcal{B}(H)$ and $W \subseteq \mathcal{B}(K)$, the corresponding mapping*

$$V \check{\otimes} W \hookrightarrow \mathcal{B}(H \otimes K)$$

is completely isometric.

Proof Owing to Proposition 8.1.5, it suffices to prove that

$$\mathcal{B}(H) \otimes_\vee \mathcal{B}(K) \hookrightarrow \mathcal{B}(H \otimes K)$$

is completely isometric. This follows from Proposition 8.1.2 and Proposition 7.2.1 since we may factor this inclusion through the identifications

$$\begin{aligned}\mathcal{B}(H) \otimes_\vee \mathcal{B}(K) &\hookrightarrow \mathcal{CB}(\mathcal{B}(H)_*, \mathcal{B}(K)) \\ &= (\mathcal{B}(H)_* \hat{\otimes} \mathcal{B}(K)_*)^* = \mathcal{B}(H \otimes K) .\end{aligned}$$

□

Corollary 8.1.7 *Given operator spaces V, W, and X, we have the completely isometric isomorphisms*

$$V \check{\otimes} W \cong W \check{\otimes} V \tag{8.1.9}$$

and

$$(V \check{\otimes} W) \check{\otimes} X \cong V \check{\otimes} (W \check{\otimes} X) . \tag{8.1.10}$$

Proof Given any Hilbert spaces H, K, and L, we have the natural isometries

$$(H \otimes K) \otimes L \cong H \otimes (K \otimes L)$$

and

$$H \otimes K \cong K \otimes H.$$

Thus, if we assume that $V \subseteq \mathcal{B}(H)$, $W \subseteq \mathcal{B}(K)$, and $X \subseteq \mathcal{B}(L)$, the result follows from Proposition 8.1.6. □

Given $p, q \in \mathbb{N}$, and matrices $v \in M_p(V)$ and $w \in M_q(W)$, the Kronecker product $v \otimes w \in M_{p\times q}(V \otimes W)$ satisfies

$$\|v \otimes w\|_\vee = \|v\|\|w\|, \tag{8.1.11}$$

since if we are given completely isometric inclusions

$$V \hookrightarrow \mathcal{B}(H) \text{ and } W \hookrightarrow \mathcal{B}(K),$$

the mapping

$$M_{p\times q}(V \otimes W) \hookrightarrow M_{p\times q}(\mathcal{B}(H \otimes K)) \cong \mathcal{B}(H^p \otimes K^q)$$

is isometric, and we may apply (1.3.2). It particular, the bilinear mapping

$$V \times W \to V \otimes_\vee W : (v, w) \mapsto v \otimes w$$

is completely contractive, and thus determines a complete contraction

$$\Phi : V \hat{\otimes} W \to V \check{\otimes} W\,. \tag{8.1.12}$$

Again, we shall refer to this as the *canonical* complete contraction between these tensor products. Given $v \in M_p(V)$ and $w \in M_q(W)$, we have from (8.1.11) and the fact that Φ is completely contractive that

$$\|v \otimes w\|_\wedge = \|v\|\|w\|\,. \tag{8.1.13}$$

Thus, in the terminology of §7.1, both $\|\cdot\|_\vee$ and $\|\cdot\|_\wedge$ are cross matrix norms.

As in Banach space theory, the kernel of Φ is often zero, and the determination of when that is true is of considerable importance (see §11.2). In what follows we see that this question is directly related to our discussion of the normal spatial and Fubini products (see §7.2).

Proposition 8.1.8 *For any operator spaces V and W, we have the complete isometry*

$$\left(V \hat{\otimes} W/\ker\Phi\right)^* \cong V^* \overline{\otimes} W^*,$$

where $\Phi : V \hat{\otimes} W \to V \check{\otimes} W$ is the canonical mapping.

Proof We have from Corollary 8.1.4 that if $u \in V \hat{\otimes} W$, then $\Phi(u) = 0$ if and only if the image of $\Phi(u)$ in $V \otimes^\lambda W$ is zero, or equivalently

$$(f \otimes g)(u) = (f \otimes g)(\Phi(u)) = 0$$

for all $f \in V^*$ and $g \in W^*$. It follows from the bipolar theorem and Theorem 7.2.3 that

$$\left(V \hat{\otimes} W/\ker\Phi\right)^* = (\ker\Phi)^\perp = (V^* \otimes W^*)^{\perp\perp} = V^* \overline{\otimes} W^*.$$

□

Corollary 8.1.9 *For any two weak* closed operator spaces* $V^* \hookrightarrow \mathcal{B}(H)$ *and* $W^* \hookrightarrow \mathcal{B}(K)$, $V^* \overline{\otimes} W^* = V^* \overline{\otimes}_{\mathcal{F}} W^*$ *if and only if* $\ker \Phi = 0$. □

There is a subtle *tensor interchange* generalization of this result, which will be useful in later chapters (see §12.3, 13.2, and 13.3). We have included the classical analogue in order to motivate the involved technical argument for operator spaces.

Theorem 8.1.10 *Suppose that* E, F, *and* G *are Banach spaces and that* V, W, *and* X *are operator spaces. Then the natural mappings*

$$E \otimes_\gamma (F \otimes_\lambda G) \to (E \otimes_\gamma F) \otimes_\lambda G$$

and

$$V \otimes_\wedge (W \otimes_\vee X) \to (V \otimes_\wedge W) \otimes_\vee X$$

are contractive and completely contractive, respectively.

Proof For any $u \in E \otimes_\gamma (F \otimes_\lambda G)$ and $\varepsilon > 0$, we may find a decomposition

$$u = \sum_i x_i \otimes u_i$$

where $x_i \in E$ and $u_i \in F \otimes G$ satisfy

$$\sum \|x_i\| \, \|u_i\|_{F \otimes_\lambda G} \leq \|u\|_{E \otimes_\gamma (F \otimes_\lambda G)} + \varepsilon .$$

We let $u_i = \sum y_{i,j} \otimes z_{i,j}$, where $y_{i,j} \in F$ and $z_{i,j} \in G$. From (7.1.1),

$$\|u\|_{(E \otimes_\gamma F) \otimes_\lambda G} = \sup \{|(h \otimes g)(u)|\} ,$$

where the supremum is taken over all $h \in (E \otimes_\gamma F)^*$, and $g \in G^*$ with $\|h\| \leq 1$ and $\|g\| \leq 1$. Let us fix such a pair h and g. Then h determines a contraction $H : E \to F^*$, where $H(x)(y) = h(x \otimes y)$, and in particular, $f_i = H(x_i)$ satisfies $\|f_i\| \leq \|x_i\|$. Thus,

$$|f_i \otimes g(u_i)| \leq \|x_i\| \, \|u_i\|_{F \otimes_\lambda G} ,$$

and

$$\begin{aligned} |h \otimes g(u)| &= \left| \sum_{i,j} h(x_i \otimes y_{i,j}) g(z_{i,j}) \right| \\ &\leq \sum_i \left| \sum_j H(x_i)(y_{i,j}) g(z_{i,j}) \right| = \sum_i |(f_i \otimes g)(u_i)| \\ &\leq \sum_i \|x_i\| \, \|u_i\|_{F \otimes_\lambda G} \leq \|u\|_{E \otimes_\gamma (F \otimes_\lambda G)} + \varepsilon. \end{aligned}$$

Since h and g are arbitrary contractions, we conclude that

$$\|u\|_{(E \otimes_\gamma F) \otimes_\lambda G} \leq \|u\|_{E \otimes_\gamma (F \otimes_\lambda G)} .$$

Now let us suppose that V, W, and X are operator spaces. We let $Z = W \otimes X$. Given $u \in M_n(V \otimes Z)$ and $\varepsilon > 0$, we may assume that

$$u = \alpha(v \otimes z)\beta = \sum \alpha_{i,k}(v_{i,j} \otimes z_{k,l})\beta_{j,l}\,, \tag{8.1.14}$$

where $v \in M_p(V)$, $z \in M_q(Z)$, $\alpha \in M_{n,p\times q}$, and $\beta \in M_{p\times q,n}$ satisfy

$$\|\alpha\|\,\|v\|\,\|z\|_\vee\,\|\beta\| < \|u\|_\wedge + \varepsilon.$$

We let $z = [z_{k,l}]$, where

$$z_{k,l} = \sum_t w^{(t)}_{k,l} \otimes x^{(t)}_{k,l}$$

with $w^{(t)}_{k,l} \in W$ and $x^{(t)}_{k,l} \in X$.

From (8.1.7),

$$\|u\|_{(V\otimes_\wedge W)\otimes_\vee X} = \sup\{\|(e\otimes h)_n(u)\|\}, \tag{8.1.15}$$

where the supremum is taken over all

$$e = [e^{a,b}] \in M_r((V\otimes_\wedge W)^*) = \mathcal{CB}(V\otimes_\wedge W, M_r)$$

and

$$h = [h^{c,d}] \in M_s(X^*) = \mathcal{CB}(X, M_s)$$

with $\|e\|_{cb}, \|h\|_{cb} \leq 1$. If we fix such elements e and h, then e determines a complete contraction

$$E \in \mathcal{CB}(V, \mathcal{CB}(W, M_r)),$$

where

$$E(v_0)(w_0) = e(v_0 \otimes w_0)$$

for arbitrary $v_0 \in V$ and $w_0 \in W$. Thus, if $f_{i,j} = E(v_{i,j})$, then

$$f = [f_{i,j}] \in M_p(\mathcal{CB}(W, M_r)) = \mathcal{CB}(W, M_{p\times r})$$

satisfies $\|f\|_{cb} \leq \|v\|$.

The matrix

$$(f\otimes h)_q(z) = \left[f^{a,b}_{i,j} \otimes h^{c,d}(z_{k,l})\right] \in M_{p\times r\times q\times s}$$

satisfies

$$\|(f\otimes h)_q(z)\| \leq \|f\|\,\|h\|_{cb}\,\|z\|_{M_q(W\otimes_\vee X)} \leq \|v\|\,\|z\|_{M_q(W\otimes_\vee X)}\,.$$

We have

$$\begin{aligned}(e\otimes h)_n(u) &= \left[(e\otimes h)\left(\sum_{i,j,k,l,t} \alpha_{g,(i,k)}(v_{i,j}\otimes w^{(t)}_{k,l}\otimes x^{(t)}_{k,l})\beta_{(j,l),h}\right)\right]_{g,h\in n}\\ &= \left[\sum_{i,j,k,l,t} \alpha_{g,(i,k)}\left(E(v_{i,j})(w^{(t)}_{k,l})\otimes h(x^{(t)}_{k,l})\right)\beta_{(j,l),h}\right]\end{aligned}$$

$$= \left[\sum_{i,j,k,l} \alpha_{g,(i,k)} \left(\sum_t f_{i,j}(w_{k,l}^{(t)}) \otimes h(x_{k,l}^{(t)}) \right) \beta_{(j,l),h} \right]$$
$$= \left[\sum_{i,j,k,l} \alpha_{g,(i,k)} (f_{i,j} \otimes h)\,(z_{k,l})\, \beta_{(j,l),h} \right]$$
$$= \alpha(f \otimes h)_q(z)\beta\,,$$

and thus

$$\begin{aligned} \|(e \otimes h)_n(u)\|_{M_{n \times r \times s}} &\leq \|\alpha\|\, \|(f \otimes h)_q(z)\|\, \|\beta\| \\ &\leq \|\alpha\|\, \|v\|\, \|z\|_\vee\, \|\beta\| \\ &\leq \|u\|_\wedge + \varepsilon. \end{aligned}$$

It follows that

$$\|u\|_{V \otimes_\vee (W \otimes_\wedge X)} \leq \|u\|_{V \otimes_\wedge (W \otimes_\vee X)} + \varepsilon,$$

or since $\varepsilon > 0$ is arbitrary, we obtain the desired inequality. □

8.2 RELATING BANACH AND OPERATOR SPACE TENSOR PRODUCTS

Given a normed space E and an operator space V, we have the isometric mappings

$$\begin{array}{ccc} \min E \check{\otimes} V & \hookrightarrow & \mathcal{CB}(V^*, \min E) \\ & & \| \\ E \otimes^\lambda V & \hookrightarrow & \mathcal{B}(V^*, E) \end{array}$$

(see (3.3.7)), and since $E \otimes V$ is dense in both spaces on the left, we obtain the *isometric* identification

$$\min E \check{\otimes} V \cong E \otimes^\lambda V\,. \tag{8.2.1}$$

In particular, if E and F are normed spaces, we have the isometry

$$\min E \check{\otimes} \min F \cong E \otimes^\lambda F\,. \tag{8.2.2}$$

If V and W are minimal operator spaces, then $V \check{\otimes} W$ is minimal since if we choose embeddings $V \hookrightarrow \mathcal{Y}$ and $W \hookrightarrow \mathcal{Z}$ with $\mathcal{Y}$ and $\mathcal{Z}$ commutative C^*-algebras, then we have a corresponding embedding of $V \check{\otimes} W$ into the commutative C^*-algebra $\mathcal{Y} \check{\otimes} \mathcal{Z}$. If $V = \min E$ and $W = \min F$, then it follows from (8.2.2) that

$$\min E \check{\otimes} \min F \cong \min (E \otimes^\lambda F) \tag{8.2.3}$$

for any normed spaces E and F.

Given a normed space E and an operator space W, we have the natural contraction

$$\varphi : E \otimes_\gamma W \to \max E \otimes_\wedge W$$

and a commutative diagram

$$\begin{array}{ccc} (\max E \otimes_\wedge W)^* & \xrightarrow{\varphi^*} & (E \otimes_\gamma W)^* \\ \| & & \| \\ \mathcal{CB}(\max E, W^*) & \longrightarrow & \mathcal{B}(E, W^*) \end{array}.$$

But from (3.3.9), the bottom row is an isometric isomorphism; hence the same is true for both φ^* and φ, and we have the natural isometry

$$\max E \mathbin{\widehat{\otimes}} W \cong E \otimes^\gamma W. \tag{8.2.4}$$

In particular, for any normed space F, we have the isometry

$$\max E \mathbin{\widehat{\otimes}} \max F \cong E \otimes^\gamma F. \tag{8.2.5}$$

If V and W are maximal operator spaces, then we claim that $V \widehat{\otimes} W$ is maximal. To see this, let us fix the complete quotient mappings

$$Y = \max \ell_1(\mathfrak{s}_1) \twoheadrightarrow V$$

and

$$Z = \max \ell_1(\mathfrak{s}_2) \twoheadrightarrow W,$$

where we regard Y and Z as the operator preduals of the von Neumann algebras $\mathcal{Y} = \ell_\infty(\mathfrak{s}_1)$ and $\mathcal{Z} = \ell_\infty(\mathfrak{s}_2)$, respectively. From Theorem 7.2.4, $Y \widehat{\otimes} Z$ is the predual of the commutative von Neumann algebra

$$\mathcal{Y} \mathbin{\overline{\otimes}} \mathcal{Z} \cong \ell_\infty(\mathfrak{s}_1 \times \mathfrak{s}_2).$$

Thus,

$$Y \mathbin{\widehat{\otimes}} Z \cong \max \ell_1(\mathfrak{s}_1 \times \mathfrak{s}_2),$$

and from Theorem 7.1.7 we have a corresponding complete quotient mapping $Y \mathbin{\widehat{\otimes}} Z \twoheadrightarrow V \mathbin{\widehat{\otimes}} W$. If $V = \max E$ and $W = \max F$, then we conclude from (8.2.5) that for any normed spaces E and F,

$$\max E \mathbin{\widehat{\otimes}} \max F \cong \max(E \otimes^\gamma F). \tag{8.2.6}$$

The following shows that there are some other interesting situations in which the operator space and the Banach space tensor products are isometric.

Proposition 8.2.1 *Suppose that H and K are Hilbert spaces. Then we have the natural isometries*

$$(K_c)^* \mathbin{\check{\otimes}} H_c \cong K^* \otimes^\lambda H \cong \mathcal{K}(K, H)$$

and

$$(K_c)^* \mathbin{\widehat{\otimes}} H_c \cong K^* \otimes^\gamma H \cong \mathcal{T}(K, H).$$

Proof We have a diagram of mappings

$$\begin{array}{ccc} (K_c)^* \check{\otimes} H_c & \hookrightarrow & \mathcal{CB}(K_c, H_c) \\ & & \| \\ K^* \otimes^\lambda H & \hookrightarrow & \mathcal{B}(K, H) \end{array},$$

in which the top row and the column mappings are completely isometric, and the bottom row is isometric (see Proposition 8.1.2 and Theorem 3.4.1). Since $K^* \otimes H$ is dense in each of these spaces, we conclude that we may isometrically identify the spaces in the first column.

The adjoint of the canonical contraction

$$\varphi : \mathcal{T}(K, H) = K^* \otimes^\gamma H \to (K_c)^* \hat{\otimes} H_c$$

is the natural mapping

$$\mathcal{CB}(H_c, K_c) \to \mathcal{B}(H, K).$$

From Theorem 3.4.1, this mapping is an isometric surjection, and thus φ is an isometry. □

8.3 NOTES AND REFERENCES

The operator space injective tensor product is the most elementary of the various tensor products. For C^*-algebras it coincides with the usual spatial tensor product, which was introduced by Turumaru (1952), and which has played an important role in operator algebra theory.

The abstract characterization of this tensor product for operator spaces was considered by Blecher and Paulsen (1991*a*), who used the Banach space theory as a model. In §8.1 Theorem 8.1.1–Corollary 8.1.7 may be found in that paper. Proposition 8.1.8 and Corollary 8.1.9 are due to Effros *et al.* (1993). Theorem 8.1.10 is a new result.

The min and max relations studied in §8.2 are due to Blecher and Paulsen (1991*a*), and Blecher (1992*a*). Proposition 8.2.1 is due to Effros and Ruan (1991*b*), and Blecher (1992*b*). Given Hilbert spaces H and K, the isometric isomorphism $K^* \otimes^\gamma H = \mathcal{T}(H, K)$ is a well-known fact.

9
The Haagerup tensor product

Soon after the decomposition theorem (Theorem 5.3.3) was proved, researchers attempted to find multilinear analogues of that result. This work was inspired by the fact that multilinear mappings arise naturally in various contexts, such as cohomology theory and the theory of modules. The success of this endeavour was in large part due to the introduction of the Haagerup tensor product, which is functorially linked to the multilinear mappings that have suitable decompositions.

In order to introduce this theory, let us suppose that we have a bilinear mapping

$$\varphi : V \times W \to \mathcal{B}(H),$$

where V and W are operator spaces, and H is a Hilbert space. In order to obtain a decomposition of φ, it is not sufficient to assume that φ is a completely bounded bilinear function, i.e. that $\|\varphi\|_{cb} < \infty$ (see §7.1.10). We must instead introduce the larger, and thus more restrictive, *multiplicative bound* $\|\varphi\|_{mb}$. As we shall see below, for any Hilbert spaces H and K, the operator multiplication mapping

$$\mathbf{m} : \mathcal{B}(K, H) \times \mathcal{B}(H, K) \to \mathcal{B}(H) : (a, b) \mapsto ab$$

satisfies $\|\mathbf{m}\|_{mb} \leq 1$. In fact, this example is prototypical in the sense that $\|\varphi\|_{mb} < \infty$ if and only if φ has a *decomposition*

$$\varphi : V \times W \to \mathcal{B}(H) : (v, w) \mapsto \psi_1(v)\psi_2(w),$$

where K is a second Hilbert space, and

$$\psi_1 : V \to \mathcal{B}(K, H) \text{ and } \psi_2 : W \to \mathcal{B}(H, K)$$

are completely bounded mappings. If V and W are C^*-algebras, then this factorization can be refined by applying the single variable decomposition theorem (Theorem 5.3.3) to both ψ_1 and ψ_2.

We define the multiplicative bound in §9.1. We use the convenient 'matrix inner product' notation for this purpose. We define the Haagerup tensor product in §9.2, and we show how it may be used to linearize multiplicatively bounded mappings. We characterize the various Haagerup tensor products of row and column Hilbert spaces in §9.3. Finally, in §9.4

we prove the desired decomposition result for bilinear mappings, and we indicate how it can be easily generalized to the multivariable case.

9.1 MULTIPLICATIVELY BOUNDED MAPPINGS

Let us suppose that V, W, and X are operator spaces, and that we have a bilinear mapping

$$\varphi : V \times W \to X. \tag{9.1.1}$$

We recall that for each $n \in \mathbb{N}$, we have a corresponding bilinear mapping

$$\varphi_{n;n} : M_n(V) \times M_n(W) \to M_{n\times n}(X),$$

where

$$\varphi_{n;n}((\alpha \otimes v_0), (\beta \otimes w_0)) = (\alpha \otimes \beta) \otimes \varphi(v_0, w_0),$$

and that the completely bounded norm of φ is defined by

$$\|\varphi\|_{cb} = \sup\{\|\varphi_{n;n}\| : n \in \mathbb{N}\}$$

(this might be infinite; see (7.1.10)).

On the other hand, we can define another bilinear mapping

$$\varphi_{(n)} : M_n(V) \times M_n(W) \to M_n(X)$$

by incorporating matrix multiplication. If $\alpha, \beta \in M_n$, $v_0 \in V$, and $w_0 \in W$, then we let

$$\varphi_{(n)}((\alpha \otimes v_0), (\beta \otimes w_0)) = \alpha\beta \otimes \varphi(v_0, w_0), \tag{9.1.2}$$

and we then linearly extend in each variable to obtain the desired mapping. For general $v = [v_{i,j}] \in M_n(V)$ and $w = [w_{k,l}] \in M_n(W)$,

$$\varphi_{(n)}(v, w) = \sum_{i,j,k,l} \varepsilon_{i,j}\varepsilon_{k,l} \otimes \varphi(v_{i,j}, w_{k,l}) = \left[\sum_k \varphi(v_{i,k}, w_{k,l})\right]_{i,l\in n}. \tag{9.1.3}$$

We define the *multiplicatively bounded norm* of φ by

$$\|\varphi\|_{mb} = \sup\left\{\left\|\varphi_{(n)}\right\| : n \in \mathbb{N}\right\}, \tag{9.1.4}$$

and we say that φ is *multiplicatively bounded* (respectively, *multiplicatively contractive*) if $\|\varphi\|_{mb} < \infty$ (respectively, $\|\varphi\|_{mb} \leq 1$).

Turning to the prototypical example discussed above, the operator composition mapping

$$\mathbf{m} : \mathcal{B}(K, H) \times \mathcal{B}(H, K) \longrightarrow \mathcal{B}(H) : (a, b) \longmapsto ab \tag{9.1.5}$$

is multiplicatively contractive since we have the commutative diagram

$$\begin{array}{ccc} M_n(\mathcal{B}(K,H)) \times M_n(\mathcal{B}(H,K)) & \xrightarrow{\mathbf{m}_{(n)}} & M_n(\mathcal{B}(H)) \\ \downarrow & & \downarrow \\ \mathcal{B}(K^n, H^n) \times \mathcal{B}(H^n, K^n) & \xrightarrow{\mathbf{m}} & \mathcal{B}(H^n, H^n) \end{array}, \tag{9.1.6}$$

where the bilinear mapping in the bottom row is contractive.

We let $\mathcal{MB}(V \times W, X)$ denote the linear space of all multiplicatively bounded bilinear mappings $\varphi : V \times W \to X$ with the norm $\|\cdot\|_{mb}$. In order to linearize these mappings, we replace the Kronecker product used in the definition of the projective tensor product by the *matrix inner product.* For elementary tensors $v = \alpha \otimes v_0 \in M_{m,r}(V)$ and $w = \beta \otimes w_0 \in M_{r,n}(W)$, the inner matrix product $v \odot w \in M_{m,n}(V \otimes W)$ is given by

$$v \odot w = \alpha\beta \otimes v_0 \otimes w_0, \tag{9.1.7}$$

and for general $v \in M_{m,r}(V)$ and $w \in M_{r,n}(W)$, we let

$$(v \odot w)_{i,j} = \sum_{k=1}^{r} v_{i,k} \otimes w_{k,j} \,. \tag{9.1.8}$$

If $\varphi : V \times W \to X$ is bilinear and $\tilde{\varphi} : V \otimes W \to X$ is its linear extension, then from (9.1.3),

$$\varphi_{(n)}(v, w) = \tilde{\varphi}_n(v \odot w),$$

and thus

$$\|\varphi\|_{mb} = \sup\{\|\tilde{\varphi}_n(v \odot w)\| : v \in M_n(V)_{\|\cdot\|\leq 1}, \\ w \in M_n(W)_{\|\cdot\|\leq 1}, n \in \mathbb{N}\}. \tag{9.1.9}$$

Given $v \in M_p(V)$ and $w \in M_q(W)$, we claim that $v \otimes I_q \in M_{p\times q}(V)$ and $I_p \otimes w \in M_{p\times q}(W)$ satisfy

$$v \otimes w = (v \otimes I_q) \odot (I_p \otimes w); \tag{9.1.10}$$

i.e. we can express the Kronecker product in terms of the matrix inner product. To see this, let us suppose that

$$v = \alpha \otimes v_0 \in M_p \otimes V$$

and

$$w = \beta \otimes w_0 \in M_q \otimes W.$$

Since we may write $\alpha \otimes \beta$ as a matrix product,

$$\alpha \otimes \beta = (\alpha \otimes I_q)(I_p \otimes \beta),$$

we have from (9.1.7)

$$\begin{aligned} v \otimes w &= (\alpha \otimes \beta) \otimes (v_0 \otimes w_0) \\ &= ((\alpha \otimes I_q) \otimes v_0) \odot ((I_p \otimes \beta) \otimes w_0) \\ &= (v \otimes I_q) \odot (I_p \otimes w). \end{aligned}$$

It follows from (9.1.10) and (9.1.9) that if $\varphi : V \times W \to X$ is bilinear, then

$$\|\varphi\|_{cb} \leq \|\varphi\|_{mb} \tag{9.1.11}$$

since

$$\|\varphi_{p;q}(v, w)\| = \|\tilde{\varphi}_{p\times q}((v \otimes I_n) \odot (I_n \otimes w))\|$$

$$\leq \|\varphi\|_{mb} \|v\| \|w\|.$$

We conclude this section by recording some of the elementary properties of the matrix inner product. If we let $V = \mathbb{C}$, and we choose $\alpha \in M_{m,n}(\mathbb{C})$ and $w \in M_{n,p}(W)$, then

$$\alpha \odot w = \alpha w \tag{9.1.12}$$

since, in particular, if $w = \gamma \otimes w_0$ we have from (9.1.7) that

$$\alpha \odot w = \alpha\gamma \otimes w_0.$$

Similarly, if $v \in M_{n,p}(V)$ and $\beta \in M_{p,q}$, then

$$v \odot \beta = v\beta.$$

Finally, if $\alpha \in M_{m,n}$ and $\beta \in M_{n,p}$, then the matrix inner product is just the usual matrix product:

$$\alpha \odot \beta = \alpha\beta = \mathbf{m}(\alpha, \beta). \tag{9.1.13}$$

It is easy to verify that for any $v \in M_{m,r}(V)$, $w \in M_{r,s}(W)$, and $x \in M_{s,n}(X)$,

$$(v \odot w) \odot x = v \odot (w \odot x). \tag{9.1.14}$$

In particular, if $\alpha \in M_{r,s}$, then

$$(v\alpha) \odot x = v \odot (\alpha x). \tag{9.1.15}$$

Given $\alpha \in M_{g,m}$, $\beta \in M_{n,h}$, $v \in M_{m,r}(V)$, and $w \in M_{r,n}(W)$,

$$\alpha(v \odot w)\beta = (\alpha v) \odot (w\beta). \tag{9.1.16}$$

If for any $v \in \mathbb{M}_{m,r}(V)$, $v' \in \mathbb{M}_{n,s}(V)$, $w \in \mathbb{M}_{r,m}(W)$, and $w' \in \mathbb{M}_{s,n}(W)$ we let $v'' = v \oplus v'$ and $w'' = w \oplus w'$, then

$$v'' \odot w'' = (v \oplus v') \odot (w \oplus w') \tag{9.1.17}$$

since

$$((v \oplus v') \odot (w \oplus w'))_{i,j} = \begin{cases} \sum_{1 \leq k \leq r} v_{i,k} \otimes w_{k,j} & \text{if } 1 \leq i,j \leq m, \\ \sum_{r+1 \leq k \leq r+s} v'_{i,k} \otimes w'_{k,j} & \text{if } m+1 \leq i,j \leq m+n, \\ 0 & \text{for all other } i,j. \end{cases}$$

Lemma 9.1.1 *Given linear spaces V and W and $u \in \mathbb{M}_n(V \otimes W)$, there exist $r \in \mathbb{N}$ and elements $v \in \mathbb{M}_{n,r}(V)$ and $w \in \mathbb{M}_{r,n}(W)$, for which*

$$u = v \odot w = \left[\sum_k v_{i,k} \otimes w_{k,j} \right].$$

Proof It suffices to show that the set

$$U_n \subseteq \mathbb{M}_n(V \otimes W) = \mathbb{M}_n \otimes V \otimes W$$

of elements of the form $u = v \odot w$ with v and w as above contains the elementary tensors $\varepsilon_{i,j}^{[n]} \otimes v \otimes w$ with $v \in V$ and $w \in W$, and that U_n is closed under addition. We have from (9.1.7) that

$$\varepsilon_{i,j}^{[n]} \otimes v \otimes w = (\varepsilon_{i,1}^{[n,1]} \otimes v) \odot (\varepsilon_{1,j}^{[1,n]} \otimes w) \in U_n.$$

On the other hand, given $u_i = v_i \otimes w_i$ $(i = 1, 2)$,

$$u_1 + u_2 = [v_1 \ v_2] \odot \begin{bmatrix} w_1 \\ w_2 \end{bmatrix} \in U_n. \tag{9.1.18}$$

□

9.2 THE TENSOR PRODUCT AND ITS ELEMENTARY PROPERTIES

Given operator spaces V and W and an element $u \in \mathbb{M}_n(V \otimes W)$, we define

$$\|u\|_h = \inf\{\|v\|\|w\| : u = v \odot w, \quad v \in M_{n,r}(V), w \in M_{r,n}(W), r \in \mathbb{N}\}. \tag{9.2.1}$$

We note from Lemma 9.1.1 that the set in brackets is non-empty.

Theorem 9.2.1 *For any operator spaces V and W, $\|\cdot\|_h$ is an operator space matrix norm on $V \otimes W$, and for any $u \in \mathbb{M}_n(V \otimes W)$,*

$$\|u\|_\vee \leq \|u\|_h \leq \|u\|_\wedge . \tag{9.2.2}$$

Proof Let us suppose that $u \in \mathbb{M}_m(V \otimes W)$, $u' \in \mathbb{M}_n(V \otimes W)$, and $\varepsilon > 0$. By definition, we may find $v \in M_{m,r}(V)$ and $w \in M_{r,m}(W)$ such that $u = v \odot w$, $\|w\| = 1$, and $\|v\| \leq \|u\|_h + \varepsilon$. Similarly, we let $u' = v' \odot w'$, where $v' \in M_{n,s}(V)$, $w' \in M_{s,n}(W)$, $\|w'\| = 1$, and $\|v'\| \leq \|u'\|_h + \varepsilon$. If $v'' = v \oplus v'$ and $w'' = w \oplus w'$, then from (9.1.17),

$$u \oplus u' = v'' \odot w'',$$

and thus

$$\|u \oplus u'\|_h \leq \|v''\|\|w''\| = \max\{\|v\|, \|v'\|\} \leq \max\{\|u\|_h, \|u'\|_h\} + \varepsilon.$$

Since $\varepsilon > 0$ is arbitrary, we obtain condition **M1**$'$ of Proposition 2.3.6:

$$\|u \oplus u'\|_h \leq \max\{\|u\|_h, \|u'\|_h\}.$$

For any $u \in \mathbb{M}_n(V \otimes W)$ and $\varepsilon > 0$, we may choose $v \in M_{n,r}(V)$ and $w \in M_{r,n}(W)$ such that $u = v \odot w$ and

$$\|v\| \, \|w\| < \|u\|_h + \varepsilon .$$

If $\alpha, \beta \in M_n$, then

$$\alpha u \beta = (\alpha v) \odot (w \beta);$$

and hence

$$\|\alpha u \beta\|_h \leq \|\alpha v\|\|w\beta\| \leq \|\alpha\|\|v\|\|w\|\|\beta\| < \|\alpha\|(\|u\|_h + \varepsilon)\|\beta\|.$$

Since $\varepsilon > 0$ is arbitrary, we obtain condition **M2**:

$$\|\alpha u \beta\|_h \leq \|\alpha\| \|u\|_h \|\beta\|.$$

If $u \in \mathbb{M}_n(V \otimes W)$, let us suppose that $\varepsilon > 0$, and that

$$u = v \odot w,$$

where $v = [v_{i,k}] \in M_{n,r}(V)$ and $w = [w_{k,j}] \in M_{r,n}(W)$ satisfy

$$\|v\| \, \|w\| \leq \|u\|_h + \varepsilon.$$

Let us suppose that $f \in M_p(V^*)$ and $g \in M_q(W^*)$ are complete contractions. We have

$$(f \otimes g)_n(v \odot w) = \left[\sum_k f(v_{i,k}) \otimes g(w_{k,j}) \right] = [f(v_{i,k}) \otimes I_q] \, [I_p \otimes g(w_{k,j})] \, ,$$

where we are using a product of matrices over $M_p \otimes M_q$, and since matrix multiplication is a contractive bilinear function on $M_p \otimes M_q$,

$$\|(f \otimes g)_n(v \odot w)\| \leq \|f_{n,r}(v)\| \, \|g_{r,n}(w)\| \leq \|v\| \, \|w\| \, .$$

It follows from (8.1.7) that

$$\|u\|_\vee \leq \|v\| \, \|w\| \leq \|u\|_h + \varepsilon,$$

from which the first inequality follows. In particular, the matrix seminorms $\| \cdot \|_h$ are non-degenerate, and from Proposition 2.3.6 they determine an operator space structure on $V \otimes W$.

For any matrices $v \in M_m(V)$ and $w \in M_n(W)$, we have from (9.1.10),

$$v \otimes w = (v \otimes I_n) \odot (I_m \otimes w),$$

and thus by definition,

$$\|v \otimes w\|_h \leq \|v\| \, \|w\| \, .$$

Thus, the bilinear mapping

$$V \times W \to V \otimes_h W : (v, w) \mapsto v \otimes w$$

is completely contractive. Since the linear extension of this mapping is just the identity mapping on $V \otimes W$, the second inequality follows from the first identification in Proposition 7.1.2. □

We let

$$V \otimes_h W = (V \otimes W, \| \cdot \|_h),$$

and we define the *Haagerup tensor product* $V \overset{h}{\otimes} W$ to be the completion of the operator space $V \otimes_h W$. The following is an immediate consequence of (9.1.9).

Proposition 9.2.2 *For any operator spaces V, W, and X, we have a natural isometry*

$$\mathcal{MB}(V \times W, X) \cong \mathcal{CB}(V \overset{h}{\otimes} W, X). \tag{9.2.3}$$

$\square$

For any $n \in \mathbb{N}$, we have the isometric identification

$$\mathcal{MB}(V \times W, M_n(X)) \cong \mathcal{CB}(V \overset{h}{\otimes} W, M_n(X)) = M_n(\mathcal{CB}(V \overset{h}{\otimes} W, X)).$$

In particular, if we use the definition

$$M_n(\mathcal{MB}(V \times W, X)) = \mathcal{MB}(V \times W, M_n(X)),$$

that is to say, we let the first space have the given norm on the second space, then (9.2.3) is a complete isometry.

For any $u \in V \otimes W$, we have by definition that $\|u\|_h < 1$ if and only if there exist $v = [v_1 \dots v_r] \in M_{1,r}(V)$ and

$$w = \begin{bmatrix} w_1 \\ \vdots \\ w_r \end{bmatrix} \in M_{r,1}(W)$$

such that

$$u = v \odot w = \sum_{j=1}^{r} v_j \otimes w_j \tag{9.2.4}$$

and $\|v\|, \|w\| < 1$. The following refinement of this statement is in some respects analogous to (4.1.2).

Lemma 9.2.3 *Suppose that V and W are operator spaces. For any u in $V \otimes_h W$ with $\|u\|_h < 1$, there exists a representation*

$$u = v \odot w = \sum_{j=1}^{r} v_j \otimes w_j$$

with $\|v\| \, \|w\| < 1$ for which $v_1, \dots, v_r$ are linearly independent in V, and $w_1, \dots, w_r$ are linearly independent in W.

Proof Given a representation (9.2.4), we have $u \in V_0 \otimes W_0$, where V_0 and W_0 are the linear subspaces of V and W spanned by the vectors v_j and w_j $(1 \le j \le r)$, respectively. We may assume that (V_0, W_0) is a minimal pair of subspaces for which u has a representation (9.2.4) with $v_j \in V_0$, $w_j \in W_0$, and $\|v\|$, $\|w\| < 1$. If we let $p = \dim V_0$ and $q = \dim W_0$, then we can assume that $q \le p$ (if $q > p$, we can interchange the roles of V_0 and W_0 in the argument). Owing to the minimality of the pair, v_j and w_j span V_0 and W_0, respectively. Permuting the vectors if necessary, we may assume that $\{w_1, \dots, w_n\}$ $(n \le q)$ is a basis for W_0. We let

$$v = [v_1 \dots v_r],$$

$$w = \begin{bmatrix} w_1 \\ \vdots \\ w_r \end{bmatrix},$$

and

$$w' = \begin{bmatrix} w_1 \\ \vdots \\ w_n \end{bmatrix}.$$

We may choose scalars $\beta_{i,j}$ with

$$w_i = \sum_{j=1}^{n} \beta_{i,j} w_j$$

for $1 \le i \le r$, or if we let $\beta = [\beta_{i,j}] \in M_{r,n}$, then

$$w = \beta w'.$$

Since the vectors w_i $(1 \le i \le n)$ are linearly independent, $\beta_{i,j} = \delta_{i,j}$ for $i, j \le n$, and in particular, $\beta : \mathbb{C}^n \to \mathbb{C}^r$ is an injective linear mapping.

We let $\beta = \tau |\beta|$ be the polar decomposition of β, where $|\beta| \in M_n$ is invertible, and

$$\tau \in M_{r,n} = \mathcal{B}(\mathbb{C}^n, \mathbb{C}^r)$$

is an isometry, i.e. $\tau^* \tau = I_n$. It follows that

$$u = v \odot \tau |\beta| w' = v\tau \odot |\beta| w',$$

and $u = \tilde{v} \odot \tilde{w}$, where

$$\tilde{v} = v\tau \in M_{1,n}(V)$$

and

$$\tilde{w} = |\beta| w' \in M_{n,1}(W).$$

We have

$$\|\tilde{v}\| = \|v\tau\| \le \|v\|$$

and

$$\|\tilde{w}\| = \||\beta| w'\| = \|\tau^* \beta w'\| = \|\tau^* w\| \le \|w\|.$$

Since $|\beta|$ is non-singular, the entries $\tilde{w}_j$ of $\tilde{w}$ are linearly independent. Owing to the minimality of V_0 and W_0, we again infer that the elements $\tilde{v}_j$ $(1 \le j \le n)$ span V_0, and since dim $V_0 \ge q \ge n$, they must also be linearly independent.

Replacing r, v, and w by $n, \tilde{v}$ and $\tilde{w}$, respectively, we obtain the desired formula. □

The next result often enables one to avoid tedious matrix calculations.

Lemma 9.2.4 *For any operator spaces V and W, we have a natural isometry*

$$M_{m,n}(V \overset{h}{\otimes} W) \cong M_{m,1}(V) \overset{h}{\otimes} M_{1,n}(W). \tag{9.2.5}$$

Proof We have natural isometries

$$M_{m,r}(V) \cong M_{1,r}(M_{m,1}(V))$$

and

$$M_{r,n}(W) \cong M_{r,1}(M_{1,n}(W)).$$

Given $v = [v_{i,j}] \in M_{m,r}(V)$ and $w = [w_{j,k}] \in M_{r,n}(W)$, the corresponding matrices are given by

$$\tilde{v} = [\tilde{v}_1 \ldots \tilde{v}_r] \text{ and } \tilde{w} = \begin{bmatrix} \tilde{w}_1 \\ \vdots \\ \tilde{w}_r \end{bmatrix},$$

where

$$\tilde{v}_j = \begin{bmatrix} v_{1,j} \\ \vdots \\ v_{m,j} \end{bmatrix} \in M_{m,1}(V) \text{ and } \tilde{w}_j = [w_{j,1}, \ldots, w_{j,n}] \in M_{1,n}(W).$$

We have

$$\tilde{v} \odot \tilde{w} = \sum_k \tilde{v}_k \otimes \tilde{w}_k,$$

and thus

$$(\tilde{v} \odot \tilde{w})_{i,j} = \sum_k v_{i,k} \otimes w_{k,j} = (v \odot w)_{i,j}.$$

We conclude that if $u \in \mathbb{M}_{m,n}(V \otimes W)$, then

$$\begin{aligned} \|u\|_{M_{m,n}(V \overset{h}{\otimes} W)} &= \inf\{\|v\|\,\|w\| : u = v \odot w,\ v \in M_{m,r}(V), w \in M_{r,n}(W)\} \\ &= \inf\{\|\tilde{v}\|\,\|\tilde{w}\| : u = \tilde{v} \odot \tilde{w}, \\ &\qquad \tilde{v} \in M_{1,r}(M_{m,1}(V)),\ \tilde{w} \in M_{r,1}(M_{1,n}(W)\} \\ &= \|u\|_{M_{m,1}(V) \overset{h}{\otimes} M_{1,n}(W)}. \end{aligned}$$

□

We next consider the properties of the Haagerup tensor product. We note that if $\varphi : V \to V'$ and $\psi : W \to W'$ are linear mappings, and $u = v \odot w \in M_n(V \otimes W)$, where $v \in M_{n,r}(V)$ and $w \in M_{r,n}(W)$, then

$$\begin{aligned} (\varphi \otimes \psi)_n(v \odot w) &= \left[\sum_k \varphi(v_{i,k}) \otimes \psi(w_{k,j})\right]_{i,j \in n} \\ &= \varphi_{n,r}(v) \odot \psi_{r,n}(w). \end{aligned} \tag{9.2.6}$$

Proposition 9.2.5 *Let V, V', W, and W' be operator spaces. For any complete contractions $\varphi : V \to V'$ and $\psi : W \to W'$, the corresponding mapping*

$$\varphi \otimes \psi : V \overset{h}{\otimes} W \to V' \overset{h}{\otimes} W' \tag{9.2.7}$$

is a complete contraction. If φ and ψ are complete isometries (respectively, complete quotient mappings), then the same is true for $\varphi \otimes \psi$.

Proof We have from the identification (9.2.5) the commutative diagram

$$\begin{array}{ccc} M_n(V \otimes_h W) & \xrightarrow{(\varphi\otimes\psi)_n} & M_n(V' \otimes_h W') \\ \downarrow & & \downarrow \\ M_{n,1}(V) \otimes_h M_{1,n}(W) & \xrightarrow{\varphi_{n,1}\otimes\psi_{1,n}} & M_{n,1}(V') \otimes_h M_{1,n}(W') \end{array}.$$

We note that if φ and ψ are completely contractive, isometric, or complete quotient mappings, then that is also the case for the mappings $\varphi_{n,1}$ and $\psi_{1,n}$. Thus, it suffices to show that the mapping $\varphi \otimes \psi$ is contractive, isometric, or a quotient map, if φ and ψ satisfy the corresponding operator space condition.

For any $u \in V \otimes W$, and representation $u = v \odot w$ with $v \in M_{1,r}(V)$ and $w \in M_{r,1}(W)$, we have from (9.2.6) that

$$\begin{aligned} \|(\varphi \otimes \psi)(u)\|_h &= \|\varphi_{1,r}(v) \odot \psi_{r,1}(w)\|_h \\ &\leq \|\varphi_{1,r}(v)\|\|\psi_{r,1}(w)\| \\ &\leq \|\varphi\|_{cb}\|\psi\|_{cb}\|v\|\|w\|. \end{aligned}$$

Taking the infimum for all such representations $u = v \odot w$,

$$\|\varphi \otimes \psi(u)\|_h \leq \|\varphi\|_{cb}\|\psi\|_{cb}\|u\|_h.$$

This shows that

$$\|\varphi \otimes \psi\| \leq \|\varphi\|_{cb}\|\psi\|_{cb} \leq 1.$$

If φ and ψ are complete isometries, then we may use them to identify V and W with operator subspaces of V' and W', respectively. To show that the inclusion is isometric, it suffices to show that if $u \in \mathbb{M}_n(V \otimes W)$ has norm less than 1 in $M_n(V' \otimes_h W')$, then it has norm less than 1 in $M_n(V \otimes_h W)$.

From Lemma 9.2.3 we may suppose that $u = v' \odot w'$, where

$$v' = [v'_1 \ldots v'_r] \in M_{1,r}(V')$$

and

$$w' = \begin{bmatrix} w'_1 \\ \vdots \\ w'_r \end{bmatrix} \in M_{r,1}(W')$$

satisfy $\|v'\| < 1$ and $\|w'\| < 1$, and that the entries v'_j and w'_j are linearly independent in V' and W', respectively. We claim that $u \in V \otimes W$ implies that $v'_j \in V$ and $w'_j \in W$.

We let V'_0 and W'_0 be the linear spans of the vectors $\{v'_j\}$ and $\{w'_j\}$ in V' and W', respectively. The sets $\{v'_j\}$ and $\{w'_j\}$ are vector bases for these subspaces, and we may select biorthogonal bases f_i and g_i in the dual spaces $V_0'^*$ and $W_0'^*$, i.e. linear functionals satisfying $f_i(v'_j) = \delta_{i,j}$ and $g_i(w'_j) = \delta_{i,j}$, respectively. From the Hahn–Banach theorem, we may extend these to contractions in V'^* and W'^*, which we again denote by f_i and g_i. By hypothesis, $u \in V \otimes W$, and thus if $u = \sum v_k \otimes w_k$, where $v_k \in V$ and $w_k \in W$, then

$$w'_i = (f_i \otimes id_{W'})(u) = \sum f_i(v_k) w_k \in W$$

and similarly,

$$v'_i = id_{V'} \otimes g_i)(u) = \sum v_k g_i(w_k) \in V.$$

Thus, the norm of u in $V \otimes_h W$ is less than or equal to the norm in $V' \otimes_h W'$, and the mapping $\varphi \otimes \psi$ is isometric.

Finally, given $u' \in V' \otimes_h W'$ with $\|u'\|_h < 1$, there exist $v' \in M_{1,r}(V')$ and $w' \in M_{r,1}(W')$ such that

$$u = v' \odot w'$$

and $\|v'\|, \|w'\| < 1$. If φ and ψ are complete quotient mappings, then there exist $v \in M_{n,r}(V)$ with $\|v\| < 1$ and $w \in M_{r,n}(W)$ with $\|w\| < 1$ such that

$$v' = (\varphi)_{1,r}(v) \quad \text{and} \quad w' = (\psi)_{r,1}(w).$$

It follows that $u = v \odot w \in V \otimes_h W$ satisfies $\|u\|_h < 1$ and $\varphi \otimes \psi(u) = u'$. Thus, $\varphi \otimes \psi$ is a quotient mapping. □

Our conclusion that the Haagerup tensor product preserves both quotient mappings and complete isometries, or in other words that it is both 'projective' and 'injective', is related to the remarkable fact that it is 'self-dual' (see Theorem 9.4.7). Another surprising property of the Haagerup tensor norm is that for any u in the algebraic tensor product $V \otimes W$, the infimum in the definition of $\|u\|_h$ (9.2.1) is actually attained by matrices v and w. We note that the argument does not use our standing assumption that V and W are complete.

Proposition 9.2.6 *Suppose that V and W are operator spaces and that $n \in \mathbb{N}$. Then for any $u \in M_n(V \otimes_h W)$ there exist $v \in M_{n,r}(V)$ and $w \in M_{r,n}(W)$ such that*

$$u = v \odot w$$

and

$$\|u\|_h = \|v\| \|w\| .$$

Proof We may suppose that $u \in V \otimes_h W$ since the general case then follows from Lemma 9.2.4. Since u lies in the algebraic tensor product, it lies in $V_0 \otimes W_0$ for finite-dimensional subspaces $V_0 \subseteq V$ and $W_0 \subseteq W$. From Theorem 9.2.5 we may identify $V_0 \otimes_h W_0$ with an operator subspace of $V \otimes_h W$. Thus, replacing V and W by V_0 and W_0, we may initially suppose that V and W are finite-dimensional. We let $r = \min\{\dim V, \dim W\}$.

For any $m \in \mathbb{N}$, it follows from Lemma 9.2.3 that there exist

$$v^m = [v_1^m, \ldots, v_r^m] \in M_{1,r}(V) \quad \text{and} \quad w^m = \begin{bmatrix} w_1^m \\ \vdots \\ w_r^m \end{bmatrix} \in M_{r,1}(W)$$

such that

$$u = v^m \odot w^m = \sum v_j^m \otimes w_j^m$$

and

$$\|v^m\|\|w^m\| < \|u\|_h + 1/m.$$

Without loss of generality, we may assume that

$$\|v^m\| = \|w^m\| < (\|u\|_h + 1/m)^{1/2}.$$

Since v^m and w^m are bounded sequences in the finite-dimensional spaces $M_{1,r}(V)$ and $M_{r,1}(W)$, respectively, it follows from the Bolzano–Weierstrass theorem that there are subsequences v^{m_k} and w^{m_k} converging in the norm topology to $\tilde{v} \in M_{n,r}(V)$ and $\tilde{w} \in M_{r,n}(W)$, respectively. It follows by continuity that $u = \tilde{v} \odot \tilde{w}$ and

$$\|u\|_h = \|\tilde{v}\|\|\tilde{w}\|.$$

□

The following proposition shows that the Haagerup tensor product is associative.

Proposition 9.2.7 *Given operator spaces V, W, and X, we have the complete isometry*

$$(V \overset{h}{\otimes} W) \overset{h}{\otimes} X \cong V \overset{h}{\otimes} (W \overset{h}{\otimes} X).$$

Proof It suffices to prove this for the algebraic tensor products. Given an element $u \in M_n(V \otimes_h W) \otimes_h X$,

$$\begin{aligned}\|u\|_h &= \inf\{\|t\|_h\|x\| : u = t \odot x : t \in M_{n,q}(V \otimes_h W), \\ &\qquad\qquad x \in M_{q,n}(X), q \in \mathbb{N}\} \\ &= \inf\{\|v\|\|w\|\|x\| : u = (v \odot w) \odot x,\ v \in M_{n,p}(V), \\ &\qquad\qquad w \in M_{p,q}(W),\ x \in M_{q,n}(X),\ p, q \in \mathbb{N}\}.\end{aligned}$$

The usual linear space identification

$$(V \otimes W) \otimes X \cong V \otimes (W \otimes X)$$

carries $(v \odot w) \odot x$ onto $v \odot (w \odot x)$. Thus, it is evident that u has the same norm in $M_n(V \otimes_h (W \otimes_h X))$. □

Identifying the above tensor products with $V \otimes W \otimes X$, we obtain the *multiple Haagerup tensor product* $V \otimes_h W \otimes_h X$ and its completion $V \overset{h}{\otimes} W \overset{h}{\otimes} X$. For the former,

$$\|u\|_h = \inf \{\|v\|\|w\|\|z\| : u = v \odot w \odot z\}, \tag{9.2.8}$$

where the infimum is taken over arbitrary decompositions using finite matrices.

Finally, we remark that unlike the projective and injective tensor products, the Haagerup tensor product is not 'commutative'. Given operator spaces V and W, the linear isomorphism $V \otimes_h W \to W \otimes_h V$ determined by $v \otimes w \mapsto w \otimes v$ need not be a complete isometry (see Proposition 9.3.4).

9.3 SOME TENSOR PRODUCT COMPUTATIONS

Among the novelties of operator space theory are the remarkable manipulations that can be performed with the Haagerup tensor product. Nothing like this seems to exist in Banach space theory. Many of these calculations involve row and column Hilbert spaces, and we shall consider some examples in this section. As in §3.4, we identify the dual K^* of a Hilbert space K with the conjugate space $\overline{K}$, and the dual operator space $(K_c)^*$ with $\overline{K}_r$.

Proposition 9.3.1 *Given an operator space V and a Hilbert space H, we have the natural complete isometries*

$$H_c \overset{h}{\otimes} V \cong H_c \check{\otimes} V \tag{9.3.1}$$

and

$$V \overset{h}{\otimes} H_r \cong V \check{\otimes} H_r. \tag{9.3.2}$$

Proof We only need to prove (9.3.1) for column Hilbert spaces since the argument for (9.3.2) is essentially the same. From (9.2.2) it suffices to show that for any $u \in \mathbb{M}_n(H_c \otimes V)$ $(n \in \mathbb{N})$,

$$\|u\|_h \leq \|u\|_\vee .$$

If H is a finite-dimensional Hilbert space with $\dim H = p$, we may identify H_c with $\mathbb{C}^p_c = M_{p,1}$, and we have the completely isometric identification

$$H_c \check{\otimes} V = M_{p,1} \check{\otimes} V = M_{p,1}(V).$$

For any

$$v \in M_n(H_c \check{\otimes} V) = M_{np,n}(V),$$

we have $v = I_{np} v = I_{np} \odot v$ (see (9.1.12)), and thus

$$\|v\|_h \leq \|I_{np}\| \, \|v\| = \|v\|_\vee .$$

Turning to the infinite-dimensional case, any element of $H_c \otimes V$ lies in $H_0 \otimes V$ for some finite-dimensional subspace H_0. Thus, any element $u \in \mathbb{M}_n(H_c \otimes V)$ lies in $\mathbb{M}_n((H_0)_c \otimes V)$ with H_0 finite-dimensional. The result follows from the commutative diagram

$$\begin{array}{ccc} (H_0)_c \otimes_\vee V & \xrightarrow{\cong} & (H_0)_c \otimes_h V \\ \downarrow & & \downarrow \\ H \otimes_\vee V & \longrightarrow & H \otimes_h V \end{array},$$

in which the vertical mappings are completely isometric embeddings since both the injective and Haagerup tensor products are injective. $\square$

Interchanging the positions of V and H_c (respectively, V and H_r) in the Haagerup tensor product in Proposition 9.3.1 leads to a very different result.

Proposition 9.3.2 *Given an operator space V and a Hilbert space H, we have the natural complete isometries*

$$V \overset{h}{\otimes} H_c \cong V \hat{\otimes} H_c \tag{9.3.3}$$

and

$$H_r \overset{h}{\otimes} V \cong H_r \hat{\otimes} V\,. \tag{9.3.4}$$

Proof We only need to prove (9.3.3) since the proof for (9.3.4) is similar. Owing to (9.2.2), it suffices to show that

$$\|u\|_\wedge \leq \|u\|_h$$

for all $u \in \mathbb{M}_n(V \otimes H_c)$.

Given $u \in M_n(V \otimes_h H_c)$ with $\|u\|_h < 1$, we may choose an $m \in \mathbb{N}$, and $v = [v_{i,j}] \in M_{n,m}(V)$ and $\xi = [\xi_{k,l}] \in M_{m,n}(H_c)$ such that $u = v \odot \xi$ and $\|v\|, \|\xi\|_c < 1$. We may suppose that

$$\xi = \sum_{l=1}^{r} \alpha_l \otimes e_l,$$

where $e_l \in H$ are orthonormal, and $\alpha_l = [\alpha_l(i,j)] \in M_{m,n}$, are such that

$$\alpha = \begin{bmatrix} \alpha_1 \\ \vdots \\ \alpha_r \end{bmatrix} \in M_{rm,n}$$

satisfies $\|\alpha\| = \|\xi\|_c$ (see (3.4.4)). If we let $e = [e_1 \ldots e_r] \in M_{1,n}(H_c)$, then

$$u = v \odot \xi = \left[\sum_k v_{i,k} \otimes \xi_{k,j}\right]_{i,j} = \left[\sum_{k,l} v_{i,k} \otimes \alpha_l(k,j) e_l\right]$$

$$= \sum_l (v \otimes e_l)\alpha_l = (v \otimes e)\alpha,$$

and since $\|e\|_c = 1$ (see (3.4.8)),

$$\|u\|_\wedge \leq \|v\| \, \|e\|_c \, \|\alpha\| < 1,$$

and we are done. □

Proposition 9.3.3 *Given an operator space V and Hilbert spaces H and K, we have the complete isometry*

$$((K_c)^* \overset{h}{\otimes} V \overset{h}{\otimes} H_c)^* \cong \mathcal{CB}(V, \mathcal{B}(H,K)) . \tag{9.3.5}$$

Proof From the associativity and commutativity of the projective tensor product and Proposition 9.3.2, we have the complete isometries

$$\begin{aligned}(K_c)^* \overset{h}{\otimes} V \overset{h}{\otimes} H_c &= \overline{K}_r \overset{h}{\otimes} V \overset{h}{\otimes} H_c \\ &\cong \overline{K}_r \,\widehat{\otimes}\, V \,\widehat{\otimes}\, H_c \\ &\cong V \,\widehat{\otimes}\, (H_c \,\widehat{\otimes}\, (K_c)^*) .\end{aligned}$$

Taking the operator dual, we obtain the complete isometries

$$\begin{aligned}((K_c)^* \overset{h}{\otimes} V \overset{h}{\otimes} H_c)^* &\cong (V \,\widehat{\otimes}\, (H_c \,\widehat{\otimes}\, (K_c)^*))^* \\ &\cong \mathcal{CB}(V, \ (H_c \,\widehat{\otimes}\, (K_c)^*)^*) \\ &\cong \mathcal{CB}(V, \ \mathcal{CB}(H_c, \ (K_c)^{**})) \\ &\cong \mathcal{CB}(V, \ \mathcal{B}(H, \ K)),\end{aligned}$$

where we have used Theorem 3.4.1 in the last step. This completes the proof. □

Proposition 9.3.4 *For any Hilbert spaces H and K, we have the natural complete isometries*

$$H_c \overset{h}{\otimes} (K_c)^* \cong \mathcal{K}(K, H) \tag{9.3.6}$$

and

$$(K_c)^* \overset{h}{\otimes} H_c \cong \mathcal{T}(K, H) . \tag{9.3.7}$$

Proof From Proposition 9.3.1 we have the complete isometries

$$H_c \overset{h}{\otimes} (K_c)^* \cong H_c \overset{h}{\otimes} \overline{K}_r \cong H_c \check{\otimes} \overline{K}_r \cong H_c \check{\otimes} (K_c)^* .$$

Hence to prove (9.3.6), it suffices to show that we have a complete isometry

$$\theta : H_c \check{\otimes} (K_c)^* \cong \mathcal{K}(K, H).$$

We proved that this is an isometry in Proposition 8.2.1. If we apply this to the Hilbert spaces H_c^n and K_c^n, we have the isometries

$$M_n(H_c \check{\otimes} (K_c)^*) \cong M_{n,1}(H_c) \check{\otimes} M_{1,n}(\overline{K}_r)$$

$$\cong H_c^n \check{\otimes} \overline{K}_r^n \cong H_c^n \check{\otimes} (K_c^n)^* \cong \mathcal{K}(K^n, H^n)$$

(see (3.4.3)). If one carefully traces what happens to elementary tensors, then one finds that the isometry of the first space onto the last is given by θ_n, and thus θ is indeed a complete isometry. □

In particular, if $H = \mathbb{C}^m$ and $K = \mathbb{C}^n$, we have

$$\mathbb{C}_c^m \overset{h}{\otimes} (\mathbb{C}_c^n)^* \cong \mathbb{C}_c^m \check{\otimes} (\mathbb{C}_c^n)^* \cong M_{m,n}$$

and

$$(\mathbb{C}_c^n)^* \overset{h}{\otimes} \mathbb{C}_c^m \cong (\mathbb{C}_c^n)^* \hat{\otimes} \mathbb{C}_c^m \cong T_{n,m}\,.$$

There are similar computations for row Hilbert spaces, which we leave to the reader.

Proposition 9.3.5 *Given Hilbert spaces H and K, we have the complete isometries*

$$H_c \hat{\otimes} K_c \cong H_c \overset{h}{\otimes} K_c \cong H_c \check{\otimes} K_c \cong (H \otimes K)_c \tag{9.3.8}$$

and

$$H_r \hat{\otimes} K_r \cong H_r \overset{h}{\otimes} K_r \cong H_r \check{\otimes} K_r \cong (H \otimes K)_r. \tag{9.3.9}$$

Proof It follows from Propositions 9.3.3 and 9.3.4 that we only need to prove the last complete isometries in (9.3.8) and (9.3.9). Choose an orthonormal basis $\{e_s\}_{s\in\mathfrak{s}}$ for H and an orthonormal basis $\{f_t\}_{t\in\mathfrak{t}}$ for K. If we identify H_c and K_c with $M_{\mathfrak{s},1}$ and $M_{\mathfrak{t},1}$, respectively, then we obtain the complete isometries

$$H_c \check{\otimes} K_c \cong M_{\mathfrak{s},1} \check{\otimes} M_{\mathfrak{t},1} \cong M_{\mathfrak{s},1}(M_{\mathfrak{t},1})) \cong M_{\mathfrak{s}\times\mathfrak{t},1} \cong (H \otimes K)_c\,.$$

Similarly, if we identify H_r and K_r with $M_{1,\mathfrak{s}}$ and $M_{1,\mathfrak{t}}$, respectively, then we obtain the complete isometries

$$H_r \check{\otimes} K_r \cong M_{1,\mathfrak{s}} \check{\otimes} M_{1,\mathfrak{t}} \cong M_{1,\mathfrak{s}}(M_{1,\mathfrak{t}})) \cong M_{1,\mathfrak{s}\times\mathfrak{t}} \cong (H \otimes K)_r\,.$$ □

It is clear from the above calculations that the Haagerup tensor product is not commutative, and in fact given operator spaces V and W, there need not exist a complete isometry (or even an isometry) of $V \overset{h}{\otimes} W$ onto $W \overset{h}{\otimes} V$.

9.4 MULTILINEAR DECOMPOSITIONS

We begin by proving a bilinear analogue of the GNS theorem.

Theorem 9.4.1 *If $\mathcal{A}$ and $\mathcal{B}$ are two unital C^*-algebras, then a linear functional*

$$F : \mathcal{A} \overset{h}{\otimes} \mathcal{B} \to \mathbb{C}$$

is bounded if and only if there exist unital $*$-representations $\pi_{\mathcal{A}} : \mathcal{A} \to \mathcal{B}(H)$ and $\pi_{\mathcal{B}} : \mathcal{B} \to \mathcal{B}(K)$, vectors $\xi \in H$ and $\eta \in K$, and a bounded linear operator $T : K \to H$ such that

$$F(a \otimes b) = \langle \pi_{\mathcal{A}}(a) T \pi_{\mathcal{B}}(b) \eta \mid \xi \rangle . \tag{9.4.1}$$

In this case, we can choose ξ, η, and T for which

$$\|F\| = \|\xi\| \|\eta\| \|T\| .$$

Proof If $F : \mathcal{A} \otimes \mathcal{B} \to \mathbb{C}$ is a linear functional satisfying (9.4.1), then for any

$$a = [a_1, \ldots, a_r] \in M_{1,r}(\mathcal{A}) \quad \text{and} \quad b = \begin{bmatrix} b_1 \\ \vdots \\ b_r \end{bmatrix} \in M_{r,1}(\mathcal{B}),$$

we have

$$\begin{aligned} |F(a \odot b)| &= \left| \sum F(a_i \otimes b_i) \right| \\ &= \left| \sum \langle \pi_{\mathcal{A}}(a_i) T \pi_{\mathcal{B}}(b_i) \eta \mid \xi \rangle \right| \\ &= |\langle (\pi_{\mathcal{A}})_{1,r}(a)(I_r \otimes T)(\pi_{\mathcal{B}})_{r,1}(b) \eta \mid \xi \rangle| \\ &\leq \|a\| \, \|T\| \, \|b\| \, \|\eta\| \, \|\xi\| . \end{aligned}$$

It follows that $F : \mathcal{A} \overset{h}{\otimes} \mathcal{B} \to \mathbb{C}$ is bounded with

$$\|F\| \leq \|\xi\| \|T\| \|\eta\| .$$

Conversely, it suffices to show that if $F : \mathcal{A} \overset{h}{\otimes} \mathcal{B} \to \mathbb{C}$ is a linear functional with $\|F\| = 1$, then we have a decomposition (9.4.1) with T, η, and ξ contractive. We claim that there exist states $p_0 \in \mathcal{A}^*$ and $q_0 \in \mathcal{B}^*$ such that

$$|F(a \otimes b)| \leq p_0(aa^*)^{1/2} q_0(b^* b)^{1/2} \tag{9.4.2}$$

for all $a \in \mathcal{A}$ and $b \in \mathcal{B}$. The proof of (9.4.2) is essentially the same as that for Lemma 2.3.2. It is enough to prove that there exist states $p_0 \in \mathcal{A}^*$ and $q_0 \in \mathcal{B}^*$ such that

$$p_0(aa^*) + q_0(b^* b) \geq 2\mathrm{Re}\, F(a \otimes b) \tag{9.4.3}$$

for all $a \in \mathcal{A}$ and $b \in \mathcal{B}$.

If we let $\mathbf{S}(\mathcal{A})$ (respectively, $\mathbf{S}(\mathcal{B})$) denote the set of states on $\mathcal{A}$ (respectively, on $\mathcal{B}$), then $\mathbf{S}(\mathcal{A}) \times \mathbf{S}(\mathcal{B})$ is a weak* compact convex subset of $(\mathcal{A} \oplus \mathcal{B})^*$, where $\mathcal{A} \oplus \mathcal{B}$ is the C^*-algebra direct sum of $\mathcal{A}$ and $\mathcal{B}$. Given matricies

$$a = [a_1, \ldots, a_n] \in M_{1,n}(\mathcal{A}) \quad \text{and} \quad b = \begin{bmatrix} b_1 \\ \vdots \\ b_n \end{bmatrix} \in M_{n,1}(\mathcal{B}),$$

we have a corresponding continuous affine function $e_{a,b}$ on $\mathbf{S}(\mathcal{A}) \times \mathbf{S}(\mathcal{B})$ defined by

$$e_{a,b}(p,q) = \sum p(a_j a_j^*) + q(b_j^* b_j) - 2\mathrm{Re}\, F(a_j \otimes b_j) .$$

It is evident that the collection $\mathcal{E}$ of all such functions $e_{a,b}$ is a convex cone of real-valued continuous affine functions on $\mathbf{S}(\mathcal{A}) \times \mathbf{S}(\mathcal{B})$. Furthermore, fixing a and b, we may choose $p_1 \in \mathbf{S}(\mathcal{A})$ and $q_1 \in \mathbf{S}(\mathcal{B})$ such that

$$p_1\left(\sum a_j a_j^*\right) = \left\|\sum a_j a_j^*\right\| \quad \text{and} \quad q_1\left(\sum b_j^* b_j\right) = \left\|\sum b_j^* b_j\right\|.$$

It follows that

$$\begin{aligned} e_{a,b}(p_1,q_1) &= p_1\left(\sum a_j a_j^*\right) + q_1\left(\sum b_j^* b_j\right) - 2\sum \mathrm{Re}\, F(a_j \otimes b_j) \\ &\geq \left\|\sum a_j a_j^*\right\| + \left\|\sum b_j^* b_j\right\| - 2\left|\sum F(a_j \otimes b_j)\right| \\ &\geq 2\left(\left\|\sum a_j a_j^*\right\|\right)^{1/2}\left(\left\|\sum b_j^* b_j\right\|\right)^{1/2} - 2\left|\sum F(a_j \otimes b_j)\right| \\ &= 2(\|a\|\|b\| - |F(a \odot b)|) \geq 0 . \end{aligned}$$

We conclude from Lemma 2.3.1 that there exist $p_0 \in \mathbf{S}(\mathcal{A})$ and $q_0 \in \mathbf{S}(\mathcal{B})$ such that

$$p_0(aa^*) + q_0(b^*b) \geq 2\mathrm{Re}\, F(a \otimes b)$$

for all $a \in \mathcal{A}$ and $b \in \mathcal{B}$. This proves (9.4.3), and thus (9.4.2).

Let $\pi_{\mathcal{A}} : \mathcal{A} \to \mathcal{B}(H)$ and $\pi_{\mathcal{B}} : \mathcal{B} \to \mathcal{B}(K)$ be the corresponding GNS representations of p_0 and q_0 with unit cyclic vectors $\xi \in H$ and $\eta \in K$. Then F determines a sesquilinear form B on $\pi_{\mathcal{B}}(\mathcal{B})\eta \times \pi_{\mathcal{A}}(\mathcal{A})\xi$ by

$$B(\pi_{\mathcal{B}}(b)\eta, \pi_{\mathcal{A}}(a)\xi) = F(a^* \otimes b),$$

which is contractive since

$$\begin{aligned} |B(\pi_{\mathcal{B}}(b)\eta, \pi_{\mathcal{A}}(a)\xi)| &= |F(a^* \otimes b)| \\ &\leq p_0(a^*a)^{1/2} q_0(b^*b)^{1/2} = \|\pi_{\mathcal{B}}(b)\eta\|\|\pi_{\mathcal{A}}(a)\xi\| . \end{aligned}$$

By continuity, B may be extended to a contractive sesquilinear form on $K \times H$. The usual argument (see §A.3) shows that there is a contractive linear operator $T \in \mathcal{B}(K,H)$ for which

$$B(\pi_{\mathcal{B}}(b)\eta, \pi_{\mathcal{A}}(a)\xi) = \langle T\pi_{\mathcal{B}}(b)\eta \mid \pi_{\mathcal{A}}(a)\xi\rangle .$$

It follows that for all $a \in \mathcal{A}$ and $b \in \mathcal{B}$, we have

$$F(a \otimes b) = B(\pi_{\mathcal{B}}(b)\eta, \pi_{\mathcal{A}}(a^*)\xi) = \langle \pi_{\mathcal{A}}(a) T \pi_{\mathcal{B}}(b)\eta \mid \xi\rangle,$$

and

$$\|\xi\|\|T\|\|\eta\| \leq \|F\| .$$

□

Corollary 9.4.2 *Given operator spaces V and W, a linear functional*

$$F : V \overset{h}{\otimes} W \to \mathbb{C}$$

is bounded if and only if there exist a Hilbert space H and completely bounded mappings

$$\varphi : V \to (H_c)^* = \mathcal{B}(H, \mathbb{C}) \quad \text{and} \quad \psi : W \to H_c = \mathcal{B}(\mathbb{C}, H)$$

(see (3.4.9)) such that

$$F(v \otimes w) = \varphi(v)\psi(w) \,. \tag{9.4.4}$$

In this case, we can choose φ and ψ such that

$$\|F\| = \|\varphi\|_{cb}\|\psi\|_{cb} \,.$$

Proof If there exist completely bounded mappings $\varphi : V \to (H_c)^*$ and $\psi : W \to H_c$ satisfying (9.4.4), then F is just the composition

$$V \otimes_h W \xrightarrow{\varphi\otimes\psi} \mathcal{B}(H, \mathbb{C}) \otimes_h \mathcal{B}(\mathbb{C}, H) \xrightarrow{\mathbf{m}} \mathbb{C},$$

and since $\mathbf{m}$ and $\varphi \otimes \psi$ are completely contractive on the Haagerup tensor products, that is also the case for F.

Conversely, we may assume that V and W are operator subspaces of unital C^*-algebras $\mathcal{A}$ and $\mathcal{B}$. From the injectivity of the Haagerup tensor product and the Hahn–Banach theorem, the linear functional F has a norm-preserving extension

$$\tilde{F} : \mathcal{A} \overset{h}{\otimes} \mathcal{B} \to \mathbb{C} \,.$$

From Theorem 9.4.1, there exist $*$-representations $\pi_{\mathcal{A}} : \mathcal{A} \to \mathcal{B}(H)$ and $\pi_{\mathcal{B}} : \mathcal{B} \to \mathcal{B}(K)$, vectors $\xi \in H$ and $\eta \in K$, and a bounded linear operator $T \in \mathcal{B}(K, H)$ such that

$$\tilde{F}(a \otimes b) = \langle \pi_{\mathcal{A}}(a) T \pi_{\mathcal{B}}(b)\eta \mid \xi \rangle \tag{9.4.5}$$

and

$$\|\tilde{F}\| = \|\xi\|\|T\|\|\eta\| \,.$$

If $\xi^* \in H^*$ is the bounded linear functional corresponding to ξ, and we restrict the representation of $\tilde{F}$ in (9.4.5) to $V \overset{h}{\otimes} W$, then we obtain that

$$F(v \otimes w) = \xi^* \pi_{\mathcal{A}}(v) T \pi_{\mathcal{B}}(w)\eta \,. \tag{9.4.6}$$

If we let $\varphi : V \to (H_c)^*$ and $\psi : W \to H_c$ be the completely bounded mappings given by

$$\varphi(v) = \xi^* \pi_{\mathcal{A}}(v) \quad \text{and} \quad \psi(w) = T \pi_{\mathcal{B}}(w)\eta \,,$$

then we have

$$F(v \otimes w) = \varphi(v)\psi(w)$$

for all $v \in V$ and $w \in W$, and

$$\|\varphi\|_{cb}\|\psi\|_{cb} \leq \|\xi\|\|T\|\|\eta\| = \|F\|.$$

This completes the proof. □

Theorem 9.4.3 *Given operator spaces V_1 and V_2 and Hilbert spaces H_0 and H_2, a linear mapping*

$$\varphi : V_1 \overset{h}{\otimes} V_2 \to \mathcal{B}(H_2, H_0)$$

is completely bounded if and only if there exists a Hilbert space H_1 and completely bounded mappings $\psi_s : V_s \to \mathcal{B}(H_s, H_{s-1})$ $(s = 1, 2)$ such that

$$\varphi(v_1 \otimes v_2) = \psi_1(v_1)\psi_2(v_2). \tag{9.4.7}$$

In this case we can choose ψ_s $(s = 1, 2)$ such that

$$\|\varphi\|_{cb} = \|\psi_1\|_{cb}\|\psi_2\|_{cb}.$$

Proof If $\psi_s : V_s \to \mathcal{B}(H_s, H_{s-1})$ $(s = 1, 2)$ are completely bounded mappings satisfying (9.4.7), then φ coincides with the composition

$$V_1 \overset{h}{\otimes} V_2 \xrightarrow{\psi_1 \otimes \psi_2} \mathcal{B}(H_1, H_0) \overset{h}{\otimes} \mathcal{B}(H_2, H_1) \xrightarrow{\mathbf{m}} \mathcal{B}(H_2, H_0),$$

and thus it is evident that $\varphi : V_1 \overset{h}{\otimes} V_2 \to \mathcal{B}(H_2, H_0)$ satisfies

$$\|\varphi\|_{cb} \leq \|\psi_1\|_{cb}\|\psi_2\|_{cb}.$$

On the other hand, if we use the identification

$$\mathcal{CB}(V_1 \overset{h}{\otimes} V_2, \mathcal{B}(H_2, H_0)) \cong ((H_{0,c})^* \overset{h}{\otimes} V_1 \overset{h}{\otimes} V_2 \overset{h}{\otimes} H_{2,c})^*$$

(see (9.3.5)), we see that φ corresponds to a bounded linear functional

$$F_\varphi : ((H_{0,c})^* \overset{h}{\otimes} V_1) \overset{h}{\otimes} (V_2 \overset{h}{\otimes} H_{2,c}) \to \mathbb{C}$$

satisfying $\|F_\varphi\| = \|\varphi\|_{cb}$ and

$$F_\varphi(\xi^* \otimes v_1 \otimes v_2 \otimes \eta) = \xi^* \varphi(v_1 \otimes v_2)\eta,$$

where we let ξ^* be the linear functional determined by a vector $\xi \in H_{0,c}$, $v_s \in V_s$ and $\eta \in H_{2,c}$. It follows from Corollary 9.4.2 that there exists a Hilbert space H_1 and completely bounded mappings

$$\tilde{\psi}_1 : (H_{0,c})^* \overset{h}{\otimes} V_1 \to (H_{1,c})^*$$

and

$$\tilde{\psi}_2 : V_2 \overset{h}{\otimes} H_{2,c} \to H_{1,c}$$

such that

$$F_\varphi(\xi^* \otimes v_1 \otimes v_2 \otimes \eta) = \tilde{\psi}_1(\xi^* \otimes v_1)\tilde{\psi}_2(v_2 \otimes \eta)$$

for $\xi \in H_{0,c}$ and $\eta \in H_{2,c}$, and

$$\|\tilde{\psi}_1\|_{cb}\|\tilde{\psi}_2\|_{cb} = \|F_\varphi\| = \|\varphi\|_{cb}.$$

Since $(H_{0,c})^* = (H_0^*)_r$, we have from (9.3.4) that

$$\begin{aligned}\mathcal{CB}((H_{0,c})^* \overset{h}{\otimes} V_1,\ (H_{1,c})^*) &\cong ((H_{0,c})^* \widehat{\otimes} V_1 \widehat{\otimes} H_{1,c})^* \\ &\cong \mathcal{CB}(V_1, ((H_{0,c})^* \widehat{\otimes} H_{1,c})^*) \\ &\cong \mathcal{CB}(V_1, \mathcal{B}(H_1, H_0)),\end{aligned}$$

$\tilde{\psi}_1$ corresponds to a completely bounded mapping $\psi_1 : V_1 \to \mathcal{B}(H_1, H_0)$ determined by

$$\langle \psi_1(v_1)(\zeta) \mid \xi \rangle = \tilde{\psi}_1(\xi^* \otimes v_1)(\zeta)$$

for $\zeta \in H_{1,c}$ and $\xi \in H_{0,c}$, or simply,

$$\xi^* \psi_1(v_1) = \tilde{\psi}_1(\xi^* \otimes v_1)$$

for $\xi \in H_{0,c}$. Similarly, since

$$\begin{aligned}\mathcal{CB}(V_2 \overset{h}{\otimes} H_{2,c},\ H_{1,c}) &\cong ((H_{1,c})^* \widehat{\otimes} V_2 \widehat{\otimes} H_{2,c})^* \\ &\cong \mathcal{CB}(V_2, \mathcal{B}(H_2, H_1)).\end{aligned}$$

$\tilde{\psi}_2$ corresponds to a completely bounded mapping $\psi_2 : V_2 \to \mathcal{B}(H_2, H_1)$ given by

$$\psi_2(v_2)\zeta = \tilde{\psi}_2(v_2 \otimes \zeta)$$

for $\zeta \in H_2$. Thus,

$$\xi^* \varphi(v_1 \otimes v_2)(\zeta) = \xi^* \psi_1(v_1)\psi_2(v_2)\eta$$

and

$$\varphi(v_1 \otimes v_2) = \psi_1(v_1)\psi_2(v_2),$$

where

$$\|\varphi\|_{cb} = \|\psi_1\|_{cb}\|\psi_2\|_{cb}.$$

□

It is a simple matter to define the notion of a multiplicatively bounded linear mapping

$$\varphi : V_1 \times \cdots \times V_n \to W,$$

or we may instead simply demand that the completely bounded norm of the corresponding linear mapping

$$\varphi : V_1 \overset{h}{\otimes} \cdots \overset{h}{\otimes} V_n \to W$$

be finite. Iterating the previous result, we obtain the following representation theorem for multiplicatively bounded multilinear mappings.

Theorem 9.4.4 *Given operator spaces $V_1, \ldots, V_n$, a linear mapping*

$$\varphi : V_1 \overset{h}{\otimes} \cdots \overset{h}{\otimes} V_n \to \mathcal{B}(H_n, H_0)$$

is completely bounded if and only if there exist Hilbert spaces $H_1, \ldots, H_{n-1}$ *and completely bounded mappings*

$$\psi_s : V_s \to \mathcal{B}(H_s, H_{s-1}) \quad (s = 1, \ldots, n)$$

such that

$$\varphi(v_1 \otimes \cdots \otimes v_n) = \psi_1(v_1) \cdots \psi_n(v_n) . \tag{9.4.8}$$

We can suppose that $\|\varphi\|_{cb} = \|\psi_1\|_{cb} \cdots \|\psi_n\|_{cb}$. □

The following refinement of Theorem 9.4.4 is often useful in algebraic contexts.

Corollary 9.4.5 *Suppose that* $V_1, \ldots, V_n$ *are operator spaces,* H_0 *and* H_{n+1} *are Hilbert spaces, and that*

$$\varphi : V_1 \overset{h}{\otimes} \cdots \overset{h}{\otimes} V_n \to \mathcal{B}(H_{n+1}, H_0)$$

is a completely bounded mapping. Then there exist Hilbert spaces $H_1, \ldots, H_n$, *completely isometric injections* $\theta_s : V_s \to \mathcal{B}(H_s)$ $(s = 1, \ldots, n)$, *and a diagram of bounded linear operators*

$$H_{n+1} \xrightarrow{T_n} H_n \xrightarrow{T_{n-1}} \cdots \xrightarrow{T_0} H_0 \tag{9.4.9}$$

such that

$$\varphi(v_1 \otimes \cdots \otimes v_n) = T_0 \theta_1(v_1) T_1 \cdots T_{n-1} \theta_n(v_n) T_n \tag{9.4.10}$$

and $\|\varphi\|_{cb} = \|T_0\| \cdots \|T_n\|$.

Proof Let us first assume that $n = 1$, and that $\varphi : V \to \mathcal{B}(H_2, H_0)$ satisfies $\|\varphi\|_{cb} = 1$. Then the mapping

$$\tilde{\varphi} : V \to \mathcal{B}(H_0 \oplus H_2) : v \longmapsto \begin{bmatrix} 0 & \varphi \\ 0 & 0 \end{bmatrix}$$

is also a complete contraction. We fix an arbitrary completely isometric injection $\Phi : V \to \mathcal{B}(K)$ for some Hilbert space K, and we may define

$$H_1 = H_0 \oplus H_2 \oplus K,$$
$$\theta = \tilde{\varphi} \oplus \Phi : V \to \mathcal{B}(H_1),$$
$$T_0 = \begin{bmatrix} I_{H_0} & 0 & 0 \end{bmatrix} \quad \text{and} \quad T_1 = \begin{bmatrix} 0 \\ I_{H_2} \\ 0 \end{bmatrix}.$$

It is immediate that $\theta : V \to \mathcal{B}(H_1)$ is a completely isometric injection and that $T_1 : H_2 \to H_1$ and $T_0 : H_1 \to H_0$ are bounded linear operators of norm 1 such that

$$\varphi = \begin{bmatrix} I_{H_0} & 0 & 0 \end{bmatrix} \begin{bmatrix} 0 & \varphi & 0 \\ 0 & 0 & 0 \\ 0 & 0 & \Phi \end{bmatrix} \begin{bmatrix} 0 \\ I_{H_2} \\ 0 \end{bmatrix} = T_0 \theta T_1 \tag{9.4.11}$$

and

$$\|\varphi\|_{cb} = \|T_0\|\|T_1\|.$$

Turning to the general case, it follows from Theorem 9.4.4 that there exist Hilbert spaces H_s and completely bounded mappings

$$\psi_s : V_s \to \mathcal{B}(H_s, H_{s-1})$$

such that

$$\varphi(v_1 \otimes \cdots \otimes v_n) = \psi_1(v_1) \cdots \psi_n(v_n)$$

and

$$\|\varphi\|_{cb} = \|\psi_1\|_{cb} \cdots \|\psi_n\|_{cb}.$$

We obtain the desired result by applying the above argument to each ψ_s $(s = 1, \ldots, n)$. □

Corollary 9.4.6 *Given C^*-algebras $\mathcal{A}_1, \ldots, \mathcal{A}_n$ and a completely bounded mapping*

$$\varphi : \mathcal{A}_1 \overset{h}{\otimes} \cdots \overset{h}{\otimes} \mathcal{A}_n \to \mathcal{B}(H_{n+1}, H_0),$$

there exist Hilbert spaces H_s $(s = 1, \ldots, n)$, faithful $*$-representations

$$\pi_s : V_s \to \mathcal{B}(H_s) \qquad (s = 1, \ldots, n),$$

and a diagram of bounded linear operators (9.4.9) *such that*

$$\varphi(a_1 \otimes \cdots \otimes a_n) = T_0 \pi_1(a_1) T_1 \cdots T_{n-1} \pi_n(a_n) T_n \tag{9.4.12}$$

and

$$\|\varphi\|_{cb} = \|T_0\| \cdots \|T_n\|.$$

Proof From Corollary 9.4.5 we may assume that

$$\varphi(a_1 \otimes \cdots \otimes a_n) = T_0 \theta_1(a_1) T_1 \theta_2(a_2) \cdots T_{n-1} \theta_n(a_n) T_n,$$

where each $\theta_s : \mathcal{A}_s \to \mathcal{B}(H_s)$ is completely isometric. For each s, we may use Theorem 5.3.3 to find a Hilbert space K_s, a $*$-representation

$$\pi_s : \mathcal{A} \to \mathcal{B}(K_s),$$

and complete contractions

$$H_s \xrightarrow{R_s} K_s \xrightarrow{S_s} H_s$$

with $\theta_s(a_s) = S_s \pi_s(a_s) R_s$. It is immediate that π_s is faithful, and we have

$$\varphi(a_1 \otimes \cdots \otimes a_n) = (T_0 S_1) \pi_1(a_1) (R_1 T_1 S_2) \pi_2(a) \cdots \pi_n(a_n) (R_n T_n),$$

which is the desired decomposition. □

Let us suppose that V and W are operator spaces. We can use the natural duality between $V \otimes W$ and $V^* \otimes W^*$ and the Haagerup tensor norm $\|\cdot\|_h$ on $V \otimes W$ to define a corresponding *dual* tensor norm $\|\cdot\|_{h'}$

on $V^* \otimes W^*$ (see (3.2.11)). In what follows we shall show that in fact this coincides with the Haagerup tensor norm on $V^* \otimes W^*$.

Theorem 9.4.7 *Given operator spaces V and W, the natural embedding*

$$V^* \overset{h}{\otimes} W^* \hookrightarrow (V \overset{h}{\otimes} W)^* \tag{9.4.13}$$

is completely isometric.

Proof We let $\theta : V^* \otimes_h W^* \to (V \overset{h}{\otimes} W)^*$ be the natural linear mapping

$$\theta(f \otimes g)(v \otimes w) = f(v)g(w)$$

for $f \in V^*, g \in W^*, v \in V$, and $w \in W$. Given $F \in M_n(V^* \otimes_h W^*)$, it follows from Proposition 9.2.6 that there exist $r \in \mathbb{N}$, $f \in M_{n,r}(V^*)$, and $g \in M_{r,n}(W^*)$ such that

$$F = f \odot g \quad \text{and} \quad \|F\|_h = \|f\|\|g\| .$$

Taking the supremum over arbitrary $v \in M_{p,s}(V)$ and $w \in M_{s,p}(W)$ with $\|v\| \leq 1$ and $\|w\| \leq 1$, we have

$$\|\theta_n(F)\|_{cb} = \sup \{\|\theta_n(F)_p(v \odot w)\|\} .$$

But we have

$$\begin{aligned}
\theta_n(F)_p(v \odot w) &= \left[\theta\left(\sum_c f_{a,c} \otimes g_{c,b}\right)\left(\sum_k v_{i,k} \otimes w_{k,j}\right)\right]_{(a,i),(b,j)} \\
&= \left[\sum_{c,k} f_{a,c}(v_{i,k}) g_{c,b}(w_{k,j})\right] \\
&= f_{p,r}(v) g_{r,p}(w),
\end{aligned}$$

where we use the matrix product of $f_{p,r}(v) \in M_{pn,rs}$ and $g_{r,p}(w) \in M_{rs,pn}$, and thus

$$\begin{aligned}
\|\theta_n(F)\|_{cb} &= \sup \{\|f_{p,r}(v)\|\|g_{r,p}(w)\|\} \\
&\leq \|f\|\|g\| = \|F\|_h .
\end{aligned}$$

This proves that $\|\theta_n(F)\|_{cb} \leq \|F\|_h$.

To prove the reverse inequality, we reduce the calculation to the finite-dimensional case. Suppose that we are given

$$F = \left[\sum_l f^l_{i,j} \otimes g^l_{i,j}\right]_{i,j \in n} \in \mathbb{M}_n(V^* \otimes W^*) .$$

We have $F \in \mathbb{M}_n(P \otimes Q)$, where P and Q are the finite-dimensional subspaces generated by coefficients $f^l_{i,j}$ and $g^l_{i,j}$, respectively. Since these spaces are weak* closed, we may identify them with operator duals of the quotients $V/P_\perp$ and $W/Q_\perp$, respectively. We have the commutative diagram

$$\begin{array}{ccc} P \overset{h}{\otimes} Q & \hookrightarrow & V^* \overset{h}{\otimes} W^* \\ \theta \downarrow & & \theta \downarrow \\ (V/P_\perp \overset{h}{\otimes} W/Q_\perp)^* & \hookrightarrow & (V \overset{h}{\otimes} W)^* \end{array},$$

in which the top and bottom rows are completely isometric (see Theorem 9.2.5), and we have from the above that the columns are complete contractions. Since F lies in the image of $P \overset{h}{\otimes} Q$, it suffices to show that the first column is completely isometric.

Therefore, we may change the notation and assume that both V and W are finite-dimensional. Our task is to show that

$$\|F\|_h \leq \|\theta_n(F)\|_{cb}\,.$$

Since $\theta_n(F) \in M_n((V \overset{h}{\otimes} W)^*) = \mathcal{CB}(V \overset{h}{\otimes} W, M_n)$, it follows from Theorem 9.4.3 that there exists a Hilbert space H and completely bounded mappings

$$\varphi : V \to \mathcal{B}(H, \mathbb{C}^n) \quad \text{and} \quad \psi : W \to \mathcal{B}(\mathbb{C}^n, H)$$

such that

$$\theta_n(F)(v \otimes w) = \varphi(v)\psi(w)$$

and

$$\|\theta_n(F)\|_{cb} = \|\varphi\|_{cb}\|\psi\|_{cb}\,.$$

Let H_1 be the subspace of H spanned by $\psi(w)\eta$ for all $w \in W$ and $\eta \in \mathbb{C}^n$. Since W and $\mathbb{C}^n$ are finite-dimensional spaces, H_1 is a finite-dimensional Hilbert space which we may identify with $\mathbb{C}^r$ for some $r \in \mathbb{N}$. If we let $e : H \to H_1$ be the orthogonal projection from H onto H_1, and we define f and g by

$$f(v) = \varphi(v)_{|H_1} \quad \text{and} \quad g(w) = e\psi(w),$$

then we may regard f and g as elements of

$$\mathcal{CB}(V, \mathcal{B}(\mathbb{C}^r, \mathbb{C}^n)) = M_{n,r}(V^*) \quad \text{and} \quad \mathcal{CB}(W, \mathcal{B}(\mathbb{C}^n, \mathbb{C}^r)) = M_{r,n}(W^*),$$

respectively, for which

$$\theta_n(F)(v \otimes w) = f(v)g(w) = \theta_n(f \odot g)(v \otimes w)$$

for all $v \in V$ and $w \in W$. This implies that $F = f \odot g$ and thus

$$\|F\|_h \leq \|f\|\|g\| \leq \|\varphi\|_{cb}\|\psi\|_{cb} = \|\theta_n(F)\|_{cb}\,.$$

This completes the proof. □

Corollary 9.4.8 *Given operator spaces V and W, if either V or W is finite-dimensional, then we have the complete isometry*

$$V^* \overset{h}{\otimes} W^* \cong (V \overset{h}{\otimes} W)^*. \qquad \square$$

The remarkable self-dual properties of the Haagerup tensor product are illustrated by the proof of the following result.

Proposition 9.4.9 *Given operator spaces V and W, and $u \in M_n(V \otimes W)$, there exist contractive elements $f \in M_{n,r}(V^*)$ and $g \in M_{r,n}(W^*)$ such that*

$$\|u\|_h = \|(f \odot g)_n(u)\|. \tag{9.4.14}$$

Thus,

$$\|u\|_h = \sup \{\|(f \odot g)_n(u)\|\}, \tag{9.4.15}$$

where the supremum is taken over all $f \in M_{n,r}(V^*)$ and $g \in M_{r,n}(W^*)$ with $\|f\|, \|g\| \leq 1$ and $r \in \mathbb{N}$.

Proof Given operator spaces V and W, and $u \in M_n(V \otimes W)$, we may find finite-dimensional subspaces $V_1 \subseteq V$ and $W_1 \subseteq W$ with $u \in \mathbb{M}_n(V_1 \otimes W_1)$. From the injectivity property for the Haagerup tensor product, u has the same norm in $M_n(V_1 \overset{h}{\otimes} W_1)$. From Lemma 2.3.4, there exists a complete contraction $\varphi : V_1 \overset{h}{\otimes} W_1 \to M_n$ for which $\|u\|_h = \|\varphi_n(u)\|$.

Since V_1 and W_1 are finite-dimensional operator spaces, it follows from Corollary 9.4.8 that we have the complete isometry

$$V_1^* \overset{h}{\otimes} W_1^* \cong (V_1 \overset{h}{\otimes} W_1)^*,$$

and thus we may regard φ as a contractive element in $M_n(V_1^* \overset{h}{\otimes} W_1^*)$. From Proposition 9.2.6, there exist contractive elements $f \in M_{n,r}(V_1^*)$ and $g \in M_{r,n}(W_1^*)$ such that $\varphi = f \odot g$. From the Arveson–Wittstock–Hahn–Banach extension theorem there exist $\tilde{f} \in M_{n,r}(V^*)$ and $\tilde{g} \in M_{r,n}(W^*)$ of f and g such that $\|f\|, \|g\| \leq 1$ and

$$\|(\tilde{f} \odot \tilde{g})_n(u)\| = \|(f \odot g)_n(u)\| = \|\varphi_n(u)\| = \|u\|_h . \qquad \square$$

9.5 NOTES AND REFERENCES

The Haagerup tensor product was essentially introduced by Haagerup in an unpublished set of notes (Haagerup 1980). Although the explicit definition does not occur in this discussion, Haagerup described the notion in another informal set of notes. To the lucky few who had access to this work, Haagerup's ideas proved to be of great importance in the period preceding Ruan's characterization of the operator spaces. Effros and Kishimoto (1987) used the Haagerup tensor product to prove Theorem 9.4.1. They employed that result, together with Haagerup–Pisier version of the Grothendieck inequality, to reprove the Connes–Haagerup theorem that nuclearity and amenability are equivalent for C^*-algebras (see Connes 1978, Haagerup 1983, and §17.2). The trilinear version of Theorem 9.4.1 was proved in Effros (1987*b*).

Subsequently, Christensen and Sinclair (1987) used the difficult set-valued techniques of Wittstock to prove Corollary 9.4.5 for C^*-algebras. The increasing complications of these arguments were overcome in the brilliant paper of Paulsen and Smith (1987), who in particular discovered the fact that the Haagerup tensor product is injective (Proposition 9.2.5). They used this to prove Theorem 9.4.3–Corollary 9.4.5. We have followed the exposition in Blecher (1992*b*) to prove Theorem 9.4.3.

Christensen and Sinclair were motivated by the Johnson–Kadison–Ringrose cohomology theory of operator algebras. Together with Effros (Christensen *et al.* 1987), they showed that certain cohomological questions could be reduced to the case of completely bounded chains, which in turn could be handled owing to the explicit 'algebraic' nature of Corollary 9.4.5. These techniques have played a major part in the solution of some of the most intractable problems in this area (see Sinclair and Smith 1995 for the details and further references).

The projective property of the Haggerup tensor product and Proposition 9.2.6 are due to Effros and Ruan (1991*b*). Theorem 9.4.7 was proved in Blecher and Paulsen (1991a) (see also Effros and Ruan 1991*b*).

The Haagerup tensor product of the column and row Hilbert operator spaces in §9.3 were studied in Blecher and Paulsen (1991*a*), Effros and Ruan (1991*b*), and Blecher (1992*b*). Blecher and Smith (1992) gave a very detailed investigation of the dual of the Haagerup tensor product. Related results can be found in Effros *et al.* (1993).

Theorem 9.4.7–Proposition 9.4.9 are due to Effros and Ruan (1991*b*).

10

Infinite matrices and asymptotic constructions

Sooner or later one must come to terms with several important infinite-dimensional techniques in order to understand the more advanced theory of operator spaces. We have placed two of these in this chapter. The reader may wish to skip this material until it is needed.

Infinite matrices play the role of infinite sequences in Banach space theory. In §10.1 we use infinite matrices to give explicit formulae for elements of projective tensor products. This is crucial to the theory of completely nuclear mappings (see §12.2).

The notion of an ultraproduct is easily generalized to the context of operator spaces (see §10.3). As in the case of Banach space theory, it provides an elegant tool for factoring various kinds of mappings (see Theorems 13.2.3 and 15.4.1).

10.1 INFINITE MATRICES OVER AN OPERATOR SPACE

In order to simplify the exposition, we shall restrict our attention to matrices over countable index sets. Nevertheless, it should be clear that the entire development is valid for arbitrary index sets $\mathfrak{s}$. The appropriate notions are obtained by considering the subalgebras $\mathbb{M}_{\mathfrak{f}}$ of $\mathbb{M}_{\mathfrak{s}}$ for finite subsets $\mathfrak{f}$ of $\mathfrak{s}$ and the corresponding truncations.

Let us suppose that V is an operator space. We recall from §1.1 that $\mathbb{M}_\infty(V)$ is the linear space of all infinite matrices $[v_{i,j}]$ with $v_{i,j} \in V$. We identify the matrix spaces $\mathbb{M}_{r,s}(V)$ and $\mathbb{M}_r(V)$, where $1 \leq r, s < \infty$, with subspaces of $\mathbb{M}_\infty(V)$ in the usual manner, and we let v^r denote the truncation of $v \in \mathbb{M}_\infty(V)$ to $\mathbb{M}_r(V)$. If $r \leq s$, then

$$\|v^r\| = \|(I_r \oplus 0_{s-r})v^s(I_r \oplus 0_{s-r})\| \leq \|v^s\|.$$

If $v \in \mathbb{M}_\infty(V)$, then we define

$$\|v\| = \sup\{\|v^r\| : r \in \mathbb{N}\} = \lim_{r\to\infty} \|v^r\|, \tag{10.1.1}$$

and we let $M_\infty(V)$ be the space of all $v \in \mathbb{M}_\infty(V)$ for which $\|v\| < \infty$. It is evident that with these conventions, $M_\infty(\mathbb{C}) = \mathcal{B}(\ell_2) = M_\infty$.

If $\mathfrak{s}$ is an arbitrary countable set, then we may apply the same definitions to $\mathbb{M}_{\mathfrak{s}}(V)$ and the corresponding space $M_{\mathfrak{s}}(V)$ by using a bijection between $\mathfrak{s}$ and $\infty = \mathbb{N}$. Each finite subset $\mathfrak{f}$ of $\mathfrak{s}$ then determines a subspace $M_{\mathfrak{f}}(V)$ and a corresponding truncation $v^{\mathfrak{f}} \in M_{\mathfrak{f}}(V)$. Given an increasing sequence of finite subsets $\mathfrak{f}_r$ of $\mathfrak{s}$ with the union $\mathfrak{s}$, it is easy to see that

$$\|v\| = \sup\{\|v^{\mathfrak{f}_r}\| : r \in \mathbb{N}\} = \lim_{r\to\infty} \|v^{\mathfrak{f}_r}\|.$$

For any $m \in \mathbb{N}$, we can use (1.1.14) to identify $\mathbb{M}_m(M_\infty(V))$ with $M_{m\times\infty}(V)$. We let $M_m(M_\infty(V))$ denote $\mathbb{M}_m(M_\infty(V))$ with the corresponding norm. The truncation mapping

$$P^r : M_\infty(V) \to M_r(V) : v \mapsto v^r$$

is contractive, and applying this to the mapping

$$P_n^r = id_{M_n} \otimes P^r : M_n(M_\infty(V)) = M_{n\times\infty}(V) \to M_{n\times r}(V),$$

it follows that P^r is completely contractive.

Proposition 10.1.1 *If V is an operator space, then $M_\infty(V)$ is an operator space.*

Proof If $v = [v_{i,j}]_{i,j\in m}$ is a matrix in $\mathbb{M}_m(\mathbb{M}_\infty(V))$, then

$$v^{m\times r} = [v_{i,j}^r] \in \mathbb{M}_m(\mathbb{M}_r(V)).$$

For any $v = [v_{i,j}] \in M_m(M_\infty(V))$ and $w = [w_{k,l}] \in M_n(M_\infty(V))$,

$$(v \oplus w)^{(m+n)\times r} = [v_{i,j}^r] \oplus [w_{k,l}^r] = v^{m\times r} \oplus w^{m\times r},$$

and thus

$$\begin{aligned}\|(v\oplus w)^{(m+n)\times r}\| &= \|v^{m\times r} \oplus w^{n\times r}\| \\ &= \max\{\|v^{m\times r}\|, \|w^{m\times r}\|\} \\ &\le \max\{\|v\|, \|w\|\}.\end{aligned}$$

Since the sets $(m+n)\times r$ have union $(m+n)\times\infty$, we obtain condition **M1**′:

$$\|v \oplus w\| = \sup\{\|(v\oplus w)^{(m+n)\times r}\|\} \le \max\{\|v\|, \|w\|\}.$$

Similarly, given $\alpha = [\alpha_{k,i}] \in M_{n,m}$, $\beta = [\beta_{j,l}] \in M_{m,n}$, we have

$$(\alpha v \beta)^{n\times r} = \alpha v^{m\times r}\beta;$$

hence

$$\|(\alpha v\beta)^{n\times r}\| = \|\alpha v^{m\times r}\beta\| \le \|\alpha\|\,\|v^{m\times r}\|\,\|\beta\| \le \|\alpha\|\,\|v\|\,\|\beta\|,$$

$$\|\alpha v\beta\| = \sup_r\{\|(\alpha v\beta)^{m\times r}\|\} \le \|\alpha\|\|v\|\|\beta\|,$$

and **M2** is also satisfied. From Theorem 2.3.6, $M_\infty(V)$ is an operator space.

Let us suppose that $(v(k))_{k\in\mathbb{N}}$ is a Cauchy sequence in $M_\infty(V)$. Then it is bounded, i.e. we may find a constant $K > 0$ such that $\|v(k)\| \leq K$ for all k. For every $r \in \mathbb{N}$,

$$v^r(k) = P^r(v(k))$$

is Cauchy. Since V is assumed complete, that is also the case for $M_r(V)$ (see the discussion after (2.1.8)). Thus, $v^r(k)$ converges in norm to an element $v^r \in M_r(V)$ with $\|v^r\| \leq K$. If $r \leq s$ and $i, j \leq r$, then from (1.1.10),

$$v^r_{i,j}(k) = E^r_i P^r(v(k)) E^{r*}_j = E^s_i P^s(v(k)) E^{s*}_j = v^s_{i,j}(k).$$

If we let $k \to \infty$, we see that $v^r_{i,j} = v^s_{i,j}$, and thus the matrices v^r $(r \in \mathbb{N})$ are the truncations of a matrix $v \in \mathbb{M}_\infty(V)$. Since $\|v^r\| \leq K$, it follows that $\|v\| \leq K$. If $\varepsilon > 0$, and we choose k_0 with $\|v(k) - v(l)\| \leq \varepsilon$ for $k, l \geq k_0$, then it follows that $\|v(k)^r - v(l)^r\| \leq \varepsilon$ for each $r \in \mathbb{N}$, and if we let $l \to \infty$, then $\|v(k)^r - v^r\| \leq \varepsilon$. Thus, $\|v(k) - v\| \leq \varepsilon$, and v is the limit of the sequence $v(k)$. □

We define $K_\infty(V)$ to be the closure of $\mathbb{M}^{\text{fin}}_\infty(V)$ in $M_\infty(V)$. The column mappings in the diagram

$$\begin{array}{ccc} M_r(V) & \cong & M_r \check{\otimes} V \\ \downarrow & & \downarrow \\ K_\infty(V) & & K_\infty \check{\otimes} V \end{array}$$

are isometric, and in each case the union of their ranges is dense. It follows that we have the isometry

$$K_\infty(V) \cong K_\infty \check{\otimes} V. \tag{10.1.2}$$

For any $v \in K_\infty(V)$,

$$\lim_{r\to\infty} \|v - P^r(v)\| = 0. \tag{10.1.3}$$

This is trivial if $v \in M_{r_0}(V)$ for some r_0 since, in this case, $P^r(v) = v$ for $r \geq r_0$. Given $v \in K_\infty(V)$ and $\varepsilon > 0$, we may choose a $v_0 \in M_{r_0}(V)$ with $\|v - v_0\| < \varepsilon$. For any $r \geq r_0$ we have $P^r(v_0) = v_0$, and thus

$$\|v - P^r(v)\| \leq \|v - v_0\| + \|v_0 - P^r(v_0)\| + \|P^r(v_0) - P^r(v)\| \leq 2\varepsilon.$$

Since $\varepsilon > 0$ is arbitrary, (10.1.3) is evident. We may identify $M_n(K_\infty(V))$ with $K_{n\times\infty}(V)$, and thus we see that

$$\lim_{r\to\infty} \|v - P^r_n(v)\| = 0$$

for any $v \in M_n(K_\infty(V))$.

If $n < \infty$, then for any matrices $\alpha \in M_{n,\infty}$ and $\beta \in M_{\infty,n}$,

$$\lim_{r\to\infty} \|\alpha - \alpha^r\| = \lim_{r\to\infty} \|\beta - \beta^r\| = 0. \tag{10.1.4}$$

To see the first of these relations, we note that since $\alpha\alpha^* \in M_n$,

$$\|\alpha\alpha^*\| \leq \operatorname{trace}(\alpha\alpha^*) \leq n\,\|\alpha\alpha^*\|.$$

It follows that

$$\|\alpha\|^2 \leq \sum |\alpha_{i,j}|^2 \leq n\,\|\alpha\|^2,$$

and in particular the operator and Hilbert–Schmidt norm topologies coincide on $M_{n,\infty}$. Thus, (10.1.4) is a simple consequence of l_2-convergence. The same argument applies to $M_{\infty,n}$, and we see that $M_{n,\infty} = K_{n,\infty}$ and $M_{\infty,n} = K_{\infty,n}$.

If $v \in M_\infty(V)$ and $\alpha \in \mathbb{M}_\infty^{\text{fin}}$, then we define $\alpha v \in \mathbb{M}_\infty$ by

$$(\alpha v)_{i,j} = \sum_k \alpha_{i,k} v_{k,j}.$$

We have, for example, that if $a, b, c, d \in \mathbb{C}$, then

$$\begin{bmatrix} a & b & 0 & \dots \\ c & d & 0 & \dots \\ 0 & 0 & 0 & \dots \\ \vdots & \vdots & \vdots & \end{bmatrix} \begin{bmatrix} v_{1,1} & v_{1,2} & v_{1,3} & v_{1,4} & \dots \\ v_{2,1} & v_{2,2} & v_{2,3} & v_{2,4} & \dots \\ v_{3,1} & v_{3,2} & \dots & \dots & \dots \\ \vdots & \vdots & \vdots & & \end{bmatrix}$$

$$= \begin{bmatrix} av_{1,1} + bv_{2,1} & av_{1,2} + bv_{2,2} & av_{1,3} + bv_{2,3} & av_{1,4} + bv_{2,4} & \dots \\ cv_{1,1} + dv_{2,1} & cv_{1,2} + dv_{2,2} & cv_{1,3} + dv_{2,3} & cv_{1,4} + dv_{2,4} & \dots \\ 0 & 0 & 0 & 0 & \dots \\ \vdots & \vdots & \vdots & \vdots & \end{bmatrix}.$$

It is apparent from this calculation that if $\alpha \in M_n \subseteq \mathbb{M}_\infty^{\text{fin}}(V)$, then

$$\alpha v = \begin{bmatrix} \alpha\tilde{v}_1 & \alpha\tilde{v}_2 & \dots \\ 0 & 0 & \dots \\ \vdots & \vdots & \end{bmatrix},$$

where $\tilde{v}_k \in M_n(V)$ is defined by

$$(\tilde{v}_k)_{i,j} = [v_{i,nk+j}].$$

If $r \in \mathbb{N}$, then we have a corresponding truncation

$$(\alpha v)^{nr} = \begin{bmatrix} \alpha\tilde{v}_1 & \dots & \alpha\tilde{v}_r \\ 0 & \dots & 0 \\ 0 & \dots & 0 \end{bmatrix} = \begin{bmatrix} \alpha & 0 & \dots \\ 0 & 0 & \\ \vdots & \vdots & \end{bmatrix} v^{nr}.$$

It follows that

$$\|(\alpha v)^{nr}\| \leq \|\alpha\|\,\|v^{nr}\| \leq \|\alpha\|\,\|v\|,$$

and in particular, $\alpha v \in M_\infty(V)$.

Given $\alpha \in K_\infty$ and $r \leq s$,

$$\|\alpha^s v - \alpha^r v\| \leq \|\alpha^s - \alpha^r\|\,\|v\|,$$

and thus $\alpha^r v$ is Cauchy. We let

$$\alpha v = \lim_{r\to\infty} \alpha^r v.$$

Similarly, we define

$$v\beta = \lim_{r\to\infty} v\beta^r.$$

With these operations, $M_\infty(V)$ is a K_∞-bimodule. To see this we must show that $\alpha(\beta v) = (\alpha\beta)v$ for $\alpha, \beta \in K_\infty$, and the corresponding right-sided condition. If we let $E^r = I_r \oplus 0 \in K_\infty$, then

$$\|\alpha(I - E^r)\|^2 = \|(I - E^r)^*\alpha^*\alpha(I - E^r)\| \to 0 \tag{10.1.5}$$

as $r \to \infty$ since this is evident for $\alpha \in M_{r_0}$, and the general case follows from the usual density argument (see above). Thus,

$$\|(\alpha\beta)^r - \alpha^r\beta^r\| = \|E^r\alpha(I - E^r)\beta E^r\| \to 0$$

and

$$(\alpha\beta)v = \lim_{r\to\infty} \alpha^r\beta^r v.$$

But we have

$$\begin{aligned}\|\alpha(\beta(v)) - \alpha^r(\beta^r v)\| &\leq \|\alpha(\beta - \beta^r)v\| + \|(\alpha - \alpha^r)\beta^r v\| \\ &\leq \|\alpha\|\,\|\beta - \beta^r\|\,\|v\| + \|\alpha - \alpha^r\|\,\|\beta\|\,\|v\| \to 0,\end{aligned}$$

and associativity follows. The same argument can be used on the right side.

The above theory applies, as well, to the spaces $M_{n,\infty}(V)$ and $M_{\infty,n}(V)$. If $v \in M_{n,\infty}(V)$, then we interpret the truncation v^r as an element of $M_r(V)$ for $r \leq n$, and as an element of $M_{n,r}(V)$ for $r \geq n$, and we use a similar convention for $M_{\infty,n}(V)$. From the previous arguments, we see that, if $\alpha \in M_{n,\infty}$, $v \in M_\infty(V)$, and $\beta \in M_{\infty,n}$, then we have a corresponding element

$$\alpha v\beta = \lim_{r\to\infty} \alpha^r v\beta^r \in M_n(V). \tag{10.1.6}$$

In §4.1 we defined a second norm $\|\cdot\|_1$ on $\mathbb{M}_r(V)$, and we let $T_r(V)$ denote the corresponding Banach space. We showed in (7.1.20) that there is a natural isometry $T_r(V) \cong T_r \hat{\otimes} V$, and using the projective tensor product norm, this enabled us to regard $T_r(V)$ as an operator space. If $r < s \in \mathbb{N}$, the inclusion $T_r(V) \hookrightarrow T_s(V)$ is completely isometric, since the canonical inclusion $T_r \hookrightarrow T_s$ is completely isometric, and the truncation projection $P^{s,r} : T_s \to T_r$ is completely contractive (see (7.1.27)).

If $v \in T_s(V)$, then $\|v^r\| \leq \|v^s\|$ since the mapping

$$P^{s,r} \otimes id : T_s \hat{\otimes} V \to T_r \hat{\otimes} V$$

is contractive. We infer that if $v \in \mathbb{M}_\infty(V)$, then the sequence $\|v^r\|_1$ $(r \in \mathbb{N})$ is increasing, and we may let

$$\|v\|_1 = \sup\{\|v^r\|_1 : r \in \mathbb{N}\} = \lim_{r \to \infty} \|v^r\|_1.$$

We define $T_\infty(V)$ to be the space of all $v \in \mathbb{M}_\infty(V)$ for which $\|v\|_1 < \infty$. For each $r \in \mathbb{N}$, $T_\infty(V)$ contains $T_r(V)$ as a subspace. The matrix norms on the $T_r(V)$ determine a complete matrix norm on $T_\infty(V)$. The proof is very similar to that given in Proposition 10.1.1, and so we leave the details to the reader.

By contrast to $M_\infty(V)$, we shall next show that the finite truncations are norm dense in $T_\infty(V)$. Given $r < s$ in $\mathbb{N}$, we define the s by r (respectively, r by s) rectangular truncation $v^{s,r}$ (respectively, $v^{r,s}$) of an element $v \in T_\infty(V)$. We let $T_{s,r}(V)$ (respectively, $T_{r,s}(V)$) denote the subspace of all these truncated elements.

Lemma 10.1.2 *Given $v \in T_\infty(V)$ and $r < s$ in $\mathbb{N}$, we have*

$$\|v^s - v^r\|_1^2 \leq 2(\|v^s\|_1^2 - \|v^r\|_1^2).$$

Proof If $v^s = 0$, then this is trivial. Thus, we may assume that $\|v^s\|_1 \neq 0$. Given $\delta > 0$, we can choose $\alpha \in HS_{s,p}$, $\beta \in HS_{p,s}$, and $x \in M_p(V)$ with

$$v^s = \alpha x \beta$$

and such that $\|x\| = 1$ and $\|\alpha\|_2 \|\beta\|_2 < \|v^s\|_1(1+\delta)$. If we let $\beta = [\beta_1 \, \beta_2]$ with $\beta_1 \in HS_{p,r}$ and $\beta_2 \in HS_{p,s-r}$, then we have

$$v^{s,r} = \alpha x \beta_1$$

and

$$v^s - v^{s,r} = \alpha x \beta_2.$$

It follows that

$$\begin{aligned} \|v^{s,r}\|_1^2 + \|v^s - v^{s,r}\|_1^2 &\leq \|\alpha\|_2^2 \|\beta_1\|_2^2 + \|\alpha\|_2^2 \|\beta_2\|_2^2 \\ &= \|\alpha\|_2^2 \|\beta\|_2^2 < \|v^s\|_1^2 (1+\delta)^2. \end{aligned}$$

Since $\delta > 0$ is arbitrary,

$$\|v^{s,r}\|_1^2 + \|v^s - v^{s,r}\|_1^2 \leq \|v^s\|_1^2$$

and

$$\|v^s - v^{s,r}\|_1^2 \leq \|v^s\|_1^2 - \|v^{s,r}\|_1^2.$$

A similar argument shows that

$$\|v^{s,r} - v^r\|_1^2 \leq \|v^{s,r}\|_1^2 - \|v^r\|_1^2,$$

and thus

$$\begin{aligned} \|v^s - v^r\|_1^2 &\leq (\|v^s - v^{s,r}\|_1 + \|v^{s,r} - v^r\|_1)^2 \\ &\leq 2\left(\|v^s - v^{s,r}\|_1^2 + \|v^{s,r} - v^r\|_1^2\right) \end{aligned}$$

$$\leq 2\left(\|v^s\|_1^2 - \|v^{s,r}\|_1^2 + \|v^{s,r}\|_1^2 - \|v^r\|_1^2\right)$$
$$= 2\left(\|v^s\|_1^2 - \|v^r\|_1^2\right).$$

□

Theorem 10.1.3 *Let V be an operator space. Then the union $\bigcup_{r\in\mathbb{N}} T_r(V)$ is norm dense in $T_\infty(V)$. Moreover, we have the complete isometry*

$$T_\infty(V) \cong T_\infty \hat{\otimes} V. \tag{10.1.7}$$

Proof If $v \in T_\infty(V)$, then $\|v^r\|_1$ is an increasing sequence such that

$$\|v\|_1 = \sup\{\|v^r\|_1 : r \in \mathbb{N}\} = \lim_{r\to\infty} \|v^r\|_1.$$

Then $\|v^r\|_1$ is a Cauchy sequence. From Lemma 10.1.2 we infer that v^r is a Cauchy sequence in the (complete) operator space $T_\infty(V)$, and thus it converges to an element v' of $T_\infty(V)$. The finite truncations of v' and v coincide, and thus

$$v = \lim v^r = v'.$$

This shows that $\bigcup_{r\in\mathbb{N}} T_r(V)$ is norm dense in $T_\infty(V)$. Since the column mappings in the diagram

$$\begin{array}{ccc} T_r(V) & \cong & T_r \hat{\otimes} V \\ \downarrow & & \downarrow \\ T_\infty(V) & & T_\infty \hat{\otimes} V \end{array}$$

are completely isometric, and in each case the union of their ranges is dense, we have the complete isometry

$$T_\infty(V) \cong T_\infty \hat{\otimes} V.$$

□

If $v \in \mathbb{M}_r(V)$, then we have from (4.1.6) that $\|v\| \leq \|v\|_1$. It follows that if $v \in T_\infty(V)$, then the truncations v^r are Cauchy in $M_\infty(V)$ and they converge to an element $\tilde{v} \in K_\infty(V)$. The mapping $v \mapsto \tilde{v}$ is one-to-one since if $\tilde{v} = 0$, then each of the truncations v^r is zero. Thus, we may identify $T_\infty(V)$ with a linear subspace of $K_\infty(V)$.

Theorem 10.1.4 *For any operator space V, we have the natural completely isometric identifications*

$$T_\infty(V)^* \cong M_\infty(V^*) \tag{10.1.8}$$

and

$$K_\infty(V)^* \cong T_\infty(V^*). \tag{10.1.9}$$

Proof From Corollary 7.1.5 we have the completely isometric identification

$$T_\infty(V)^* \cong (T_\infty \hat{\otimes} V)^* \cong \mathcal{CB}(V, M_\infty),$$

and thus to prove (10.1.8) it suffices to describe the natural complete isometry

$$\theta : M_\infty(V^*) \to \mathcal{CB}(V, M_\infty).$$

For any $f \in [f_{i,j}] \in M_\infty(V^*)$, we define

$$\theta(f) : V \to \mathbb{M}_\infty : \theta(f)(v) = [f_{i,j}(v)].$$

If $r \in \mathbb{N}$, then

$$(\theta(f)(v))^r = \theta(f^r)(v),$$

where

$$f^r \in M_r(V^*) = \mathcal{CB}(V, M_r),$$

and thus

$$\|\theta(f)^r\|_{cb} = \|f^r\|_{M_r(V^*)} \leq \|f\|.$$

In particular, if $v \in V$, then

$$\|\theta(f)(v)\| = \lim_{r\to\infty} \|\theta(f)^r(v)\| \leq \|f\| \, \|v\| < \infty$$

and $\theta(f)(v) \in M_\infty$. We have from (10.1.1), (3.2.2), and a routine argument for the last equality that

$$\|f\| = \lim_{r\to\infty} \|f^r\| = \lim_{r\to\infty} \|\theta(f)^r\|_{cb} = \|\theta(f)\|_{cb},$$

and θ is an isometric injection. Since we may identify θ_n with the isometric injection

$$\theta : M_{n\times\infty}(V^*) \to \mathcal{CB}(V, M_{n\times\infty}),$$

it follows that θ is a completely isometric injection.

If $\varphi \in \mathcal{CB}(V, M_\infty)$, then we define

$$f_{i,j} = E_i \varphi E_j^* \in V^*,$$

where $E_j = [0 \ldots 1_j \, 0 \ldots]$, and we let

$$f = [f_{i,j}] \in \mathbb{M}_\infty(V^*).$$

The truncation $f^r \in M_r(V^*)$ corresponds to the function $\varphi^r \in \mathcal{CB}(V, M_r)$, and thus

$$\|f^r\|_{M_r(V^*)} = \|\varphi^r\|_{cb} \leq \|\varphi\|.$$

It follows that $f \in M_\infty(V^*)$, and it is evident that $\theta(f) = \varphi$. We conclude that θ is a surjection, and we have proved (10.1.8).

For any r, we have a diagram of completely contractive mappings

$$\begin{array}{ccc} T_r(V^*) & \longrightarrow & M_r(V)^* \\ \downarrow & & \downarrow \\ T_\infty(V^*) & & K_\infty(V)^* \end{array},$$

where the top row is a complete isometry (see (7.1.23)) and the two columns are isometric injections. Since $\bigcup_{r\in\mathbb{N}} T_r(V^*)$ is norm dense in $T_\infty(V^*)$, this extends to a complete isometry $\theta : T_\infty(V^*) \to K_\infty(V)^*$. On the other hand, if $f \in K_\infty(V)^*$, then from Theorem 5.3.4,

$$f = \theta(\alpha(\varepsilon \otimes g)\beta),$$

where $\alpha \in M_{1,\infty\times\infty}$ and $\beta \in M_{\infty\times\infty,1}$, and $\varepsilon = id$ is the identity mapping on K_∞. If we let $\alpha^{r\times r}$ denote the $r \times r$ row truncation of α and let $\beta^{r\times r}$ denote the $r \times r$ column truncation of β, then $f_r = \alpha^{r\times r}(\varepsilon \otimes g)\beta^{r\times r}$ is an element in $T_\infty \otimes_\wedge V^*$ such that $\|f - f_r\|_\wedge \to 0$. It follows that

$$f = \alpha(\varepsilon \otimes g)\beta \in T_\infty \widehat{\otimes} V^*,$$

from which we infer that θ is surjective. □

10.2 REPRESENTING ELEMENTS OF THE PROJECTIVE TENSOR PRODUCT

If E and F are Banach spaces, $u \in E\otimes^\gamma F$, then there exist bounded sequences $x_n \in E$ and $y_n \in F$ with

$$u = \sum_{n=1}^{\infty} \lambda_n(x_n \otimes y_n),$$

where $\|x_n\|, \|y_n\| \leq 1$, and

$$\sum \lambda_n < \|u\|_\wedge + \varepsilon.$$

Perturbing the summands, we may also assume that the sequences $\|x_n\|$ and $\|y_n\|$ converge to zero. Our task is to prove a natural operator space analogue for this result. This will play an important part in our discussion of completely nuclear mappings (see §12.2).

Let us suppose that $v \in M_\infty(V)$, $w \in M_\infty(W)$, $\alpha \in M_{n,\infty\times\infty}$, and $\beta \in M_{\infty\times\infty,n}$. We have $v \otimes w \in M_{\infty\times\infty}(V \widehat{\otimes} W)$ since

$$\left\|(v \otimes w)^{r\times s}\right\|_\wedge = \|v^r \otimes w^s\|_\wedge = \|v^r\| \, \|w^s\| \leq \|v\| \, \|w\|,$$

and thus from (10.1.6) we have a well-defined element

$$u = \alpha(v \otimes w)\beta = \lim_{r\to\infty} \alpha^{r\times r}(v \otimes w)\beta^{r\times r} \in M_n(V \widehat{\otimes} W). \tag{10.2.1}$$

If we let $u_r = \alpha^{r\times r}(v \otimes w)\beta^{r\times r}$, then

$$u_r = \alpha^{r\times r}(v^r \otimes w^r)\beta^{r\times r}, \tag{10.2.2}$$

and thus

$$\|u_r\|_\wedge \leq \|\alpha\| \, \|v\| \, \|w\| \, \|\beta\|.$$

It is evident that the entries of u_r converge to those of u in the norm topology, and thus applying (2.1.8) to the operator space matrix norm $\|\cdot\|_\wedge$,

$$\|u\|_\wedge \leq \|\alpha\| \, \|v\| \, \|w\| \, \|\beta\|.$$

Theorem 10.2.1 *Given operator spaces V and W, and $u \in M_n(V \widehat{\otimes} W)$,*

$$\|u\|_\wedge = \inf\{\|\alpha\|\,\|v\|\,\|w\|\,\|\beta\| : u = \alpha(v \otimes w)\beta\},$$

where the infimum is taken over all such representations with $v \in M_\infty(V)$, $w \in M_\infty(W)$, $\alpha \in M_{n,\infty\times\infty}$, and $\beta \in M_{\infty\times\infty,n}$. Furthermore, we may assume that $v \in K_\infty(V)$ and $w \in K_\infty(W)$.

Proof If $u \in M_n(V \widehat{\otimes} W)$ and $\|u\|_\wedge < 1$, then there exists a sequence of elements $\{u_k\}$ in $M_n(V \otimes_\wedge W)$ such that

$$u = \sum_{k=1}^{\infty} u_k \quad \text{and} \quad \sum_{k=1}^{\infty} \|u_k\|_\wedge < 1.$$

Let ε be any number with $0 < \varepsilon < (1 - \sum_{k=1}^{\infty} \|u_k\|_\wedge)$. For each u_k in $M_n(V \otimes_\wedge W)$ we can choose $v_k \in M_{p_k}(V)$, $w_k \in M_{q_k}(W)$, $\alpha_k \in M_{n,p_k\times q_k}$, and $\beta_k \in M_{p_k\times q_k,n}$ such that

$$u_k = \alpha_k(v_k \otimes w_k)\beta_k$$

and

$$\|\alpha_k\|\|v_k\|\|w_k\|\,\|\beta_k\| < \|u_k\|_\wedge + \varepsilon/2^k\,.$$

Without loss of generality, we can suppose that $\|v_k\| = \|w_k\| = 1$ and $\|\alpha_k\| = \|\beta_k\| < (\|u_k\|_\wedge + \varepsilon/2^k)^{1/2}$. We have

$$\sum \|\alpha_k\|\,\|\beta_k\| < 1.$$

We may choose constants $c_k \geq 1$ such that $c_k \to \infty$, and yet we still have $\sum c_k^2\,\|\alpha_k\|\,\|\beta_k\| < 1$. The matrices

$$v = \sum\nolimits^{\oplus} c_k^{-1} v_k \in K_\infty(V),$$

$$w = \sum\nolimits^{\oplus} c_k^{-1} v_k \in K_\infty(W),$$

$$\alpha = [c_1\alpha_1\ 0_{12}\ \ldots\ 0_{21}\ c_2\alpha_2\ 0_{23}\ \ldots\,] \in M_{n,\infty\times\infty},$$

and the matrix $\beta \in M_{\infty\times\infty,n}$ which is the transpose of

$$[c_1\beta_1\ 0_{12}\ \ldots\ 0_{21}\ c_2\beta_2\ 0_{23}\ \ldots\,],$$

have norm < 1 and satisfy $u = \alpha(v \otimes w)\beta$. □

10.3 ULTRAPRODUCTS

Given an indexed family of operator spaces $(V_s)_{s\in\mathfrak{s}}$, and a free ultrafilter $\mathcal{U}$ on $\mathfrak{s}$ we define the *ultraproduct operator space* to be the quotient operator space

$$\prod_{s\in\mathfrak{s}} V_s/\mathcal{U} = \left(\prod_{s\in\mathfrak{s}} V_s\right)\Big/ J_\mathcal{U},$$

where

$$J_{\mathcal{U}} = \left\{ (v_s)_{s \in \mathfrak{s}} \in \prod_{s \in \mathfrak{s}} V_s : \lim_{\mathcal{U}} \|v_s\| = 0 \right\}.$$

Equivalently, we may identify this with the Banach space ultraproduct (see §A.6), together with the matrix norms determined by the identifications

$$M_n\left(\prod_{s \in \mathfrak{s}} V_s/\mathcal{U}\right) \cong \prod_{s \in \mathfrak{s}} M_n(V_s)/\mathcal{U}.$$

We let $\pi_{\mathcal{U}}$ denote the complete quotient mapping. It follows that if $v = (v_s) \in M_n\left(\prod_{s \in \mathfrak{s}} V_s\right)$, then

$$\|(\pi_{\mathcal{U}})_n((v_s))\| = \lim_{\mathcal{U}} \|v_s\|.$$

As in the case of Banach spaces, if $V_s = V$ for some fixed operator space V, then we write $V^{\mathcal{U}}$ for the *ultrapower operator space* $\prod_{s \in \mathfrak{s}} V_s/\mathcal{U}$. We have a natural embedding

$$\Delta : V \hookrightarrow V^{\mathcal{U}}$$

defined by $\Delta(v) = \pi_{\mathcal{U}}((v_s))$, where we let $v_s = v$ for all $s \in \mathfrak{s}$. The second statement in the following result represents one of the key distinctions between asymptotic products and ultraproducts.

Lemma 10.3.1 *For any operator space V and a free ultrafilter $\mathcal{U}$ on an index set $\mathfrak{s}$, the mapping $\Delta : V \hookrightarrow V^{\mathcal{U}}$ is a completely isometric injection. If V is a finite-dimensional operator space, then Δ is also surjective.*

Proof Again we can restrict our attention to the case $n = 1$. If $v \in V$, then

$$\|\Delta(v)\| = \lim_{\mathcal{U}} \|v_s\| = \lim_{\mathcal{U}} \|v\| = \|v\|.$$

The mapping $\Delta : V \hookrightarrow V^{\mathcal{U}}$ is an isometric injection.

Let us suppose that V is finite-dimensional. If we have an element $\pi_{\mathcal{U}}((v_s)) \in V^{\mathcal{U}}$, where $\|v_s\| \leq 1$, then since $V_{\|\cdot\| \leq 1}$ is compact, the limit $v_0 = \lim_{\mathcal{U}} v_s$ exists in that set (see §A.6). We have

$$\|\pi_{\mathcal{U}}((v_s)) - \Delta(v_0)\| = \lim_{\mathcal{U}} \|v_s - v_0\| = 0,$$

and thus $\Delta(v_0) = \pi_{\mathcal{U}}((v_s))$. □

Let us suppose that $(V_s)_{s \in \mathfrak{s}}$ and $(W_s)_{s \in \mathfrak{s}}$ are two family of operator spaces, and that we are given a family of linear mappings $\varphi_s : V_s \to W_s$ with $\|\varphi_s\|_{cb}$ uniformly bounded. Then we have a corresponding mapping $(\varphi_s)_{s \in \mathfrak{s}} : \prod_{s \in \mathfrak{s}} V_s \to \prod_{s \in \mathfrak{s}} W_s$ given by

$$(\varphi_s)_{s \in \mathfrak{s}}(v_s) = (\varphi_s(v_s))$$

with completely bounded norm

$$\|(\varphi_s)_{s \in \mathfrak{s}}\|_{cb} = \sup\{\|\varphi_s\|_{cb} : s \in \mathfrak{s}\}.$$

Thus, we have an isometric injection

$$\prod_{s\in\mathfrak{s}} \mathcal{CB}(V_s, W_s) \hookrightarrow \mathcal{CB}\left(\prod_{s\in\mathfrak{s}} V_s, \prod_{s\in\mathfrak{s}} W_s\right).$$

In fact, this is a completely isometric inclusion since the operator space matrix norm on $\prod_{s\in\mathfrak{s}} \mathcal{CB}(V_s, W_s)$ is given by

$$M_n\left(\prod_{s\in\mathfrak{s}} \mathcal{CB}(V_s, W_s)\right) = \prod_{s\in\mathfrak{s}} M_n(\mathcal{CB}(V_s, W_s)) = \prod_{s\in\mathfrak{s}} \mathcal{CB}(V_s, M_n(W_s)).$$

Proposition 10.3.2 *Let us suppose that V_s and W_s ($s \in \mathfrak{s}$) are operator spaces. Each '$\mathfrak{s}$-tuple' $(\varphi_s)_{s\in\mathfrak{s}} \in \prod_{s\in\mathfrak{s}} \mathcal{CB}(V_s, W_s)$ determines a mapping*

$$(\varphi_s)_{\mathcal{U}} : \prod_{s\in\mathfrak{s}} V_s/\mathcal{U} \to \prod_{s\in\mathfrak{s}} W_s/\mathcal{U} : \; \pi_{\mathcal{U}}((v_s)) \mapsto \pi_{\mathcal{U}}((\varphi_s(v_s))),$$

which satisfies

$$\|(\varphi_s)_{\mathcal{U}}\|_{cb} \le \lim_{\mathcal{U}} \|\varphi_s\|_{cb}\,.$$

If each $\varphi_s : V_s \to W_s$ is a completely isometric injection (respectively, a complete quotient mapping), then $(\varphi_s)_{\mathcal{U}}$ is a completely isometric injection (respectively, a complete quotient mapping).

Proof For any

$$v = (v_s) \in M_n\left(\prod_{s\in\mathfrak{s}} V_s\right) = \prod_{s\in\mathfrak{s}} M_n(V_s),$$

we have

$$\lim_{\mathcal{U}} \|(\varphi_s)_n(v_s)\| \le \lim_{\mathcal{U}} \|\varphi_s\|_{cb}\|v_s\| = \left(\lim_{\mathcal{U}} \|\varphi_s\|_{cb}\right)\left(\lim_{\mathcal{U}} \|v_s\|\right).$$

This implies

$$\|(\pi_{\mathcal{U}})_n(((\varphi_s)_n(v_s)))\| \le \lim_{\mathcal{U}} \|\varphi\|_{cb} \|(\pi_{\mathcal{U}})_n((v_s))\|$$

for all $(v_s) \in M_n(\prod_{s\in\mathfrak{s}} V_s)$, and

$$\|(\varphi_s)_{\mathcal{U}}\|_{cb} \le \lim_{\mathcal{U}} \|\varphi_s\|_{cb}\,.$$

If each φ_s is a completely isometric injection, then for any

$$\pi_{\mathcal{U}}((v_s)) \in \prod_{s\in\mathfrak{s}} V_s/\mathcal{U}$$

we have

$$\|\pi_{\mathcal{U}}((\varphi_s(v_s)))\| = \lim_{\mathcal{U}} \|\varphi_s(v_s)\| = \lim_{\mathcal{U}} \|v_s\| = \|\pi_{\mathcal{U}}((v_s))\|\,.$$

Therefore, $(\varphi_s)_{\mathcal{U}}$ is an isometric injection from $\prod_{s\in\mathfrak{s}} V_s/\mathcal{U}$ into $\prod_{s\in\mathfrak{s}} W_s/\mathcal{U}$. A simple matrix argument shows that $(\varphi_s)_{\mathcal{U}}$ is a completely isometric injection.

Let us suppose that each φ_s is a complete quotient mapping. Given an element

$$\tilde{w} = (\pi_{\mathcal{U}})_n((w_s)) \in M_n\left(\prod_{s\in\mathfrak{s}} W_s/\mathcal{U}\right)$$

with $\|\tilde{w}\| < 1$, we may assume that $\|w_s\| < 1$ for all $s \in \mathfrak{s}$. By hypothesis we can choose an element $v_s \in M_n(V_s)$ with $\|v_s\| < 1$ and $w_s = (\varphi_s)_n(v_s)$. If we let $v = (v_s)$, then

$$\tilde{w} = ((\varphi_s)_{\mathcal{U}})_n((\pi_{\mathcal{U}})_n(v))$$

with $\|(\pi_{\mathcal{U}})_n(v)\| < 1$. □

Given operator spaces V_s and W_s ($s \in \mathfrak{s}$), it follows from Proposition 10.3.2 that we have a well-defined (complete) contraction

$$\prod_{s\in\mathfrak{s}} \mathcal{CB}(V_s, W_s)/\mathcal{U} \to \mathcal{CB}\left(\prod_{s\in\mathfrak{s}} V_s/\mathcal{U}, \prod_{s\in\mathfrak{s}} W_s/\mathcal{U}\right) : \pi_{\mathcal{U}}((\varphi_s)) \mapsto (\varphi_s)_{\mathcal{U}}. \tag{10.3.1}$$

However, in contrast to Banach space theory, this mapping need not be a completely isometric injection. This represents a marked difference between Banach spaces and operator spaces, and deserves further comment.

If N is a finite-dimensional operator space, and $V_s = N$ for all $s \in \mathfrak{s}$, then we obtain the complete contraction

$$\prod_{s\in\mathfrak{s}} \mathcal{CB}(N, W_s)/\mathcal{U} \to \mathcal{CB}\left(N^{\mathcal{U}}, \prod_{s\in\mathfrak{s}} W_s/\mathcal{U}\right) = \mathcal{CB}\left(N, \prod_{s\in\mathfrak{s}} W_s/\mathcal{U}\right),$$

or letting $L = N^*$, we have the corresponding complete contraction

$$\theta : \prod_{s\in\mathfrak{s}} (L \check{\otimes} W_s)/\mathcal{U} \to L \check{\otimes} \prod_{s\in\mathfrak{s}} W_s/\mathcal{U}.$$

Pisier proved that θ is completely isometric (for arbitrary families W_s) if and only if L is an *exact* finite-dimensional operator space (we shall prove this in Theorem 14.4.3).

Proposition 10.3.3 *If $\{V_s\}_{s\in\mathfrak{s}}$ is a family of operator spaces then for each $n \in \mathbb{N}$, the mapping*

$$\pi_{\mathcal{U}}((\varphi_s)) \in \prod_{s\in\mathfrak{s}} \mathcal{CB}(V_s, M_n)/\mathcal{U} \mapsto (\varphi_s)_{\mathcal{U}} \in \mathcal{CB}\left(\prod_{s\in\mathfrak{s}} V_s/\mathcal{U}, M_n\right)$$

is a (completely) isometric injection.

Proof Let us choose $n \in \mathbb{N}$, and an element $\pi_{\mathcal{U}}((\varphi_s)) \in \prod_{s\in\mathfrak{s}} \mathcal{CB}(V_s, M_n)/\mathcal{U}$. Given $s \in \mathfrak{s}$, we have from Proposition 2.2.2 that

$$\|\varphi_s\|_{cb} = \|(\varphi_s)_n\|.$$

For arbitrary $\varepsilon > 0$, there exists a $v_s \in M_n(V_s)$ with $\|v_s\| = 1$ such that

$$\|(\varphi_s)_n(v_s)\| > \|(\varphi_s)_n\| - \varepsilon = \|\varphi_s\|_{cb} - \varepsilon .$$

Therefore,

$$\|(\varphi_s)_\mathcal{U}\|_{cb} \geq \lim_\mathcal{U} \|(\varphi_s)_n(v_s)\| \geq \lim_\mathcal{U} \|\varphi_s\|_{cb} - \varepsilon .$$

Since $\varepsilon > 0$,

$$\|(\varphi_s)_\mathcal{U}\|_{cb} \geq \lim_\mathcal{U} \|\varphi_s\|_{cb} .$$

Thus, the mapping

$$\prod_{s\in\mathfrak{s}} \mathcal{CB}(V_s, M_n)/\mathcal{U} \ni \pi_\mathcal{U}((\varphi_s)) \mapsto (\varphi_s)_\mathcal{U} \in \mathcal{CB}\Big(\prod_{s\in\mathfrak{s}} V_s/\mathcal{U}, M_n\Big)$$

is an isometric injection. It is also a complete isometry since for each $m \in \mathbb{N}$ the mapping

$$\begin{aligned} M_m\left(\prod_{s\in\mathfrak{s}} \mathcal{CB}(V_s, M_n))/\mathcal{U}\right) &= \prod_{s\in\mathfrak{s}} \mathcal{CB}(V_s, M_{mn})/\mathcal{U} \\ &\to \mathcal{CB}(V_s/\mathcal{U}, M_{mn}) \\ &= M_m\left(\mathcal{CB}\left(\prod_{s\in\mathfrak{s}} V_s/\mathcal{U}, M_n\right)\right) \end{aligned}$$

is isometric. □

Corollary 10.3.4 *For any indexed family of operator spaces V_s $(s \in \mathfrak{s})$, the natural mapping*

$$\pi_\mathcal{U}((\varphi_s)) \in \prod_{s\in\mathfrak{s}} V_s^*/\mathcal{U} \to (\varphi_s)_\mathcal{U} \in \left(\prod_{s\in\mathfrak{s}} V_s/\mathcal{U}\right)^*,$$

determined by $\langle(\varphi_s)_\mathcal{U}, \pi_\mathcal{U}((v_s))\rangle = \lim_\mathcal{U}\langle\varphi_s, v_s\rangle$, is a completely isometric injection. □

Corollary 10.3.5 *For any indexed family of operator spaces V_s $(s \in \mathfrak{s})$, we have a completely isometric inclusion*

$$\prod_{s\in\mathfrak{s}} V_s^*/\mathcal{U} \hookrightarrow \left(\prod_{s\in\mathfrak{s}} V_s\right)^* .$$

Proof The complete quotient mapping

$$\pi_\mathcal{U} : \prod_{s\in\mathfrak{s}} V_s \to \prod_{s\in\mathfrak{s}} V_s/\mathcal{U}$$

determines a completely isometric inclusion

$$\pi_\mathcal{U}^* : \left(\prod_{s\in\mathfrak{s}} V_s/\mathcal{U}\right)^* \hookrightarrow \left(\prod_{s\in\mathfrak{s}} V_s\right)^* .$$

and the result follows from Corollary 10.3.4. □

An important fact in Banach space theory states that an ultraproduct of L_1-spaces is again an L_1-space. Groh (1984) has proved that an ultrapower of von Neumann algebraic preduals is again the predual of a von Neumann algebra. Since we are more concerned with ultrapowers of *injective* von Neumann algebras, this general result will not play a role in the subsequent theory. We shall need only a very special case of this (see the proof of Theorem 15.4.1).

Proposition 10.3.6 *Suppose that $\mathcal{U}$ is a free ultrafilter on an index set $\mathfrak{s}$. Then the ultrapower $T_\infty^{\mathcal{U}}$ is completely isometric to the predual of a von Neumann algebra.*

Proof We have a natural completely isometric injection

$$\theta : T_\infty^{\mathcal{U}} \to \left(\prod_{s\in\mathfrak{s}} M_\infty\right)^* \tag{10.3.2}$$

defined by the pairing

$$\langle \theta(\pi_{\mathcal{U}}((\omega_s)), (y_s)\rangle = \lim_{\mathcal{U}} \langle \omega_s, y_s\rangle .$$

We may regard $(\prod_{s\in\mathfrak{s}} M_\infty)^*$ as a bimodule over $\prod_{s\in\mathfrak{s}} M_\infty$ or $(\prod_{s\in\mathfrak{s}} M_\infty)^{**}$ in the usual manner. The subspace $\mathcal{T} = \theta(T_\infty^{\mathcal{U}})$ is a norm closed two-sided $\prod_{s\in\mathfrak{s}} M_\infty$ submodule since if $f = \pi_{\mathcal{U}}((f_s)) \in T_\infty^{\mathcal{U}}$ and $x = (x_s) \in \prod_{s\in\mathfrak{s}} M_\infty$, then we have $(x_s f_s) \in \prod_{s\in\mathfrak{s}} T_\infty$ and thus

$$xf = \theta(\pi_{\mathcal{U}}((x_s f_s))) \in \mathcal{T}.$$

The same argument shows that $fx \in \mathcal{T}$. We conclude that the annihilator of $\mathcal{T}$ is a weak* closed two-sided ideal in the von Neumann algebra $(\prod_{s\in\mathfrak{s}} M_\infty)^{**}$, and in particular there is a central projection e in $(\prod_{s\in\mathfrak{s}} M_\infty)^{**}$ for which

$$\mathcal{T} = \left(\prod_{s\in\mathfrak{s}} M_\infty\right)^* e = \left[\left(\prod_{s\in\mathfrak{s}} M_\infty\right)^{**} e\right]_* . \tag{10.3.3}$$

□

Corollary 10.3.7 *If $(V_s)_{s\in\mathfrak{s}}$ is a family of n-dimensional operator spaces, then $\prod_{s\in\mathfrak{s}} V_s/\mathcal{U}$ is again an n-dimensional operator space.*

Proof Let us suppose that V is an n-dimensional operator space. From Corollary 2.2.5 there exist inverse linear isomorphisms $\varphi_s : V_s \to V$ and $\psi_s : V \to V_s$ with $\|\varphi_s\|_{cb} \le n$ and $\|\psi_s\|_{cb} \le n$. It follows that

$$\varphi = (\varphi_s)_{\mathcal{U}} : \prod_{s\in\mathfrak{s}} V_s/\mathcal{U} \to V^{\mathcal{U}} \cong V$$

satisfies $\|\varphi\|_{cb} \leq n$, and it has the inverse mapping $\psi = (\psi_s)_{\mathcal{U}}$, which satisfies $\|\psi\|_{cb} \leq n$. □

Turning to the projective tensor product, the following is often useful.

Lemma 10.3.8 *Given an index set $\mathfrak{s}$, a family of operator spaces $\{V_s\}_{s\in\mathfrak{s}}$, and an ultrafilter $\mathcal{U}$ on $\mathfrak{s}$, we have a natural complete isometry*

$$T_n \widehat{\otimes} \prod_{s\in\mathfrak{s}} V_s/\mathcal{U} \cong \prod_{s\in\mathfrak{s}} (T_n \widehat{\otimes} V_s)/\mathcal{U}.$$

Proof Let us first show that this is an isometric identification. Given $\dot{v} \in \prod_{s\in\mathfrak{s}}(T_n \widehat{\otimes} V_s)/\mathcal{U}$ with $\|\dot{v}\| < 1$, we may suppose that $\dot{v} = \pi_{\mathcal{U}}(v)$, where $v = (v(s))_{s\in\mathfrak{s}}$, and

$$v(s) = [v(s)_{i,j}] \in T_n \widehat{\otimes} V_s$$

satisfies

$$\lim_{\mathcal{U}} \|[v_{i,j}(s)]\|_{T_n \widehat{\otimes} V_s} < 1.$$

Since we have the norm identification $T_n \widehat{\otimes} V_s \cong T_n(V_s)$ (see (7.1.20)), we may choose $a(s), b(s) \in HS_n$ and $\tilde{v}(s) \in M_n(V_s)$ with $v(s) = a(s)\tilde{v}(s)b(s)$, $\|a(s)\|_2 = \|b(s)\|_2 = 1$, and

$$\lim_{\mathcal{U}} \|\tilde{v}(s)\|_{M_n(V_s)} < 1.$$

If $\tilde{v} = (\tilde{v}(s))$, then $(\pi_{\mathcal{U}})_n(\tilde{v}) \in M_n(\prod_{s\in\mathfrak{s}} V_s/\mathcal{U})$ satisfies $\|(\pi_{\mathcal{U}})_n(\tilde{v})\| < 1$. Since the unit ball of HS_n is compact, we may define $a_0 = \lim_{\mathcal{U}} a(s) \in HS_n$. If we similarly let $b_0 = \lim_{\mathcal{U}} b(s) \in HS_n$, then

$$\lim_{\mathcal{U}} \|a_0\tilde{v}(s)b_0 - a(s)\tilde{v}(s)b(s)\|_{T_n \widehat{\otimes} V_s} = 0;$$

hence

$$\begin{aligned} \dot{v} &= (\pi_{\mathcal{U}})_n((v(s)) = (\pi_{\mathcal{U}})_n((a(s)\tilde{v}(s)b(s))) \\ &= (\pi_{\mathcal{U}})_n(a_0(\tilde{v}(s))b_0) = a_0(\pi_{\mathcal{U}})_n(\tilde{v})b_0 \end{aligned}$$

and

$$\|\dot{v}\|_{T_n \widehat{\otimes} \prod_{s\in\mathfrak{s}} V_s/\mathcal{U}} \leq 1.$$

On the other hand, if

$$\dot{v} = [\dot{v}_{i,j}] \in T_n \widehat{\otimes} \prod_{s\in\mathfrak{s}} V_s/\mathcal{U} = T_n\left(\prod_{s\in\mathfrak{s}} V_s/\mathcal{U}\right)$$

satisfies $\|\dot{v}\| < 1$, then we may choose $a, b \in HS_n$ with $\|a\|_2, \|b\|_2 < 1$ and an element $\tilde{v} \in M_n(\prod_{s\in\mathfrak{s}} V_s/\mathcal{U})$ with $\|\tilde{v}\| < 1$ and $\dot{v} = a\tilde{v}b$. We may suppose that $\tilde{v} = (\pi_{\mathcal{U}})_n(w)$, where $w \in M_n(\prod_{s\in\mathfrak{s}} V_s)$ satisfies

$$\lim_{\mathcal{U}} \|w(s)\| < 1.$$

If we let $v(s) = aw(s)b$, then it follows that

$$\dot{v} = (\pi_{\mathcal{U}})_n((v(s)),$$

where $\lim_{\mathcal{U}} \|v(s)\|_{T_n \hat{\otimes} V_s} \leq 1$, and thus

$$\|\dot{v}\|_{\prod_{s\in\mathfrak{s}}(T_n \hat{\otimes} V_s)/\mathcal{U}} \leq 1.$$

Finally, it follows from the above result that for any $r \in \mathbb{N}$, we have the natural isometry

$$T_r \hat{\otimes} T_n \hat{\otimes} \prod_{s\in\mathfrak{s}} V_s/\mathcal{U} \cong \prod_{s\in\mathfrak{s}}(T_r \hat{\otimes} T_n \hat{\otimes} V_s)/\mathcal{U} \cong T_r \hat{\otimes} \prod_{s\in\mathfrak{s}}(T_n \hat{\otimes} V_s)/\mathcal{U}.$$

Hence, from Theorem 4.1.8 our identification is a complete isometry. □

Corollary 10.3.9 *Suppose that Q is a complete quotient of T_n for some $n \in \mathbb{N}$. Then given an index set $\mathfrak{s}$, a family of operator spaces $\{V_s\}_{s\in\mathfrak{s}}$, and an ultrafilter $\mathcal{U}$ on $\mathfrak{s}$, we have a natural complete isometry*

$$Q \hat{\otimes} \prod_{s\in\mathfrak{s}} V_s/\mathcal{U} \cong \prod_{s\in\mathfrak{s}}(Q \hat{\otimes} V_s)/\mathcal{U}.$$

Proof Let $q : T_n \to Q$ be a complete quotient mapping, and consider the commutative diagram

$$\begin{array}{ccc} T_n \hat{\otimes} \prod_{s\in\mathfrak{s}} V_s/\mathcal{U} & \longrightarrow & \prod_{s\in\mathfrak{s}}(T_n \hat{\otimes} V_s)/\mathcal{U} \\ {\scriptstyle q\otimes id}\downarrow & & \downarrow \\ Q \hat{\otimes} \prod_{s\in\mathfrak{s}} V_s/\mathcal{U} & \longrightarrow & \prod_{s\in\mathfrak{s}}(Q \hat{\otimes} V_s)/\mathcal{U} \end{array} .$$

From Proposition 10.3.8, the top row is completely isometric, and from Propositions 7.1.7 and 10.3.2, the columns are complete quotient mappings. Thus, the bottom row is completely isometric. □

10.4 NOTES AND REFERENCES

Infinite matrix calculations for operator spaces were first considered in Effros and Ruan (1990). The ultraproduct theory of §10.3 is due to Pisier (1996*b*, 1998).

Part III

The Grothendieck Programme

11
The approximation property

Perhaps the most straightforward approach to understanding the structure of Banach spaces is to seek analogues of the orthonormal bases in a Hilbert space. The existence of a basis of one sort or another for a Banach space E generally implies that one can 'strongly' approximate E by its finite-dimensional subspaces. Grothendieck made a concerted effort to understand various notions of approximability, including what are now known as the 'approximation property' and the more restrictive 'metric approximation property'.

Enflo provided the first example of a separable Banach space without the approximation property. Owing to the seemingly *ad hoc* constructions of separable Banach spaces which do not satisfy the approximation property or the metric approximability property, the further study of such spaces might seem premature. What would be desirable is a context in which approximability provides an invariant related to some phenomenon outside of Banach space theory. It seems likely that the theory of operator spaces will shed some light on these problems. Szankowski (1981) proved that M_∞ is an example of a (non-separable) Banach space which does not have the approximation property, and it has been conjectured that the same might be true for the reduced group C^*-algebra of $SL(3, \mathbb{Z})$, which is separable.

We are, of course, more concerned with the operator space analogues of these approximation properties. The appropriate version of the metric approximation property for operator spaces was first considered by Haagerup. In contrast to the situation in Banach space theory, this notion provides crucial invariants for distinguishing *natural* C^*-algebras. Cowling and Haagerup showed that one can calculate precise approximation invariants for various important group C^*-algebras by relating them to the harmonic analysis on the groups. These provide new methods for studying non-amenable groups.

In this chapter we consider the operator space version of the approximability property. Although this theory is not as well understood as Haagerup's metric version, it is more interesting from the operator space point of view. A step toward proving the conjecture mentioned above would be to prove that the reduced group C^*-algebra of $SL(3, \mathbb{Z})$ does not satisfy the operator space approximation property.

11.1 THE GROTHENDIECK APPROXIMATION PROPERTY

A Banach space E is said to have the *approximation property* if, for every compact set K of E and every $\varepsilon > 0$, there exists a finite-rank mapping $\varphi : E \to E$ such that

$$\|\varphi(x) - x\| < \varepsilon$$

for all $x \in K$. Since the analogue of compactness for subsets of operator spaces is not well understood at this time, we begin by considering an equivalent property. We say that a sequence $x_n \in E$ $(n \in \mathbb{N})$ is a *null sequence* if $\lim x_n = 0$. Owing to the following result, it suffices to verify uniform convergence on sets of the form $S = \{x_n : n \in \mathbb{N}\}$, with x_n a null sequence.

Lemma 11.1.1 *Suppose that E is a Banach space. Then a subset K of E is compact if and only if K is contained in the closed convex hull of a null sequence.*

Proof Let us assume that K is compact. For each $n \in \mathbb{N}$, we may choose a finite subset F_n of K which is 4^{-n-1} dense in K, by which we mean that

$$d(x, F_n) \leq \frac{1}{4^{n+1}}$$

for all $x \in K$. Since $S = \bigcup F_n$ is dense in K, it suffices to show that S is contained in the closed convex hull of a null sequence.

For each $y \in F_{n+1}$, we may select an element $y' \in F_n$ such that $\|y - y'\| \leq \frac{1}{4^{n+1}}$. It follows that if $y_{n+1} \in F_{n+1}$, then we may inductively choose a sequence $y_n, \ldots, y_1$ with $y_k \in F_k$ and $\|y_{k+1} - y_k\| \leq \frac{1}{4^{k+1}}$ for $1 \leq k \leq n$. Then

$$y_{n+1} = y_1 + (y_2 - y_1) + \cdots + (y_{n+1} - y_n),$$

and we may write this as a convex combination:

$$y_{n+1} = \frac{1}{2} z_0 + \frac{1}{2^2} z_1 + \cdots + \frac{1}{2^{n+1}} z_n + \frac{1}{2^{n+1}} 0,$$

where $z_0 = 2y_1$ and

$$z_k = 2^{k+1}(y_{k+1} - y_k)$$

for $k \geq 1$. For each $k \geq 1$, $\|z_k\| \leq \frac{1}{2^{k+1}}$, and thus z_k lies in the finite set

$$A_k = \left(2^{k+1}\left(F_{n+1} - F_n\right)\right) \cap E_{\|\cdot\| \leq \frac{1}{2^{k+1}}}.$$

It follows that S is in the convex hull of the countable set

$$A = \{0\} \cup 2F_1 \cup A_1 \cup A_2 \cup \cdots.$$

If we let $\{a_1, a_2, \ldots\}$ be the distinct elements in A, it is evident that a_k is either a finite or a null sequence.

Conversely, if z_n is a null sequence, then $Z = \{z_n : n \in \mathbb{N}\} \cup \{0\}$ is norm compact. It follows that Z is totally bounded, i.e. if $\varepsilon > 0$ is given, then there exists a finite set $F \subseteq Z$ with $Z \subseteq F + B_\varepsilon$, where we let $B_\varepsilon = E_{\|\cdot\| < \varepsilon}$. But that implies that the convex hull $\mathrm{co}(Z)$ is totally bounded. To see this, let us suppose that $\varepsilon > 0$, and choose a finite set F with $Z \subseteq F + B_{\varepsilon/2}$. Since $\mathrm{co}(F)$ is compact, we may find a finite set $F_1 \subseteq \mathrm{co}(F)$ with

$$\mathrm{co}(F) \subseteq F_1 + B_{\varepsilon/2}.$$

Z is a subset of the convex set $\mathrm{co}(F) + B_{\varepsilon/2}$, and thus

$$\mathrm{co}(Z) \subseteq \mathrm{co}(F) + B_{\varepsilon/2} \subseteq F_1 + B_{\varepsilon/2} + B_{\varepsilon/2} \subseteq F_1 + B_\varepsilon.$$

Since E is complete, the closure of $\mathrm{co}(Z)$ is compact. □

The space $c_0(E) = c_0 \otimes^\lambda E$ consists of the null sequences $x = (x_n)$ in E, together with the norm

$$\|(x_n)\|_\lambda = \sup\{\|x_n\|\}.$$

Any bounded linear contraction $\varphi : E \to E$ determines a corresponding linear contraction

$$\varphi^\infty = id \otimes \varphi : c_0(E) \to c_0(E),$$

and for any sequence $x = (x_n) \in c_0(E)$,

$$\|\varphi^\infty(x) - x\| = \sup\{\|\varphi(x_n) - x_n\| : n \in \mathbb{N}\}.$$

Corollary 11.1.2 *If E is a Banach space, then E has the approximation property if and only if for each $\varepsilon > 0$ and element $x \in c_0(E)$, there exists a finite-rank mapping $\varphi : E \to E$ such that*

$$\|\varphi^\infty(x) - x\| < \varepsilon.$$

Equivalently, there is a net of finite-rank mappings $\varphi_\nu : E \to E$ for which the mappings

$$\varphi_\nu^\infty : c_0(E) \to c_0(E)$$

converge to the identity mapping in the point-norm topology.

Proof We note that if we have null sequences $x = (x_n)$ and $y = (y_n)$, then $z = (x_1, y_1, x_2, \ldots)$ is a null sequence, and if we approximate the identity mapping uniformly on the entries of z, we shall simultaneously approximate the entries of x and y. The remainder of the argument is routine. □

11.2 THE OPERATOR SPACE APPROXIMATION PROPERTY

Turning to operator spaces, we replace the Banach spaces E and $c_0(E)$ by the operator spaces V and $K_\infty(V) = K_\infty \check{\otimes} V$. We let $\varphi_\infty = id_{K_\infty} \otimes \varphi$. We recall that we assume that all operator spaces are complete unless otherwise indicated.

We define the *stable point-norm topology* $\mathcal{T}_n$ on $\mathcal{CB}(V, W)$ to be the weakest topology in which the seminorms

$$\varphi \mapsto \|\varphi_\infty(v)\| \qquad (v \in K_\infty(V))$$

are continuous. We say that V has the *operator space approximation property* if there exists a net of finite-rank mappings $\varphi_\nu \in \mathcal{CB}(V,V)$ converging to the identity mapping id_V in this topology. Equivalently, for each finite set $v_1,\ldots,v_n \in K_\infty(V)$, there is a finite-rank mapping $\varphi \in \mathcal{CB}(V,W)$ such that

$$\|\varphi_\infty(v_k) - v_k\| < \varepsilon \tag{11.2.1}$$

for $k = 1,\ldots,n$. Since we may identify $K_\infty \oplus \cdots \oplus K_\infty$ with a subalgebra of $M_n(K_\infty)$ and we may identify the latter with K_∞, it suffice to find approximations (11.2.1) for a single element $v \in K_\infty(V)$.

We identify the algebraic tensor product $V^* \otimes V$ with the finite-rank operators in $\mathcal{CB}(V,V)$.

Lemma 11.2.1 *Suppose that V is a complete operator space. Then the following are equivalent:*

(i) *V has the operator space approximation property;*
(ii) *$V^* \otimes V$ is dense in $(\mathcal{CB}(V,V), \mathcal{T}_n)$;*
(iii) *for all operator spaces W, $V^* \otimes W$ is dense in $(\mathcal{CB}(V,W), \mathcal{T}_n)$;*
(iv) *for all operator spaces W, $W^* \otimes V$ is dense in $(\mathcal{CB}(W,V), \mathcal{T}_n)$.*

Proof The equivalence (i)$\Leftrightarrow$(ii) is immediate from the definitions. If φ_ν is a net in $V^* \otimes V$ converging to $id : V \to V$ in the $\mathcal{T}_n$ topology, then for any $\psi \in \mathcal{CB}(V,W)$, $\psi \circ \varphi_\nu \in V^* \otimes W$ converges to ψ in that topology. Similarly, if $\psi \in \mathcal{CB}(W,V)$, then $\varphi_\nu \circ \psi \in W^* \otimes V$ converges to ψ in that topology. This proves (i)$\Rightarrow$(iii) and (i)$\Rightarrow$ (iv). We obtain (iii)$\Rightarrow$(ii) and (iv)$\Rightarrow$(ii) by letting $W = V$. □

If an operator space V has the operator space approximation property, then the underlying Banach space has the classical approximation property. This follows since we may identify $c_0(V)$ with the diagonal matrices in $K_\infty(V)$. As we remarked at the beginning of this chapter, there exist separable C^*-algebras that do not have the Banach space approximation property.

We define the *stable point-weak* topology $\mathcal{T}_w$ on $\mathcal{CB}(V,W)$ to be the weakest topology in which the seminorms

$$\mathcal{CB}(V,W) \to [0,\infty) : \varphi \mapsto \omega(\varphi_\infty(v)) \tag{11.2.2}$$

are continuous for all $v \in K_\infty(V)$ and $\omega \in K_\infty(W)^*$. Although the $\mathcal{T}_w$ topology is weaker than the $\mathcal{T}_n$ topology, we can use a standard convexity argument to show that these topologies have the same continuous linear functionals.

Proposition 11.2.2 *Suppose that V and W are operator spaces and F is a linear functional on $\mathcal{CB}(V,W)$. Then F is stably point-norm continuous if and only if it is stably point-weakly continuous.*

Proof It is trivial that stable point-weak continuity implies stable point-norm continuity. Conversely, given $F \in (\mathcal{CB}(V,W),\mathcal{T}_n)^*$, there exists a constant C and finitely many elements $v_k \in K_\infty(V)$ $(1 \le k \le n)$ such that

$$|F(\varphi)| \le C \max\{\|\varphi_\infty(v_k)\| : k = 1,\ldots,n\} \tag{11.2.3}$$

for all $\varphi \in \mathcal{CB}(V,W)$. Since it is easy to construct an embedding

$$K_\infty \oplus \cdots \oplus K_\infty \hookrightarrow K_\infty,$$

we may find a single element $v_1 \in K_\infty(V)$ with

$$|F(\varphi)| \le C\|\varphi_\infty(v_1)\|.$$

If we define $\theta : \mathcal{CB}(V,W) \to K_\infty(W)$ by $\theta(\varphi) = \varphi_\infty(v_1)$, then it follows that the relation

$$\omega(\varphi_\infty(v_1)) = F(\varphi)$$

determines a bounded linear functional ω on $\theta(\mathcal{CB}(V,W))$. We use the Hahn–Banach theorem to extend ω to a bounded linear functional $\tilde{\omega}$ on $K_\infty(W)$, and we then have

$$F(\varphi) = \langle \varphi_\infty(v_1), \tilde{\omega}\rangle,$$

and thus F is continuous in the stable point-weak topology. □

Corollary 11.2.3 *An operator space V satisfies the operator space approximation property if and only if $V^* \otimes V$ is dense in $(\mathcal{CB}(V,V),\mathcal{T}_w)$.*

Proof From Proposition 11.2.2 and the geometric Hahn–Banach theorem, a convex set in $\mathcal{CB}(V,V)$ is closed in the $\mathcal{T}_n$ topology if and only if it is closed in the $\mathcal{T}_w$ topology. If we apply this to $V^* \otimes V$, then the result follows from Theorem 11.2.1. □

We next give an important characterization of the $\mathcal{T}_w$ topology.

Lemma 11.2.4 *Suppose that V and W are operator spaces and consider the natural embedding*

$$\mathcal{CB}(V,W) \subseteq \mathcal{CB}(V,W^{**}) = (V \hat{\otimes} W^*)^*. \tag{11.2.4}$$

The stable point-weak topology on $\mathcal{CB}(V,W)$ is just the relative weak topology determined by $V \hat{\otimes} W^*$, and thus each stable point-weakly continuous functional on $\mathcal{CB}(V,W)$ is determined by an element of $V \hat{\otimes} W^*$.*

Proof From Theorem 5.3.4 the functionals

$$\omega \in K_\infty(W)^* = T_\infty \hat{\otimes} W^*$$

are given by

$$\omega(\gamma \otimes w) = \alpha(\gamma \otimes g(w))\beta,$$

where $g \in M_{\infty\times\infty}(W^*)$ and $\alpha \in M_{1,\infty\times\infty}$ and $\beta \in M_{\infty\times\infty,1}$. If $v = \gamma \otimes v_0$, where $\gamma \in K_\infty$ and $v_0 \in V$, then

$$\langle \varphi_\infty(v), \omega\rangle = \alpha(\gamma \otimes g(\varphi(v_0)))\beta$$

$$= \sum \alpha_{1,(i,k)}(\gamma_{i,j} \otimes g_{k,l}(\varphi(v_0)) \beta_{(j,l),1}$$
$$= \left\langle \varphi, \sum \alpha_{1,(i,k)}(\gamma_{i,j} \otimes v_0 \otimes g_{k,l}) \beta_{(j,l),1} \right\rangle$$
$$= \langle \varphi, \alpha(v \otimes g)\beta \rangle.$$

By continuity this extends to all $v \in K_\infty(V)$; hence, letting

$$u = \alpha(v \otimes g)\beta \in V \widehat{\otimes} W^*,$$

we see that

$$\langle \varphi_\infty(v), \omega \rangle = \langle \varphi, u \rangle.$$

Conversely, given an element $u \in V \widehat{\otimes} W^*$, we have

$$u = \alpha(v \otimes g)\beta,$$

where $\alpha \in M_{1,\infty\times\infty}$, $\beta \in M_{\infty\times\infty,1}, v \in K_\infty(V)$, and $g \in K_\infty(W^*)$. Reversing the above calculation,

$$\langle \varphi, u \rangle = \langle \varphi_\infty(v), \omega \rangle,$$

where $\omega = \alpha(id_{K_\infty} \otimes g)\beta$. It follows that $\mathcal{T}_w$ convergence in $\mathcal{CB}(V, W)$ is equivalent to convergence with respect to the $V \widehat{\otimes} W^*$ topology. It is wel-known that this implies the last statement. □

We can use the inclusion (11.2.4) to define a completely contractive pairing

$$\langle \cdot, \cdot \rangle : V \widehat{\otimes} W^* \times \mathcal{CB}(V, W) : (v \otimes f, \varphi) \mapsto f(\varphi(v)). \qquad (11.2.5)$$

From Lemma 11.2.4 this determines the $\mathcal{T}_w$ or, equivalently, the $\mathcal{T}_n$ continuous functionals on $\mathcal{CB}(V, W)$. On the other hand, if we let $V = W$, then the mapping $id_V \in \mathcal{CB}(V, V)$ determines the trace functional on $V \widehat{\otimes} V^*$ (see (7.1.12)) in the sense that

$$\operatorname{trace}(u) = \langle u, id_V \rangle,$$

since if $v \in V$ and $f \in V^*$,

$$\langle v \otimes f, id_V \rangle = f(v) = \operatorname{trace}(v \otimes f).$$

The natural mapping

$$\Phi_0 : V \otimes V^* \to \mathcal{CB}(V, V)$$

is one-to-one and its image is the subspace $\mathcal{FCB}(V, V)$ of finite-rank mappings. Thus, we may define

$$\operatorname{trace} : \mathcal{FCB}(V, V) \to \mathbb{C}$$

by letting

$$\operatorname{trace}(\Phi_0(u)) = \operatorname{trace}(u).$$

On the other hand, the canonical mapping

$$\Phi : V \widehat{\otimes} V^* \to \mathcal{CB}(V, V)$$

(see (8.1.12)) need not be one-to-one, and hence this procedure does not determine a well-defined linear functional on the range of Φ. As we shall see in the next result, this phenomenon is closely related to the operator approximation property.

In order to include one more condition, it is necessary to introduce another notion. A dual operator space V^* is said to have the *dual slice mapping property* if

$$V^* \overline{\otimes} W^* = V^* \overline{\otimes}_{\mathcal{F}} W^*$$

for all operator spaces W.

Theorem 11.2.5 *Suppose that V is an operator space. Then the following are equivalent:*

(i) *V has the operator space approximation property;*

(ii) *for any operator space W, the canonical mapping*

$$\Phi : V \widehat{\otimes} W \to V \check{\otimes} W \subseteq \mathcal{CB}(V^*, W) \tag{11.2.6}$$

is one-to-one;

(iii) *$\Phi : V \widehat{\otimes} V^* \to V \check{\otimes} V^*$ is one-to-one;*

(iv) *if $u \in V \widehat{\otimes} V^*$ is such that $\Phi(u) = 0$, then* $\operatorname{trace}(u) = 0$*;*

(v) *V^* has the dual slice mapping property.*

Proof (i)$\Rightarrow$(ii). Given a contractive element $u \in V \widehat{\otimes} W$ with $\Phi(u) = 0$, we may suppose that

$$u = \alpha(v \otimes w)\beta,$$

where $v \in K_\infty(V)$, $w \in K_\infty(W)$, and all the matrices are of norm < 1. Given $f_0 \in V^*$ and $v_0 \in V$,

$$\varphi_0 = f_0 \otimes v_0 : V \to V$$

has the property that

$$\varphi_0 \otimes id(u) \in V \otimes W,$$

since

$$\begin{aligned} \varphi_0 \otimes id(u) &= \alpha(f_0(v_{i,j})v_0 \otimes w_{k,l})\beta \\ &= \sum \alpha_{i,k}(f_0(v_{i,j})v_0 \otimes w_{k,l})\beta_{j,l} \\ &= v_0 \otimes \left[\sum_{i,j,k,l} \alpha_{i,k} f_0(v_{i,j}) w_{k,l} \beta_{j,l}\right] = v_0 \otimes w_0, \end{aligned}$$

where $f_0(v) \in K_\infty(\mathbb{C})$ implies that

$$w_0 = \alpha\langle\langle f_0(v) \otimes w\rangle\rangle\beta \in \mathbb{C} \widehat{\otimes} W = W.$$

Thus, if $\varphi : V \to V$ is any finite-rank mapping, then $\varphi \otimes id(u) \in V \otimes W$.

For any $\varepsilon > 0$, we may, by assumption, choose a finite-rank mapping $\varphi_\varepsilon : V \to V$ such that

$$\|(\varphi_\varepsilon)_\infty(v) - v\| < \varepsilon.$$

It follows that if

$$u_\varepsilon = \varphi_\varepsilon \otimes id(u) = \alpha((\varphi_\varepsilon)_\infty(v) \otimes w)\beta,$$

then

$$\|u - u_\varepsilon\|_\wedge \leq \|\alpha\| \|v - (\varphi_\varepsilon)_\infty(v)\| \|w\| \|\beta\| < \varepsilon.$$

We may regard $\Phi(u)$ as an element of $\mathcal{CB}(V^*, W)$ (see (11.2.6)). For any $\varphi \in \mathcal{CB}(V, V)$, we have

$$\Phi((\varphi \otimes id)(u)) = \Phi(u) \circ \varphi^*,$$

since if $f \in V^*$ and $g \in W^*$, then

$$\begin{aligned} \langle \Phi((\varphi \otimes id)(u))(f), g \rangle &= \langle (\varphi \otimes id)(u), f \otimes g \rangle \\ &= \langle u, \varphi^*(f) \otimes g \rangle = \langle \Phi(u)(\varphi^*(f)), g \rangle. \end{aligned}$$

It follows that $\Phi(u_\varepsilon) = \Phi(u) \circ \varphi_\varepsilon^* = 0$, and since Φ restricts to the identity on $V \otimes W$, $u_\varepsilon = 0$. Thus, $\|u\|_\wedge < \varepsilon$, and since $\varepsilon > 0$ was arbitrary, $u = 0$.

(ii)$\Rightarrow$(iii) and (iii)$\Rightarrow$(iv) are trivial.

(iv)$\Rightarrow$(i) From Corollary 11.2.3 it suffices to show that the identity mapping id_V is contained in the closure of $V^* \otimes V$ in $(\mathcal{CB}(V, V), \mathcal{T}_w)$. If that is not the case, then there exists a continuous linear functional $F \in (\mathcal{CB}(V, V), \mathcal{T}_w)^*$ such that $F(id_V) \neq 0$ but $F_{|V^* \otimes V} = 0$. From Lemma 11.2.4, F is determined by an element $u \in V \hat{\otimes} V^* \subseteq \mathcal{CB}(V, V)^*$. We have

$$\operatorname{trace}(u) = \langle id_V, u \rangle = F(id_V) \neq 0.$$

On the other hand, $\Phi(u) \in V \check{\otimes} V^* \subseteq \mathcal{CB}(V, V)$ satisfies

$$\langle \Phi(u)(v), f \rangle = \langle u, f \otimes v \rangle = \langle F, f \otimes v \rangle = 0$$

for all $v \in V$ and $f \in V^*$. Thus, $\Phi(u) = 0$ which contradicts (iv).

(i)$\Leftrightarrow$(v) is an immediate consequence of Corollary 8.1.9. □

Corollary 11.2.6 *If an operator space V is such that V^* has the operator space approximation property, then V has the operator space approximation property.*

Proof We have from (7.1.28) that the top row of the diagram

$$\begin{array}{ccc} V \hat{\otimes} V^* & \longrightarrow & V^{**} \hat{\otimes} V^* \\ {\scriptstyle \Phi} \big\downarrow & & \big\downarrow {\scriptstyle \Phi} \\ V \check{\otimes} V^* & \longrightarrow & V^{**} \check{\otimes} V^* \end{array}$$

is a completely isometric injection.If V^* has the operator space approximation property, then the second column is one-to-one. Since the diagram commutes, it follows that the first column is also one-to-one. □

11.3 TOMIYAMA'S SLICE MAPPING PROPERTY

We have seen in the previous section that one can use the properties of the normal spatial tensor products $V^* \overline{\otimes} W^*$ to characterize the approximation property for V. In this section we show that one can use the injective tensor products $V \check{\otimes} W$ for the same purpose. This result will subsequently be important for relating the approximation property to exactness and local reflexivity. As in our foregoing discussion, Tomiyama's notions of slice mappings and Fubini products are the key techniques.

Given operator spaces Z and W, we recall from Proposition 8.1.1 that there is a complete isometric embedding

$$\theta : Z \check{\otimes} W \hookrightarrow \mathcal{CB}(Z^*, W)$$

given by

$$\theta(x \otimes w)(g) = g(x)w = (g \otimes id_W)(x \otimes w).$$

Therefore, given $g \in Z^*$, we can define a completely bounded mapping $R_g : Z \check{\otimes} W \to W$ by

$$R_g(u) = (g \otimes id_W)(u)$$

for all $u \in Z \check{\otimes} W$. We call R_g the *right slice mapping* of $Z \check{\otimes} W$ (this should be compared with the corresponding notion in §7.2). Similarly, given $f \in W^*$, we can define the *left slice mapping* $L_f : Z \check{\otimes} W \to Z$ by

$$L_f(u) = (id_Z \otimes f)(u).$$

Given operator spaces $X \subseteq Z$ and $V \subseteq W$, we define the *relative Fubini product* of X and V with respect to Z and W to be the operator space

$$\mathcal{F}(X, V; Z \check{\otimes} W) = \{u \in Z \check{\otimes} W : \; L_f(u) \in X, \; R_g(u) \in V \quad \text{for all } g \in Z^*, f \in W^*\}.$$

In this section we shall be particularly interested in the case $V = W$. It is evident that for any operator spaces $X \subseteq Z$,

$$X \check{\otimes} V \subseteq \mathcal{F}(X, V; Z \check{\otimes} V).$$

An operator space V is said to have the *slice mapping property for subspaces of Z* if for any operator space $X \subseteq Z$,

$$\mathcal{F}(X, V; Z \check{\otimes} V) = X \check{\otimes} V.$$

Theorem 11.3.1 *An operator space V has the slice mapping property for subspaces of Z if and only if for every $u \in Z \check{\otimes} V$ and $\varepsilon > 0$, there exists a finite-rank mapping $\varphi : V \to V$ such that*

$$\|(id_Z \otimes \varphi)(u) - u\| < \varepsilon.$$

Proof Let us assume the second condition. Given a finite-rank mapping $\varphi : V \to V$, we may assume that

$$\varphi = \sum f_i \otimes v_i$$

where $f_i \in V^*$ and $v_i \in V$ $(i = 1, \ldots, n)$. For any $u \in Z \check{\otimes} V$ we have that

$$(id_Z \otimes \varphi)(u) = \sum (id_Z \otimes f_i)(u) \otimes v_i$$

since this is immediate for $u \in Z \otimes V$, and the general result follows by continuity. If we are given $X \subseteq Z$, $u \in \mathcal{F}(X, V; Z \check{\otimes} V)$, and $\varepsilon > 0$, then by assumption there exists a finite-rank mapping $\varphi : V \to V$ such that

$$\|(id_Z \otimes \varphi)(u) - u\| < \varepsilon.$$

We have

$$(id_Z \otimes \varphi)(u) = \sum (id_Z \otimes f_i)(u) \otimes v_i = \sum L_{f_i}(u) \otimes v_i \in X \otimes_\vee V.$$

Thus, u is a limit of a net of elements in $X \otimes_\vee V$, and $u \in X \check{\otimes} V$.

Conversely, suppose that V has the slice mapping property for subspaces of Z. Given $u \in Z \check{\otimes} V$, we let X_u denote the norm closed subspace of Z spanned by $\{L_f(u) : \ f \in V^*\}$. By hypothesis,

$$u \in \mathcal{F}(X_u, V; Z \check{\otimes} V) = X_u \check{\otimes} V.$$

For every $\varepsilon > 0$, there exist $f_i \in V^*$ and $v_i \in V$ such that

$$u_\varepsilon = \sum (L_{f_i})(u) \otimes v_i \in X_u \otimes V$$

satisfies $\|u_\varepsilon - u\| < \varepsilon$. The corresponding finite-rank mapping

$$\varphi = \sum f_i \otimes v_i : V \to V$$

satisfies

$$\|(id_Z \otimes \varphi)(u) - u\| = \|u_\varepsilon - u\| < \varepsilon.$$

□

We recall from the previous section that an operator space V is said to have the operator space approximation property if for any

$$v_1, \ldots, v_r \in K_\infty(V) = K_\infty \check{\otimes} V$$

and $\varepsilon > 0$ we may find a finite-rank mapping $\varphi : V \to V$ such that

$$\|(id_{K_\infty} \otimes \varphi)(v_k) - v_k\| < \varepsilon. \tag{11.3.1}$$

We say that V has the *strong operator space approximation property* if for any $v_1, \ldots, v_r \in M_\infty \check{\otimes} V$ and $\varepsilon > 0$, we may find a finite-rank mapping $\varphi : V \to V$ such that

$$\|(id_{M_\infty} \otimes \varphi)(v_k) - v_k\| < \varepsilon. \tag{11.3.2}$$

The following is an immediate consequence of Theorem 11.3.1.

Corollary 11.3.2 *An operator space V has the operator space approximation property (respectively, strong operator space approximation property) if and only if it has the slice mapping property for subspaces of K_∞ (respectively, for subspaces of M_∞).* □

The relations between slice mapping properties and appoximation properties have obvious Banach space analogues. The Banach space theory is simpler since it is not necessary to distinguish between the approximation property and the strong approximation property. This follows since if E is a Banach space, then E has the slice mapping property for subspaces of $c_0(\mathbb{N})$ if and only if it has the slice mapping property for subspaces of $\ell_\infty(\mathbb{N})$. We will not give the simple argument for this result.

There are other approximation properties of interest. A Banach space E has the *bounded approximation property* (respectively, the *contractive approximation property*) if there exists a net of finite-rank mappings $\varphi_\nu : V \to V$ such that $\|\varphi_\nu\| \leq K$ for some constant K (respectively, $\|\varphi_\nu\| \leq 1$) and for every $v \in V$, $\|\varphi_\nu(v) - v\| \to 0$.

By analogy, we say that an operator space V has the *completely bounded approximation property* (respectively, *completely contractive approximation property*) if there exists a net of finite-rank mappings $\varphi_\nu : V \to V$ such that $\|\varphi_\nu\|_{cb} \leq K$ for some constant K (respectively, $\|\varphi_\nu\|_{cb} \leq 1$) and for every $v \in V$, $\|\varphi_\nu(v) - v\| \to 0$.

Theorem 11.3.3 *If V has the completely bounded approximation property, then V has the strong operator space approximation property.*

Proof Let us assume that $\{\varphi_\nu\}$ is a net of finite-rank mappings on V such that $\|\varphi_\nu\|_{cb} \leq K$ for some constant K and for every $v \in V$,

$$\|\varphi_\nu(v) - v\| \to 0.$$

We claim that for every $u \in M_\infty \check{\otimes} V$,

$$\|(\varphi_\nu)_\infty(u) - u\| \to 0.$$

To see this, let us first consider $u = \alpha \otimes v \in M_\infty \otimes V$. It is clear that

$$\|(\varphi_\nu)_\infty(\alpha \otimes v) - \alpha \otimes v\| = \|\alpha \otimes \varphi_\nu(v) - \alpha \otimes v\| \leq \|\alpha\| \, \|\varphi_\nu(v) - v\| \to 0.$$

It follows that $\|(\varphi_\nu)_\infty(u) - u\| \to 0$ for every $u \in M_\infty \otimes V$. Given a contractive element $u \in M_\infty \check{\otimes} V$ and $\varepsilon > 0$, there exists a contractive element $u_\varepsilon \in M_\infty \otimes V$ such that $\|u - u_\varepsilon\| < \varepsilon/(K+1)$. Then

$$\begin{aligned}\|(\varphi_\nu)_\infty(u) - u\| &\leq \|(\varphi_\nu)_\infty(u) - (\varphi_\nu)_\infty(u_\varepsilon)\| \\ &\qquad + \|(\varphi_\nu)_\infty(u_\varepsilon) - u_\varepsilon\| + \|u_\varepsilon - u\| \\ &\leq \|\varphi_\nu\|_{cb} \|u - u_\varepsilon\| + \|(\varphi_\nu)_\infty(u_\varepsilon) - u_\varepsilon\| + \|u - u_\varepsilon\| \\ &\leq \varepsilon + \|(\varphi_\nu)_\infty(u_\varepsilon) - u_\varepsilon\|.\end{aligned}$$

This implies that

$$\|(\varphi_\nu)_\infty(u) - u\| \to 0,$$

which completes the proof. □

One of the most important applications of the Fubini product is that it can be used to calculate kernels.

Proposition 11.3.4 *Suppose that V and W are operator spaces and that J is a closed subspace of W. If $\pi : W \to W/J$ is the complete quotient mapping, then the kernel of the mapping*

$$id_V \otimes \pi : V \check{\otimes} W \to V \check{\otimes} (W/J)$$

is $\mathcal{F}(V, J)$.

Proof Since the natural mapping $V \check{\otimes} (W/J) \to V \otimes^\lambda (W/J)$ is one-to-one (see Corollary 8.1.4), we have that $id_V \otimes \pi(u) = 0$ if and only if $(f \otimes \bar{g})(u) = 0$ for all $f \in V^*$ and $\bar{g} \in (W/J)^*$. Thus, $id_V \otimes \pi(u) = 0$ if and only if

$$g((f \otimes id)u) = 0$$

for all $f \in V^*$ and $g \in J^\perp$, or equivalently, $(f \otimes id)(u) \in J$. □

11.4 NOTES AND REFERENCES

Grothendieck's remarkable theory of the approximation property and Enflo's counter-example may be found in Grothendieck (1955) and Enflo (1973), respectively (see also Lindenstrauss and Tzafriri 1977, I, §I.e). Grothendieck proved that a Banach space E has the approximation property if and only if $\mathcal{FB}(F, E)$ is norm dense in $\mathcal{K}(F, E)$ for all Banach spaces Banach spaces F (see §A.2).

Pisier's monograph on tensor products of Banach spaces (Pisier, 1986) remains a key resource for those interested in generalizing Banach space techniques to operator spaces.

The formulation of Grothendieck's approximation theory for operator spaces was first considered in Effros and Ruan (1990). The slice mapping techniques were introduced by Tomiyama (1970), who used them to study a wide range of operator algebraic questions.

Owing to work of Haagerup, Cowling, and de Canniere (see Haagerup and Kraus 1994 for the references) it is known that if G is a discrete subgroup of a (compact or non-compact) simple Lie group of rank 1, then the reduced group C^*-algebra of G has the completely bounded approximation property. They computed the smallest bound K, and they gave examples showing that K can be any odd integer. Whether or not other values of K can occur remains an open problem. Haagerup and Kraus (1994) have shown that if G is a discrete group, then $C^*_\lambda(G)$ has the operator space approximation property if and only if it has the strong operator space approximation property.

12

Mapping spaces

Mapping spaces naturally arise in both the theory and applications of functional analysis. To cite two examples in Grothendieck's work, he used absolutely summing mappings to investigate the Dvoretzky–Rogers theory, and he introduced the nuclear mappings to study differential operators. In this chapter and the next we shall explore the operator space analogues of some of the most important mapping spaces. We begin with an examination of the Banach space theory.

12.1 NUCLEAR AND INTEGRAL MAPPINGS OF BANACH SPACES

One of the fundamental ideas in Grothendieck's theory of Banach spaces is the relation between various tensor products and mapping spaces. The latter theory is most conveniently formulated in terms of *mapping ideals* (the more common terminology 'operator ideal' would be too confusing in this discussion). A (Banach space) mapping ideal $\mathcal{O}$ is an assignment to each pair of Banach spaces (E, F) of a linear space of mappings $\mathcal{O}(E, F)$, together with a norm $\|\cdot\|_{\mathcal{O}}$, such that for each $\varphi \in \mathcal{O}(E, F)$

(a) $\|\varphi\| \leq \|\varphi\|_{\mathcal{O}}$ and
(b) for any linear mappings $r : D \to E$ and $s : F \to G$,

$$\|s \circ \varphi \circ r\|_{\mathcal{O}} \leq \|s\| \, \|\varphi\|_{\mathcal{O}} \, \|r\| \, .$$

We say that a Banach space mapping ideal $\mathcal{O}$ is (left) *local* if for each linear mapping $\varphi : E \to F$,

$$\|\varphi\|_{\mathcal{O}} = \sup \{\|\varphi|_L\|\},$$

where the supremum is taken over all finite-dimensional subspaces $L \subseteq E$.

Perhaps the first important example of a mapping ideal is Grothendieck's *space of nuclear mappings* $\mathcal{N}^B(E, F)$ between Banach spaces E and F. He defined this to be the image of the canonical mapping

$$\Phi : E^* \otimes^\gamma F \to E^* \otimes^\lambda F \subseteq \mathcal{B}(E, F), \tag{12.1.1}$$

equipped with the quotient norm ν^B determined by the linear isomorphism

$$\mathcal{N}^B(E, F) \cong \frac{E^* \otimes^\gamma F}{\ker \Phi}.$$

Given a sequence $d \in \ell_1$, the *multiplication mapping*

$$M(d) : \ell_\infty \to \ell_1 : (\alpha_i) \mapsto (d_i \alpha_i) \tag{12.1.2}$$

is nuclear since $M(d) = \Phi(u)$, where

$$u = \sum d_i (e_i \otimes e_i) \in \ell_1 \otimes^\gamma \ell_1 \subseteq (\ell_\infty)^* \otimes^\gamma \ell_1.$$

Mappings of this form may be regarded as the *prototypes* for all nuclear mappings. By this we mean that for any Banach spaces E and F, a mapping $\varphi : E \to F$ is nuclear if and only if there is a commutative diagram

$$\begin{array}{ccc} \ell_\infty & \xrightarrow{M(d)} & \ell_1 \\ r \uparrow & & \downarrow s \\ E & \xrightarrow{\varphi} & F \end{array}, \tag{12.1.3}$$

and one has that

$$\nu^B(\varphi) = \inf \{\|s\| \, \|d\|_1 \, \|r\|\},$$

with the infimum taken over all such diagrams. It is customary to refer to the diagram (12.1.3) as a *factorization* of φ through the multiplication mapping. It is easy to verify that the assignment

$$\mathcal{N}^B : (E, F) \mapsto (\mathcal{N}^B(E, F), \nu^B)$$

is a mapping ideal.

Nuclearity is a very strong condition, and this is illustrated by the following observation (see §A.2).

Proposition 12.1.1 *If $\varphi : E \to F$ is a nuclear linear mapping of Banach spaces, then it is a compact linear mapping.*

Proof It suffices to show that if φ is nuclear, then it is a uniform limit of finite-rank mappings. If $\varphi = \Phi(u)$, where $u \in E^* \otimes^\gamma F$, then for any $\varepsilon > 0$, we may choose an element $u_0 \in E^* \otimes F$ such that $\|u - u_0\|_\gamma < \varepsilon$. It is immediate that $\varphi_0 = \Phi(u_0)$ is finite-rank and

$$\|\varphi - \varphi_0\| \le \nu^B(\varphi - \varphi_0) < \varepsilon.$$

□

Examining the prototypical nuclear mapping (12.1.2), it is natural to wonder what the situation is for continuous measure spaces. We may identify (12.1.2) with the inclusion mapping

$$\mathrm{i} : \ell_\infty(\mathbb{N}, \mu) \to \ell_1(\mathbb{N}, \mu),$$

where $\mu(\{n\}) = d_n$. We are thus led to consider inclusion mappings of the form

$$\mathrm{i} : L_\infty(X, \mu) \to L_1(X, \mu), \tag{12.1.4}$$

where μ is a probability measure. Such mappings are generally not nuclear since they need not be compact. To see this, let us consider the example

$X = [-\pi, \pi]$ with $\mu = \frac{1}{2\pi}\,\mathrm{d}x$ the corresponding normalized Lebesgue measure on the Borel sets of X. It suffices to prove that $\mathrm{i}\left(L_\infty(X,\mu)_{\|\cdot\|\leq 1}\right)$ does not have norm compact closure in $L_1(X,\mu)$. The functions e^{inx} $(n \in \mathbb{N})$ lie in this image, but they do not have any norm limit points in $L_1(X,\mu)$ owing to the fact that they have constant distance from each other. If $m \neq n$, then

$$\begin{aligned}\left\|\mathrm{e}^{inx} - \mathrm{e}^{imx}\right\|_1 &= \frac{1}{2\pi}\int_0^{2\pi}\left|\mathrm{e}^{inx} - \mathrm{e}^{imx}\right|\,\mathrm{d}x = \frac{1}{2\pi}\int_0^{2\pi}\left|\mathrm{e}^{\mathrm{i}(n-m)x} - 1\right|\mathrm{d}x \\ &= \frac{1}{2\pi(n-m)}\int_0^{2\pi(n-m)}\left|\mathrm{e}^{\mathrm{i}u} - 1\right|\,\mathrm{d}u \\ &= \frac{1}{2\pi}\int_0^{2\pi}\left|\mathrm{e}^{\mathrm{i}u} - 1\right|\,\mathrm{d}u = \frac{1}{2\pi}\int_0^{2\pi}\left|\mathrm{e}^{\mathrm{i}u/2}\right|\left|\mathrm{e}^{\mathrm{i}u/2} - \mathrm{e}^{-\mathrm{i}u/2}\right|\,\mathrm{d}u \\ &= \frac{1}{\pi}\int_0^{2\pi}|\sin(u/2)|\,\mathrm{d}u = \frac{1}{\pi}\int_0^{2\pi}\sin(u/2)\,\mathrm{d}u = \frac{4}{\pi}.\end{aligned}$$

Both this example and the fact that $\mathcal{N}^B$ is not a local operator ideal lead us to consider a larger mapping ideal. Given Banach spaces E and F, the *integral norm* of a linear mapping $\varphi : E \to F$ is given by

$$\iota^B(\varphi) = \sup\left\{\nu_B(\varphi|_L) : L \subseteq E \text{ finite-dimensional}\right\}, \qquad (12.1.5)$$

and φ is said to be *integral* if $\iota^B(\varphi) < \infty$. With this norm, the space of integral mappings $\mathcal{I}^B(E,F)$ is again a Banach space, and it is easy to verify that

$$\mathcal{I}^B : (E,F) \mapsto (\mathcal{I}^B(E,F), \iota^B)$$

is a mapping ideal. If μ is a probability measure on a measure space $(X, \mathcal{S})$, then Grothendieck showed that the inclusion mapping

$$\mathrm{i} : L^\infty(X,\mu) \to L^1(X,\mu) \qquad (12.1.6)$$

satisfies $\iota^B(\mathrm{i}) = 1$. Furthermore, he proved that $\varphi : E \to F$ satisfies $\iota^B(\varphi) \leq 1$ if and only if there exists a commutative diagram

$$\begin{array}{ccccc} L^\infty(\mu) & \xrightarrow{\ \mathrm{i}\ } & L^1(\mu) & & \\ {\scriptstyle r}\uparrow & & & \searrow{\scriptstyle s} & \\ E & \xrightarrow{\ \varphi\ } & F & \subseteq & F^{**} \end{array}\ , \qquad (12.1.7)$$

where r and s are contractions. Owing to this result, i is said to be the 'prototype' for the integral mappings.

Grothendieck also showed that one may use a duality result to characterize the injective norm. Given Banach spaces E and F, we have the natural isomorphism

$$\mathcal{I}^B(E,F^*) \cong (E \otimes^\lambda F)^*. \qquad (12.1.8)$$

The operator space version of this result is not always true (see §14.2). As we shall see, this phenomenon is related to some of the most interesting results in the subject.

12.2 COMPLETELY NUCLEAR MAPPINGS

We now turn our attention to linear mappings of operator spaces $\varphi : V \to W$. As before, we identify a matrix $\varphi = [\varphi_{i,j}]$ of linear mappings $\varphi_{i,j} : V \to W$ with a linear mapping $\varphi : V \to M_n(W)$, and conversely given such a mapping, we let $\varphi_{i,j} = E_i \varphi E_j^*$ (see (1.1.1)). An *(operator space) mapping ideal* $\mathcal{O}$ is an assignment to each pair of operator spaces V, W of a linear space $\mathcal{O}(V, W)$ of completely bounded mappings $\varphi : V \to W$, together with an operator space matrix norm $\|\cdot\|_{\mathcal{O}}$, such that for each $\varphi \in M_n(\mathcal{O}(V, W))$,

(a) $\|\varphi\|_{cb} \leq \|\varphi\|_{\mathcal{O}}$ and
(b) for any linear mappings $r : U \to V$ and $s : W \to X$,

$$\|s_n \circ \varphi \circ r\|_{\mathcal{O}} \leq \|s\|_{cb} \|\varphi\|_{\mathcal{O}} \|r\|_{cb} . \tag{12.2.1}$$

Guided by classical theory, we define the *completely nuclear mappings* $\mathcal{N}(V, W)$ to be the image of the mapping

$$\Phi : V^* \widehat{\otimes} W \to V^* \check{\otimes} W \subseteq \mathcal{CB}(V, W)$$

with the quotient operator space structure determined by the identification

$$\mathcal{N}(V, W) \cong \frac{V^* \widehat{\otimes} W}{\ker \Phi}. \tag{12.2.2}$$

We let ν_n be the corresponding matrix norm on $M_n(\mathcal{N}(V, W))$. These matrix norms must be handled with care since we need not have that $\nu_n(\varphi)$ is equal to the ν_1 norm of the mapping $\varphi : V \to M_n(W)$; that is, in general

$$M_n(\mathcal{N}(V, W)) \neq \mathcal{N}(V, M_n(W)).$$

The correct relations are given by

$$T_n(\mathcal{N}(V, W)) \cong \mathcal{N}(V, T_n(W)) \cong \mathcal{N}(M_n(V), W), \tag{12.2.3}$$

which is evident from the diagram

$$\begin{array}{ccccc} T_n \widehat{\otimes} (V^* \widehat{\otimes} W) & \cong & V^* \widehat{\otimes} T_n(W) & \cong & M_n(V)^* \widehat{\otimes} W \\ \downarrow & & \downarrow & & \downarrow \\ T_n \widehat{\otimes} \mathcal{N}(V, W) & & \mathcal{N}(V, T_n(W)) & & \mathcal{N}(M_n(V), W) \end{array},$$

in which the column mappings are complete quotient mappings, and their null spaces are the same.

Given a linear mapping $\varphi : V \to W$, we write $\nu(\varphi) = \infty$ if φ is not in the range of Φ. Since Φ is (completely) contractive, we have

$$\|\varphi\|_{cb} \leq \nu(\varphi) \tag{12.2.4}$$

for any linear mapping $\varphi : V \to W$.

Proposition 12.2.1 *If $\varphi : V \to W$ is a completely nuclear mapping of operator spaces, then it is a compact (Banach space) mapping.*

Proof Any $u \in V^* \hat{\otimes} W$ is the norm limit of a sequence u_n in $V^* \otimes W$. The finite-rank mappings $\Phi(u_n)$ satisfy $\|\Phi(u) - \Phi(u_n)\|_{cb} \to 0$. □

As in Banach space theory, there are simple prototypes for the completely nuclear mappings. Let us suppose that we have Hilbert–Schmidt matrices $a \in HS_{\infty,\infty}$ and $b \in HS_{\infty,\infty}$ with $\|a\|_2 = \|b\|_2 \leq 1$. We claim that the *multiplication mapping*

$$M(a,b) : M_\infty \to T_\infty : x \mapsto axb = \left[\sum_{k,l} a_{i,k} x_{k,l} b_{l,j}\right] \tag{12.2.5}$$

is completely nuclear with $\nu(M(a,b)) \leq 1$. To see this we observe that

$$\begin{aligned}(axb)_{i,j} &= \sum_{k,l} \varepsilon_{i,j} \otimes a_{i,k} x_{k,l} b_{l,j}\\ &= \sum_{k,l} a_{i,k}(\varepsilon_{i,j} \otimes x_{k,l}) b_{l,j}\\ &= \sum_{k,l} \alpha_{1,(i,k)}(\varepsilon_{i,j} \otimes \varepsilon_{k,l}(x))\beta_{(j,l),1},\end{aligned}$$

where the row and column matrices $\alpha \in M_{1,\infty\times\infty}$ and $\beta \in M_{\infty\times\infty,1}$ defined by $\alpha_{1,(i,k)} = a_{i,k}$ and $\beta_{(j,l),1} = b_{l,j}$ satisfy $\|\alpha\| = \|a\|_2$ and $\|\beta\| = \|b\|_2$. We recall that the matrix of matrix units

$$\varepsilon = [\varepsilon_{i,j}] \in M_\infty(T_\infty) = \mathcal{CB}(K_\infty, M_\infty)$$

is just the inclusion mapping, and thus $\|\varepsilon\|_{M_\infty(T_\infty)} = 1$. It follows that $\alpha(\varepsilon \otimes \varepsilon)\beta \in T_\infty \hat{\otimes} T_\infty$ and

$$\|\alpha(\varepsilon \otimes \varepsilon)\beta\|_\wedge \leq \|\alpha\| \, \|\beta\|.$$

From the above calculation,

$$M(a,b)(x) = \Phi(\alpha(\varepsilon \otimes \varepsilon)\beta)(x);$$

hence $M(a,b) = \Phi(\alpha(\varepsilon \otimes \varepsilon)\beta)$ and

$$\nu(M(a,b)) \leq \|a\|_2 \, \|b\|_2 .$$

We have just proved the first part of the following result.

Proposition 12.2.2 *For any Hilbert–Schmidt matrices $a \in HS_{\infty,\infty}$, and $b \in HS_{\infty,\infty}$ with $\|a\|_2 = \|b\|_2 \leq 1$, the mapping*

$$M(a,b) : M_\infty \to T_\infty : x \mapsto axb = \left[\sum_{i,j} a_{k,i} x_{i,j} b_{j,l}\right] \tag{12.2.6}$$

satisfies $\nu(M(a,b)) \leq 1$. *More generally, for any* $a = [a_{g,(k,i)}] \in M_{n,\infty\times\infty}$ *and* $b = [b_{(j,l),h}] \in M_{\infty\times\infty,n}$ *with* $\|a\|, \|b\| \leq 1$, *the mapping*

$$M(a,b) : M_\infty \to M_n(T_{\infty\times\infty}) : x \mapsto axb = \left[\sum_{k,l} a_{g,(i,k)} x_{k,l} b_{(l,j),h} \right]$$

satisfies

$$\nu_n(M(a,b)) \leq 1. \tag{12.2.7}$$

Proof If a and b are given as above, then

$$M(a,b) = [\Phi(a_g(\varepsilon_{i,j} \otimes \varepsilon_{k,l})b_h)] = \Phi_n(\alpha(\varepsilon \otimes \varepsilon)\beta),$$

where $\alpha = a$, and $\beta_{(j,l),h} = b_{(l,j),h}$ determines a matrix with $\|\beta\| = \|b\|$ (since we are just permuting rows). It follows that

$$\|\alpha(\varepsilon \otimes \varepsilon)\beta\|_\wedge \leq 1,$$

and thus we have (12.2.7). □

To see that $\mathcal{N}$ is an operator space mapping ideal, let us suppose that we are given $\varphi \in M_n(\mathcal{N}(V,W))$ and linear mappings $r : U \to V$ and $s : W \to X$. If we choose $u \in M_n(V^* \hat{\otimes} W)$ with $\varphi = \Phi_n(u)$, it follows that

$$s_n \circ \varphi \circ r = \Phi_n(u'),$$

where

$$u' = (r^* \otimes s)_n(u) \in M_n(U^* \hat{\otimes} X),$$

and thus

$$\nu_n(s_n \circ \varphi \circ r) \leq \|u'\|_\wedge \leq \|s\|_{cb} \|u\| \|r\|_{cb}. \tag{12.2.8}$$

Taking the infimum over all u with $\varphi = \Phi_n(u)$, we conclude that

$$\nu_n(s_n \circ \varphi \circ r) \leq \|s\|_{cb} \, \nu_n(\varphi) \, \|r\|_{cb}.$$

Proposition 12.2.3 *Given operator spaces* V *and* W, *a linear mapping* $\varphi : V \to M_n(W)$ *is completely nuclear with* $\nu_n(\varphi) < 1$ *if and only if there is a commutative diagram*

$$\begin{array}{ccc} M_\infty & \xrightarrow{M(a,b)} & M_n(T_\infty) \\ r \uparrow & & \downarrow s_n \\ V & \xrightarrow{\varphi} & M_n(W) \end{array}, \tag{12.2.9}$$

where $r : V \to M_\infty$ *and* $s : T_\infty \to W$ *are complete contractions, and* $a \in M_{n,\infty\times\infty}$ *and* $b \in M_{\infty\times\infty,n}$ *satisfy* $\|a\|, \|b\| < 1$.

Proof If we have such a diagram, then it is immediate from (12.2.7) and (12.2.8) that $\nu_n(\varphi) < 1$.

Conversely, if φ satisfies the hypothesis, $\varphi = \Phi_n(u)$, where $\|u\|_\wedge < 1$, then we can assume that $u = \alpha(f \otimes w)\beta$, where $f \in K_\infty(V^*)$, $w \in K_\infty(W)$,

$\alpha \in M_{n,\infty\times\infty}$, $\beta \in M_{\infty\times\infty,n}$, and each of these elements has norm less than 1. From the inclusions

$$K_\infty(V^*) = K_\infty \check{\otimes} V^* \subseteq \mathcal{CB}(V, K_\infty) \tag{12.2.10}$$

and

$$K_\infty(W) = K_\infty \check{\otimes} W \subseteq \mathcal{CB}(T_\infty, W), \tag{12.2.11}$$

we infer that f and w determine the linear mappings

$$r = f : V \to K_\infty : v \mapsto [f_{i,j}(v)] \tag{12.2.12}$$

and

$$s : T_\infty \to W : t \mapsto \sum t_{k,l} w_{k,l}, \tag{12.2.13}$$

respectively. Since

$$\begin{aligned}\varphi(v) &= \left[\sum_{i,j,k,l} a_{g,(i,k)}(\Phi(f_{i,j} \otimes w_{k,l})(v)) b_{(l,j),h}\right]_{g,h\in n} \\ &= \left[\sum_{i,j,k,l} a_{g,(i,k)}(f_{i,j}(v) b_{(l,j),h} w_{k,l})\right]_{g,h\in n} \\ &= \left[s(M(a,b)(r(v))_{g,h})\right]_{g,h\in n} \\ &= s_n(M(a,b)(r(v))),\end{aligned}$$

we have proved (12.2.9). □

The following result shows that there are relatively few completely bounded mappings between row and column Hilbert operator spaces.

Proposition 12.2.4 *For any Hilbert spaces H and K, we have*

$$\mathcal{CB}(H_c, K_r) = \mathcal{N}(H_c, K_r)$$

and

$$\mathcal{CB}(H_r, K_c) = \mathcal{N}(H_r, K_c).$$

Proof If we let $\overline{H}$ denote the Hilbert space dual of H, then

$$\begin{aligned}\mathcal{CB}(H_c, K_r) &= \mathcal{CB}(H_c, (\overline{K}_c)^*) = (H_c \hat{\otimes} \overline{K}_c)^* \\ &= ((H \otimes \overline{K})_c)^* = (\overline{H} \otimes K)_r \\ &= \overline{H}_r \hat{\otimes} K_r = (H_c)^* \hat{\otimes} K_r = \mathcal{N}(H_c, K_r),\end{aligned}$$

where in the last step we used the fact that H_c satisfies the complete metric approximation property. □

The following result will be useful in our discussion of integrality.

Proposition 12.2.5 *If V and W are operator spaces and $\varphi : V \to W$ is a completely bounded mapping, then the mapping $\varphi^* : W^* \to V^*$ satisfies*

$$\nu(\varphi^*) \leq \nu(\varphi).$$

If V or W is finite-dimensional, then

$$\nu(\varphi^*) = \nu(\varphi).$$

Proof If we let $S(\varphi) = \varphi^*$, it is evident from the commutative diagram

$$\begin{array}{ccc} V^* \widehat{\otimes} W & \xrightarrow{id \otimes \iota_W} & V^* \widehat{\otimes} W^{**} \\ \Phi_1 \downarrow & & \downarrow \Phi_2 \\ \mathcal{N}(V, W) & \xrightarrow{S} & \mathcal{N}(W^*, V^*) \end{array}$$

that S is contractive. Even though the top row is isometric (see (7.1.28)) and the columns are quotient mappings, it need not follow that S is isometric since we might have

$$\ker \Phi_2 \cap (V^* \widehat{\otimes} W) \neq \ker \Phi_1 .$$

If either V or W is finite-dimensional, then the columns are isometric, and thus that is also the case for S. □

Corollary 12.2.6 *Suppose that L is a finite-dimensional operator space. Then for any operator space W, the natural injection*

$$\mathcal{N}(L, W) \to \mathcal{N}(L, W^{**})$$

is completely isometric.

Proof Consider the diagram

$$\begin{array}{ccc} L^* \widehat{\otimes} W & \longrightarrow & L^* \widehat{\otimes} W^{**} \\ \downarrow & & \downarrow \\ \mathcal{N}(L, W) & \longrightarrow & \mathcal{N}(L, W^{**}) \end{array} .$$

Since L is finite-dimensional, the columns have no kernel, and thus they are isometric. We have seen in (7.1.28) that the top row is completely isometric, and thus the bottom row of the diagram is completely isometric. □

The following result will play an important role in our discussion of local reflexivity and the notion of integral mappings. We let $\mathcal{N}_1(V, W)$ denote the closed unit ball of $\mathcal{N}(V, W)$.

Lemma 12.2.7 *Suppose that V and W are operator spaces with V finite-dimensional. Then $\mathcal{N}_1(V, W)$ is a closed subset of $\mathcal{N}(V, W)$ in the point-weak topology.*

Proof We first observe that the point-weak closure of $\mathcal{N}_1(V, W)$ coincides with the point-norm closure. To see this, let us suppose that $\varphi : V \to W$

is a point-weak limit of a net $\psi_\beta \in \mathcal{N}_1(V,W)$. Given $v_1,\ldots,v_r \in V$, we define a mapping $T : \mathcal{N}_1(V,W) \to W \oplus \cdots \oplus W$ by

$$T(\psi) = (\psi(v_1),\ldots,\psi(v_r)).$$

We give $W \oplus \cdots \oplus W$ the norm $\|w\|_\infty = \max\{\|w_i\|\}$. It is evident that

$$(\psi_\beta(v_1),\ldots,\psi_\beta(v_r)) \to (\varphi(v_1),\ldots,\varphi(v_r))$$

weakly, and thus $(\varphi(v_1),\ldots,\varphi(v_r))$ is in the weak closure of $T(\mathcal{N}_1(V,W))$. Since $T(\mathcal{N}_1(V,W))$ is convex, it follows from the (classical) Hahn–Banach theorem that its weak and norm closures coincide. Thus, for any $\varepsilon > 0$, there is a $\psi \in \mathcal{N}_1(V,W)$ that satisfies

$$\|\psi(v_j) - \varphi(v_j)\| < \varepsilon$$

for $j = 1,\ldots,r$. It follows that φ is in the point-norm closure of $\mathcal{N}_1(V,W)$.

Let us suppose that $\varphi : V \to W$ is a point-norm limit of a net ψ_β in $\mathcal{N}_1(V,W)$. If we fix a vector basis $x_j \in V$ $(1 \le j \le n)$, then the corresponding dual basis $f_i \in V^*$ is determined by

$$f_i\left(\sum_{j=1}^{n} c_j x_j\right) = c_i.$$

If we use the algebraic identification $\mathcal{CB}(V,W) = V^* \otimes W$, it follows that

$$\psi_\beta = \sum_{i=1}^{n} f_i \otimes y_i^\beta$$

and

$$\varphi = \sum_{i=1}^{n} f_i \otimes y_i,$$

where $y_i^\beta = \psi_\beta(x_i)$ and $y_i = \varphi(x_i) \in W$. By assumption,

$$\left\|y_i^\beta - y_i\right\| = \|\psi_\beta(x_i) - \varphi(x_i)\| \to 0,$$

and since the operator space projective tensor norm is a cross norm in the Banach space sense,

$$\begin{aligned}\nu(\varphi - \psi_\beta) &\le \left\|\sum_{i=1}^{n} f_i \otimes (y_i^\beta - y_i)\right\|_{V^* \hat{\otimes} W} \\ &\le \sum_{i=1}^{n} \|f_i\|\,\|y_i^\beta - y_i\| \to 0.\end{aligned}$$

Since $\nu(\psi_\beta) \le 1$ and

$$\nu(\varphi) \le \nu(\varphi - \psi_\beta) + \nu(\psi_\beta),$$

we conclude that $\nu(\varphi) \le 1$. □

12.3 COMPLETELY INTEGRAL MAPPINGS

Given operator spaces V and W, we define a mapping $\varphi : V \to W$ to be *completely integral* if

$$\iota(\varphi) = \sup \left\{ \nu(\varphi_{|L}) : L \subseteq V \text{ finite-dimensional} \right\} < \infty.$$

We let $\mathcal{I}(V,W)$ denote the set of all completely integral mappings, and given

$$\varphi = [\varphi_{i,j}] \in M_n(\mathcal{I}(V,W)),$$

we define

$$\iota_n(\varphi) = \sup \left\{ \nu_n(\varphi_{|L}) : L \subseteq V \text{ finite-dimensional} \right\}. \tag{12.3.1}$$

It is a routine exercise to check that $\mathcal{I}(V,W)$ is a linear space, and ι is an operator space matrix norm on $\mathcal{I}(V,W)$.

Given a linear mapping $\varphi : V \to W$ and a finite-dimensional subspace L of V we have

$$\varphi_{|L} = \varphi \circ r,$$

where $r : L \to V$ is the inclusion mapping, and thus

$$\left\| \varphi_{|L} \right\|_{cb} \leq \nu(\varphi_{|L}) \leq \nu(\varphi) \left\| r \right\|_{cb} = \nu(\varphi).$$

From this we infer that

$$\left\| \varphi \right\|_{cb} \leq \iota(\varphi) \leq \nu(\varphi). \tag{12.3.2}$$

If V is finite-dimensional, then from the definition, $\nu(\varphi) \leq \iota(\varphi)$, i.e. $\nu(\varphi) = \iota(\varphi)$. The same argument shows that $\nu_n(\varphi) = \iota_n(\varphi)$, and thus for any finite-dimensional space V,

$$\mathcal{I}(V,W) = \mathcal{N}(V,W). \tag{12.3.3}$$

$\mathcal{I}$ is an operator space mapping ideal. To see this let us suppose that $\varphi \in M_n(\mathcal{I}(V,W))$ and we are given the linear mappings $r : U \to V$ and $s : W \to X$. If K is a finite subspace of U and we let $L = r(U)$, then

$$\nu_n(s_n \circ \varphi \circ r_{|K}) \leq \left\| s \right\|_{cb} \nu_n(\varphi_{|L}) \left\| r \right\|_{cb} \leq \left\| s \right\|_{cb} \iota_n(\varphi) \left\| r \right\|_{cb},$$

and thus

$$\iota_n(s_n \circ \varphi \circ r) \leq \left\| s \right\|_{cb} \iota_n(\varphi) \left\| r \right\|_{cb}. \tag{12.3.4}$$

Lemma 12.3.1 *Given operator spaces V and W, and a linear mapping $\varphi : V \to M_n(W)$, the following are equivalent:*

(i) $\iota_n(\varphi) \leq 1$;

(ii) *for all finite-dimensional operator spaces L,*

$$\left\| id_L \otimes \varphi : L \check{\otimes} V \to M_n(L \hat{\otimes} W) \right\| \leq 1; \tag{12.3.5}$$

(iii) *for all operator spaces U,*

$$\left\| id_U \otimes \varphi : U \check{\otimes} V \to M_n(U \hat{\otimes} W) \right\| \leq 1; \tag{12.3.6}$$

(iv) *there is a net* $\psi_\alpha \in M_n(\mathcal{N}(V,W))$ *which satisfies* $\nu_n(\psi_\alpha) \leq 1$ *and which converges to* φ *in the point-norm topology;*

(v) *there is a net* $\psi_\alpha \in M_n(\mathcal{N}(V,W))$ *which satisfies* $\nu_n(\psi_\alpha) \leq 1$ *and which converges to* φ *in the point-weak topology.*

It follows from Theorem 8.1.10 that we may use the completely bounded norms in conditions (ii) and (iii). We have, for example, that (iii) and that result imply that the top row of the following commutative diagram is completely contractive:

$$\begin{array}{ccc} (M_p \check{\otimes} U)\check{\otimes} V \to M_n((M_p \check{\otimes} U) \hat{\otimes} W) & \to & M_n(M_p \check{\otimes} (U \hat{\otimes} W)) \\ \| & & \| \\ M_p(U \check{\otimes} V) & \longrightarrow & M_p(M_n(U \hat{\otimes} W)) \end{array}.$$

Proof (i)$\Leftrightarrow$(ii). Given a finite-dimensional operator space L, we have the commutative diagram

$$\begin{array}{ccc} L \check{\otimes} V & \xrightarrow{id_L \otimes \varphi} & M_n(L \hat{\otimes} W) \\ \| & & \| \\ \mathcal{CB}(L^*, V) & \xrightarrow{r \mapsto \varphi \circ r} & M_n(\mathcal{N}(L^*, W)) \end{array}.$$

From (i) and (12.2.8), we have for any $r \in \mathcal{CB}(L^*, V)$

$$\nu_n(\varphi \circ r) \leq \nu_n(\varphi_{|r(L^*)}) \|r\|_{cb} \leq \|r\|_{cb}, \tag{12.3.7}$$

and thus the top row is contractive. Conversely, if the top row is contractive for each inclusion mapping $r : L^* \to V$, where L^* is a finite-dimensional subspace of V, then $\nu_n(\varphi_{|L^*}) \leq 1$ for each such subspace, and by definition, $\iota_n(\varphi) \leq 1$.

(ii)$\Leftrightarrow$ (iii). Given an element $u \in U \otimes_\vee V$, we have that $u \in U_0 \otimes V$ where U_0 a finite-dimensional subspace of U. Using the natural complete contractions

$$U_0 \otimes_\vee V \xrightarrow{id \otimes \varphi} M_n(U_0 \otimes_\wedge W) \xrightarrow{(\iota \otimes id)_n} M_n(U \otimes_\wedge W),$$

we have

$$\|id \otimes \varphi(u)\|_{M_n(U \otimes_\wedge W)} \leq \|u\|_{U_0 \otimes_\vee V} = \|u\|_{U \otimes_\vee V}.$$

(iii) follows by continuity. Again the converse is trivial.

(i)$\Rightarrow$(iv). Given $v_1, \ldots, v_r \in V$ and $\varepsilon > 0$, we let L be the subspace of V spanned by the v_j. We have from (i) that $\nu_n(\varphi_{|L}) \leq 1$, and thus from Proposition 12.2.3 we have the commutative rectangle of the diagram

$$\begin{array}{ccccc} & & M_\infty & \xrightarrow{M(a,b)} & M_n(T_\infty) \\ & {}^{r}\nearrow & \uparrow{\scriptstyle r_0} & & \downarrow{\scriptstyle s_n} \\ V & \supseteq & L & \xrightarrow{\varphi} & M_n(W) \end{array},$$

where r_0 and s are completely contractive and, in addition, $a \in M_{n,\infty\times\infty}$ and $b \in M_{\infty\times\infty,n}$ satisfy $\|a\|, \|b\| < 1$. If we use the injectivity of M_∞, then we can extend r_0 to a complete contraction $r : V \to M_\infty$, and from Proposition 12.2.3 we have that

$$\psi(v) = s_n \circ M(a,b) \circ r$$

satisfies $\nu_n(\psi) \leq 1 + \varepsilon$ and $\psi(v_i) = \varphi(v_i)$. It follows that φ is in the point-norm limit closure of the completely nuclear mappings $\psi : V \to M_n(W)$ satisfying $\nu(\psi) \leq 1$.

(iv)⇒(v) is trivial.

(v)⇒(i). Given a finite-dimensional subspace $L \subseteq V$, it follows that $\psi_\alpha|_L$ converges to $\varphi_{|L}$ in the point-weak topology. From Lemma 12.2.7, $\nu_n(\varphi_{|L}) \leq 1$ and (i) follows by definition of ι_n. □

The following 'global' analogue of Proposition 12.2.6 will play an important role in Proposition 14.2.2.

Lemma 12.3.2 *Given operator spaces V and W, the natural mapping*

$$\mathcal{I}(V,W) \to \mathcal{I}(V,W^{**})$$

is completely isometric.

Proof Since $\mathcal{I}$ is a mapping ideal, this mapping is a complete contraction. On the other hand, letting $J : W \to W^{**}$ be the canonical injection, let us suppose that $\iota_n(J_n \circ \varphi) \leq 1$. Given a finite-dimensional subspace $L \subseteq V$, it follows from Proposition 12.2.6 that

$$\nu_n(\varphi_{|L}) = \nu_n(J_n \circ \varphi_{|L}) \leq 1,$$

and thus $\iota_n(\varphi) \leq 1$. □

As we indicated before, the analogue of Grothendieck's fundamental result (12.1.8) is false for operator spaces. Nevertheless, it is frequently correct, and there is much one can say in general. We recall that we have a complete isometry

$$S : \mathcal{CB}(V, W^*) \cong (V \hat{\otimes} W)^*,$$

where, by definition, if $\varphi : V \to W^*$ is completely bounded, then

$$S(\varphi) : V \otimes W \to \mathbb{C} : v \otimes w \mapsto \varphi(v)(w).$$

Let us consider the commutative diagram of complete contractions

$$\begin{array}{ccccc} \mathcal{N}(V,W^*) \subseteq & \mathcal{I}(V,W^*) & \subseteq & \mathcal{CB}(V,W^*) & \\ \hat{\Phi}\uparrow & \searrow S_{nuc} & & \downarrow S & , \\ V^* \hat{\otimes} W^* \xrightarrow{\theta} & (V \check{\otimes} W)^* & \overset{\Phi^*}{\hookrightarrow} & (V \hat{\otimes} W)^* & \end{array} \qquad (12.3.8)$$

where

$$\Phi : V \hat{\otimes} W \to V \check{\otimes} W$$

and

$$\widehat{\Phi} : V^* \widehat{\otimes} W^* \to \mathcal{N}(V, W^*)$$

are the canonical mappings, and θ is determined by the completely contractive bilinear mapping

$$V^* \times W^* \to (V \check{\otimes} W)^* : (f, g) \mapsto f \otimes g.$$

Since Φ maps onto a dense subspace, Φ^* is injective. Thus, if $\theta(u) = 0$, then $\widehat{\Phi}(u) = 0$, and S induces a complete contraction

$$S_{nuc} : \mathcal{N}(V, W^*) \to (V \check{\otimes} W)^*.$$

Now let us suppose that $\iota(\varphi) \leq 1$. Given a contractive $u \in V \otimes_\vee W$, we may assume that $u \in V_0 \otimes W_0$, where V_0 and W_0 are finite-dimensional. Then $S(\varphi)(u) = S(\varphi \circ j)(u)$, where $j : V_0 \to V$ is the inclusion mapping, and thus

$$\|S(\varphi) : V \otimes_\vee W \to \mathbb{C}\| \leq \nu(\varphi \circ j) \leq \iota(\varphi) \leq 1.$$

We conclude that S determines a contractive mapping

$$S_{int} : \mathcal{I}(V, W^*) \to (V \check{\otimes} W)^*. \tag{12.3.9}$$

A similar argument shows that this is a complete contraction. We have the commutative diagram

$$\begin{array}{ccc} \mathcal{I}(V, W^*) & \subseteq & \mathcal{CB}(V, W^*) \\ {\scriptstyle S_{int}} \downarrow & & \downarrow {\scriptstyle S} \\ (V \check{\otimes} W)^* & \stackrel{\Phi^*}{\hookrightarrow} & (V \widehat{\otimes} W)^* \end{array} . \tag{12.3.10}$$

Although S_{int} need not be completely isometric, a closely related mapping does have that property.

Lemma 12.3.3 *For any operator spaces V and W, the composition*

$$S_0 : \mathcal{I}(V, W) \hookrightarrow \mathcal{I}(V, W^{**}) \xrightarrow{S_{int}} (V \check{\otimes} W^*)^*$$

is completely isometric.

Proof Let us suppose that $\varphi \in M_n(\mathcal{I}(V, W))$ satisfies $\|(S_0)_n(\varphi)\| \leq 1$. Then

$$(S_0)_n(\varphi) \in M_n((V \check{\otimes} W^*)^*) = \mathcal{CB}(V \check{\otimes} W^*, M_n).$$

Since

$$V \check{\otimes} W^* \cong W^* \check{\otimes} V \hookrightarrow \mathcal{CB}(W, V^{**}) \cong (V^* \widehat{\otimes} W)^*$$

are completely isometric, we may identify $V \check{\otimes} W^*$ with an operator subspace of $(V^* \widehat{\otimes} W)^*$. It follows from the Arveson–Wittstock–Hahn–Banach theorem that $(S_0)_n(\varphi)$ has a completely contractive extension

$$F_\varphi \in \mathcal{CB}((V^* \widehat{\otimes} W)^*, M_n) = M_n((V^* \widehat{\otimes} W)^{**}).$$

From the bipolar theorem, we may choose a net of elements

$$u_\lambda \in M_n(V^* \hat{\otimes} W)$$

such that

$$\|u_\lambda\|_{M_n(V^* \hat{\otimes} W)} < 1$$

and u_λ converges to F_φ in the point-norm topology on $\mathcal{CB}((V^* \hat{\otimes} W)^*, M_n)$. It follows that

$$\varphi_\lambda = \Phi_n(u_\lambda) \in M_n(\mathcal{N}(V, W))$$

is a net with $\nu_n(\varphi_\lambda) < 1$, and for each $v \in V$ and $g \in W^*$,

$$((\varphi_\lambda)(v))(g) = (u_\lambda)(v \otimes g) \to F_\varphi(v \otimes g) = S_0(\varphi)_n(v \otimes g) = \varphi(v)(g)$$

in the norm on M_n. Therefore, φ_λ converges to φ in the point-weak topology, and thus $\iota_n(\varphi) \leq 1$. □

Corollary 12.3.4 *If L is a finite-dimensional operator space, then for any operator space V we have the complete isometry*

$$S_{int} : \mathcal{I}(V, L^*) \cong (V \check{\otimes} L)^*.$$

□

Corollary 12.3.5 *For each $n \in \mathbb{N}$, we have the natural complete isometries*

$$T_n \hat{\otimes} \mathcal{I}(V, W) \cong \mathcal{I}(V, T_n(W)) \cong \mathcal{I}(M_n(V), W).$$

Proof If we are given a matrix $\varphi = [\varphi_{i,j}] \in T_n \hat{\otimes} \mathcal{I}(V, W)$ and an element $v \in V$, then $\| \sum_{i,j} \varepsilon_{i,j} \otimes \varphi_{i,j}(v) \|_{T_n \hat{\otimes} W} \leq \sum_{i,j} \|\varphi_{i,j}(v)\|_W < \infty$, and thus φ determines a linear mapping

$$\tilde{\varphi} : V \to T_n(W) : v \mapsto [\varphi_{i,j}(v)].$$

For each i, j, the mapping

$$s_{i,j} : W \to T_n(W) : w \mapsto \varepsilon_{i,j} \otimes w$$

is completely contractive since

$$s_{i,j} = t_{i,j} \otimes id : \mathbb{C} \hat{\otimes} W \to T_n \hat{\otimes} W,$$

where $t_{i,j} : \mathbb{C} \to T_n : \alpha \to \alpha \varepsilon_{i,j}$ is completely contractive. Thus, since $\mathcal{I}$ is an operator ideal,

$$\iota(\tilde{\varphi}) \leq \sum \iota(s_{i,j} \circ \varphi_{i,j}) \leq \sum \iota(\varphi_{i,j}) < \infty,$$

and $\tilde{\varphi} \in \mathcal{I}(V, T_n(W))$. We have placed the linear mapping $\varphi \mapsto \tilde{\varphi}$ in the first column of the following diagram:

$$\begin{array}{ccccc}
T_n \hat{\otimes} \mathcal{I}(V, W) & \hookrightarrow & T_n \hat{\otimes} \mathcal{I}(V, W^{**}) & \xrightarrow{id \otimes S_{int}} & T_n \hat{\otimes} (V \check{\otimes} W^*)^* \\
\downarrow & & & & \| \\
\mathcal{I}(V, T_n(W)) & \hookrightarrow & \mathcal{I}(V, T_n(W^{**})) & \xrightarrow{S_{int}} & (V \check{\otimes} M_n(W^*))^*
\end{array}.$$

It is a simple exercise to verify that this diagram is commutative. From Lemma 12.3.3 the composition of each row mapping is completely isometric. The right column is obviously completely isometric, and thus the same is true for the first column. A similar argument may be used for the second complete isometry. □

It is natural to seek a *prototypical completely integral mapping.* A possible candidate for this designation is suggested by the corollary of the following result.

Proposition 12.3.6 *Given operator spaces V and W and a completely bounded mapping $\varphi : V \to W$, $\iota(\varphi) \leq 1$ if and only if there exist Hilbert spaces H and K, a contractive functional $\omega \in B(H \otimes K)^*$, and completely contractive mappings $r : V \to B(H)$ and $t : W^* \to B(K)$ such that for all $v \in V$ and $g \in W^*$,*

$$\langle \varphi(v), g \rangle = \langle \omega, r(v) \otimes t(g) \rangle. \tag{12.3.11}$$

Proof Let us suppose that $\iota(\varphi) \leq 1$. We fix completely isometric embeddings $r : V \to B(H)$ and $s : W^* \to B(K)$. Since

$$\|S_0(\varphi) : V \otimes_\vee W^* \to \mathbb{C}\| \leq 1,$$

we may extend $S_0(\varphi)$ to an element $\omega \in B(H \otimes K)^*$ with $\|\omega\| \leq 1$. It follows that

$$\langle \varphi(v), g \rangle = S_0(\varphi)(v \otimes g) = \langle \omega, r(v) \otimes t(g) \rangle.$$

Conversely, given such a factorization with $\|\omega\| \, \|r\|_{cb} \, \|t\|_{cb} \leq 1$, we have for any $u \in V \otimes W^*$,

$$\begin{aligned} |S_0(\varphi)(u)| &= |\langle \omega, (r \otimes t)(u) \rangle| \\ &\leq \|\omega\| \, \|(r \otimes t)(u)\|_{B(H \otimes K)} \\ &\leq \|\omega\| \, \|r\|_{cb} \, \|t\|_{cb} \, \|u\|_{V \otimes_\vee W^*} \leq \|u\|_{V \otimes_\vee W^*} . \end{aligned}$$

Therefore, from Lemma 12.3.3 we have $\iota(\varphi) = \|S_0(\varphi)\| \leq 1$. □

Given a bounded linear functional $\omega : B(H \otimes K) \to \mathbb{C}$, we define a linear mapping $M(\omega) : B(H) \to B(K)^*$ by

$$M(\omega)(b)(g) = \omega(b \otimes g). \tag{12.3.12}$$

Corollary 12.3.7 *Let us suppose that V and W are operator spaces, and that $\varphi : V \to W$ is a linear mapping. We have $\iota(\varphi) \leq 1$ if and only if there is a commutative diagram*

$$\begin{array}{ccccc} \mathcal{B}(H) & \xrightarrow{M(\omega)} & \mathcal{B}(K)^* & & \\ r\uparrow & & & \searrow s & \\ V & \xrightarrow{\varphi} & W & \overset{\iota_W}{\hookrightarrow} & W^{**} \end{array} , \tag{12.3.13}$$

where $\omega \in B(H \otimes K)^*$ *satisfies* $\|\omega\| \leq 1$, r *and* s *are complete contractions,* $\iota_W : W \to W^{**}$ *is the canonical embedding, and* $s : B(K)^* \to W^{**}$ *is weak* continuous.*

Proof If we let $s = t^*$, then this follows from Proposition 12.3.6. □

Given a von Neumann algebra $\mathcal{R}$ on a Hilbert space H, each unit vector $\xi \in H$ determines a complete contraction

$$\theta_\xi : \mathcal{R} \to \mathcal{R}'_*$$

by

$$\theta_\xi(r)(r') = \langle rr'\xi \mid \xi \rangle.$$

This mapping may be regarded as a non-commutative analogue of the mapping (12.1.4), and thus the following result is of particular interest. It is conceivable that completely integral mappings of this type might provide a more satisfactory prototype for the completely integral mappings.

Theorem 12.3.8 *Suppose that* $\mathcal{R}$ *is a factor on a Hilbert space* H. *If* $\mathcal{R}$ *is injective, then each mapping* θ_ξ *with* ξ *a unit vector in* H, *has integral norm less than or equal to one. Conversely, if there is a single unit vector* ξ *for which* θ_ξ *is completely integral, then* $\mathcal{R}$ *is injective.*

Proof Let us suppose that $\mathcal{R}$ is injective. Replacing $\mathcal{R}$ by $\mathcal{R} \otimes \mathbb{C}I_K$ for some Hilbert space K, we may assume that all the normal states are vector states. If $\mathcal{R}$ is injective, then it is semidiscrete, and in particular the multiplication mapping

$$\mathbf{m} : \mathcal{R} \check{\otimes} \mathcal{R}' \to \mathcal{B}(H) : r \otimes r' \mapsto rr'$$

is completely contractive. Thus, we may extend the linear functional

$$\omega : \mathcal{R} \check{\otimes} \mathcal{R}' \to \mathbb{C} : r \otimes r' \mapsto \langle rr'\xi \mid \xi \rangle$$

to a contractive linear functional

$$\overline{\omega} : \mathcal{R} \overline{\otimes} \mathcal{R}' \to \mathbb{C}.$$

Since this is in

$$(\mathcal{R} \overline{\otimes} \mathcal{R}')^* = (\mathcal{R}_* \hat{\otimes} \mathcal{R}'_*)^{**},$$

we may use the bipolar theorem to approximate $\overline{\omega}$ by a net of contractive functionals

$$\omega_\nu \in \mathcal{R}_* \hat{\otimes} \mathcal{R}'_* \subseteq \mathcal{R}^* \hat{\otimes} \mathcal{R}'_*$$

(see (7.1.28)). By definition, if we let

$$\Phi : \mathcal{R}^* \hat{\otimes} \mathcal{R}'_* \to \mathcal{R}^* \check{\otimes} \mathcal{R}'_* \subseteq \mathcal{CB}(\mathcal{R}, \mathcal{R}'_*)$$

be the fundamental mapping, the images

$$\varphi_\nu = \Phi(\omega_\nu) : \mathcal{R} \to \mathcal{R}'_*$$

satisfy $\nu(\varphi_\nu) \leq 1$. On the other hand,

$$\varphi_\nu(r)(r') = \omega_\nu(r \otimes r') \to \omega(r \otimes r') = \langle rr'\xi \mid \xi\rangle = \theta_\xi(r)(r'),$$

i.e. the mappings φ_ν converge to θ_ξ in the point-weak topology. Thus, it follows from (v) of Theorem 12.3.1 that $\iota(\theta_\xi) \leq 1$.

Conversely, if there is a unit vector ξ with $\iota(\theta_\xi) < \infty$, we may assume that $\iota(\theta_\xi : \mathcal{R} \to \mathcal{R}'_*) \leq 1$. Then

$$\iota\left(\theta_\xi : \mathcal{R} \to \mathcal{R}'^*\right) \leq 1$$

and therefore

$$\|S(\theta_\xi) : \mathcal{R} \otimes_\vee \mathcal{R}' \to \mathbb{C}\| \leq 1.$$

It follows from Effros and Lance (1977) that $\mathcal{R}$ is injective. □

12.4 NOTES AND REFERENCES

Grothendieck introduced the theory of nuclear and integral mappings of locally convex spaces in Grothendieck (1955). Nuclear mappings play an important part in the theory of Schwartz spaces and their applications to differential equations and quantum field theory. There is an operator analogue of this theory that is based on the fact that one can regard Frechet spaces as projective limits of Banach spaces (see Effros and Webster 1997).

The material in this chapter has been taken from Effros and Ruan (1994*a*,*b*, and 1997) and Effros *et al* (2000). The fact that the operator space analogue of (12.1.8) need not be true was pointed out to the authors by G. Pisier.

13
Absolutely summing mappings

It has been said that Grothendieck's theory of absolutely summing mappings is 'in some sense at the heart of Banach space theory'. Grothendieck began by expressing the notions of absolute and unconditional summability in terms of tensor products. Since this represented one of the earliest non-trivial applications of tensor products to the study of concrete analytic questions, we feel that a brief sketch of the theory is in order. We have included this in the first section of this chapter. We then consider the corresponding operator space versions of absolutely summing and Hilbert factorable mappings in the following sections, as well as an operator space analogue of the Dvoretzky–Rogers theorem.

With regard to the first section, we note that the idea of 'unconditionally summable matrices' has yet to be explored, and this has provided us with an additional motivation for our initial digression.

13.1 1-SUMMING MAPPINGS OF BANACH SPACES

Let us suppose that we are given a sequence $(x_n)_{n\in\mathbb{N}}$ in a Banach space E. We recall that the sequence x_n is said to be *absolutely summable* if

$$\sum_{n=1}^{\infty} \|x_n\| < \infty.$$

It follows from our introduction to the projective tensor products of Banach spaces that the absolutely summable sequences comprise the Banach space $\ell_1 \otimes^\gamma E$.

Ordering the finite sets $F \subseteq \mathbb{N}$ by inclusion, we say that a sequence x_n in a Banach space E is *unconditionally summable* if the net of partial sums

$$S_F = \sum_{n\in F} x_n$$

converges in the norm topology, in which case we use the notation

$$\sum_n x_n = \lim S_F.$$

The usual argument from real variable theory shows that an absolutely summable sequence in a Banach space is unconditionally convergent. If E

is finite-dimensional, then these concepts coincide. Reviewing the proof for $E = \mathbb{C}$, let us suppose that we are given an unconditional sequence $\alpha_n \in \mathbb{C}$, and that $\varepsilon > 0$. By assumption, we may choose a finite set $F_\varepsilon \subseteq \mathbb{N}$ such that for arbitrary finite set $F \subseteq \mathbb{N}$,

$$F \cap F_\varepsilon = \emptyset \Rightarrow |S_F| = \left|\sum_{n\in F} \alpha_n\right| < \varepsilon. \tag{13.1.1}$$

We have

$$\left|\sum_{n\in F} \mathrm{Re}\ \alpha_n\right| = \left|\mathrm{Re} \sum_{n\in F} \alpha_n\right| < \varepsilon$$

and

$$\left|\sum_{n\in F} \mathrm{Im}\ \alpha_n\right| = \left|\mathrm{Im} \sum_{n\in F} \alpha_n\right| < \varepsilon$$

for any such F. The subsets $F_+ = \{n \in F : \mathrm{Re}\ \alpha_n \geq 0\}$ and $F_- = F \backslash F_+$ are also disjoint from F_ε, and thus

$$\sum_{n\in F} \left|\mathrm{Re}\ \alpha_n\right| = \left|\sum_{n\in F_+} \mathrm{Re}\ \alpha_n\right| + \left|\sum_{n\in F_-} \mathrm{Re}\ \alpha_n\right| < 2\varepsilon.$$

Similarly,

$$\sum_{n\in F} \left|\mathrm{Im}\ \alpha_n\right| < 2\varepsilon,$$

and we conclude that

$$F \cap F_\varepsilon = \emptyset \Rightarrow \sum_{n\in F} |\alpha_n| < 4\varepsilon. \tag{13.1.2}$$

For infinite-dimensional Banach spaces E, this converse is generally false. We have, for example, that if $e_n \in \ell_2$ is the usual orthonormal basis, then $\sum \frac{1}{n} e_n$ is unconditionally summable, but not absolutely summable. In fact, this is a general phenomenon: the Dvoretzky–Rogers theorem states that the two notions of summability agree in a Banach space E if and only if E is finite-dimensional.

Our next task is to show that unconditional convergence can also be formulated in terms of a tensor product. If we are given Banach spaces E and F, and a bounded linear mapping $f : E \to F$, it is immediate that f maps any unconditionally summable sequence $(x_n) \in E$ into an unconditionally summable sequence $(f(x_n)) \in F$. In particular, if $F = \mathbb{C}$, then $(f(x_n))$ is absolutely summable, or equivalently, $(f(x_n)) \in \ell_1$.

Proposition 13.1.1 *Given an unconditionally summable sequence*

$$x = (x_n)_{n\in\mathbb{N}} \in E,$$

we may define a function $\theta(x)$ from E^* into ℓ_1 by letting

$$\theta(x)(f) = (f(x_n)).$$

In this case, $\theta(x)$ can be indentified with an element in $\ell_1 \otimes^\lambda E$, and all elements of the latter tensor product arise in this fashion.

Proof Given an unconditionally convergent sequence (x_n) in E, we choose a finite set $F_\varepsilon \subseteq \mathbb{N}$ such that

$$F \cap F_\varepsilon = \emptyset \Rightarrow \left\| \sum_{n \in F} x_n \right\| < \varepsilon.$$

Then for each $f \in E^*$ with $\|f\| \leq 1$, the sequence $\alpha_n = f(x_n)$ satisfies (13.1.1). It follows from (13.1.2) that

$$F \cap F_\varepsilon = \emptyset \Rightarrow \sum_{n \in F} |f(x_n)| < 4\varepsilon. \tag{13.1.3}$$

If we let $N = \max F_\varepsilon$, then we conclude that

$$\sum_{n=1}^{\infty} |f(x_n)| \leq \left(\sum_{n=1}^{N} \|x_n\| + 4\varepsilon \right), \tag{13.1.4}$$

and thus $\theta(x) \in \mathcal{B}(E^*, \ell_1)$. In fact, we have $\theta(x) \in \ell_1 \otimes^\lambda E$, i.e. $\theta(x)$ is a norm limit of finite-rank mappings. To see this consider the truncation

$$\tilde{x} = (\tilde{x}_n) \in \ell_1 \otimes E \subseteq \mathcal{B}(E^*, \ell_1)$$

defined by

$$\tilde{x}_n = \begin{cases} x_n & \text{if } n \leq N, \\ 0 & \text{if } n > N. \end{cases}$$

If $F \cap \{1, \ldots, N\} = \emptyset$, then

$$\left\| \sum_{n \in F} (x_n - \tilde{x}_n) \right\| = \left\| \sum_{n \in F} x_n \right\| < \varepsilon,$$

and thus the above calculation shows that for $\|f\| \leq 1$,

$$\|\theta(x)(f) - \theta(\tilde{x})(f)\|_{\ell_1} = \sum_{n=1}^{\infty} |f(x_n - \tilde{x}_n)| = \sum_{n=N+1}^{\infty} |f(x_n)| \leq 4\varepsilon.$$

Conversely, given an element

$$u \in \ell_1 \otimes^\lambda E \subseteq \mathcal{B}(c_0, E),$$

we claim that $x_n = u(e_n)$ is an unconditionally summable sequence. To see this let us suppose that $\varepsilon > 0$. Since $\ell_1 \otimes^\lambda E$ is the completion of the algebraic tensor product, we may find an element

$$u_\varepsilon = \sum f_i \otimes v_i \in \ell_1 \otimes E$$

with $\|u - u_\varepsilon\| < \varepsilon$.

Since we may approximate each f_i by a function which vanishes off of a finite set, we may also suppose that all of the f_i vanish off of a common finite set F_ε. Given a finite set F with $F \cap F_\varepsilon = \emptyset$, we let χ_F be the characteristic function of F. Since $\|\chi_F\|_{c_0} \leq 1$, we have

$$F \cap F_\varepsilon = \emptyset \Rightarrow \left\| \sum_{n \in F} x_n \right\| = \|u(\chi_F)\| = \|(u - u_\varepsilon)(\chi_F)\| \leq \varepsilon.$$

□

Given Banach spaces E and F and a linear mapping $\varphi : E \to F$, we define the 1-*absolutely summing norm* of φ by

$$\begin{aligned} \pi_1^B(\varphi) &= \|id_{\ell_1} \otimes \varphi : \ell_1 \otimes_\lambda E \to \ell_1 \otimes_\gamma F\| \\ &= \sup_{n \in \mathbb{N}} \left\| id_{\ell_1^n} \otimes \varphi : \ell_1^n \otimes_\lambda E \to \ell_1^n \otimes_\gamma F \right\|. \end{aligned}$$

We say that φ is 1-*absolutely summing* or simply 1-*summing* if $\pi_1^B(\varphi) < \infty$. If this is the case, then $id \otimes \varphi$ extends to a mapping

$$id \otimes \varphi : \ell_1 \otimes^\lambda E \to \ell_1 \otimes^\gamma F,$$

with the norm $\pi_1^B(\varphi)$.

It is evident that π_1^B is a norm on the space $\Pi_1^B(E, F)$ of all 1-summing mappings, and in fact the isometric embedding

$$\Pi_1^B(E, F) \hookrightarrow \mathcal{B}(\ell_1 \otimes^\lambda E, \ell_1 \otimes^\gamma F) : \varphi \mapsto id \otimes \varphi$$

may be used to see that it is a Banach space. From our earlier discussion, we see that 1-absolutely summing mappings send unconditionally convergent sequences into absolutely summable sequences, a fact which accounts for the terminology.

We note that if we are given a diagram

$$D \xrightarrow{r} E \xrightarrow{\varphi} F \xrightarrow{s} G,$$

then we have a corresponding diagram

$$\ell_1 \otimes^\lambda D \xrightarrow{id \otimes r} \ell_1 \otimes^\lambda E \xrightarrow{id \otimes \varphi} \ell_1 \otimes^\gamma F \xrightarrow{id \otimes s} \ell_1 \otimes^\gamma G,$$

from which it follows that

$$\pi_1^B(s \circ \varphi \circ r) \leq \|s\| \, \pi_1^B(\varphi) \, \|r\|,$$

and thus

$$\Pi_1^B : (E, F) \mapsto (\Pi_1^B(E, F), \pi_1)$$

is a mapping ideal. If s is an isometric injection, then so is the mapping $id \otimes s$ (see the argument we gave for (7.1.25)), and thus

$$\pi_1^B(s \circ \varphi) = \pi_1^B(\varphi).$$

Thus, enlarging the range does not affect the absolutely summing norm.

Proposition 13.1.2 *For any Banach spaces E and F, a linear mapping $\varphi : E \to F$ satisfies $\pi_1^B(\varphi) \le 1$ if and only if for each $n \in \mathbb{N}$ and contraction $\theta : \ell_\infty^n \to E$, $\nu^B(\varphi \circ \theta) \le 1$.*

Proof This is apparent from the commutative diagram

$$\begin{array}{ccc} \ell_1^n \otimes^\lambda E & \longrightarrow & \ell_1^n \otimes^\gamma F \\ \downarrow & & \downarrow \\ \mathcal{B}(\ell_\infty^n, E) & \longrightarrow & \mathcal{N}^B(\ell_\infty^n, F) \end{array}$$

□

Corollary 13.1.3 *Π_1^B is a local mapping ideal, and for any linear mapping $\varphi : E \to F$, $\pi_1^B(\varphi) \le \iota^B(\varphi)$.*

Proof Given a contraction $\theta : \ell_\infty^n \to E$, we have $\varphi \circ \theta = \varphi|_L \circ \theta$, where $L = \theta(\ell_\infty^n)$. Thus, it is immediate that Π_1^B is a local mapping ideal. If $\nu^B(\varphi) \le 1$, then for each contraction $\theta : \ell_\infty^n \to E$,

$$\nu^B(\varphi \circ \theta) \le \nu^B(\varphi)\, \|\theta\| \le 1,$$

and in general,

$$\pi_1^B(\varphi) \le \nu^B(\varphi).$$

Since the mapping ideal is local,

$$\begin{aligned} \pi_1^B(\varphi) &= \sup\left\{\pi_1^B(\varphi_{|L}) : L \text{ finite-dimensional in } E\right\} \\ &\le \sup\left\{\nu^B(\varphi_{|L}) : L \text{ finite-dimensional in } E\right\} = \iota^B(\varphi). \end{aligned}$$

□

Proposition 13.1.4 *Given a compact Hausdorff space Ω and a bounded Borel measure $\mu \ge 0$ on Ω, the natural mapping*

$$J : C(\Omega) \to L^1(\Omega, \mu) \tag{13.1.5}$$

satisfies

$$\pi_1^B(J) \le \|\mu\|.$$

Proof To see this we note that if $(x_j) \in \ell_1^n \otimes C(\Omega)$, then

$$\begin{aligned} \|(id \otimes J)((x_j))\|_{\ell_1^n \otimes^\gamma L_1(\Omega,\mu)} &= \sum \|J(x_j)\|_{L_1(\Omega,\mu)} \\ &= \int \sum |x_j| \, \mathrm{d}\mu \le \|\mu\| \sup_{\omega\in\Omega} \left\{\sum |x_j(\omega)|\right\} \\ &\le \|\mu\| \sup_{f \in C(\Omega)^*} \left\{\sum |f(x_j)| : \|f\| \le 1\right\} \\ &= \|\mu\|\, \|(x_j)\|_{\ell_1^n \otimes^\lambda C(\Omega)}. \end{aligned}$$

□

The following result of Grothendieck shows that the mappings (13.1.5) are the prototypes for the 1-summing mappings.

Theorem 13.1.5 *Given Banach spaces E and F and a linear mapping $\varphi : E \to F$, $\pi_1(\varphi) \leq 1$ if and only if there exists a compact Hausdorff space Ω, a probability measure μ on Ω, and a commutative diagram*

$$\begin{array}{ccc} C(\Omega) & \xrightarrow{J} & L^1(\Omega, \mu) \\ \cup & & \cup \\ E_\infty & \xrightarrow{J_0} & F_1 \\ {\scriptstyle r}\uparrow & & \downarrow{\scriptstyle s} \\ E & \xrightarrow{\varphi} & F \end{array}, \tag{13.1.6}$$

where r and s are contractions, J is the usual inclusion mapping, E_∞ is the closure of $r(E)$, J_0 is the restriction of J to E_∞, and E_1 is the closure of $J_0(r(E))$. □

Given such a diagram of mappings, we have from the above that

$$\pi_1(\varphi) \leq \|s\| \, \pi_1(J_0) \, \|r\| \leq \pi_1(J) \leq 1.$$

The converse is proved by a simple application of the Hahn–Banach separation theorem. We do not pursue this theory since we take a rather different approach for operator spaces.

13.2 COMPLETELY 1-SUMMING MAPPINGS

If $\varphi : V \to W$ is a linear mapping of operator spaces, then we define $\pi_1(\varphi)$ in $[0, \infty]$ by

$$\begin{aligned} \pi_1(\varphi) &= \left\| id_{T_\infty} \otimes \varphi : T_\infty \check{\otimes} V \to T_\infty \hat{\otimes} W \right\| \\ &= \sup \left\{ \left\| id_{T_r} \otimes \varphi : T_r \check{\otimes} V \to T_r \hat{\otimes} W \right\| : r \in \mathbb{N} \right\}. \end{aligned} \tag{13.2.1}$$

This definition is 'stable' in the sense that we may replace the bounded norms with completely bounded norms. To see this, let us suppose that $\pi_1(\varphi) \leq 1$. Let us fix r. We have from Theorem 4.1.8 that

$$\|id_{T_r} \otimes \varphi\|_{cb} = \sup_{p \in \mathbb{N}} \left\| id_{T_p} \otimes id_{T_r} \otimes \varphi : T_p \hat{\otimes} (T_r \check{\otimes} V) \to T_p \hat{\otimes} (T_r \hat{\otimes} V) \right\|.$$

From Theorem 8.1.10 and (13.2.1) (with r replaced by $p \times r$), the two mappings in the diagram

$$T_p \hat{\otimes} (T_r \check{\otimes} V) \longrightarrow T_{p \times r} \check{\otimes} V \longrightarrow T_{p \times r} \hat{\otimes} V = T_p \hat{\otimes} (T_r \hat{\otimes} V)$$

are contractions, and thus $\|id_{T_r} \otimes \varphi\|_{cb} \leq 1$. If we let $r = 1$, then $\|\varphi\|_{cb} \leq 1$, and thus $\|\varphi\|_{cb} \leq \pi_1(\varphi)$.

If $\pi_1(\varphi) < \infty$, we say that φ is *completely* 1-*summing* and we refer to $\pi_1(\varphi)$ as the *completely* 1-*summing norm* of φ. We let $\Pi_1(V, W)$ denote the space of all completely 1-summing mappings from V into W. This is again an operator space since we may use the embedding

$$\Pi_1(V,W) \hookrightarrow \mathcal{CB}(T_\infty \check{\otimes} V, T_\infty \hat{\otimes} W) : \varphi \mapsto id \otimes \varphi.$$

The reader may prefer to consider only the norm on $\Pi_1(V,W)$ since we shall not have any particular use for matrices of such mappings. Nevertheless, in order to show that we have an operator space mapping ideal, we shall briefly avail ourselves of the operator space matrix norm structure. For any matrix $\varphi = [\varphi_{i,j}] \in M_m(\Pi_1(V,W))$,

$$\pi_{1,m}(\varphi) = \left\| id \otimes \varphi = [id \otimes \varphi_{i,j}] : T_\infty \check{\otimes} V \to M_m(T_\infty \hat{\otimes} W) \right\|_{cb}.$$

Let us suppose that we are given mappings $r : U \to V$, $s : W \to X$, and $\varphi : V \to M_m(W)$ (again we suggest that the reader consider only the case $m = 1$). Then it is apparent from the diagram

$$T_\infty \check{\otimes} U \xrightarrow{id \otimes r} T_\infty \check{\otimes} V \xrightarrow{id \otimes \varphi} M_m(T_\infty \hat{\otimes} W) \xrightarrow{(id \otimes s)_m} M_m(T_\infty \hat{\otimes} X)$$

that

$$\pi_{1,m}(s_m \circ \varphi \circ r) \le \|s\|_{cb} \, \pi_{1,m}(\varphi) \, \|r\|_{cb}.$$

In the remainder of this section we confine our attention to the *normed* space $\Pi_1(V,W)$.

Proposition 13.2.1 *For any operator spaces V and W, a linear mapping $\varphi : V \to W$ satisfies $\pi_1(\varphi) \le 1$ if and only if for each $n \in \mathbb{N}$ and complete contraction $\theta : M_n \to V$, $\nu(\varphi \circ \theta) \le 1$.*

Proof This is apparent from the commutative diagram

$$\begin{array}{ccc} T_n \check{\otimes} V & \xrightarrow{id \otimes \varphi} & T_n \hat{\otimes} W \\ \downarrow & & \downarrow \\ \mathcal{CB}(M_n, V) & \longrightarrow & \mathcal{N}(M_n, W) \end{array}.$$

□

Corollary 13.2.2 *The bifunctor $\Pi_1 : (V,W) \mapsto (\Pi_1(V,W), \pi_1)$ is a local mapping ideal, and for any linear mapping $\varphi : V \to W$,*

$$\pi_1(\varphi) \le \iota(\varphi).$$

If $s : W \hookrightarrow G$ is a completely isometric injection, then

$$\pi_1(s \circ \varphi) = \pi_1(\varphi);$$

in other words, if we enlarge the range, then the completely 1-summing norm remains the same.

Proof We may use the argument for the Banach 1-summing norm. □

Theorem 13.2.3 *Suppose that V and W are operator spaces, and that $\varphi : V \to W$ is a linear mapping. Then the following are equivalent:*

(i) $\pi_1(\varphi) \le 1$;

(ii) *there are indexed families* $(a_\alpha)_{\alpha\in I}$ *and* $(b_\alpha)_{\alpha\in I}$ *of Hilbert–Schmidt contractions, an ultrafilter* $\mathcal{U}$ *on* I*, and a commutative diagram*

$$\begin{array}{ccc} \prod_I M_\infty & \xrightarrow{\mathcal{M}} & \prod_I T_\infty/\mathcal{U} \\ \cup\!| & & \cup\!| \\ V_\infty & \longrightarrow & V_1 \\ {\scriptstyle r}\big\uparrow & & \big\downarrow{\scriptstyle s} \\ V & \xrightarrow{\varphi} & W \end{array} , \qquad (13.2.2)$$

where r *and* s *are complete contractions,* $\mathcal{M}$ *is determined by the complete contractions*

$$M(a_\alpha, b_\alpha) : M_\infty \to T_\infty : x \mapsto a_\alpha x b_\alpha,$$

and where $V_\infty = \overline{r(V)}$ *and* $V_1 = \overline{\mathcal{M}(V_\infty)}$.

Proof (i)⇒(ii). Let us suppose that $W \subseteq \mathcal{B}(H)$ for some Hilbert space H. For each finite-dimensional subspace L of V, we let $\iota_L : L \hookrightarrow V$ be the inclusion mapping, and we let $\mathcal{L}$ denote the collection of all finite-dimensional subspaces L of V. For each finite-dimensional subspace F of H, we let P_F denote the projection of H onto F, and we define

$$\tau_F : \mathcal{B}(H) \to \mathcal{B}(H) : x \mapsto P_F x P_F.$$

We let $\mathcal{F}$ denote the collection of all finite-dimensional subspaces F of H. We may identify $P_F\mathcal{B}(H)P_F$ with $\mathcal{B}(F)$, or with $M_{n(F)}$, where $n(F)$ is the dimension of F.

By assumption, $\pi_1(\varphi) \le 1$. Hence, given $L \in \mathcal{L}$,

$$\varphi_L = \varphi \circ \iota_L : L \to W$$

satisfies

$$\pi_1(\varphi_L) \le \pi_1(\varphi)\, \|\iota_L\|_{cb} \le 1,$$

and thus for each $n \in \mathbb{N}$,

$$\left\| id \otimes \varphi_L : T_n \check{\otimes} L \to T_n \hat{\otimes} W \right\| \le 1.$$

Let us consider the adjoint of this mapping. We have

$$(T_n \hat{\otimes} W)^* \cong M_n \check{\otimes} W^*,$$

and since any finite-dimensional operator space is reflexive,

$$(T_n \check{\otimes} L)^* \cong (M_n \hat{\otimes} L^*)^{**} \cong M_n \hat{\otimes} L^*.$$

It follows that

$$\left\| id \otimes \varphi_L^* : M_n \check{\otimes} W^* \to M_n \hat{\otimes} L^* \right\| \le 1.$$

We may identify $M_n \check{\otimes} W^*$ and $M_n \hat{\otimes} L^*$ with the spaces $\mathcal{CB}(W, M_n)$ and $\mathcal{N}(L, M_n)$ of completely bounded and completely nuclear mappings, respectively, and φ_L^* with the mapping

$$\mathcal{CB}(W, M_n) \to \mathcal{N}(L, M_n) : \tau \mapsto \tau \circ \varphi_L.$$

Thus, for each $F \in \mathcal{F}$, $\nu(\tau_F \circ \varphi_L) \le 1$. It follows that we have a commutative diagram

$$\begin{array}{ccc} M_\infty & \xrightarrow{M(a_{(L,F)}, b_{(L,F)})} & T_\infty \\ {\scriptstyle r_{(L,F)}}\uparrow & & \downarrow{\scriptstyle s_{(L,F)}} \\ L & \xrightarrow{\tau_F \circ \varphi} & M_{n(F)} \subseteq \mathcal{B}(H) \end{array},$$

where $r_{(L,F)}$ and $s_{(L,F)}$ are complete contractions, and the Hilbert–Schmidt norms of the operators $a_{(L,F)}, b_{(L,F)} \in \mathcal{B}(\ell_2)$ satisfy

$$\left\|a_{(L,F)}\right\|_2, \left\|b_{(L,F)}\right\|_2 \le 1 + \frac{1}{\dim F}.$$

Using the injectivity of M_∞, we may extend $r_{(L,F)}$ to a completely contractive mapping $r'_{(L,F)} : V \to M_\infty$, from which we obtain the commutative diagram

$$\begin{array}{ccccc} & & M_\infty & \xrightarrow{M(a_{(L,F)}, b_{(L,F)})} & T_\infty \\ & {\scriptstyle r'_{(L,F)}}\nearrow & \uparrow{\scriptstyle r_{(L,F)}} & & \downarrow{\scriptstyle s_{(L,F)}} \\ V & \xleftarrow{\iota_L} & L & \xrightarrow{\tau_F \circ \varphi} & M_{n(F)} \subseteq \mathcal{B}(H) \end{array}.$$

We let $\mathcal{L} \times \mathcal{F}$ have the partial ordering $(L, F) \preceq (L', F')$ if $L \subseteq L'$ and $F \subseteq F'$. The family $\mathcal{U}_0$ of intervals

$$I(L, F) = \{(L', F') : (L, F) \preceq (L', F')\}$$

is a filter on $\mathcal{L} \times \mathcal{F}$, and we let $\mathcal{U}$ be an ultrafilter containing $\mathcal{U}_0$. The mappings

$$r'_{(L,F)} : V \to M_\infty$$

determine a complete contraction

$$r : V \to \prod_{\mathcal{L} \times \mathcal{F}} M_\infty : v \to (r'_{(L,F)}(v)),$$

and the uniformly completely bounded mappings

$$M(a_{(L,F)}, b_{(L,F)}) : M_\infty \to T_\infty$$

determine a mapping

$$\mathcal{M} = \pi_{\mathcal{U}} \circ (M(a_{(L,F)}, b_{(L,F)})) : \prod_{\mathcal{L} \times \mathcal{F}} M_\infty \to \prod_{\mathcal{L} \times \mathcal{F}} T_\infty / \mathcal{U}.$$

Let us denote a typical element of $\prod_{\mathcal{L}\times\mathcal{F}} T_\infty$ by $(t_{(L,F)})$ and its quotient image in $\prod_{\mathcal{L}\times\mathcal{F}} T_\infty/\mathcal{U}$ by $\pi_\mathcal{U}((t_{(L,F)}))$. Since weak* closed bounded subsets of $\mathcal{B}(H)$ are weak* compact, we may define a mapping

$$s : \prod_{\mathcal{L}\times\mathcal{F}} T_\infty/\mathcal{U} \to \mathcal{B}(H)$$

by using the weak* limit

$$s(\pi_\mathcal{U}((t_{(L,F)}))) = \lim_\mathcal{U} s_{(L,F)}(t_{(L,F)}).$$

In this manner, we obtain a diagram of mappings

$$\begin{array}{ccccc} & \prod_{\mathcal{L}\times\mathcal{F}} M_\infty & \xrightarrow{\mathcal{M}} & \prod_{\mathcal{L}\times\mathcal{F}} T_\infty/\mathcal{U} & \\ {}^{r}\nearrow & & & & \searrow^{s} \\ V & & \xrightarrow{\varphi} & W & \subseteq \mathcal{B}(H) \end{array}.$$

Given a point $v \in V$, we may suppose that $v \in L_0 \in \mathcal{L}$. Then for $L \supseteq L_0$,

$$r'_{(L,F)}(v) = r_{(L,F)}(v),$$

and thus

$$s\left(\mathcal{M}(r(v))\right) = \lim_\mathcal{U} a_{(L,F)} r_{(L,F)}(v) b_{(L,F)} = \lim_\mathcal{U} P_F \varphi(v) P_F = \varphi(v).$$

If we let $V_\infty = \overline{r(V)}$ and $V_1 = \overline{\mathcal{M}(V_\infty)}$, then it follows that $s(V_1) = \varphi(V) \subseteq W$, and thus if we let $I = \mathcal{L}\times\mathcal{F}$, then we obtain the commutative diagram (13.2.2). Finally, since

$$\lim_\mathcal{U} \dim F = \infty,$$

it is evident that the diagram still commutes if we replace $a_{(L,F)}$ and $b_{(L,F)}$ by $a_{(L,F)}/\left\|a_{(L,F)}\right\|_2$ and $b_{(L,F)}/\left\|b_{(L,F)}\right\|_2$, respectively; that is, we may initially suppose that $a_{(L,F)}$ and $b_{(L,F)}$ are Hilbert–Schmidt contractions.

(ii)⇒(i). Let us suppose that we are given a diagram (13.2.2) with r and s complete contractions. Since

$$\begin{aligned} \pi_1(\varphi) &\le \|s\|_{cb}\, \pi_1(\mathcal{M}\circ r : V \to V_1) \\ &\le \pi_1\left(\mathcal{M}\circ r : V \to \prod_{\mathcal{L}\times\mathcal{F}} T_\infty/\mathcal{U}\right) \end{aligned}$$

(see Corollary 13.2.2), it suffices to show that the latter quantity is ≤ 1. Given $\alpha \in I$, we have from Proposition 12.2.2 that

$$\pi_1(M(a_\alpha, b_\alpha)) \le \nu(M(a_\alpha, b_\alpha)) \le 1.$$

Thus, if we are given a matrix $v = [v_{i,j}] \in T_n \check{\otimes} V$, then

$$id \otimes r(v) = [r(v_{i,j})] \in T_n \check{\otimes} M_\infty$$

and

$$\|[M(a_\alpha, b_\alpha)(r(v_{i,j}))]\|_{T_n \hat{\otimes} T_\infty} \le \|id \otimes r(v)\|_{T_n \check{\otimes} M_\infty} \le \|v\|_{T_n \check{\otimes} V}.$$

It follows from Lemma 10.3.8 that

$$\begin{aligned}\|[M(r(v_{i,j})]\|_{T_n \hat{\otimes} \prod_{\mathcal{L}\times\mathcal{F}} T_\infty/\mathcal{U}} &= \lim_{\mathcal{U}} \|[M(a_\alpha, b_\alpha)(r(v_{i,j}))]\|_{T_n \hat{\otimes} T_\infty} \\ &\le \|v\|_{T_n \check{\otimes} V},\end{aligned}$$

which completes the proof. □

13.3 HILBERT SPACE FACTORABLE MAPPINGS

Mappings that factor through Hilbert spaces play an important role in Banach space theory. The corresponding operator space theory is more complex since there is a multiplicity of quantizations for Hilbert spaces. We shall confine our attention to column and row Hilbert operator spaces. Given operator spaces V and W, we say that a mapping $\varphi : V \to W$ *factors through a column Hilbert operator space* if there is a Hilbert space K and a commutative diagram

$$\begin{array}{ccccc} & & K_c & & \\ & {}^{r}\nearrow & & \searrow^{s} & \\ V & & \xrightarrow{\varphi} & & W \end{array},$$

where r and s are completely bounded. We let $\Gamma_2^c(V,W)$ be the linear space of all such mappings. We identify the matrices

$$\varphi = [\varphi_{i,j}] \in M_n(\Gamma_2^c(V,W))$$

with the mappings $\varphi : V \to M_n(W)$, and given such a matrix, it is easy to see that there must exist complete contractive mappings

$$r : V \to M_{1,n}(H_c) \text{ and } s : H_c \to M_{n,1}(W)$$

for which the diagram

$$\begin{array}{ccccc} & & M_{1,n}(H_c) & & \\ & {}^{r}\nearrow & & \searrow^{s_{1,n}} & \\ V & & \xrightarrow{\varphi} & & M_n(W) \end{array} \qquad (13.3.1)$$

commutes. We define

$$\gamma_{2,n}^c(\varphi) = \inf\left\{\|r\|_{cb}\|s\|_{cb}\right\},$$

where the infimum is taken over all such diagrams (13.3.1). As we shall see below, this determines an operator space structure on $\Gamma_2^c(V,W)$.

It is immediate from (13.3.1) that Γ_2^c is a mapping ideal. On the other hand, if $\psi : W \to X$ is a completely isometric injection, then

$$\gamma_{2,n}^c(\psi \circ \varphi) = \gamma_{2,n}^c(\varphi). \qquad (13.3.2)$$

To see this, let us suppose that we are given a commutative diagram

$$\begin{array}{ccccc} & & M_{1,n}(H_c) & & \\ & {}^{r}\nearrow & & \searrow^{s_{1,n}} & \\ V & \longrightarrow & M_n(W) & \subseteq & M_n(X) \end{array} .$$

If we let

$$r = [r_1 \ldots r_n] \quad \text{and} \quad s = \begin{bmatrix} s_1 \\ \vdots \\ s_n \end{bmatrix},$$

then it follows that $\varphi_{i,j}(v) = s_i(r_j((v))$. If we let $\tilde{H}$ denote the linear span of the subspaces $r_j(V)$ $(1 \leq j \leq n$), then we have $r(V) \subseteq M_{1,n}(\tilde{H}_c)$. On the other hand, since $s_i(r_j(v)) = \varphi_{i,j}(v) \in W$ and W is closed in X (it is assumed complete), $s_i(\xi) \in W$ for all $\xi \in \tilde{H}$. Thus, we have the desired commutative diagram

$$\begin{array}{ccc} & M_{1,n}(\tilde{H}_c) & \\ {}^{r}\nearrow & & \searrow^{s_{1,n}} \\ V & \xrightarrow{\ \varphi\ } & M_n(W) \end{array} .$$

Lemma 13.3.1 *Given operator spaces V and W, we have a natural complete isometry*

$$\Gamma_2^c(V, W^*) \cong (W \overset{h}{\otimes} V)^*.$$

Proof We define $S_0 : \Gamma_2^c(V, W^*) \to (W \otimes_h V)^*$ by letting

$$S_0(\varphi)(w \otimes v) = \varphi(v)(w).$$

If we let S_Γ be the composition

$$\Gamma_2^c(V, W^*) \xrightarrow{j} \mathcal{CB}(V, W^*) \longrightarrow (W \widehat{\otimes} V)^*,$$

where $j : \Gamma_2^c(V, W^*) \to \mathcal{CB}(V, W^*)$ is the natural inclusion mapping, and the second mapping is the natural complete isometry, then we have

$$S_0(\varphi) = S_\Gamma(\varphi)_{|W \otimes V}.$$

It follows that S_0 and thus $(S_0)_n$ are one-to-one. Thus, it suffices to show that if $\varphi : V \to M_n(W^*)$ satisfies $\gamma_{2,n}^c(\varphi) < 1$, then

$$F = (S_0)_n(\varphi) : W \otimes_h V \to M_n$$

is a complete contraction, and conversely that any $F \in M_n((W \otimes_h V)^*)$ with $\|F\|_{cb} < 1$ has the form $(S_0)_n(\varphi)$ for a mapping $\varphi \in M_n(\Gamma_2^c(V, W^*))$ satisfying $\gamma_{2,n}^c(\varphi) \leq 1$.

If we are given $\gamma^c_{2,n}(\varphi) < 1$ as above, we have a corresponding commutative diagram

$$\begin{array}{ccccc} & & M_{1,n}(H_c) & & \\ & {}^{r}\nearrow & & \searrow^{s_{1,n}} & \\ V & & \xrightarrow{\varphi} & & M_n(W^*) \end{array} \qquad (13.3.3)$$

with r and s completely contractive. If we use the identification

$$\mathcal{B}(H, \mathbb{C}^n) = \mathcal{CB}(H_c, \mathbb{C}^n_c),$$

then we see that

$$\mathcal{CB}(H_c, \mathcal{CB}(W, \mathbb{C}^n_c)) = \mathcal{CB}(W \widehat{\otimes} H_c, \mathbb{C}^n_c) = \mathcal{CB}(W, \mathcal{B}(H, \mathbb{C}^n)). \qquad (13.3.4)$$

It follows that the complete contraction $s : H_c \to \mathcal{CB}(W, M_{n,1})$ determines a complete contraction $\overline{s} : W \to \mathcal{B}(H, \mathbb{C}^n)$, where if $\xi \in H_c$, the matrix product $\overline{s}(w)\xi$ is equal to $s(\xi)(w)$. Given $v \in V$ and $w \in W$, we have

$$\begin{aligned} F(w \otimes v) &= \varphi(v)(w) = s_{1,n}(r(v))(w) \\ &= [s(r_1(v))(w) \ \dots \ s(r_n(v))(w)] \\ &= [\overline{s}(w)r_1(v) \ \dots \ \overline{s}(w)r_n(v)] \\ &= \overline{s}(w)r(v), \end{aligned}$$

where for fixed v and w, the last expression is a composition of Hilbert space operators $\overline{s}(w) \in \mathcal{B}(H, \mathbb{C}^n)$ and $r(v) \in \mathcal{B}(\mathbb{C}^n, H)$. It follows from Theorem 9.4.3 that $F : W \otimes_h V \to M_n$ is completely contractive.

Conversely, given a complete contraction $F : W \otimes_h V \to M_n$, we have from Theorem 9.4.3 that there exists a Hilbert space H and complete contractions $r : V \to \mathcal{B}(\mathbb{C}^n, H)$ and $\overline{s} : W \to \mathcal{B}(H, \mathbb{C}^n)$ with

$$F(w \otimes v) = \overline{s}(w)r(v).$$

Reversing the above argument, we have from (13.3.4) that $\overline{s}$ determines a complete contraction

$$s : H_c \to \mathcal{CB}(W, \mathbb{C}^n_c) = M_{n,1}(W^*),$$

where $s(\xi)(w) = \overline{s}(w)(\xi)$. It follows that $\varphi = s_{1,n} \circ r : V \to M_n(W^*)$ satisfies $\gamma^c_{2,n}(\varphi) \leq 1$, and it is easy to check that $S_0(\varphi) = F$. □

Corollary 13.3.2 *For any operator spaces V and W, $\Gamma^c_2(V, W)$ is an operator space.*

Proof This is immediate from Lemma 13.3.1, and (13.3.2) applied to the inclusion $W \to W^{**}$. □

Theorem 13.3.3 *Suppose that V and W are operator spaces, and that $\varphi : V \to M_n(W)$ is a linear mapping. Then the following are equivalent:*

(i) $\gamma^c_{2,n}(\varphi) \leq 1$;

(ii) *for any column Hilbert operator space H_c,*

$$\left\| id \otimes \varphi : H_c \check{\otimes} V \to M_n(H_c \widehat{\otimes} W) \right\|_{cb} \le 1;$$

(iii) *for any index set I (respectively, any finite set I or $I = \mathbb{N}$),*

$$\left\| id \otimes \varphi : T_I \overset{h}{\otimes} V \to M_n(T_I \widehat{\otimes} W) \right\|_{cb} \le 1.$$

Proof The various alternatives in (iii) are equivalent since for any index set I, $\bigcup_F T_F \overset{h}{\otimes} X$ is dense in $T_I \overset{h}{\otimes} V$, where the union is taken over finite subsets F of I.

(i)⇒(ii). If we assume (i), we may factor the mapping

$$id \otimes \varphi : H_c \check{\otimes} V \to M_n(H_c \widehat{\otimes} W)$$

as follows:

$$\begin{aligned} H_c \check{\otimes} V &\overset{id\otimes r}{\longrightarrow} H_c \check{\otimes} M_{1,n} \check{\otimes} K_c = M_{1,n} \check{\otimes} (H_c \check{\otimes} K_c) = M_{1,n}(H_c \widehat{\otimes} K_c) \\ &\overset{id\otimes(id\otimes s)}{\longrightarrow} M_{1,n}(H_c \widehat{\otimes} M_{n,1}(W)) = M_{1,n}(M_{n,1}(H_c \widehat{\otimes} W)), \end{aligned}$$

where we have used the identifications

$$H_c \widehat{\otimes} M_{n,1}(W) = M_{n,1} \overset{h}{\otimes} W \overset{h}{\otimes} H_c = M_{n,1}(H_c \widehat{\otimes} W).$$

(ii)⇒(iii). Given an index set I, we let $H = \ell_2(I)$. Then

$$T_I = (H_c)^* \overset{h}{\otimes} H_c,$$

and we have the complete contractions

$$\begin{array}{c} T_I \overset{h}{\otimes} V = \overline{H}_r \overset{h}{\otimes} (\overline{H}_r \overset{h}{\otimes} V) \\ \Big\downarrow {\scriptstyle id\otimes(id\otimes\varphi)} \\ \overline{H}_r \overset{h}{\otimes} M_n(H_c \widehat{\otimes} W) = \overline{H}_r \widehat{\otimes} M_n(H_c \widehat{\otimes} W) = (M_n \check{\otimes} (H_c \widehat{\otimes} W)) \widehat{\otimes} \overline{H}_r \\ \Big\downarrow \\ M_n \check{\otimes} (H_c \widehat{\otimes} W \widehat{\otimes} \overline{H}_r) = M_n(T_I \widehat{\otimes} W), \end{array}$$

where the second mapping was shown to be completely contractive in Theorem 8.1.10.

(iii)⇒(i). Let us first suppose that W is the dual of some operator space W_1. We may choose an index set I and a complete quotient mapping $\pi : T_I \to W_1$. The mapping $F_\varphi : W_1 \otimes V \to M_n$ defined by

$$F_\varphi(w_1 \otimes v) = \varphi(v)(w_1)$$

is factored in the bottom row of the commutative diagram

$$\begin{array}{ccccc} T_I \otimes_h V & \xrightarrow{id \otimes \varphi} & M_n(T_I \widehat{\otimes} W_1^*) & & \\ \Big\downarrow{\scriptstyle \pi\otimes id} & & \Big\downarrow{\scriptstyle (\pi\otimes id)_n} & & \\ W_1 \otimes_h V & \xrightarrow{id\otimes\varphi} & M_n(W_1 \widehat{\otimes} W_1^*) & \xrightarrow{\mu_n} & M_n \end{array}.$$

By assumption, the top row is completely contractive, and thus

$$F_\varphi \circ (\pi \otimes id) : T_I \otimes_h V \to M_n$$

is a complete contraction. From Proposition 9.2.5, $\pi \otimes id$ is a complete quotient mapping, and thus F_φ is completely contractive. We conclude from Lemma 13.3.1 that $\gamma_{2,n}^c(\varphi) \leq 1$.

Now let us suppose that W is a general operator space. Then

$$\left\| id \otimes (J \circ \varphi) : H_c \check{\otimes} V \to M_n(H_c \widehat{\otimes} W^{**}) \right\|_{cb} \leq 1,$$

and thus

$$\gamma_{2,n}^c(\varphi) = \gamma_{2,n}^c(J \circ \varphi) \leq 1.$$

□

Corollary 13.3.4 *If V and W are operator spaces, then*

$$\Pi_1(V, W) \subseteq \Gamma_2^c(V, W),$$

and $\gamma_2^c \leq \pi_1$.

Proof If $\varphi : V \to M_n(W)$ satisfies $\pi_{1,n}(\varphi) \leq 1$, then the composition

$$id \otimes \varphi : T_I \overset{h}{\otimes} V \to T_I \check{\otimes} V \to M_n(T_I \widehat{\otimes} W)$$

is completely contractive, and from (iii) of Theorem 13.3.3 we conclude that $\gamma_{2,n}^c(\varphi) \leq 1$. □

By symmetry, there is an analogous theory for mappings $\varphi : V \to W$ which can be factored through row Hilbert operator spaces. We denote the corresponding mapping space by $\Gamma_2^r(V, W)$, and the matrix norms by $\gamma_{2,n}^r(\varphi)$. Although the arguments are identical, it is necessary to take some care with the notation since, for example,

$$\Gamma_2^r(V, W^*) \cong (V \overset{h}{\otimes} W)^*.$$

13.4 THE DVORETZKY–ROGERS THEOREM FOR OPERATOR SPACES

An elementary version of the Dvoretzky–Rogers theorem states that if the unconditionally summable sequences in a Banach space V all converge absolutely, then V must be finite-dimensional. Grothendieck realized that this could be understood in terms of absolutely summing mappings. If the identity mapping $id : V \to V$ is 1-absolutely summing, then V must be finite-dimensional.

Turning to operator spaces, we shall use the following.

Proposition 13.4.1 *Given operator spaces V, W, X, and a diagram of linear mappings*

$$V \xrightarrow{\varphi} W \xrightarrow{\psi} X,$$

we have

$$\nu(\psi \circ \varphi) \leq \gamma_2^c(\psi)\gamma_2^r(\varphi),$$

and

$$\nu(\psi \circ \varphi) \leq \gamma_2^r(\psi)\gamma_2^c(\varphi).$$

Proof We have from Proposition 12.2.4 the isometric identification

$$\mathcal{CB}(H_c, K_r) = \mathcal{N}(H_c, K_r).$$

Given $\gamma_2^r(\psi) < 1$ and $\gamma_2^c(\varphi) < 1$, we have a diagram of completely contractive mappings

$$\begin{array}{ccccccccc} & & H_c & & \xrightarrow{\theta} & & H_r & & \\ & {}^r\nearrow & & {}^s\searrow & & {}^t\nearrow & & {}^u\searrow & , \\ V & & \xrightarrow{\varphi} & & W & & \xrightarrow{\psi} & & X \end{array}$$

where $\theta = t \circ s$, and thus

$$\nu(\psi \circ \varphi) \leq \nu(\theta) = \|\theta\|_{cb} \leq 1.$$

□

Corollary 13.4.2 *Under the hypotheses of Proposition* 13.4.1, *we have*

$$\nu(\psi \circ \varphi) \leq \pi_1(\psi)\pi_1(\varphi).$$

Proof This follows from Proposition 13.4.1 and Corollary 13.3.4. □

Theorem 13.4.3 *If V is an operator space for which the identity mapping $id : V \to V$ satisfies $\pi_1(id) < \infty$, then V must be finite-dimensional.*

Proof If $\pi_1(id) < \infty$, then from Corollary 13.4.2,

$$\nu(id) = \nu(id \circ id) \leq \pi_1(id)^2 < \infty,$$

and id is completely nuclear. It follows from Proposition 12.2.1 that id is compact in the usual Banach space sense, and from classical theory, V must be finite-dimensional. □

13.5 NOTES AND REFERENCES

Absolutely summing mappings were introduced in Grothendieck (1955), and the quotation in the introduction of this chapter may be found in Defant and Floret (1993). Grothendieck's theory was subsequently generalized to a notion of p-absolutely summing operators by Pietsch and Kwapien (see Pisier (1986) for further references).

Completely 1-summing mappings first appeared in Effros and Ruan (1994*b*), as did the material in §13.3 and §13.4. The factorization of the completely 1-summing mappings through ultraproducts is due to Pisier (1998). Pisier and Junge have succeeded in generalizing the Pietsch–Kwapien theory to operator spaces (see Junge 1996 and Pisier 1998). This remarkable work depends upon Pisier's operator space versions of interpolation theory, a topic we have not included in this text. The reader should note that Pisier has another notion of Hilbert-factorable mappings that is closer to the Banach space concept (see Pisier 1996*b*).

Part IV

Local Theory and Integrality

14

Local reflexivity, exactness, and nuclearity

Archbold and Batty were the first to recognize an unexpected complication that can occur for non-nuclear C^*-algebras. It is now understood that they had found a common tensorial context for two of the most important *trans-nuclear* properties in the subject, namely Kirchberg's notion of exactness, and the Effros–Haagerup theory of local reflexivity.

Once again we begin by considering the classical theory, in which these new phenomena are always precluded.

14.1 THE LOCAL STRUCTURE OF BANACH SPACES

In principle, the structure of a Banach space is determined by the nature and relative positions of its finite-dimensional subspaces. In order to exploit this obvious proposition, it is necessary to have quantitative estimates on the distance between finite-dimensional subspaces.

If $r : E \to F$ is a linear isomorphism of finite-dimensional linear spaces, then we shall write $E \overset{r}{\cong} F$. Given finite-dimensional Banach spaces, we define the *Banach–Mazur distance* $d(E, F)$ by

$$d(E,F) = \inf \{\|r\| \, \|r^{-1}\| : E \overset{r}{\cong} F\}. \tag{14.1.1}$$

If E and F are not isomorphic, then we let $d(E,F) = \infty$. It is a simple exercise to verify that for any three spaces E, F, and G, we have $d(E,E)=1$, $d(E,F) = d(F,E)$, and

$$d(E,G) \leq d(E,F)\, d(F,G).$$

It follows that $\ln d$ is a semimetric on any fixed collection of isomorphic spaces.

A fundamental principle of the Banach space theory is that *any* finite-dimensional normed space can be approximated by subspaces of ℓ^n_∞. We include two proofs. The first is simpler, but the second illustrates an argument that is frequently used in operator space theory.

Theorem 14.1.1 *For any finite-dimensional Banach space L and $\varepsilon > 0$, there exists an $n \in \mathbb{N}$ and a subspace S of ℓ^n_∞ such that*

$$d(L, S) \leq 1 + \varepsilon.$$

Proof It suffices to show that if we let $\delta = \varepsilon(1+\varepsilon)^{-1}$, we may find a contractive linear mapping $r : L \to \ell^n_\infty$ such that

$$\|r(x)\|_\infty \geq (1-\delta)\|x\|. \tag{14.1.2}$$

To see this we note that the inequality implies that r is a linear isomorphism of L onto $S = r(L)$, for which

$$\left\|r^{-1}r(x)\right\| = \|x\| \leq (1-\delta)^{-1}\|r(x)\|_\infty = (1+\varepsilon)\|r(x)\|_\infty;$$

hence, $\left\|r^{-1}\right\| \leq (1+\varepsilon)$ and $d(L,S) \leq (1+\varepsilon)$.

First proof. Since $K = L^*_{\|\cdot\|\leq 1}$ is norm compact, we can choose open balls $B_k = B_\delta(f_k)$ with $f_k \in K$ $(k = 1, \ldots, n)$ and

$$K \subseteq \bigcup_{k=1}^n B_k.$$

We define a mapping $r : L \to \ell^n_\infty$ by

$$\tilde{r}(x) = (f_1(x), \ldots, f_n(x)).$$

Since $\|f_k\| \leq 1$, $\|r(x)\|_\infty \leq \|x\|$. On the other hand, for any $x \in L$, we can choose an element $f \in K$ with $|f(x)| = \|x\|$. If we choose k such that $\|f - f_k\| < \delta$, then we have

$$\|r(x)\|_\infty \geq |f_k(x)| \geq |f(x)| - |(f - f_k)(x)| \geq (1-\delta)\|x\|.$$

Second proof. We may suppose that L is a subspace of ℓ_∞. Since $L_{\|\cdot\|_\infty = 1}$ is compact, we have

$$L_{\|\cdot\|_\infty = 1} \subseteq \bigcup_{k=1}^n B_k,$$

where $B_k = B_{\delta/2}(x_k)$ and $x_k \in L$ satisfies $\|x_k\|_\infty = 1$. For each $n \in \mathbb{N}$, we let

$$\rho_n : \ell_\infty \to \ell^n_\infty$$

be the truncation mapping. It is evident that we can find an $n \in \mathbb{N}$ such that

$$\|\rho_n(x_k)\|_\infty \geq \|x_k\|_\infty - \delta/2 = 1 - \delta/2$$

for all k. We let $r = (\rho_n)_{|L}$. If $x \in L$ and $\|x\|_\infty = 1$, then we may choose a k such that $\|x - x_k\|_\infty \leq \delta/2$. It follows that

$$\|r(x)\|_\infty \geq \|r(x_k)\|_\infty - \|r(x - x_k)\|_\infty \geq 1 - 2\delta/2,$$

and we obtain (14.1.2). □

Corollary 14.1.2 *Let us suppose that L is a finite-dimensional Banach space. Then for any Banach space E we have the isometry*

$$L^* \otimes^\gamma E^* \cong (L \otimes^\lambda E)^*.$$

Proof If $L = \ell_\infty^n$, then $L^* = \ell_1^n$ and

$$\ell_1^n \otimes^\gamma E^* = \ell_1^n(E^*) \cong \ell_\infty^n(E)^* = (\ell_\infty^n \otimes^\lambda E)^*.$$

If $L \subseteq \ell_\infty^n$, then we may identify L^* with $\ell_1^n / L^\perp$. We have a commutative diagram

$$\begin{array}{ccc} \ell_1^n \otimes^\gamma E^* & \xrightarrow{\theta_{\ell_\infty^n}} & (\ell_\infty^n \otimes^\lambda E)^* \\ \downarrow & & \downarrow \\ L^* \otimes^\gamma E^* \cong (\ell_1^n / L^\perp) \otimes^\gamma E^* & \xrightarrow{\theta_L} & (L \otimes^\lambda E)^* \end{array},$$

where the contraction θ_L is determined by the contractive bilinear function $(f, g) \mapsto f \otimes g$. Since $\otimes^\gamma$ is projective and $\otimes^\lambda$ is injective, it is evident that the columns are quotient mappings, and thus θ_L is an isometry.

If L is a general finite-dimensional Banach space and $\varepsilon > 0$, then we may use Theorem 14.1.1 to find a linear isomorphism $r : L \to S$, where $S \subseteq \ell_n^\infty$ for some $n \in \mathbb{N}$, such that $\|r\|\|r^{-1}\| < 1 + \varepsilon$. If we let $s = r^{-1}$ and define θ_S and θ_L as above, then we obtain a diagram of mappings

$$\begin{array}{ccc} L^* \otimes^\gamma E^* & \xrightarrow{\theta_L} & (L \otimes^\lambda E)^* \\ {\scriptstyle s^* \otimes id}\downarrow & & \uparrow{\scriptstyle (r \otimes id)^*} \\ S^* \otimes^\gamma E^* & \xrightarrow{\theta_S} & (S \otimes^\lambda E)^* \end{array}.$$

This commutes since if we are given $f \in L^*$, $g \in E^*$, $x \in S$, and $y \in E$, then we have

$$\begin{aligned} \langle (r \otimes id)^* \circ \theta_S \circ (s^* \otimes id)(f \otimes g), x \otimes y \rangle &= \langle \theta_S (s^* \otimes id)(f \otimes g), r(x) \otimes y \rangle \\ &= \langle s^*(f) \otimes g, r(x) \otimes y \rangle \\ &= \langle f \otimes g, sr(x) \otimes y \rangle \\ &= \langle \theta_L (f \otimes g), x \otimes y \rangle. \end{aligned}$$

It follows that

$$\|\theta_L^{-1}\| = \left\|(s^* \otimes id)^{-1} \theta_S ((r \otimes id)^*)^{-1}\right\| \leq 1 + \varepsilon,$$

and since ε is arbitrary, we see that θ_L is isometric. □

The following result is a simple version of the *principle of local reflexivity,* a property that holds for all Banach spaces. As we shall see, the corresponding property is not universally true for operator spaces. Indeed, operator space local reflexivity becomes one of the key properties that distinguishes certain classes of C^*-algebras. Perhaps the most surprising discovery along these lines is that *all* C^*-algebraic duals are operator space locally reflexive (see Theorem 15.3.1).

Corollary 14.1.3 *Suppose that L and E are Banach spaces with L finite-dimensional. Every linear contraction $\varphi : L \to E^{**}$ may be approximated in the point-weak* topology by linear contractions $\psi : L \to E$.*

Proof Each element $F \in \mathcal{B}(L, E)^{**}$ determines a mapping

$$\varphi_F : L \to E^{**}$$

by the relation

$$\langle \varphi_F(x), f \rangle = \langle F, \theta(x \otimes f) \rangle, \tag{14.1.3}$$

where

$$\theta : L \otimes^\gamma E^* \to (L^* \otimes^\lambda E)^* \cong \mathcal{B}(L, E)^* \tag{14.1.4}$$

is the natural isometric isomorphism (see Corollary 14.1.2). It follows that the adjoint mapping

$$\theta^* : \mathcal{B}(L, E)^{**} \to \mathcal{B}(L, E^{**}) : F \mapsto \varphi_F$$

(see (14.1.3)) is an isometric isomorphism. We have a commutative diagram

$$\begin{array}{ccc} \mathcal{B}(L,E)^{**} & \xrightarrow{\theta^*} & \mathcal{B}(L,E^{**}) \\ \cup\! \mid & & \cup\! \mid \\ \mathcal{B}(L,E) & \xrightarrow{id} & \mathcal{B}(L,E) \end{array},$$

where the columns are the obvious isometries, and the top mapping is continuous in the weak* and point-weak* topologies. The desired approximation is immediate since the unit ball of $\mathcal{B}(L, E)$ is weak* dense in that of its second dual. □

In the following result and its proof, it is notationally convenient to use the incomplete tensor products $\otimes_\gamma$ and $\otimes_\lambda$.

Corollary 14.1.4 *For any Banach spaces E and F, the inclusion*

$$E^{**} \otimes F^{**} \hookrightarrow (E \otimes_\lambda F)^{**} \tag{14.1.5}$$

induces the Banach space injective tensor norm $\|\cdot\|_\lambda$ on $E^{**} \otimes F^{**}$.

Proof We first prove that the natural inclusion

$$E \otimes_\lambda F^{**} \hookrightarrow (E \otimes_\lambda F)^{**} \tag{14.1.6}$$

is isometric. Given an element $u \in E \otimes F^{**}$, it follows that $u \in L \otimes F^{**}$ for some finite-dimensional space L. From Corollary 14.1.2,

$$(L \otimes_\lambda F)^{**} \cong (L^* \otimes_\gamma F^*)^* \cong L \otimes_\lambda F^{**}.$$

Thus, (14.1.6) follows from the commutative diagram

$$\begin{array}{ccc} L \otimes_\lambda F^{**} & \cong & (L \otimes_\lambda F)^{**} \\ \cap\!\!| & & \cap\!\!| \\ E \otimes_\lambda F^{**} & \to & (E \otimes_\lambda F)^{**} \end{array},$$

where the inclusion symbols indicate isometries.

If we apply (14.1.6) to the spaces E^{**} and F, then it follows that

$$E^{**} \otimes_\lambda F^{**} \subseteq (E^{**} \otimes_\lambda F)^{**} \subseteq (E \otimes_\lambda F)^{****},$$

where the inclusions again indicate isometries. Since $E^{**} \otimes F^{**}$ is situated inside the isometric image of $(E \otimes_\lambda F)^{**}$ in the fourth dual, we obtain (14.1.5). □

It can be shown that Grothendieck's theorem (12.1.8) is a consequence of Corollary 14.1.4. The reader can extract the argument from the corresponding formula for operator spaces that satisfy the analogue of (14.1.5) (see Proposition 14.2.2).

14.2 THE ARCHBOLD–BATTY CONDITIONS

There is a natural analogue of (14.1.1), due to Pisier. If V and W are operator spaces of the same finite dimension, we define the *completely bounded Banach–Mazur distance* $d_{cb}(V, W)$ by

$$d_{cb}(V, W) = \inf\left\{ \|\varphi\|_{cb} \left\|\varphi^{-1}\right\|_{cb} : V \overset{\varphi}{\cong} W \right\}$$

(see the notation of the previous section). For any three operator spaces V, W, and Z, we have $d_{cb}(V, V) = 1$, $d_{cb}(V, W) = d_{cb}(W, V)$, and

$$d_{cb}(V, Z) \leq d_{cb}(V, W)\, d_{cb}(W, Z).$$

It follows that $\ln d_{cb}$ is a semimetric on any fixed collection of isomorphic operator spaces.

If V and W are n-dimensional operator spaces, then we have from Corollary 2.2.5 that $d_{cb}(V, W) \leq n^2$, and as we indicated in §2.4, Pisier has shown that, in fact, $d_{cb}(V, W) \leq n$. Thus, the set $\mathcal{OS}_n$ of n-dimensional subspaces of $\mathcal{B}(\ell_2)$ with the metric $\ln d_{cb}$ has diameter less than or equal to $\ln n$. In contrast to the Banach space situation, $\mathcal{OS}_n$ need not be compact or even separable (see Junge and Pisier 1995). We shall not pursue these ideas below.

It is a reflection of the richer structure of operator space theory that the analogue of Theorem 14.1.1 is false. We shall see in §14.5 that there exist finite-dimensional operator spaces which cannot be approximated in the metric $\ln d_{cb}$ by finite-dimensional subspaces of M_n for any $n \in \mathbb{N}$. Furthermore, the analogues of Corollary 14.1.3 (local reflexivity), and Corollary 14.1.4 do not hold in general. These phenomena were initially studied by

Archbold and Batty, who observed that they could be used to distinguish C^*-algebras.

Given operator spaces V and W, we have a commutative diagram

$$\begin{array}{ccc} & & (V \check{\otimes} W)^{**} \\ & \overset{\tau}{\nearrow} & \\ V^{**} \otimes W^{**} & & \downarrow \theta^* \\ & \underset{\sigma}{\searrow} & \\ & & (V^* \hat{\otimes} W^*)^* \cong \mathcal{CB}(V^*, W^{**}) \end{array}, \tag{14.2.1}$$

where the diagonal mappings σ and τ are the natural inclusions, and

$$\theta : V^* \hat{\otimes} W^* \to (V \check{\otimes} W)^*$$

is determined by the completely contractive bilinear mapping

$$V^* \times W^* \to (V \check{\otimes} W)^* : (f, g) \mapsto f \otimes g.$$

It is evident that with the relative operator space matrix norm,

$$\sigma(V^{**} \otimes W^{**}) = V^{**} \otimes_{\vee} W^{**}$$

(see Proposition 8.1.2). We let $V^{**}{:}\otimes{:}_{\vee} W^{**}$ denote $\tau(V^{**} \otimes W^{**})$ with the relative operator space matrix norm in $(V \check{\otimes} W)^{**}$. Similarly, we write

$$V \otimes{:}_{\vee} W^{**} = \tau(V \otimes W^{**})$$

and

$$V^{**}{:}\otimes_{\vee} W = \tau(V^{**} \otimes W)$$

for the (incomplete) operator spaces, and we denote the norm closures of these operator subspaces by $V^{**} : \check{\otimes} : W^{**}$, $V \check{\otimes} : W^{**}$, and $V^{**} : \check{\otimes} W$, respectively. We refer to these as the *augmented, right augmented*, and *left augmented injective tensor products,* respectively. With these conventions, we have the commutative diagram

$$\begin{array}{ccccc} & & V \check{\otimes}{:} W^{**} & & \\ & \nearrow & & \searrow & \\ V \check{\otimes} W & & & & V^{**} : \check{\otimes} : W^{**} \xrightarrow{\theta_0^*} V^{**} \check{\otimes} W^{**}, \\ & \searrow & & \nearrow & \\ & & V^{**}{:}\check{\otimes} W & & \end{array} \tag{14.2.2}$$

where the diagonal mappings are completely isometric injections, and the restriction θ_0^* of θ^* is a complete contraction. It is easy to see that for any operator spaces V and W, we have the natural complete isometry

$$V \check{\otimes}{:} W^{**} \cong W^{**}{:}\check{\otimes} V. \tag{14.2.3}$$

These definitions behave well with respect to matrices. For example, it is easy to see that we may identify the mapping

$$\tau_n : M_n(V^{**} \otimes W^{**}) \to M_n((V \check{\otimes} W)^{**})$$

with the corresponding mapping

$$\tau : M_n(V)^{**} \otimes W^{**} \to (M_n(V) \check{\otimes} W)^{**},$$

and thus

$$M_n(V^{**} {:}\check{\otimes}{:} W^{**}) = M_n(V)^{**} {:}\check{\otimes}{:} W^{**}.$$

Similarly, we have the identifications

$$M_n(V^{**} {:}\check{\otimes}{:} W^{**}) = V^{**} {:}\check{\otimes}{:} M_n(W)^{**}$$

and

$$M_n(V^{**} {:}\check{\otimes} W) = M_n(V)^{**} {:}\check{\otimes} W = V^{**} {:}\check{\otimes} M_n(W).$$

We say that an operator space V satisfies *condition* C if for all operator spaces W, the mapping

$$\theta_0^* : V^{**} {:}\check{\otimes}{:} W^{**} \cong V^{**} \check{\otimes} W^{**}$$

is isometric. It is equivalent to suppose that θ_0^* is a complete isometry, since the isometric condition implies that

$$M_n(V^{**}{:}\check{\otimes}{:}W^{**}) = V^{**}{:}\check{\otimes}{:}M_n(W)^{**} = V^{**} \check{\otimes} M_n(W)^{**} = M_n(V^{**} \check{\otimes} W^{**}).$$

Similarly, we say that V satisfies *condition* C' if for all operator spaces W, θ_0^* restricts to an isometry

$$V \check{\otimes}{:} W^{**} \cong V \check{\otimes} W^{**}$$

and that V satisfies *condition* C'' if for all operator spaces W, it restricts to an isometry

$$V^{**} {:}\check{\otimes} W \cong V^{**} \check{\otimes} W.$$

Once again, these conditions are stable in the sense that if they hold, then these identifications are completely isometric.

Lemma 14.2.1 *Suppose that V is an operator space. Then V satisfies condition C if and only if it satisfies both C' and C''.*

Proof If V satisfies condition C and W is an arbitrary operator space, then the bottom row of the commutative diagram

$$\begin{array}{ccc} V \check{\otimes}{:} W^{**} & \longrightarrow & V \check{\otimes} W^{**} \\ \cap\!\mid & & \cap\!\mid \\ V^{**} {:}\check{\otimes}{:} W^{**} & \longrightarrow & V^{**} \check{\otimes} W^{**} \end{array}$$

is isometric. Hence, the same is true for the top row, and V satisfies condition C'. A similar diagram shows that V satisfies condition C''.

On the other hand, let us suppose that V satisfies conditions C' and C''. From condition C'',

$$V^{**} \otimes_{\vee} W^{**} = V^{**} :\otimes_{\vee} W^{**} \subseteq (V \otimes_{\vee} W^{**})^{**}.$$

From condition C',

$$V \otimes_{\vee} W^{**} = V \otimes:_{\vee} W^{**} \subseteq (V \otimes_{\vee} W)^{**},$$

and thus we have the isometric inclusion

$$V^{**} \otimes_{\vee} W^{**} \subseteq (V \otimes_{\vee} W)^{****}.$$

But we have

$$V^{**} \otimes_{\vee} W^{**} \subseteq (V \otimes_{\vee} W)^{**} \subseteq (V \otimes_{\vee} W)^{****},$$

where the second inclusion is isometric. It follows that the first inclusion is isometric and we are done. □

We may use the right augmented injective tensor product to study the mapping (12.3.9).

Proposition 14.2.2 *For any operator spaces V and W, the mapping*

$$S_{int} : \mathcal{I}(V, W^*) \to (V \check{\otimes} W)^* \tag{14.2.4}$$

is a (completely) isometric surjection if and only if we have the natural (completely) isometric isomorphism

$$V \check{\otimes}: W^{**} \cong V \check{\otimes} W^{**} \tag{14.2.5}$$

Proof Let us suppose that we have (14.2.5). For any $\varphi \in \mathcal{I}(V, W^*)$, $F_{\varphi} = S_{int}(\varphi) = S(\varphi)$ is determined by

$$\langle F_{\varphi}, v \otimes w \rangle = \varphi(v)(w)$$

(see (12.3.9)). In order to prove that S_{int} is isometric, we must show that $\iota(\varphi) \leq \|F_{\varphi}\|$. From Lemma 12.3.3, we have the natural complete isometry

$$S_0 : \mathcal{I}(V, W^*) \hookrightarrow (V \check{\otimes} W^{**})^*.$$

It follows that

$$\begin{aligned}\iota(\varphi) &= \sup \left\{ |\langle F_{\varphi}, u \rangle| : u \in V \otimes W^{**}, \|u\|_{V \check{\otimes} W^{**}} \leq 1 \right\} \\ &= \sup \left\{ |\langle F_{\varphi}, u \rangle| : u \in V \otimes W^{**}, \|u\|_{V \check{\otimes} : W^{**}} \leq 1 \right\}.\end{aligned}$$

Since the closed unit ball of $V \otimes_{\vee} W$ is weak* dense in the closed unit ball of $(V \check{\otimes} W)^{**}$,

$$\iota(\varphi) = \sup \left\{ |\langle F_{\varphi}, u \rangle| : u \in V \otimes W, \|u\|_{V \check{\otimes} W} \leq 1 \right\} = \|F_{\varphi}\|.$$

To prove that S_{int} is a surjection, let us suppose that $f \in (V \check{\otimes} W)^*$. Then since the mapping S in the diagram (12.3.8) is a complete isometric surjection, there is a complete contraction $\varphi : V \to W^*$ such that $S(\varphi) = \Phi^*(f)$. Restricting to the algebraic tensor product $V \otimes W$, we have $F_\varphi = f$, and thus from the above calculations we obtain $\iota(\varphi) = \|f\| < \infty$. We conclude that $\varphi \in \mathcal{I}(V, W^*)$ and $S_{int}(\varphi) = f$.

Conversely, let us suppose that (14.2.4) is an isometric bijection. Then we have the commutative diagram

$$\begin{array}{ccccc} \mathcal{I}(V,W^*) & \overset{S_{int}}{\cong} & (V \check{\otimes} W)^* & \overset{\Phi^*}{\longrightarrow} & (V \hat{\otimes} W)^* \cong \mathcal{CB}(V,W^*) \\ {\scriptstyle \tilde{J}}\big\downarrow & & & & \big\downarrow \\ \mathcal{I}(V,W^{***}) & \overset{\tilde{S}_{int}}{\longrightarrow} & (V \check{\otimes} W^{**})^* & \longrightarrow & (V \hat{\otimes} W^{**})^* \cong \mathcal{CB}(V,W^{***}) \end{array},$$

where $\tilde{J}$ is the isometry described in Lemma 12.3.2 (for V and W^*), and the right column is the obvious isometric inclusion. Thus, if we let

$$\eta = \tilde{S}_{int} \circ \tilde{J} \circ S_{int}^{-1},$$

then we obtain a diagram of contractions

$$\begin{array}{ccc} (V \check{\otimes} W)^* & \overset{\Phi^*}{\longrightarrow} & (V \hat{\otimes} W)^* \cong \mathcal{CB}(V,W^*) \\ {\scriptstyle \eta}\big\downarrow & & \big\downarrow \\ (V \check{\otimes} W^{**})^* & \longrightarrow & (V \hat{\otimes} W^{**})^* \cong \mathcal{CB}(V,W^*) \end{array}.$$

If we take the adjoints of the mappings in this diagram, then we obtain the commutative diagram

$$\begin{array}{cccc} & (V \hat{\otimes} W^{**})^{**} & \longrightarrow & (V \check{\otimes} W^{**})^{**} \\ \nearrow & & & \\ V \otimes W^{**} & \big\downarrow & & \big\downarrow{\scriptstyle \eta^*} \\ \searrow & & & \\ & (V \hat{\otimes} W)^{**} & \longrightarrow & (V \check{\otimes} W)^{**} \end{array}.$$

The bottom composition has range $V \otimes:_\vee W^{**}$. On the other hand, $V \otimes W^{**}$ inherits the matrix norm $V \check{\otimes} W^{**}$ in $(V \check{\otimes} W^{**})^{**}$, and thus the algebraic identification $V \otimes_\vee W^{**} = V \otimes:_\vee W^{**}$ is isometric.

The completely isometric case can be obtained by applying Corollary 12.3.5. □

Corollary 14.2.3 *Let V be an operator space.*

(i) *V satisfies condition C' if and only if $\mathcal{I}(V, W^*) \cong (V \check{\otimes} W)^*$ for all operator spaces W.*

(ii) *V satisfies condition C'' if and only if $\mathcal{I}(W, V^*) \cong (V \check{\otimes} W)^*$ for all operator spaces W.*

Proof This is an immediate consequence of the above result and the definitions of the two conditions. □

We can recapture Grothendieck's classical result by defining the *weak integral mappings* to be the linear space

$$\mathcal{I}^w(V, W^*) = S^{-1}((V \check{\otimes} W)^*) \subseteq \mathcal{CB}(V, W^*),$$

together with the norm determined by $(V \check{\otimes} W)^*$. From the proof of the above result, we see that this does not determine a mapping ideal, since in particular we cannot expect $\eta : \mathcal{I}^w(V, W^*) \to \mathcal{I}^w(V, W^{***})$ to be contractive, or even defined. We shall return to this notion in §15.5.

14.3 LOCAL REFLEXIVITY AND CONDITION C''

Following the classical definition, we say that an operator space W is *locally reflexive* (respectively, λ*-locally reflexive*) if for any finite-dimensional operator space L, every complete contraction $\varphi : L \to W^{**}$ is the point-weak* limit of a net of linear mappings $\varphi_\alpha : L \to W$ with $\|\varphi_\alpha\|_{cb} \leq 1$ (respectively, with $\|\varphi_\alpha\|_{cb} \leq \lambda$). It is known that the full group C^*-algebra $C^*(\mathbb{F}_2)$ is not locally reflexive (see Corollary 14.3.8). One can prove a version of the following result for λ-locally reflexivity. Since the proof is identical, we shall only consider the local reflexivity case.

Theorem 14.3.1 *Suppose that W is an operator space. Then the following are equivalent:*

(i) *W is locally reflexive;*

(ii) *for any finite-dimensional operator space L, we have the isometry* (*or complete isometry*)

$$L^* \hat{\otimes} W^* \cong (L \check{\otimes} W)^*;$$

(ii′) *for any finite-dimensional operator space L, we have the isometry* (*or complete isometry*)

$$\mathcal{N}(W, L^*) = \mathcal{I}(W, L^*);$$

(iii) *for any operator space V, we have the isometry* (*or complete isometry*)

$$\mathcal{I}(V, W^*) \cong (V \check{\otimes} W)^*;$$

(iv) *W satisfies condition C''.*

Proof Let us first prove the equivalence with the isometric conditions.

We have already proved (iii) ⇔(iv) (see Corollary 14.2.3).

(ii)⇔(ii′) is immediate from Corollary 12.3.4.

(iii)⇒(ii) is obvious since if L is finite-dimensional, then we obtain the (complete) isometries

$$L^* \hat{\otimes} W^* \cong \mathcal{N}(L, W^*) = \mathcal{I}(L, W^*) \cong (L \check{\otimes} W)^*.$$

(ii)⇒(iii). We have seen that

$$S_{int} : \mathcal{I}(V, W^*) \to (V \check{\otimes} W)^*$$

is a completely contractive injection. Let us suppose that the mapping in (ii) is completely isometric (the argument for the isometric case is the same). If we have a contractive functional

$$F \in M_n((V \check{\otimes} W)^*) = \mathcal{CB}(V \check{\otimes} W, M_n),$$

then $F = S_n(\varphi)$ for some

$$\varphi : V \to M_n(W^*).$$

For any finite-dimensional subspace L of V,

$$F_{|L \check{\otimes} W} = S_n(\varphi_{|L}),$$

and thus from (ii),

$$\left\| F_{|L \check{\otimes} W} \right\|_{cb} = \nu_n(\varphi_{|L}).$$

If we take the union over all finite-dimensional subspaces $L \subseteq V$, we conclude from (12.3.1) that

$$\iota_n(\varphi) = \sup \{\nu_n(\varphi_{|L})\} = \sup \left\{ \left\| F_{|L \check{\otimes} W} \right\|_{cb} \right\} = \|F\|_{cb} \, .$$

(ii) ⇔(i). Let us look at the isometric case first. Since

$$(L^* \hat{\otimes} W^*)^* \cong \mathcal{CB}(L, W^{**}) \cong L \check{\otimes} W^{**},$$

(ii) holds if and only if we have the natural isometric isomorphism

$$L \check{\otimes} W^{**} \cong (L \check{\otimes} W)^{**}.$$

This correspondence is explicitly given by the norm-increasing linear isomorphism

$$\tau : L \check{\otimes} W^{**} \to (L \check{\otimes} W)^{**}.$$

Thus, the relation is isometric if and only if

$$\varphi \in (L \check{\otimes} W^{**})_{\| \cdot \| \leq 1} \cong \mathcal{CB}(L^*, W^{**})_{\| \cdot \|_{cb} \leq 1}$$

implies that

$$\varphi \in (L \check{\otimes} W)^{**}_{\| \cdot \| \leq 1}.$$

From the bipolar theorem, the latter is the case if and only if φ is a weak* limit of elements in

$$(L \check{\otimes} W)_{\| \cdot \| \leq 1} = \mathcal{CB}(L^*, W)_{\| \cdot \|_{cb} \leq 1}.$$

Since it is evident that

$$\tau : \mathcal{CB}(L^*, W^{**}) \to (L \check{\otimes} W)^{**}$$

is a homeomorphism in point-weak* and weak* topologies, we are done.

If the mapping in (ii) is isometric, then we have the natural isometric identifications

$$\begin{aligned} T_n((L\check{\otimes} W)^*) &= (M_n(L\check{\otimes} W))^* = (M_n(L)\check{\otimes} W)^* \\ &= (M_n(L))^* \,\hat{\otimes}\, W^* = T_n(L^*) \,\hat{\otimes}\, W^* = T_n(L^* \,\hat{\otimes}\, W^*), \end{aligned}$$

and from Theorem 4.1.8 the identification is completely isometric. □

Corollary 14.3.2 *If W is a locally reflexive operator space, then any subspace $X \subseteq W$ is locally reflexive.*

Proof This is immediate from condition C'' and the commutative diagram

$$\begin{array}{ccc} V \otimes_{\vee} X^{**} & \longrightarrow & (V \check{\otimes} X)^{**} \\ \downarrow & & \downarrow \\ V \otimes_{\vee} W^{**} & \longrightarrow & (V \check{\otimes} W)^{**} \end{array},$$

in which the columns are automatically isometric. □

Lemma 14.3.3 *An operator space V is locally reflexive if and only if for any finite-dimensional subspaces $E \subseteq V^{**}$, $F \subseteq V^*$, and $\epsilon > 0$, there exists a mapping $\varphi : E \to V$ such that $\|\varphi\|_{cb} < 1 + \epsilon$ and*

$$\langle \varphi(x), f\rangle = \langle x, f\rangle$$

for all $x \in E$ and $f \in F$.

Proof Since V is locally reflexive, we may regard the inclusion mapping

$$\iota : E \to V^{**}$$

as a contractive element of $E^* \check{\otimes} V^{**} \cong (E^* \check{\otimes} V)^{**}$, and $L = E \otimes F$ as a finite-dimensional subspace of $E\hat{\otimes}V^* \cong (E^* \check{\otimes} V)^*$. From Helly's lemma (see §A.2), we can choose an element

$$\varphi \in E^* \check{\otimes} V \cong \mathcal{CB}(E, V)$$

such that $\|\varphi\|_{cb} < 1 + \epsilon$, and

$$\langle \varphi(x), f\rangle = \langle \varphi, x \otimes f\rangle = \langle \iota, x \otimes f\rangle = \langle x, f\rangle$$

for all $x \in E$ and $f \in F$.

To prove the converse, it is enough to consider a net of completely contractive mappings of the form $\psi_{(E,F,\varepsilon)} = \varphi/(1 + \varepsilon)$ with φ chosen as above. □

We shall next prove that an operator space is locally reflexive if and only if that is the case for each separable subspace. This is a rather subtle fact, and we must first consider the following result of Ge and Hadwin.

Lemma 14.3.4 *Let V be an operator space and E a finite-dimensional subspace of V^{**}. For any finite-dimensional subspace $F \subseteq V^*$, $\varepsilon > 0$, and $n \in \mathbb{N}$, there exists a linear isomorphism ψ from E onto a subspace $E_\psi \subseteq V$ such that*

(i) $\|\psi_n\| \le 1$, $\|\psi_n^{-1}\| \le 1 + \varepsilon$;
(ii) $\langle \psi(x), f\rangle = \langle x, f\rangle$ *for all $x \in E$ and $f \in F$;*
(iii) $\psi_{|E\cap V} = id_{E\cap V}$.

Proof Let us fix n. Since we have the isometry $M_n(V)^{**} \cong M_n(V^{**})$, we may identify $M_n(E)$ with a finite-dimensional subspace of $M_n(V)^{**}$ and identify $T_n(F)$ with a finite-dimensional subspace of the Banach dual $T_n(V^*) = M_n(V)^*$. We recall that the duality is given by

$$\langle [x_{i,j}], [f_{i,j}]\rangle = \sum_{i,j} \langle x_{i,j}, f_{i,j}\rangle.$$

From the principle of local reflexivity for Banach spaces, there exists a mapping $\tilde{\varphi} : M_n(E) \to M_n(V)$ such that

(i′) $\|\tilde{\varphi}\| \le 1, \|\tilde{\varphi}^{-1}\| \le 1 + \varepsilon$;
(ii′) $\langle \tilde{\varphi}(x), f\rangle = \langle x, f\rangle$ for all $x \in M_n(E)$ and $f \in T_n(F)$;
(iii′) $\tilde{\varphi}_{|M_n(E\cap V)} = id_{M_n(E\cap V)}$.

Since the unitary group $U(n)$ of M_n is a compact group, there exists a normalized Haar measure μ on $U(n)$. We define $\tilde{\psi} : M_n(E) \to M_n(V)$ by

$$\tilde{\psi}(x) = \int\!\!\int_{U(n)\times U(n)} \alpha\tilde{\varphi}(\alpha^* x\beta)\beta^* \,\mathrm{d}\mu(\alpha)\mathrm{d}\mu(\beta).$$

Since μ is translation invariant, we have

$$\tilde{\psi}(\alpha x\beta) = \alpha\tilde{\psi}(x)\beta$$

for $\alpha, \beta \in U(n)$, and thus for all $\alpha, \beta \in M_n$. This implies that

$$\tilde{\psi} = id_{M_n} \otimes \psi = \psi_n$$

for some linear mapping $\psi : E \to V$. It follows from (i′) that

$$\|\psi_n\| = \|\tilde{\psi}\| \le \|\tilde{\varphi}\| \le 1,$$
$$\|\psi_n^{-1}\| = \|\tilde{\psi}^{-1}\| \le \|\tilde{\varphi}^{-1}\| \le 1 + \varepsilon.$$

For any $x \in M_n(E)$ and $f \in T_n(F)$, $\alpha x\beta^* \in M_n(E)$ and $\alpha^{tr} f\overline{\beta} \in T_n(F)$ for $\alpha, \beta \in U(n)$. We have from (ii′) that

$$\begin{aligned}\langle \psi_n(x), f\rangle &= \left\langle \int\!\!\int_{U(n)\times U(n)} \alpha\tilde{\varphi}(\alpha^* x\beta)\beta^* \,\mathrm{d}\mu(\alpha)\mathrm{d}\mu(\beta), f\right\rangle \\ &= \int\!\!\int_{U(n)\times U(n)} \langle \alpha\tilde{\varphi}(\alpha^* x\beta)\beta^*, f\rangle \,\mathrm{d}\mu(\alpha)\mathrm{d}\mu(\beta)\end{aligned}$$

$$= \int\int_{U(n)\times U(n)} \langle \tilde{\varphi}(\alpha^* x\beta),\, \alpha^{tr} f\overline{\beta}\rangle \, \mathrm{d}\mu(\alpha)\mathrm{d}\mu(\beta)$$

$$= \int\int_{U(n)\times U(n)} \langle \alpha^* x\beta,\, \alpha^{tr} f\overline{\beta}\rangle \, \mathrm{d}\mu(\alpha)\mathrm{d}\mu(\beta)$$

$$= \langle x,\, f\rangle.$$

Therefore, for any $x \in E$ and $f \in F$,

$$\langle \psi(x),\, f\rangle = \langle \psi_n(\varepsilon_{1,1}^{[n]} \otimes x),\, \varepsilon_{1,1}^{[n]} \otimes f\rangle = \langle \varepsilon_{1,1}^{[n]} \otimes x,\, \varepsilon_{1,1}^{[n]} \otimes f\rangle = \langle x,\, f\rangle.$$

A similar argument based on (iii′) shows that $\psi_{|E\cap V} = id_{E\cap V}$. □

Theorem 14.3.5 *An operator space V is locally reflexive if and only if that is the case for each separable subspace.*

Proof Since any subspace of a locally reflexive operator space is locally reflexive, it is evident that the first condition implies the second. Conversely, let us suppose that each separable subspace of V is locally reflexive. It suffices to show that if $\varphi : E \to V^{**}$ is a complete contraction, and $F \subseteq V^*$ is finite-dimensional, then for each $\varepsilon > 0$ there exists a mapping $\psi_\varepsilon : E \to V$ such that

$$\|\psi_\varepsilon\|_{cb} < 1 + \varepsilon \text{ and } \langle \psi_\varepsilon(x), f\rangle = \langle \varphi(x), f\rangle \qquad (14.3.1)$$

for all $x \in E$ and $f \in F$ (see Lemma 14.3.3).

From Lemma 14.3.4 we may find a mapping $\psi^{(n)} : E \to V$ such that $\|\psi_n^{(n)}\| < 1 + 1/n$ and

$$\langle \psi^{(n)}(x), f\rangle = \langle \varphi(x), f\rangle$$

for all $x \in E$ and $f \in F$. The norm closed linear span V_0 of the union of the subspaces $\psi^{(n)}(E)$ with $n \in \mathbb{N}$ is separable in the norm topology, and we can regard $\psi^{(n)}$ as a sequence in $\mathcal{B}(E, V_0^{**})$. Since the closed ball of radius 2 is compact in the point-weak* topology on the latter space, we may choose a limit point $\psi : E \to V_0^{**}$ of the sequence $\psi^{(n)}$. If $r \leq n$, then

$$\|(\psi^{(n)})_r\| \leq \|(\psi^{(n)})_n\| \leq 1 + 1/n,$$

and thus $\|\psi_r\| \leq 1$. It follows that $\|\psi\|_{cb} \leq 1$. Furthermore,

$$\langle \psi(x), f\rangle = \langle \varphi(x), f\rangle$$

for all $x \in E$ and $f \in F$.

By assumption, V_0 is locally reflexive as an operator space. Thus, given $\varepsilon > 0$, we may find a mapping $\psi_\varepsilon : E \to V_0$ such that $\|\psi_\varepsilon\|_{cb} < 1+\varepsilon$, which satisfies (14.3.1). □

A C^*-algebraic version of the following result is due to Archbold and Batty (1980). They used it to show that the full group C^*-algebra $C^*(\mathbb{F}_n)$ is not locally reflexive (see Corollary 14.3.8).

Lemma 14.3.6 *Suppose that $\mathcal{A}$ is a locally reflexive C^*-algebra, $\mathcal{J}$ is a closed two-sided ideal in $\mathcal{A}$, and $\pi : \mathcal{A} \to \mathcal{A}/\mathcal{J}$ is the quotient mapping. Then for any finite-dimensional operator space L, the mapping*

$$\pi \otimes id_L : \mathcal{A} \check{\otimes} L \to (\mathcal{A}/\mathcal{J}) \check{\otimes} L$$

is a complete quotient mapping. Moreover, the quotient C^*-algebra $\mathcal{A}/\mathcal{J}$ is locally reflexive.

Proof We have a commutative diagram

$$\begin{array}{ccc} \mathcal{A} \check{\otimes} L & \xrightarrow{\pi \otimes id_L} & (\mathcal{A}/\mathcal{J}) \check{\otimes} L \\ \downarrow & & \downarrow \\ \mathcal{A}^{**} \check{\otimes} L & \xrightarrow{\pi^{**} \otimes id_L} & (\mathcal{A}/\mathcal{J})^{**} \check{\otimes} L \end{array},$$

where we have from Proposition 8.1.5 that the columns are completely isometric. The weak* closure $\overline{\mathcal{J}}$ of $\mathcal{J}$ is a closed two-sided ideal in the von Neumann algebra $\mathcal{A}^{**}$, and thus it has the form $\mathcal{A}^{**}e$ for some central projection e in $\mathcal{A}^{**}$. In this case, we have the completely isometric $*$-isomorphsims

$$(\mathcal{A}/\mathcal{J})^{**} \cong \mathcal{A}^{**}/\overline{\mathcal{J}} \cong \mathcal{A}^{**}(1-e),$$

and the complete quotient mapping

$$\pi^{**} : \mathcal{A}^{**} \to (\mathcal{A}/\mathcal{J})^{**}$$

has a completely contractive lifting given by the canonical inclusion

$$\mathcal{A}^{**}(1-e) \hookrightarrow \mathcal{A}^{**}.$$

It follows from Proposition 8.1.5 that

$$\pi^{**} \otimes id_L : \mathcal{A}^{**} \check{\otimes} L \longrightarrow (\mathcal{A}^{**}/\overline{\mathcal{J}}) \check{\otimes} L$$

is a complete quotient mapping. Since L is finite-dimensional,

$$\ker(\pi \otimes id_L) = \mathcal{J} \check{\otimes} L$$

and

$$\ker(\pi^{**} \otimes id_L) = \overline{\mathcal{J}} \check{\otimes} L.$$

Therefore, we obtain a complete isometry

$$(\mathcal{A}^{**} \check{\otimes} L)/(\overline{\mathcal{J}} \check{\otimes} L) \cong (\mathcal{A}/\mathcal{J})^{**} \check{\otimes} L.$$

Since $\mathcal{A}$ is locally reflexive, we have the complete isometry

$$(\mathcal{A} \check{\otimes} L)^{**} \cong \mathcal{A}^{**} \check{\otimes} L$$

and thus the complete isometries

$$\begin{aligned} ((\mathcal{A} \check{\otimes} L)/(\mathcal{J} \check{\otimes} L))^{**} &\cong ((\mathcal{J} \check{\otimes} L)^{\perp})^{*} \cong (\mathcal{A} \check{\otimes} L)^{**}/(\mathcal{J} \check{\otimes} L)^{\perp\perp} \\ &\cong (\mathcal{A} \check{\otimes} L)^{**}/\overline{(\mathcal{J} \check{\otimes} L)} \cong (\mathcal{A}^{**} \check{\otimes} L)/(\overline{\mathcal{J}} \check{\otimes} L). \end{aligned}$$

It follows that the columns in the following diagram are completely isometric injections

$$\begin{array}{ccc} (\mathcal{A}\check{\otimes} L)/(\mathcal{J}\check{\otimes} L) & \longrightarrow & (\mathcal{A}/\mathcal{J})\check{\otimes} L \\ \downarrow & & \downarrow \\ (\mathcal{A}^{**}\check{\otimes} L)/(\overline{\mathcal{J}}\check{\otimes} L) & \cong & (\mathcal{A}/\mathcal{J})^{**}\check{\otimes} L \end{array},$$

and thus the top row is a complete isometry. This implies the first assertion of the theorem.

Applying our first result, we obtain the following complete isometries:

$$\begin{aligned} ((\mathcal{A}/\mathcal{J})\check{\otimes} L)^{**} &\cong ((\mathcal{A}\check{\otimes} L)/(\mathcal{J}\check{\otimes} L))^{**} \\ &\cong (\mathcal{A}^{**}\check{\otimes} L)/(\overline{\mathcal{J}}\check{\otimes} L) \cong (\mathcal{A}/\mathcal{J})^{**}\check{\otimes} L. \end{aligned}$$

We conclude from Theorem 14.3.1 that $\mathcal{A}/\mathcal{J}$ is locally reflexive. □

We note that the above proof applies in more general situations. Thus, the theorem holds when we replace $\mathcal{A}$ with an arbitrary locally reflexive operator space V and a closed subspace J for which the complete quotient mapping $V^{**} \to (V/J)^{**}$ has a completely contractive lifting. In particular, this is the situation for a closed one-sided ideal in a locally reflexive C^*-algebra.

Proposition 14.3.7 *Suppose that $\mathcal{A}$ is a locally reflexive C^*-algebra, $\mathcal{J}$ is a closed two-sided ideal in $\mathcal{A}$, and $\pi : \mathcal{A} \to \mathcal{A}/\mathcal{J}$ is the quotient mapping. Then for any operator space V,*

$$\ker\left(\pi \otimes id : \mathcal{A}\check{\otimes} V \to (\mathcal{A}/\mathcal{J})\check{\otimes} V\right) = \mathcal{J}\check{\otimes} V. \tag{14.3.2}$$

Proof Suppose that $u \in \mathcal{A}\check{\otimes} V$ satisfies $\pi \otimes id(u) = 0$. Then given $\varepsilon > 0$, we may choose an element

$$u_0 = \sum_{i=1}^{n} h_i \otimes v_i \in \mathcal{A} \otimes_{\vee} V$$

such that $\|u - u_0\| < \varepsilon$. It follows that $u_0 \in \mathcal{A}\check{\otimes} L$, where L is a finite-dimensional subspace of V spanned by $v_1, \ldots, v_n$. Since the obvious mapping $(\mathcal{A}/\mathcal{J})\check{\otimes} L \to (\mathcal{A}/\mathcal{J})\check{\otimes} V$ is isometric,

$$\begin{aligned} \|\pi \otimes id_L(u_0)\| &= \|\pi \otimes id_V(u_0)\| \\ &\leq \|\pi \otimes id_V(u_0) - \pi \otimes id_V(u)\| + \|\pi \otimes id_V(u)\| \\ &= \|(\pi \otimes id_V)(u_0 - u)\| + 0 \leq \|u_0 - u\| < \varepsilon. \end{aligned}$$

From Lemma 14.3.6, $\pi \otimes id_L : \mathcal{A}\check{\otimes} L \to (\mathcal{A}/\mathcal{J})\check{\otimes} L$ is a quotient mapping, and thus there is an element $u_1 \in \mathcal{A}\check{\otimes} L$ with $\|u_1\| < \varepsilon$ and

$$\pi \otimes id_L(u_1) = \pi \otimes id_L(u_0).$$

We have

$$\|u - (u_0 - u_1)\| \leq \|u - u_0\| + \|u_1\| < 2\varepsilon,$$

where

$$u_0 - u_1 \in \mathcal{J}\check{\otimes} L \subseteq \mathcal{J}\check{\otimes} V,$$

and thus

$$\text{dist}(u, \mathcal{J}\check{\otimes} V) < 2\varepsilon.$$

Since $\varepsilon > 0$ is arbitrary, it follows that $u \in \mathcal{J}\check{\otimes} V$. □

Corollary 14.3.8 *For $n > 1$, the full C^*-algebra $C^*(\mathbb{F}_n)$ is not locally reflexive.*

Proof Wassermann has proved that if $n > 1$ and $\mathcal{J}$ is the kernel of the canonical $*$-homomorphism

$$\rho : C^*(\mathbb{F}_n) \to C^*_\lambda(\mathbb{F}_n),$$

then the kernel of

$$\rho \otimes id : C^*(\mathbb{F}_n)\check{\otimes} C^*(\mathbb{F}_n) \to C^*_\lambda(\mathbb{F}_n)\check{\otimes} C^*(\mathbb{F}_n) \tag{14.3.3}$$

is not equal to $\mathcal{J}\check{\otimes} C^*(\mathbb{F}_n)$. It follows from Proposition 14.3.7 that $C^*(\mathbb{F}_n)$ is not locally reflexive. □

Kirchberg proved the following result relating local reflexivity to the strong operator space approximation property (see §11.3).

Theorem 14.3.9 *Let V be an operator space. If V is locally reflexive and it has the operator space approximation property, then V has the strong operator space approximation property.*

Proof Let us suppose that V is locally reflexive. Then the natural inclusion

$$M_\infty \otimes_\vee V^{**} \hookrightarrow (M_\infty \check{\otimes} V)^{**}$$

is isometric. To prove the strong operator space approximation property, it suffices to show that for every $u \in M_\infty \check{\otimes} V$, $F \in (M_\infty \check{\otimes} V)^*$, and $\varepsilon > 0$, there exists a finite-rank mapping $\varphi : V \to V$ such that

$$|\langle (id \otimes \varphi)(u) - u,\ F\rangle| < \varepsilon.$$

Given $F \in (M_\infty \check{\otimes} V)^*$, F determines a bounded weak* continuous linear functional on $(M_\infty \check{\otimes} V)^{**}$. Since the canonical embedding

$$M_\infty \otimes V^{**} \hookrightarrow (M_\infty \check{\otimes} V)^{**}$$

extends to a completely isometric mapping on $M_\infty \check{\otimes} V^{**}$, F induces a bounded linear functional

$$\tilde{F} : M_\infty \check{\otimes} V^{**} \to \mathbb{C}$$

such that for each $b \in M_\infty$, $\overline{v} \mapsto \tilde{F}(b \otimes \overline{v})$ is weak* continuous on V^{**}.

Given $u \in M_\infty \check{\otimes} V$ and a mapping $\varphi \in \mathcal{CB}(V, V^{**})$, we have

$$(id \otimes \varphi)(u) \in M_\infty \check{\otimes} V^{**}.$$

Thus, we can define a linear functional

$$\tilde{F}_u : \mathcal{CB}(V, V^{**}) = (V^* \hat{\otimes} V)^* \to \mathbb{C}$$

by letting

$$\langle \varphi,\ \tilde{F}_u \rangle = \langle (id \otimes \varphi)(u),\ \tilde{F} \rangle .$$

We have

$$\begin{aligned} \|\tilde{F}_u\| &= \sup\{ |\langle \varphi,\ \tilde{F}_u \rangle| :\ \|\varphi\|_{cb} \leq 1\} \\ &= \sup\{ |\langle (id \otimes \varphi)(u),\ \tilde{F} \rangle| :\ \|\varphi\|_{cb} \leq 1\} \leq \|u\| \|\tilde{F}\|. \end{aligned}$$

We claim that $\tilde{F}_u$ is weak* continuous in the $V \hat{\otimes} V^*$ topology on $\mathcal{CB}(V, V^{**})$, i.e. $\tilde{F}_u \in V \hat{\otimes} V^*$.

If $u = x \otimes v \in M_\infty \otimes V$, then we have for $\varphi \in \mathcal{CB}(V, V^{**})$

$$\langle \varphi,\ \tilde{F}_u \rangle = \langle (id \otimes \varphi)(u),\ \tilde{F} \rangle = \langle x \otimes \varphi(v),\ \tilde{F} \rangle.$$

If $\varphi_\gamma \to \varphi$ in the $V \hat{\otimes} V^*$ topology, then for $f \in V^*$,

$$\langle \varphi_\gamma(v), f \rangle = \langle \varphi_\gamma, v \otimes f \rangle \to \langle \varphi, v \otimes f \rangle = \langle \varphi(v), f \rangle,$$

and $\varphi_\gamma(v) \to \varphi(v)$ in the V^* topology. It follows from the construction of $\tilde{F}$ that

$$\langle \varphi_\gamma,\ \tilde{F}_u \rangle \to \langle \varphi, \tilde{F}_u \rangle.$$

The same is true for any element u in the algebraic tensor product $M_\infty \otimes V$. Given an arbitrary element $u \in M_\infty \check{\otimes} V$, we may approximate it in the norm topology by a sequence $u_n \in M_\infty \otimes V$. If we let D be the closed unit ball of $\mathcal{CB}(V, V^{**})$ with the relative weak* topology, then the restriction $\tilde{F}_{u\,|D}$ is the uniform limit of the continuous functions $\tilde{F}_{u_n\,|D}$, and thus $\tilde{F}_{u\,|D}$ is weak* continuous. That in turn implies that $\tilde{F}_u$ is weak* continuous (see §A.2).

Since, by hypothesis, V satisfies the operator space approximation property, we may find a net of finite-rank mappings $\varphi_\gamma : V \to V$ such that $\varphi_\gamma \to id_V$ in the $V \hat{\otimes} V^*$ topology on $\mathcal{CB}(V, V)$ (see Corollary 11.2.3 and Lemma 11.2.4). It follows that for any $u \in M_\infty \check{\otimes} V$ and $F \in (M_\infty \check{\otimes} V)^*$,

$$|\langle (id \otimes \varphi_\gamma)(u) - u,\ F \rangle| = |\langle \varphi_\gamma - id_V, \tilde{F}_u \rangle| \to 0,$$

which completes the proof. □

14.4 EXACTNESS AND CONDITION C'

As we have already indicated, the operator space analogue of Theorem 14.1.1 is false. In order to study this phenomenon, we define a finite-dimensional operator space L to be λ-*exact* if for every $\varepsilon > 0$ there exist an integer $n \in \mathbb{N}$ and a subspace S of M_n such that $d_{cb}(L, S) < \lambda + \varepsilon$. It follows from the discussion in the second paragraph of §14.2 that if $n = \dim L$, then L is n-exact. We define

$$d_{ex}(L) = \inf\{\lambda : L \text{ is } \lambda\text{-exact}\} \tag{14.4.1}$$

to be the *exact approximation constant* of L. Thus, in particular,

$$d_{ex}(L) \leq \dim L.$$

We generalize this notion to arbitrary operator spaces V by letting

$$d_{ex}(V) = \sup\{d_{ex}(L) : L \subseteq V, L \text{ finite-dimensional}\}. \tag{14.4.2}$$

If $\lambda < \infty$, then we say that V is λ-*exact* if $d_{ex}(V) \leq \lambda$, and that V is *exact* if it is 1-exact. It can be shown that any nuclear C^*-algebra is exact (see §14.6), and the same is true for reduced group C^*-algebras of free groups. On the other hand, in §14.5 we shall use an embedding in the full group C^*-algebras of free groups to show that

$$d_{ex}(\max \ell_1^n) \geq \frac{n}{2\sqrt{n-1}}, \tag{14.4.3}$$

and that, in particular, the three-dimensional operator space $\max \ell_1^3$ is not exact.

Our next goal is to prove several characterizations of the exact operator spaces. A completely analogous result can be proved for the λ-exact spaces by the same arguments, but for simplicity we restrict ourselves to the case $\lambda = 1$. We begin with a few preliminaries.

We say that a diagram of operator spaces and complete contractions

$$0 \longrightarrow V \xrightarrow{\varphi} W \xrightarrow{\psi} X \longrightarrow 0 \tag{14.4.4}$$

is 1-*exact* if φ is a complete isometry, ψ is a complete quotient mapping, and $\ker \psi = \varphi(V)$.

Theorem 14.4.1 *Suppose that V is an operator space. Then the following are equivalent:*

(i) *V is exact;*
(ii) *V satisfies condition C';*
(iii) *for each finite-dimensional subspace $L \subseteq V$, and for each C^*-algebra $\mathcal{A}$ with closed ideal $\mathcal{J} \subseteq \mathcal{A}$, the natural mapping*

$$\mathcal{A} \check{\otimes} L \to (\mathcal{A}/\mathcal{J}) \check{\otimes} L \tag{14.4.5}$$

is a quotient (respectively, complete quotient) mapping;
(iv) *for any C^*-algebra $\mathcal{A}$ and closed ideal $\mathcal{J} \subseteq \mathcal{A}$,*

$$0 \longrightarrow \mathcal{J} \check{\otimes} V \xrightarrow{\iota \otimes id} \mathcal{A} \check{\otimes} V \xrightarrow{\pi \otimes id} (\mathcal{A}/\mathcal{J}) \check{\otimes} V \longrightarrow 0$$

is 1-exact.

Proof Let us first suppose that $V = L$ is finite-dimensional, and prove the equivalence of (i)–(iii).

(i)$\Rightarrow$ (ii). Let us fix an arbitrary operator space W. If L is exact, then for any $\varepsilon > 0$ we may find an $n \in \mathbb{N}$, a subspace $S \subseteq M_n$, and a linear isomorphism $r : L \to S$ for which

$$\|r\|_{cb}\,\|r^{-1}\|_{cb} < 1 + \varepsilon.$$

Since the columns of the commutative diagram

$$\begin{array}{ccc} W^{**} \otimes_\vee S & \longrightarrow & (W \otimes_\vee S)^{**} \\ \big\downarrow & & \big\downarrow \\ W^{**} \otimes_\vee M_n & & (W \otimes_\vee M_n)^{**} \\ \big\| & & \big\| \\ M_n(W^{**}) & \cong & M_n(W)^{**} \end{array}$$

are completely isometric, the top row is a complete isometry. Thus, it follows from the commutative diagram

$$\begin{array}{ccc} W^{**} \otimes_\vee L & \xrightarrow{\ \tau\ } & (W \otimes_\vee L)^{**} \\ {\scriptstyle id\otimes r}\big\downarrow & & \big\uparrow{\scriptstyle (id\otimes r^{-1})^{**}} \\ W^{**} \otimes_\vee S & \longrightarrow & (W \otimes_\vee S)^{**} \end{array}$$

that $\|\tau\|_{cb} \leq 1 + \varepsilon$. Since $\varepsilon > 0$ is arbitrary, τ is a complete isometry. This shows that L satisfies condition C'.

(ii)$\Rightarrow$(iii) We have a commutative diagram

$$\begin{array}{ccc} L\check{\otimes}\mathcal{A} & \longrightarrow & L\check{\otimes}(\mathcal{A}/\mathcal{J}) \\ \big\downarrow & & \big\downarrow \\ L\check{\otimes}\mathcal{A}^{**} & \longrightarrow & L\check{\otimes}(\mathcal{A}/\mathcal{J})^{**} \quad , \\ \big\| & & \big\| \\ (L\check{\otimes}\mathcal{A})^{**} & \longrightarrow & (L\check{\otimes}(\mathcal{A}/\mathcal{J}))^{**} \end{array}$$

where the equalities follow from condition C' for L. Since the quotient mapping $\mathcal{A}^{**} \to (\mathcal{A}/\mathcal{J})^{**}$ has a completely contractive lifting, the middle row is a complete quotient mapping (see Proposition 8.1.5). Thus, the bottom row is a complete quotient mapping, and since it is the second adjoint of the first row, it follows from (A.2.1) and (A.2.3) that the first row is a complete quotient mapping.

(iii)$\Rightarrow$(i). We may assume that L is a finite-dimensional subspace of M_∞. As in §10.1, we write $P^n : M_\infty \to M_n$ for the truncation mapping, and we let $\rho_n = (P^n)_{|L}$ and

$$S_n = \rho_n(L) \subseteq M_n.$$

If $m \leq n$, then $P^m \circ \rho_n = \rho_m$, and thus for each $v \in M_p(L)$,

$$\left\|(id_{M_p} \otimes \rho_m)(v)\right\| \leq \|P^m\|_{cb} \left\|(id_{M_p} \otimes \rho_n)(v)\right\| \leq \left\|(id_{M_p} \otimes \rho_n)(v)\right\|.$$

As in the second proof of Theorem 14.1.1, we may select an $n_0 \in \mathbb{N}$ such that $n \geq n_0$ implies that $P^n : L \to M_n$ is one-to-one, and therefore $\rho_n : L \to S_n$ is a linear isomorphism. We let $\sigma_n = \rho_n^{-1}$ for $n \geq n_0$, which we indicate in the following diagram:

$$\begin{array}{ccc} L & \subseteq & M_\infty \\ {\scriptstyle \rho_n}\big\downarrow\big\uparrow{\scriptstyle \sigma_n} & & \big\downarrow{\scriptstyle P^n} \\ S_n & \subseteq & M_n \end{array}.$$

It follows from Corollary 2.2.4 that $\sigma_{n_0} = (\rho_{n_0})^{-1}$ is completely bounded. From the above inequalities, $\|\sigma_n\|_{cb}$ is a decreasing sequence. We wish to show that

$$\lim_{n\to\infty} \|\sigma_n\|_{cb} = 1.$$

For each $n \geq n_0$, we may select a $k_n \in \mathbb{N}$ and a $b_n \in M_{k_n}(S_n)_{\|\cdot\|\leq 1}$ such that

$$\|\sigma_n\|_{cb} - \frac{1}{n} < \left\|(id_{M_{k_n}} \otimes \sigma_n)(b_n)\right\|.$$

We have

$$b = (b_n)_{n\geq n_0} \in \prod (M_{k_n} \check{\otimes} S_n)$$

and

$$h = (h_n) = ((id_{M_{k_n}} \otimes \sigma_n)(b_n))_{n\geq n_0} \in \prod_{n\geq n_0} (M_{k_n} \check{\otimes} L).$$

It suffices to show that

$$\limsup_{n\to\infty} \|h_n\| = 1,$$

and we shall do this using asymptotic products (see §A.6).

In the following calculations we assume that $n \geq n_0$. We let

$$\pi_\infty : \prod M_{k_n} \to Q = \prod M_{k_n} \Big/ \sum M_{k_n}$$

and

$$\pi_\infty^L : \prod (M_{k_n} \check{\otimes} L) \to \prod (M_{k_n} \check{\otimes} L) \Big/ \sum (M_{k_n} \check{\otimes} L)$$

be the usual quotient mappings. From Proposition A.6.1

$$\pi_\infty^L(h) = \limsup_{n\to\infty} \|h_n\|.$$

Since L is finite-dimensional,

$$\prod (M_{k_n} \check{\otimes} L) = \left(\prod M_{k_n}\right) \check{\otimes} L.$$

To see this, let $v_k \in L$ $(k = 1, \ldots, r)$ be a vector basis for L and f_k be a dual basis. We have

$$\left\| id_{M_{k_n}} \otimes f_k : M_{k_n} \check{\otimes} L \to M_{k_n} \right\| \leq \|f_k\|_{cb} = \|f_k\|,$$

and thus we have a corresponding bounded mapping

$$\theta_k = \left(id_{M_{k_n}} \otimes f_k\right) : \prod \left(M_{k_n} \check{\otimes} L\right) \to \prod M_{k_n}.$$

It is easy to verify that if $u \in \prod \left(M_{k_n} \check{\otimes} L\right)$, then

$$u = \sum \theta_k(u) \otimes v_k \in \left(\prod M_{k_n}\right) \check{\otimes} L.$$

The reverse inclusion is trivial. It also follows from this that

$$\sum \left(M_{k_n} \check{\otimes} L\right) = \left(\sum M_{k_n}\right) \check{\otimes} L.$$

We infer from (iii) that the mapping

$$\prod \left(M_{k_n} \check{\otimes} L\right) \Big/ \sum \left(M_{k_n} \check{\otimes} L\right) = \left(\prod M_{k_n}\right) \check{\otimes} L \Big/ \left(\sum M_{k_n}\right) \check{\otimes} L \to Q \check{\otimes} L$$

is completely isometric, and thus

$$\left\|\pi_\infty^L(h)\right\| = \|\pi_\infty \otimes id_L(h)\|.$$

The mappings $\rho_m : L \to M_m$ determine a completely isometric injection

$$\rho : L \to \prod M_m : v \mapsto (\rho_m(v)),$$

and thus the mapping

$$id \otimes \rho : Q \check{\otimes} L \to Q \check{\otimes} \prod M_m \subseteq \prod Q \check{\otimes} M_m$$

is isometric. We have

$$\begin{aligned} \|\pi_\infty \otimes id_L(h)\| &= \|(id \otimes \rho)(\pi_\infty \otimes id_L)(h)\| \\ &= \sup \left\{\|(id \otimes \rho_m)(\pi_\infty \otimes id_L)(h)\|\right\}. \end{aligned}$$

We must show that each of the norms in the supremum is less than or equal to one.

If we fix m, then we have the commutative diagram

$$\begin{array}{ccc} \prod M_{k_n} \check{\otimes} L & \xrightarrow{id \otimes \rho_m} & \prod M_{k_n} \check{\otimes} M_m \\ {\scriptstyle \pi_\infty \otimes id_L} \big\downarrow & & \big\downarrow {\scriptstyle \pi_\infty \otimes id} \\ Q \check{\otimes} L & \xrightarrow{id \otimes \rho_m} & Q \check{\otimes} M_m \end{array}.$$

Since the right column is a quotient of C^*-algebras, it is a complete quotient mapping, which is to say that $Q \check{\otimes} M_m$ is the asymptotic product of the $M_{k_n} \check{\otimes} M_m$. Thus,

$$\|(id \otimes \rho_m)(\pi_\infty \otimes id_L)(h)\| = \limsup_{n \to \infty} \left\|(id_{M_{k_n}} \otimes \rho_m)(h_n)\right\|.$$

If $n \geq m$, then

$$(id_{M_{k_n}} \otimes \rho_m)(h_n) = (id_{M_{k_n}} \otimes P^m)(id_{M_{k_n}} \otimes \rho_n)(id_{M_{k_n}} \otimes \sigma_n)(b_n) = (id_{M_{k_n}} \otimes P^m)(b_n),$$

and thus

$$\|(id_{M_{k_n}} \otimes \rho_m)(h_n)\| \leq \|b_n\| \leq 1.$$

This shows that $\|(id \otimes \rho_m)(\pi_\infty \otimes id_L)(h)\| \leq 1$ for all $m \in \mathbb{N}$, hence

$$\limsup \|h_n\| = \|\pi_\infty \otimes id_L(h)\| = \sup\{\|(id \otimes \rho_m)(\pi_\infty \otimes id_L)(h)\|\} \leq 1.$$

To see that (i)–(iii) hold for arbitrary operator spaces V, we note that each of these conditions is 'local' in the sense that V has the given property if and only if that is the case for each finite-dimensional subspace L of V. That is obvious for (i) from the definition of exactness, and it is also trivial for (iii). To see that it is true for (ii), let us suppose that we wish to show that

$$\tau : W^{**} \otimes_\vee V \to (W \otimes_\vee V)^{**}$$

satisfies $\|\tau\| \leq 1$. Given an element $u \in W^{**} \otimes_\vee V$, we have $u \in W^{**} \otimes_\vee L$ for some finite-dimensional subspace L of V. Since the mapping

$$(W \otimes_\vee L)^{**} \to (W \otimes_\vee V)^{**}$$

is isometric, it suffices to prove that the mapping

$$\tau_{L \otimes_\vee W^{**}} : W^{**} \otimes_\vee L \to (W \otimes_\vee L)^{**}$$

is isometric for each such L.

We also note that if the mapping in (iii) is always a quotient mapping, then it is always a complete quotient mapping, since we have the commutative diagram

$$\begin{array}{ccc} M_n(\mathcal{A} \check{\otimes} L) & \twoheadrightarrow & M_n((\mathcal{A}/\mathcal{J}) \check{\otimes} L) \\ \| & & \| \\ M_n(\mathcal{A}) \check{\otimes} L & \twoheadrightarrow & (M_n(\mathcal{A})/M_n(\mathcal{J})) \check{\otimes} L \end{array},$$

where the bottom row is a quotient mapping by hypothesis.

(iii)⇒(iv). It suffices to show that the kernel of $\pi \otimes id$ is $\mathcal{J} \check{\otimes} V$, and that $\pi \otimes id$ is a quotient mapping since the above argument will then show that the latter is a complete quotient mapping. The first statement follows immediately from the proof of Proposition 14.3.7.

To show the quotient condition, it suffices to prove that $\pi \otimes id$ maps $(\mathcal{A} \otimes_\vee V)_{\|\cdot\|<1}$ onto a dense subset of $((\mathcal{A}/\mathcal{J}) \otimes_\vee V)_{\|\cdot\|<1}$ (see §A.2). Given an element $\tilde{u}$ in the latter set, there is a finite-dimensional subspace $L \subseteq V$ with

$$\tilde{u} \in (\mathcal{A}/\mathcal{J}) \otimes_\vee L = (\mathcal{A}/\mathcal{J}) \check{\otimes} L.$$

From (iii) there exists an element $u \in \mathcal{A}\check{\otimes} L$ with $\|u\| < 1$ and $\pi\otimes id(u) = \tilde{u}$, and since we may regard u as an element of $\mathcal{A}\check{\otimes} V$, we have the desired result.

(iv)⇒(iii). We have a commutative diagram

$$\begin{array}{ccccc} \mathcal{A}\check{\otimes} L & \twoheadrightarrow & (\mathcal{A}\check{\otimes} L)/(\mathcal{J}\check{\otimes} L) & \xrightarrow{q} & (\mathcal{A}/\mathcal{J})\check{\otimes} L \\ \downarrow & & \downarrow & & \downarrow \\ \mathcal{A}\check{\otimes} V & \twoheadrightarrow & (\mathcal{A}\check{\otimes} V)/(\mathcal{J}\check{\otimes} V) & \xrightarrow{q'} & (\mathcal{A}/\mathcal{J})\check{\otimes} V \end{array},$$

where the first and third columns are isometric injections, and the second column is a contraction. In fact, the second column is also isometric. To see this, it suffices to show that if $u \in \mathcal{A}\check{\otimes} L$, then

$$\mathrm{dist}(u, \mathcal{J}\check{\otimes} L) \leq \mathrm{dist}(u, \mathcal{J}\check{\otimes} V) \tag{14.4.6}$$

(the reverse inequality is trivial). Given $\varepsilon > 0$, we may choose an element $r \in \mathcal{J} \otimes V$ for which

$$\|u - r\| < \mathrm{dist}(u, \mathcal{J}\check{\otimes} V) + \varepsilon.$$

We let $e_\lambda \in \mathcal{J}$, $0 \leq e_\lambda \leq 1$, be an approximate identity for $\mathcal{J}$. If

$$r = \sum_{i=1}^{n} h_i \otimes v_i,$$

where $h_i \in \mathcal{J}$ and $v_i \in V$, then

$$\begin{aligned} \|u - u(e_\lambda \otimes 1)\| &\leq \|(u - r)(1 \otimes 1 - e_\lambda \otimes 1)\| + \|r - r(e_\lambda \otimes 1)\| \\ &\leq \|u - r\| + \sum_{i=1}^{n} \|h_i(1 - e_\lambda)\| \|v_i\|. \end{aligned}$$

If λ is sufficiently large, then we may suppose that

$$\sum_{i=1}^{n} \|h_i(1 - e_\lambda)\| \, \|v_i\| < \varepsilon,$$

and since $u(e_\lambda \otimes 1) \in \mathcal{J}\check{\otimes} L$,

$$\mathrm{dist}(u, \mathcal{J}\check{\otimes} L) \leq \|u - u(e_\lambda \otimes 1)\| < \mathrm{dist}(u, \mathcal{J}\check{\otimes} V) + 2\varepsilon.$$

The number $\varepsilon > 0$ was arbitrary, and thus we obtain (14.4.6).

From (iv), q' is an isometry, and thus the same is true for q. It follows that the composition of the top row is a quotient mapping, which is condition (iii). □

It is instructive to consider two special cases of (iv) in the above theorem. If $V = \mathcal{B}$ is a C^*-algebra, then the mapping

$$\mathcal{A}\check{\otimes} \mathcal{B} \to (\mathcal{A}/\mathcal{J})\check{\otimes} \mathcal{B}$$

is a C^*-algebraic homomorphism, and thus it is automatically a quotient mapping. Therefore we only need to check that the mapping $\pi \otimes id$ has the 'correct' kernel. This is the criterion that Kirchberg initially used when he introduced the notion of exactness for C^*-algebras. On the other hand, as we saw in the above proof, if $V = L$ is finite-dimensional, then the kernel condition is automatic, and thus only the quotient condition is relevant.

The quotient C^*-algebra $Q_\infty = M_\infty / K_\infty$ is called the *Calkin algebra.* It follows from the early work of Busby that the exact sequence of C^*-algebras

$$0 \to K_\infty \to M_\infty \to Q_\infty \to 0 \tag{14.4.7}$$

is, in a certain precise sense, a 'prototypical' C^*-algebraic extension. Subsequently, Kirchberg showed that in order to prove that a C^*-algebra is exact, it suffices to prove (iv) of Theorem 14.4.1 for this quotient. We begin with Pisier's proof that Kirchberg's result is valid for all operator spaces. We use the terminology of Theorem 14.4.1 (iv) (see also (14.4.6)).

Theorem 14.4.2 *Suppose that V is an operator space. Then V is exact if and only if the sequence*

$$0 \longrightarrow K_\infty \check{\otimes} V \longrightarrow M_\infty \check{\otimes} V \xrightarrow{\pi \otimes id} Q_\infty \check{\otimes} V \longrightarrow 0 \tag{14.4.8}$$

is 1-exact.

Proof The necessity of the condition follows from Theorem 14.4.1.

Turning to the converse, it again suffices to prove the theorem for the case when $V = L$ is finite-dimensional. Furthermore, since we may identify $M_n(M_\infty)$ and $M_n(K_\infty)$ with M_∞ and K_∞, if (14.4.8) is a quotient mapping, then it is also a complete quotient mapping. It is apparent from the proof of (iii)⇒(i) in Theorem 14.4.1 that it suffices to prove condition (iii) for C^*-algebraic quotients of the form

$$\pi_\infty : \prod M_{k_n} \to \prod M_{k_n} \Big/ \sum M_{k_n}.$$

In other words, we must use the hypothesis to prove that

$$\pi_\infty \otimes id_L : \prod M_{k_n} \check{\otimes} L \to Q \check{\otimes} L$$

is a quotient mapping, or equivalently that the induced mapping

$$\left(\prod M_{k_n} \check{\otimes} L\right) \Big/ \left(\sum M_{k_n} \check{\otimes} L\right) \to Q \check{\otimes} L$$

is isometric. We can relate this to (14.4.8) by identifying ℓ_2 with $\bigoplus \mathbb{C}^{k_n}$, and $\prod M_{k_n}$ with the corresponding 'block diagonal' subalgebra of M_∞. Since

$$\sum M_{k_n} \check{\otimes} L = \left(\prod M_{k_n} \check{\otimes} L\right) \cap (K_\infty \check{\otimes} L),$$

we have the commutative diagram

$$\begin{array}{ccc} \left(\prod M_{k_n} \check{\otimes} L\right) / \left(\sum M_{k_n} \check{\otimes} L\right) & \longrightarrow & (M_\infty \check{\otimes} L)/(K_\infty \check{\otimes} L) \\ \downarrow \tilde{\mu}_L & & \downarrow \mu_L \\ \left(\prod M_{k_n} / \sum M_{k_n}\right) \check{\otimes} L & \longrightarrow & Q_\infty \check{\otimes} L \end{array}.$$

The top row is isometric; that is, we claim that if

$$u = \sum_{i=1}^{p} b_i \otimes v_i \in \prod M_{k_n} \check{\otimes} L,$$

then

$$\operatorname{dist}(u, K_\infty \check{\otimes} L) = \operatorname{dist}\left(u, \sum M_{k_n} \check{\otimes} L\right).$$

To see this, we note that the 'diagonalization' complete contraction

$$\Phi : M_\infty \to \prod M_{k_n}$$

maps K_∞ into $\sum M_{k_n}$, and thus for any $v \in K_\infty \check{\otimes} L$,

$$\|u - (\Phi \otimes id_L)(v)\| = \|(\Phi \otimes id_V)(u - v)\| \leq \|u - v\|.$$

The mapping $\prod M_{k_n} / \sum M_{k_n} \to Q_\infty$ is a C^*-algebraic inclusion, and thus it is completely isometric. It follows that the bottom row is also isometric. Thus, if μ_L is completely isometric, then the same is true for $\tilde{\mu}_L$. □

The following ultraproduct characterization of exactness is due to Pisier.

Theorem 14.4.3 *Suppose that L is a finite-dimensional operator space. Then L is exact if and only if for any indexed family of operator spaces $(W_i)_{i \in I}$ and a free ultrafilter $\mathcal{U}$ on I, we have a complete isometry*

$$\prod (L \check{\otimes} W_i)/\mathcal{U} \cong L \check{\otimes} \prod W_i/\mathcal{U}. \tag{14.4.9}$$

Proof Let us suppose that $L = S \subseteq M_n$ for some $n \in \mathbb{N}$. We have a commutative diagram

$$\begin{array}{ccc} \prod (S \check{\otimes} W_i)/\mathcal{U} & \longrightarrow & S \check{\otimes} \prod W_i/\mathcal{U} \\ \downarrow & & \downarrow \\ \prod (M_n \check{\otimes} W_i)/\mathcal{U} & = & M_n \check{\otimes} \prod W_i/\mathcal{U} \end{array},$$

where the bottom row is a complete isometry by definition of the operator space ultraproduct, and the columns are completely isometric since $\check{\otimes}$ is an injective tensor product.

If L is exact, then given $\varepsilon > 0$, we may find a completely bounded isomorphism $r : L \to S$, where $S \subseteq M_n$, such that

$$\|r\|_{cb} \|r^{-1}\|_{cb} < 1 + \varepsilon.$$

We have a commutative diagram of completely bounded mappings

$$\begin{array}{ccc} \prod(S\check{\otimes} W_i)/\mathcal{U} & \xleftarrow{\widehat{\theta}_S^{-1}} & S\check{\otimes}\prod W_i/\mathcal{U} \\ {\scriptstyle (r^{-1}\otimes id_{W_i})_{\mathcal{U}}}\Big\downarrow & & \Big\uparrow{\scriptstyle r\otimes id} \\ \prod(L\check{\otimes} W_i)/\mathcal{U} & \xleftarrow{\widehat{\theta}_L^{-1}} & L\check{\otimes}\prod W_i/\mathcal{U} \end{array},$$

from which we see that

$$\|\widehat{\theta}_L^{-1}\|_{cb} \leq \|r\|_{cb}\|\widehat{\theta}_S^{-1}\|_{cb}\|r^{-1}\|_{cb} < 1+\varepsilon.$$

Since ε is arbitrary, $\widehat{\theta}_L^{-1}$ is a complete contraction. We conclude that

$$\widehat{\theta}_L : \prod(L\check{\otimes} W_i)/\mathcal{U} \to L\check{\otimes}\prod W_i/\mathcal{U}$$

is a complete isometry.

Now let us suppose that we have (14.4.9). We can identify L with a subspace of M_∞, and we define

$$\rho_n : L \to S_n \subseteq M_n$$

as in the proof of (iii)$\Rightarrow$(i) in Theorem 14.4.1. We again choose $n_0 \in \mathbb{N}$ such that ρ_n is a linear isomorphism for $n \geq n_0$, and we let $\sigma_n = \rho_n^{-1}$ for $n \geq n_0$, and $\sigma_n = 0$ for $n < n_0$, and we fix a free ultrafilter $\mathcal{U}$ on the set $\mathbb{N}$. From the proof of Corollary 10.3.7, the mapping

$$\rho = (\rho_n)_{\mathcal{U}} : L \to \prod S_n/\mathcal{U}$$

is a completely isometric surjection for which the inverse mapping is

$$\sigma = (\sigma_n)_{\mathcal{U}} : \prod S_n/\mathcal{U} \to \prod L/\mathcal{U} = L. \tag{14.4.10}$$

It follows that σ is a complete isometry.

Since $\prod S_n^*/\mathcal{U}$ is also finite-dimensional, with the same dimension as L, we have from Corollary 10.3.5 the complete isometry

$$\prod S_n^*/\mathcal{U} \cong \Big(\prod S_n/\mathcal{U}\Big)^*.$$

Thus, from (14.4.9) we have the complete isometries

$$\prod(S_n^*\check{\otimes} L)/\mathcal{U} \cong \Big(\prod S_n^*/\mathcal{U}\Big)\check{\otimes} L \cong \Big(\prod S_n/\mathcal{U}\Big)^*\check{\otimes} L,$$

or equivalently, we have the natural complete isometry.

$$\prod \mathcal{CB}(S_n, L)/\mathcal{U} \cong \mathcal{CB}\Big(\prod S_n/\mathcal{U}, L\Big).$$

This implies that

$$\lim_{\mathcal{U}} \|\sigma_n\|_{cb} = \|\pi_{\mathcal{U}}((\sigma_n))\|_{cb} = \|(\sigma_n)_{\mathcal{U}}\|_{cb} = 1.$$

Given $\varepsilon > 0$, there is an integer $n(\varepsilon) \geq n_0$ such that $\|\sigma_{n(\varepsilon)}\|_{cb} < 1+\varepsilon$, and hence $d_{cb}(L, S_{n(\varepsilon)}) < 1+\varepsilon$. Since $\varepsilon > 0$ is arbitrary, it follows that L is exact. □

14.5 EXAMPLES OF NON-EXACT OPERATOR SPACES

Maximal quantizations provide some of the simplest examples of non-exact operator spaces. Pisier was the first to show that for each $n > 2$, $\max \ell_1^n$ is not exact. In order to prove Pisier's result, we use a well-known embedding of the reduced group C^*-algebra $C^*_\lambda(\mathbb{F}_n)$ into an ultraproduct of matrix algebras.

Lemma 14.5.1 *For each $g \in \mathbb{F}_n$ with $g \neq e$, there exists a finite group G and a homomorphism $\pi : \mathbb{F}_n \to G$ such that $\pi(g) \neq e$.*

Proof We may suppose that we have the reduced form

$$g = s_{i_p}^{\varepsilon_p} s_{i_{p-1}}^{\varepsilon_{p-1}} \cdots s_{i_1}^{\varepsilon_1},$$

where $\varepsilon_k = \pm 1$. It suffices to show that we can find permutations γ_k $(1 \leq k \leq n)$ of the set $J_{p+1} = \{1, \ldots, p+1\}$ such that if $1 \leq j \leq p$, then

$$\gamma_{i_j}^{\varepsilon_j}(j) = j+1, \tag{14.5.1}$$

and thus $s_k \mapsto \gamma_k$ determines a homomorphism π of $\mathbb{F}_n$ into the symmetric group of J_{p+1} with

$$\pi(g)(1) = \gamma_{i_p}^{\varepsilon_p} \gamma_{i_{p-1}}^{\varepsilon_{p-1}} \cdots \gamma_{i_1}^{\varepsilon_1}(1) = p+1.$$

The requirements for the permutations γ_κ may be understood in terms of a table. If, for example,

$$g = \cdots s_1 s_3^{-1} s_2 s_1^{-1} s_1^{-1},$$

then the corresponding table is given by

	1	2	3	4	5	...	$p+1$
γ_1	$1 \leftarrow 2$	$2 \leftarrow 3$			$5 \rightarrow 6$		
γ_2			$3 \rightarrow 4$				
γ_3				$4 \leftarrow 5$			
$\vdots$							

Owing to the fact that we cannot have a pair $s_k s_k^{-1}$ or $s_k^{-1} s_k$ in the reduced form, we cannot have either

$$j-1 \leftarrow j \quad j \rightarrow j+1$$

or

$$j-1 \rightarrow j \quad j \leftarrow j+1$$

in the kth row. As a result, the set of arrows (or more precisely, the corresponding set of ordered pairs) in the kth row is a well-defined bijective mapping γ_k^0 of a subset S_k of J_{p+1} onto another subset of J_{p+1}. Since the complements of the domain S_k and range $\gamma_k^0(S_k)$ must have the same number of elements, we may extend γ_k^0 to a permutation γ_k of J_{p+1}. It is evident that the γ_k satisfy (14.5.1). □

Corollary 14.5.2 *There is a sequence of finite groups G_k and homomorphisms $\theta_k : \mathbb{F}_n \to G_k$ such that* $\ker\theta_1 \supseteq \ker\theta_2 \supseteq \cdots$ *and* $\bigcap \ker\theta_k = \{e\}$.

Proof From above, we may find a sequence of finite groups H_k and homomorphisms $\eta_k : \mathbb{F}_n \to H_k$ with $\bigcap \ker \eta_k = \{e\}$. If we let $G_k = H_1 \times \cdots \times H_k$, then it is evident that the homomorphisms

$$\theta_k : \mathbb{F}_n \to H_1 \times \cdots \times H_k : g \mapsto (\eta_1(g), \ldots, \eta_k(g))$$

have the desired properties. □

We let λ_k be the regular representation of G_k on the Hilbert space $\mathbb{C}^{d(k)} = \ell_2(G_k)$, where $d(k)$ is the cardinality of G_k, and let $\{\delta_g : g \in G_k\}$ be the canonical orthonormal basis in that Hilbert space. If $h \neq e$, then

$$\tau_{d(k)}(\lambda_k(h)) = \frac{1}{d(k)} \sum_{h \in G_k} \langle \lambda_k(h)\delta_g \mid \delta_g \rangle = \frac{1}{d(k)} \sum_{h \in G_k} \langle \delta_{hg} \mid \delta_g \rangle = 0. \tag{14.5.2}$$

We let

$$\pi_k = \lambda_k \circ \theta_k : \mathbb{F}_n \to M_{d(k)}$$

be the corresponding unitary representations of $\mathbb{F}_n$, and we let I stand for the sequence $(d(k))$. These determine a unitary representation

$$\pi : \mathbb{F}_n \to \mathcal{M}_I = \prod_{k \in \mathbb{N}} M_{d(k)} \subseteq \mathcal{B}\Big(\bigoplus \mathbb{C}^{d(k)}\Big),$$

where

$$\pi(g) = (\pi_k(g)) \,.$$

We let $\beta\mathbb{N}$ be the spectrum of the C^*-algebra $\ell_\infty(\mathbb{N})$, and we fix an element $\omega \in \beta\mathbb{N}\backslash\mathbb{N}$, which corresponds to a free ultrafilter on $\mathbb{N}$ (see §A.6). We may regard the elements of $\ell_\infty(\mathbb{N})$ as continuous functions on $\beta\mathbb{N}$, and given a bounded sequence $\alpha = (\alpha_k) \in \ell_\infty(\mathbb{N})$, we define

$$\lim_{k \to \omega} \alpha_k = \alpha(\omega).$$

In order to construct a von Neumann algebraic ultraproduct of the matrix algebras M_m, we must initially choose a state on each such algebra. We let τ_m be the normalized trace on M_m. If $\alpha \in M_m$ and the eigenvalues of $|\alpha|$ are the non-negative numbers $\lambda_1, \ldots, \lambda_m$, then

$$\begin{aligned}\tau_m(|\alpha|) &= \frac{1}{m}\sum \lambda_i 1 \\ &\le \Big(\frac{1}{m}\sum \lambda_i^2\Big)^{1/2}\Big(\frac{1}{m}\sum 1^2\Big)^{1/2}\end{aligned}$$

$$= \tau_m(\alpha^*\alpha)^{1/2}. \tag{14.5.3}$$

Owing to the fact that $\tau_{d(k)}$ is a state on $M_{d(k)}$,

$$\left|\tau_{d(k)}(\alpha_k)\right| \leq \|\alpha_k\|.$$

We define a trace τ_0 on $\mathcal{M}_I$ by letting

$$\tau_0(\alpha) = \lim_{k\to\omega} \tau_{d(k)}(\alpha_k).$$

The set

$$\mathcal{J}_\omega = \{\alpha \in \mathcal{M}_I : \tau_0(\alpha^*\alpha) = 0\}$$

is a closed two-sided ideal in $\mathcal{M}_I$, and we let Φ denote the quotient mapping of $\mathcal{M}_I$ onto the C^*-algebra $\mathcal{M}_\omega = \mathcal{M}_I/\mathcal{J}_\omega$. It must be emphasized that this is not an ultraproduct of operator spaces, since the ideal $\mathcal{J}_\omega$ is not defined in terms of the norms. We define a faithful trace τ_ω on $\mathcal{M}_\omega$ by

$$\tau_\omega(\Phi(\alpha)) = \tau_0(\alpha).$$

τ_ω determines a corresponding GNS representation $U_\omega : \mathcal{M}_\omega \to \mathcal{B}(H_\omega)$ with a cyclic and separating vector ξ. This is a faithful representation of $\mathcal{M}_\omega$, and it has the remarkable property that $U_\omega(\mathcal{M}_\omega)$ is a factor on H_ω (see Sakai (1962), Ch. 2, §7), and thus the C^*-algebra $\mathcal{M}_\omega$ is a II_1 factor.

Let $\mathcal{R}_n$ be the von Neumann subalgebra of $\mathcal{M}_\omega$ generated by $\Phi(\pi(\mathbb{F}_n))$. We claim that this is isomorphic to the regular group von Neumann algebra $VN(\mathbb{F}_n)$. To see this we let E' be the projection on

$$K = \overline{\mathcal{R}_n\xi} \subseteq H_\pi.$$

Since K is invariant under $\mathcal{R}_n$, $E' \in \mathcal{R}_n'$. On the other hand, since ξ is separating for all of $\mathcal{M}_\omega$, it is separating for $\mathcal{R}_n$, and thus the mapping

$$\mathcal{R}_n \to \mathcal{R}_nE' : T \mapsto TE'$$

is an isomorphism.

The unitary representation

$$\lambda_0 : \mathbb{F}_n \to \mathcal{B}(K) : g \mapsto \Phi(\pi(g))E'$$

is unitarily equivalent to the regular representation of $\mathbb{F}_n$. To see this we note that for any $g \in \mathbb{F}_n$ with $g \neq e$,

$$\langle \lambda_0(g)\xi \mid \xi\rangle = \tau_\omega((\pi_k(g))) = \lim \tau_{d(k)}(\pi_k(g)) = 0,$$

since for sufficiently high k, $g_k = \theta_k(g) \neq e$, and thus $\tau_{d(k)}(\lambda_k(g_k)) = 0$ (see (14.5.2)). Thus, the elements $\lambda_0(g)\xi$ $(g \in \mathbb{F}_n)$ are orthonormal, and the mapping

$$V : \ell_2(\mathbb{F}_n) \to K : \lambda(g)\delta_e \mapsto \lambda_0(g)\xi$$

satisfies $V\lambda(g)V^* = \lambda_0(g)$. It follows that we have the isomorphisms

$$\mathcal{R}_n \cong \mathcal{R}_nE' \cong VN(\mathbb{F}_n),$$

and using the natural inclusion $\iota : C^*_\lambda(\mathbb{F}_n) \hookrightarrow VN(\mathbb{F}_n)$, we have the commutative diagram

$$\begin{array}{ccccc} & & C^*(\mathbb{F}_n) & \xrightarrow{\pi} & \mathcal{M}_I \\ & & \downarrow & & \downarrow \Phi \\ \ell^n_\infty & \xrightarrow{\varphi} & C^*_\lambda(\mathbb{F}_n) & \hookrightarrow & \mathcal{M}_\omega \end{array}, \tag{14.5.4}$$

where $\varphi(e_j) = \lambda(s_j)$.

Lemma 14.5.3 *If $\psi : \ell^n_\infty \to \mathcal{M}_I$ is a lifting of φ, i.e. we have a commutative diagram*

$$\begin{array}{ccc} & & \mathcal{M}_I \\ & \overset{\psi}{\nearrow} & \downarrow \Phi \\ \ell^n_\infty & \xrightarrow{\varphi} & \mathcal{M}_\omega \end{array},$$

then

$$\|\psi\|_{cb} \geq \frac{n}{2\sqrt{n-1}} \|\varphi\|_{cb}. \tag{14.5.5}$$

Proof Since $\mathcal{M}_I$ is injective and $\mathcal{M}_\omega$ is a type II_1 factor, we have

$$\|\psi\|_{cb} = \|\psi\|_{dec} \geq \|\varphi\|_{dec} = n.$$

On the other hand, we have verified in Theorem 5.4.7 that

$$\|\varphi\|_{cb} \leq 2\sqrt{n-1}.$$

Combining these two inequalities, we obtain (14.5.5). □

Theorem 14.5.4 *For any $n > 2$, $\max \ell^n_1$ and T_n are not exact.*

Proof If $F_n = \max \ell^n_1$ is exact, then we have the commutative diagram

$$\begin{array}{ccc} F_n \check{\otimes} \mathcal{M}_I & \longrightarrow & F_n \check{\otimes} \mathcal{M}_\omega \\ \downarrow & & \downarrow \\ \mathcal{CB}(\ell^n_\infty, \mathcal{M}_I) & \longrightarrow & \mathcal{CB}(\ell^n_\infty, \mathcal{M}_\omega) \end{array},$$

where the columns are complete isometries and the top row is a complete quotient mapping. It follows that the bottom row is a complete quotient mapping, and thus given $\varepsilon > 0$, any $\varphi \in \mathcal{CB}(\ell^n_\infty, \mathcal{M}_\omega)$ has a lifting ψ with $\|\psi\|_{cb} < \|\varphi\|_{cb} + \varepsilon$, which is impossible for $n > 2$.

Given $n \in \mathbb{N}$, we may identify $\max \ell^n_1$ with the diagonal operator subspace of T_n. To see this, let us identify $\min \ell^n_\infty$ with the diagonal operator subspace of M_n and let

$$D_n : M_n \to \min \ell^n_\infty$$

denote the diagonal truncation mapping from M_n onto $\min \ell^n_\infty$. Then D_n is a completely contractive projection, and thus

$$D^*_n : \max \ell^n_1 \hookrightarrow T_n$$

is a completely isometric inclusion. We can conclude that T_n is not exact for $n > 2$ since it contains the non-exact operator subspace $\max \ell_1^n$. □

With a little more effort, one can show that

$$d_{ex}(\max\, \ell_1^n) \geq \frac{n}{2\sqrt{n-1}}$$

(see Pisier 1995). On the other hand, we have shown in §3.3 that there is a completely isometric injection

$$\max \ell_1^n \hookrightarrow C^*(\mathbb{F}_{n-1}),$$

and, in particular, we have a completely isometric injection

$$\max \ell_1^2 = \min \ell_1^2 \hookrightarrow C^*(\mathbb{F}_1) = C^*(\mathbb{Z}).$$

As a consequence, $\max\, \ell_1^2$ is exact.

The following result is an immediate consequence of the above theorem.

Corollary 14.5.5 *For any $n \geq 2$, $C^*(\mathbb{F}_n)$ is not exact.*

Proof If $n \geq 2$, then $\max\, \ell_1^{n+1}$ is completely isometric to an operator subspace of $C^*(\mathbb{F}_n)$. Since $n + 1 > 2$, $\max\, \ell_1^{n+1}$ is not exact by Theorem 14.5.4, and thus $C^*(\mathbb{F}_n)$ is not exact. □

Theorem 14.5.6 *There exists an operator space B such that B^{**} is completely isometric to $\ell_\infty^3 \oplus M_\infty$, but B is not locally reflexive.*

Proof From Theorem 14.5.4 ℓ_1^3, or more precisely $\max\, \ell_1^3$, is not exact. It follows from Theorem 14.4.2 that the mapping

$$\ell_1^3 \check{\otimes} M_\infty \longrightarrow \ell_1^3 \check{\otimes} (M_\infty / K_\infty)$$

is not a quotient mapping. Since we have a commuting diagram

$$\begin{array}{ccc} \ell_1^3 \check{\otimes} M_\infty & \longrightarrow & \ell_1^3 \check{\otimes} (M_\infty / K_\infty) \\ \| & & \| \\ \mathcal{CB}(\ell_\infty^3, M_\infty) & \longrightarrow & \mathcal{CB}(\ell_\infty^3, M_\infty / K_\infty) \end{array},$$

it follows that there is a complete contraction $\varphi : \ell_\infty^3 \to Q_\infty = M_\infty / K_\infty$ which does not have a completely contractive lifting $\psi : \ell_\infty^3 \to M_\infty$. We may also assume that φ is completely isometric. To see this, we let K be the Hilbert space $\bigoplus_{n \in \mathbb{N}} \ell_2^3$. The diagonal $*$-representation

$$\theta : \ell_\infty^3 \to \mathcal{B}(K)$$

has the property that

$$\theta(\ell_\infty^3) \cap \mathcal{K}(K) = \{0\}\,.$$

If we identify K with ℓ_2, then it follows that we have a completely isometric mapping

$$\dot{\theta} = \pi \circ \theta : \ell_\infty^3 \to Q_\infty,$$

and the corresponding mapping

$$\ell^3_\infty \to Q_\infty \oplus Q_\infty : \alpha \to (\varphi(\alpha), \dot{\theta}(\alpha))$$

is a complete isometry which does not have a completely contractive lifting

$$\ell^3_\infty \to M_\infty \oplus M_\infty.$$

If we use an identification of $\ell_2 \oplus \ell_2$ with ℓ_2, and we regard $M_\infty \oplus M_\infty$ and $Q_\infty \oplus Q_\infty$ as C^*-subalgebras of M_∞ and Q_∞, then we obtain the desired modification of φ.

If we let $L = \varphi(\ell^3_\infty) \subseteq Q_\infty$ and $B = \pi^{-1}(L)$, then we have a commutative diagram

$$\begin{array}{ccccc} & B & \subseteq B^{**} & \subseteq M^{**}_\infty \\ & \downarrow{\scriptstyle \pi_0} & \downarrow{\scriptstyle \pi_0^{**}} & \downarrow{\scriptstyle \pi^{**}} \\ \ell^3_\infty \xrightarrow{\varphi} & L & = L & \subseteq Q^{**}_\infty \end{array} \tag{14.5.6}$$

where π_0 is the restriction of π to B. Since π is a C^*-algebraic homomorphism, there is a lifting for π^{**}, i.e. a complete contraction $p : Q^{**}_\infty \to M^{**}_\infty$ such that $\pi^{**} \circ p = id_{Q^{**}_\infty}$. More precisely, there is a central projection $e \in M^{**}_\infty$ such that $M_\infty \cong \ker \pi^{**} = M^{**}_\infty e$, and p is given by $p(b) = b(1-e)$. We have that

$$p(L) \subseteq (\pi^{**})^{-1}(L) = B^{**}.$$

To see this we note that since π is a surjection of B onto L, π^{**} is also a surjection of B^{**} onto L, and thus

$$(\pi^{**})^{-1}(L) = B^{**} + \ker \pi^{**}.$$

But we have that $K_\infty \subseteq B$ implies that $\ker \pi^{**} = \overline{K}_\infty \subseteq B^{**}$.

If B is locally reflexive, then since L is trivially locally reflexive, we have the commutative diagram

$$\begin{array}{ccccc} \ell^3_1 \check{\otimes} B & \subseteq & (\ell^3_1 \check{\otimes} B)^{**} & = & \ell^3_1 \check{\otimes} B^{**} \\ {\scriptstyle id\otimes\pi_0}\downarrow & & {\scriptstyle (id\otimes\pi_0)^{**}}\downarrow & & {\scriptstyle id\otimes\pi_0^{**}}\downarrow \\ \ell^3_1 \check{\otimes} L & \subseteq & (\ell^3_1 \check{\otimes} L)^{**} & = & \ell^3_1 \check{\otimes} L \end{array}$$

(see Theorem 14.3.6). Since there is a lifting for π^{**}, the right-hand mapping is a complete quotient mapping. But the the right-hand mapping is the second adjoint of the left-hand mapping, and thus the latter is also a complete quotient mapping. This contradicts the fact that the mapping φ in (14.5.6) cannot have a lifting $\psi : \ell^3_\infty \to B$ with $\|\psi\|_{cb} \leq 1$.

We have a commutative diagram

$$\begin{array}{ccc} B^{**} & \subseteq & M^{**}_\infty \\ {\scriptstyle \pi_0^{**}}\downarrow & & {\scriptstyle \pi^{**}}\downarrow \uparrow{\scriptstyle p}, \\ L & \subseteq & Q_\infty \end{array}$$

where

$$B^{**} = (\pi^{**})^{-1}(L) = M_\infty^{**} e \oplus p(L) \cong M_\infty \oplus \ell_\infty^3.$$

□

14.6 NUCLEAR OPERATOR SPACES

Given operator spaces V and W and a mapping $\varphi : V \to W$, we say that φ is C^*-*nuclear* if there exist diagrams of complete contractions

$$\begin{array}{ccccc} & & M_{n(\gamma)} & & \\ & {}^{r_\gamma}\nearrow & & \searrow^{s_\gamma} & \\ V & & \xrightarrow{\varphi} & & W, \end{array} \tag{14.6.1}$$

which approximately commute in the point-norm topology, by which we mean that there exist nets of complete contractions

$$V \xrightarrow{r_\gamma} M_{n(\gamma)} \xrightarrow{s_\gamma} W$$

such that for each $v \in V$, $\|s_\gamma \circ r_\gamma(v) - \varphi(v)\| \to 0$. We use the prefix '$C^*$' to distinguish this from the unrelated notion of a complete nuclear mapping. We define an operator space V to be *nuclear* if the identity mapping $id_V : V \to V$ is C^*-nuclear. This coincides with the usual notion of 'nuclearity' for C^*-algebras, and we shall continue to use this abbreviated terminology for operator spaces (but *not* for mappings).

Theorem 14.6.1 *An operator space V is nuclear if and only if it has the following property. For any 1-exact sequence of operator spaces*

$$0 \longrightarrow X \xrightarrow{\varphi} Y \xrightarrow{\psi} Z \longrightarrow 0$$

(*see* (14.4.4)) *it follows that*

$$0 \longrightarrow X \check{\otimes} V \xrightarrow{\varphi \otimes id} Y \check{\otimes} V \xrightarrow{\psi \otimes id} Z \check{\otimes} V \longrightarrow 0$$

is 1-exact.

Proof If V satisfies this property, then it is immediate from Theorem 14.4.1 that V is exact. Furthermore, given finite-dimensional operator spaces $E \subseteq F$, the top row of the commutative diagram

$$\begin{array}{ccc} F^* \check{\otimes} V & \longrightarrow & E^* \check{\otimes} V \\ \Vert & & \Vert \\ \mathcal{CB}(F, V) & \longrightarrow & \mathcal{CB}(E, V) \end{array}$$

is a quotient mapping, and thus the same is true for the bottom row. In other words, any mapping $\varphi : E \to V$ with $\|\varphi\|_{cb} < 1$ can be extended to a mapping $\psi : F \to V$ with $\|\psi\|_{cb} < 1$.

Given a finite-dimensional subspace $L \subseteq V$ and $\varepsilon > 0$, we may choose an $n \in \mathbb{N}$, a subspace $G \subseteq M_n$, and a linear isomorphism $r : L \to G$ with

$\|r\|_{cb} = 1$ and $\|r^{-1}\|_{cb} < 1 + \varepsilon$. From the previous argument, we may find a corresponding extension $s : M_n \to V$ of r^{-1} with $\|s\|_{cb} < 1 + \varepsilon$. We thus obtain a diagram

$$\begin{array}{ccccc} & & M_n & & \\ & {}^{t_0}\nearrow & \cup\!| & \searrow^{s} & \\ L & \xrightarrow{r} & G & \xrightarrow{r^{-1}} & V \end{array},$$

in which $t_0 : L \to M_n$ is just the inclusion mapping composed with r. We may extend t_0 to a complete contraction $t : V \to M_n$. From this construction it is evident that V is nuclear.

For the converse we may suppose that X is an operator subspace of Y and that Z is the operator quotient Y/X. If V is nuclear, then there exist nets of complete contractions $r_\gamma : V \to M_{n(\gamma)}$ and $s_\gamma : M_{n(\gamma)} \to V$ with $s_\gamma \circ r_\gamma : V \to V$ converging to the identity mapping id_V on V in the point-norm topology. From Theorem 11.3.3 V has the slice mapping property; hence from Proposition 11.3.4 $\ker(\pi \otimes id_V) = X \check{\otimes} V$.

We have that

$$\|(id_Y \otimes s_\gamma) \circ (id_Y \otimes r_\gamma)(u) - u\| \to 0$$

for $u \in V \check{\otimes} Z$. This is immediate if $u \in V \otimes Z$, and since the mappings $(id \otimes s_\gamma) \circ (id \otimes r_\gamma)$ are completely contractive, the same is true for elements of the completion.

Given an element $u \in Z \check{\otimes} V$ such that $\|u\| < 1$ and $\varepsilon > 0$, there exists a γ such that

$$\|(id_Y \otimes s_\gamma) \circ (id_Y \otimes r_\gamma)(u) - u\| < \varepsilon.$$

We have that $(id_Z \otimes r_\gamma)(u)$ is an element in $Z \check{\otimes} M_{n(\gamma)}$ with norm < 1. Since π is a complete quotient mapping, there is an element $u_1 \in Y \check{\otimes} M_{n(\gamma)}$ for which $\|u_1\| < 1$ and

$$(\pi \otimes id_{M_{n(\gamma)}})(u_1) = (id_Z \otimes r_\gamma)(u).$$

It follows that

$$u_2 = (id_Z \otimes s_\gamma)(u_1)$$

is a contractive element in $V \check{\otimes} Y$ with $\|u_2\| < 1$ and

$$(\pi \otimes id_V)(u_2) = (\pi \otimes s_\gamma)(u_1) = (id_Z \otimes (s_\gamma \circ r_\gamma))(u).$$

This shows that $id_V \otimes \pi$ maps the open unit ball $(V \check{\otimes} Y)_{\|\cdot\|<1}$ onto a dense subset of the open unit ball $(V \check{\otimes} Z)_{\|\cdot\|<1}$, and thus $id_V \otimes \pi$ is a quotient mapping (see §A.2). Since we can replace X by $M_n(X)$ and the same is true for Y and Z, $\pi \otimes id_V$ is a complete quotient mapping. □

The following result is an immediate consequence of the first part of the proof in Theorem 14.6.1

Corollary 14.6.2 *If V is a nuclear operator space, then it is exact.*

It is useful to single out a special class of nuclear operator spaces. We say that an operator space V is an *approximately square matrix space* (or simply, an *ASM space*) if

$$V = \lim_{n\to\infty} V_n,$$

where

$$V_1 \overset{\varphi_1}{\hookrightarrow} V_2 \overset{\varphi_2}{\hookrightarrow} \cdots$$

is a diagram with completely isometric injections φ_k and operator spaces V_k completely isometric to some matrix spaces $M_{n(k)}$ (see §3.1 for a discussion of direct limits). It is easy to see that any ASM space is nuclear. The following theorem shows that every separable exact (and thus nuclear) operator space is completely isometric to a subspace of some ASM space.

Theorem 14.6.3 *If V is a separable exact operator space, then it is completely isometric to a subspace of some ASM space.*

Proof We let

$$E_1 \subseteq E_2 \subseteq \cdots$$

be an increasing sequence of finite-dimensional operator subspaces of V such that $V = \overline{\bigcup E_k}$.

We claim that there is an increasing sequence $\{n(k)\}$ of natural numbers and a diagram of mappings

$$\begin{array}{ccccccccc} E_1 & \subseteq & \cdots & \subseteq & E_k & \subseteq & E_{k+1} & \subseteq & \cdots \\ {\scriptstyle i_1}\downarrow & & & & {\scriptstyle i_k}\downarrow & & {\scriptstyle i_{k+1}}\downarrow & & \\ M_{n_1} & \overset{\varphi_1}{\hookrightarrow} & \cdots & \overset{\varphi_{k-1}}{\hookrightarrow} & M_{n(k)} & \overset{\varphi_k}{\hookrightarrow} & M_{n(k+1)} & \overset{\varphi_{k+1}}{\hookrightarrow} & \cdots \end{array} \tag{14.6.2}$$

such that

(i) $\|i_k\|_{cb} \leq 1$ and $\left\|i_k^{-1}\right\|_{cb} \leq 1 + \frac{1}{2^k}$;
(ii) φ_k is a complete isometry;
(iii) $\left\|i_{k+1|E_k} - \varphi_k \circ i_k\right\|_{cb} < \frac{1}{2^k}$.

We proceed by induction. Since E_1 is exact, we may find an integer $n(1)$, an operator space $S_1 \subseteq M_{n(1)}$, and a linear isomorphism

$$i_1 : E_1 \to S_1$$

such that $\|i_1\|_{cb} \leq 1$ and $\left\|i_1^{-1}\right\|_{cb} < 1 + \frac{1}{2}$. Suppose that $n(k)$ and

$$i_k : E_k \to S_k \subseteq M_{n(k)}$$

are given. We shall construct $n(k+1)$, i_{k+1}, and φ_k so that they satisfy the above conditions.

Since E_{k+1} is exact, there exist an integer m, an operator space S_{k+1} in M_m, and a linear isomorphism

$$j : E_{k+1} \to S_{k+1}$$

such that $\|j\|_{cb} \leq 1$ and $\|j^{-1}\|_{cb} < 1 + \frac{1}{2^{k+1}}$. Since $M_{n(k)}$ is injective, there is a mapping

$$\alpha : E_{k+1} \to M_{n(k)}$$

such that $\alpha_{|E_k} = i_k$ and

$$\|\alpha\|_{cb} \leq \|i_k\|_{cb} \leq 1.$$

Since M_m is injective, there is a linear mapping

$$\beta : M_{n(k)} \to M_m$$

such that $\beta_{|i_k(E_k)} = j \circ i_k^{-1}$ and

$$\|\beta\|_{cb} \leq \|i_k^{-1}\|_{cb} \leq 1 + \frac{1}{2^k}.$$

We let $n(k+1) = n(k) + m$, and we define $i_{k+1} : E_{k+1} \to M_{n(k+1)}$ by

$$i_{k+1}(x) = \begin{bmatrix} \alpha(x) & 0 \\ 0 & j(x) \end{bmatrix}.$$

It is clear that $\|i_{k+1}\|_{cb} \leq 1$ and $S_{k+1} = i_{k+1}(E_{k+1})$ is a subspace of $M_{n(k+1)}$. If $r \in \mathbb{N}$ and $y \in M_r(S_{k+1})$, then there exists an $x \in M_r(E_{k+1})$ such that $y = i_{k+1}(x)$ and thus

$$\begin{aligned} \|(i_{k+1}^{-1})_r(y)\| &= \|x\| \\ &= \|j_r^{-1} j_r(x)\| \\ &\leq \|j^{-1}\|_{cb} \|j_r(x)\| \\ &\leq \left(1 + \frac{1}{2^{k+1}}\right) \|j_r(x)\| \\ &\leq \left(1 + \frac{1}{2^{k+1}}\right) \|(i_{k+1})_r(x)\|, \end{aligned}$$

from which we conclude

$$\|i_{k+1}^{-1}\|_{cb} \leq \left(1 + \frac{1}{2^{k+1}}\right).$$

If we define $\varphi_k : M_{n(k)} \to M_{n(k+1)}$ by

$$\varphi_k(x) = \begin{bmatrix} x & 0 \\ 0 & (1 + \frac{1}{2^k})^{-1} \beta(x) \end{bmatrix},$$

then it is a simple matter to verify that we also have (ii) and (iii). It is evident that the mappings i_k induce a complete isometry of V into an ASM space determined by the bottom row in (14.6.2). □

The following important result is due to Kirchberg. The proof is much more difficult than that for the above embedding theorem, and we refer the reader to the literature for its proof.

Theorem 14.6.4 *If V is a separable nuclear operator space, then there exist a UHF C^*-algebra $\mathcal{B}$, and norm closed left ideal L and right ideal R in $\mathcal{B}$ such that V is completely isometric to $\mathcal{B}/(L+R)$. In particular, any separable nuclear operator space is locally reflexive.* □

Corollary 14.6.5 *Every exact operator space is locally reflexive.*

Proof Let us suppose that V is an exact operator space. Then from the above theorem each separable subspace $V_0 \subseteq V$ is locally reflexive, and thus from Theorem 14.3.5, V is locally reflexive. □

Archbold and Batty proved that nuclear C^*-algebras have condition C, and thus such C^*-algebras are exact (condition C') and locally reflexive (condition C''). Subsequently, Kirchberg proved that for separable C^*-algebras, exactness implies local reflexivity (and thus C' is equivalent to C). From the above discussion, we see that the implications

$$\text{nuclearity} \Rightarrow \text{exactness} \Rightarrow \text{locally reflexivity}$$

are valid for all operator spaces. The equivalence of condition C and condition C' can be obtained by applying Corollary 14.6.5 and Lemma 14.2.1. In general, locally reflexive operator spaces need not be exact since the finite-dimensional operator spaces max ℓ_1^n and T_n $(n > 2)$ are trivially reflexive, but not exact. Kirchberg has raised the question of whether or not exactness and local reflexivity are equivalent for C^*-algebras. The following result provides some evidence for this possibility.

Theorem 14.6.6 *If a C^*-algebra $\mathcal{B}$ has the strong operator space approximation property, then $\mathcal{B}$ is exact.*

In particular, that is the case for C^*-algebras with the completely bounded approximation property, or for locally reflexive C^*-algebras with the operator space approximation property.

Proof Suppose that $\mathcal{B}$ has the strong operator space approximation property. Then $\mathcal{B}$ has the slice mapping property for subspaces of M_∞ by Theorem 11.3.3, and in particular,

$$\mathcal{F}(K_\infty, \mathcal{B}, M_\infty \check{\otimes} \mathcal{B}) = K_\infty \check{\otimes} \mathcal{B}.$$

It follows from Proposition 11.3.4 that

$$\ker \left\{\pi \otimes id : M_\infty \check{\otimes} \mathcal{B} \to Q_\infty \check{\otimes} \mathcal{B}\right\} = \mathcal{F}(K_\infty, \mathcal{B}, M_\infty \check{\otimes} \mathcal{B}),$$

where $Q_\infty = M_\infty / K_\infty$ is the Calkin algebra and $\pi : M_\infty \to Q_\infty$ is the quotient mapping. The result follows from Theorem 14.4.2.

The second statement follows from Theorems 11.3.3 and 14.3.9. □

We say that a dual operator space V is *semidiscrete* if there exist diagrams of weak* continuous complete contractions

$$\begin{array}{ccc} & M_{n(\alpha)} & \\ {}^{\varphi_\alpha}\nearrow & & \searrow^{\psi_\alpha} \\ V & \xrightarrow{\ id\ } & V \end{array},$$

which approximately commute in the point-weak* topology.

The following provides a 'second dual' characterization of the nuclear operator spaces.

Theorem 14.6.7 *Suppose that V is an operator space. Then the following are equivalent:*

(i) *V is nuclear;*
(ii) *V is locally reflexive, and V^{**} is semidiscrete (or equivalently, injective).*

Proof (i)$\Rightarrow$(ii). If V is nuclear, then from Corollary 14.6.5, V is locally reflexive. To see that V^{**} is semidiscrete, consider the diagrams

$$\begin{array}{ccccc} E & \subseteq & V^{**} & \longrightarrow & V^{**} \\ & & \downarrow \tilde{s} & & \\ {}^{\theta}\downarrow & & M_n & & \uparrow \\ & {}^{s}\nearrow & & \searrow^{t} & \\ V & & \longrightarrow & & V \end{array},$$

where the bottom triangle is approximately commutative in the point-norm topology and E is a finite-dimensional subspace of V^{**}. Given a finite-dimensional subspace $F \subseteq V^*$, we may choose $\theta : E \to V$ such that $\|\theta\|_{cb} < 1 + \varepsilon$ and $\langle \theta(x), f \rangle = \langle x, f \rangle$ for $x \in E$, $f \in F$ (see Lemma 14.3.3).

The composition $s \circ \theta : E \to M_n$ has a weak* continuous extension

$$\tilde{s} : V^{**} \to M_n$$

satisfying $\|\tilde{s}\|_{cb} < 1 + \varepsilon$. To see this, we note that the inclusion mapping $\iota : E \hookrightarrow V^{**}$ is continuous in the weak* topologies defined by E^* and V^*, and thus it is the adjoint of a mapping $\rho : V^* \to E^*$. Since ι is completely isometric, ρ must be a complete quotient mapping. We have the commutative diagram

$$\begin{array}{ccc} M_n(V^*) & \longrightarrow & M_n(E^*) \\ \| & & \| \\ \mathcal{CB}^\sigma(V^{**}, M_n) & \longrightarrow & \mathcal{CB}(E, M_n) \end{array},$$

in which the exponent σ indicates the weak* continuous completely bounded mappings (see §3.2), and the top and bottom rows are quotient mappings.

It follows that $s \circ \theta \in \mathcal{B}(E, M_n)$ is the restriction of a weak* continuous mapping $\tilde{s}$ in $\mathcal{CB}(V^{**}, M_n)$ with the desired norm property.

We conclude that the diagrams

$$\begin{array}{ccccc} & & M_n & & \\ & \tilde{s}\nearrow & & \searrow t & \\ V^{**} & & \xrightarrow{\varphi} & & V^{**} \end{array}$$

with $\tilde{s}$ continuous in the weak* topology, approximately commute in the point-weak* topology.

(ii)⇒(i). This is immediate from the diagrams

$$\begin{array}{ccccc} V^{**} & & \longrightarrow & & V^{**} \\ & s\searrow & & \nearrow t & \\ \cup| & & M_n & & \cup| \\ & s_0\nearrow & & \searrow t_0 & \\ V & & \longrightarrow & & V \end{array},$$

in which s_0 is the restriction of s, and we use local reflexivity to approximate t in the point-weak* topology by complete contractions t_0. □

We have already seen in Theorem 14.5.6 that the condition of local reflexivity cannot be deleted from condition (ii). By contrast, this condition is not needed for C^*-algebras.

14.7 NOTES AND REFERENCES

Although the principle of local reflexivity was already implicit in early work of Schatten (1950) and Grothendieck (1955), its importance for the local theory of Banach spaces was explicitly pointed out by Lindenstrauss and Rosenthal (1969). A history of the result and the argument that we have given may be found in Dean (1973).

Archbold and Batty's conditions C and C' first appeared in Archbold and Batty (1980). Local reflexivity and condition C'' were introduced in Effros and Haagerup (1985). These notions were used in an attempt to generalize the lifting theory of Choi and Effros (1976). The proof for Theorem 14.3.5 is due to Effros et al. (1999).

Exactness was defined by Kirchberg (1983). Subsequently, he proved that this condition is equivalent to condition C' (Kirchberg 1995). In that paper he also proved Theorem 14.6.4. A convenient reference for this theory may be found in Wassermann (1994). The operator space version of this theory was considered in great detail by Pisier (1995), who simplified Kirchberg's arguments and introduced ultraproducts into the discussion. The proof for Theorem 14.4.1 is new.

Lemma 14.5.3 was first proved by Haagerup (1980). We have given a simplified proof for Pisier's Theorem 14.5.4 (see Pisier (1995) for the original form). The counter-example in Theorem 14.5.6 is due to Kirchberg

(1995). We are indebted to Ozawa for clarifying Kirchberg's argument for us.

Nuclearity for C^*-algebras and semidiscreteness for von Neumann algebras was introduced in Lance (1973) and Effros and Lance (1977), respectively, and the C^*-algebraic version of Theorem 14.6.7 was proved by Choi and Effros (1977*b*, 1978). The notions of C^*-nuclear mappings and of nuclear operator spaces first appeared in Kirchberg (1995). Theorem 14.6.1 was discussed in Pisier (1995), who attributed the result to Kirchberg and Valliant. The material in §14.6 is due to Effros *et al.* (1999).

The weak expectation property (the 'WEP') was introduced by Lance (1973). Kirchberg found several profound characterizations of this property (Kirchberg 1993), and he investigated the C^*-algebraic quotients of WEP algebras, which he called QWEP algebras. He suggested that perhaps all separable C^*-algebras have this property. In that paper he showed that this conjecture is equivalent to a number of other open problems in functional analysis. We shall not attempt to discuss this notion, nor others that Kirchberg has introduced, such as the important 'local lifting property'.

15
Local reflexivity and exact integrality

In this chapter we consider local reflexivity and a closely related mapping space, called the exactly integral mappings. Our purpose is to prove one of the most surprising results in the theory of operator spaces. In light of the fact that C^*-algebras need not be locally reflexive, it was thought the same would be true for their dual operator spaces. It therefore came as quite a surprise to find that all such dual spaces, as well as all von Neumann algebraic preduals, are locally reflexive. Indeed, one has a strong version of local reflexivity which is generally false even for nuclear C^*-algebras. The result lends credence to the notion that the local theory of von Neumann algebraic preduals might be much better behaved than that of the von Neumann algebras themselves.

15.1 THE JUNGE APPROXIMATION THEOREM

Given operator spaces V and W, we say that a completely bounded mapping $\varphi : V^* \to W$ satisfies the *weak* approximation property* (or simply, *W*AP*) if there exists a net of finite-rank weak* continuous mappings $\varphi_\gamma : V^* \to W$ with $\|\varphi_\gamma\|_{cb} \leq \|\varphi\|_{cb}$ which converges to φ in the *point-norm* topology. If H is an infinite-dimensional Hilbert space, the identity mapping *id* on $\mathcal{B}(H)$ does not have such approximations since $\mathcal{B}(H)$ does not have the Grothendieck's approximation property (see the discussion in the introduction to Chapter 11). Our objective in this section is to show Junge's surprising result that if $\mathcal{A}$ and $\mathcal{B}$ are C^*-algebras, then *any* completely bounded mapping $\varphi : \mathcal{A}^* \to \mathcal{B}$ has the W^*AP. This will play a key role in our discussion of C^*-algebra duals and von Neumann algebra preduals.

Theorem 15.1.1 *Given C^*-algebras $\mathcal{A}$ and $\mathcal{B}$, every complete contraction $\varphi : \mathcal{A}^* \to \mathcal{B}^{**}$ can be approximated by a net of finite-rank weak* continuous complete contractions $\varphi_\gamma : \mathcal{A}^* \to \mathcal{B}$ in the point-weak* topology. Every completely bounded mapping $\varphi : \mathcal{A}^* \to \mathcal{B}$ satisfies the W^*AP.*

Proof Using the universal representations of $\mathcal{A}$ and $\mathcal{B}$, we may identify $\mathcal{A}^{**}$ and $\mathcal{B}^{**}$ with von Neumann algebras on Hilbert spaces H and K. It is evident that the $*$-subalgebra $\mathcal{A} \otimes \mathcal{B}$ is weak operator dense in the von Neumann algebra

$$\mathcal{A}^{**} \overline{\otimes} \mathcal{B}^{**} = (\mathcal{A} \otimes \mathcal{B})''$$

on $H \otimes K$. From the Kaplansky density theorem, the unit ball of the $*$-algebra $\mathcal{A} \otimes \mathcal{B}$ is weak operator dense in that of $\mathcal{A}^{**} \overline{\otimes} \mathcal{B}^{**}$.

If $\varphi : \mathcal{A}^* \to \mathcal{B}^{**}$ is a complete contraction, then we may suppose that $\varphi = \varphi_u$ for some contractive element $u \in \mathcal{A}^{**} \overline{\otimes} \mathcal{B}^{**}$ since we have the isometry

$$\mathcal{A}^{**} \overline{\otimes} \mathcal{B}^{**} \cong \mathcal{CB}(\mathcal{A}^*, \mathcal{B}^{**})$$

by (7.2.4). There exists a net of contractive elements $u_\gamma \in \mathcal{A} \otimes \mathcal{B}$ converging to u in the weak operator topology on $\mathcal{B}(H \otimes K)$. It follows that u_γ converges to u relative to the topology determined by the algebraic tensor product $\mathcal{A}^* \otimes \mathcal{B}^*$. The corresponding mappings $\varphi_\gamma = \varphi_{u_\gamma}$ comprise a net of finite-rank weak* continuous complete contractions from $\mathcal{A}^*$ into $\mathcal{B}$ which converges to $\varphi = \varphi_u$ in the point-weak* topology, i.e. for each $f \in \mathcal{A}^*$,

$$\varphi_\gamma(f) = f \otimes id(u_\gamma) \in \mathcal{B} \to \varphi(f) = f \otimes id(u) \in \mathcal{B}^{**}$$

in the weak* topology. If φ is a complete contraction from $\mathcal{A}^*$ into $\mathcal{B}$, then φ_γ converges to φ in the point-weak topology in $\mathcal{B}$. The usual convexity argument shows that we can find a net of finite-rank weak* continuous complete contractions $\psi_{\tilde{\gamma}} : \mathcal{A}^* \to \mathcal{B}$ in the convex hull of $\{\varphi_\gamma\}$, which converges to φ in the point-norm topology. □

15.2 EXACTLY INTEGRAL MAPPINGS I

Weakening the characterization in Corollary 12.3.7, we say that a linear mapping $\varphi : V \to W$ is *exactly integral* if it has a factorization

$$\begin{array}{ccccc} \mathcal{B}(H) & \xrightarrow{M(\omega)} & \mathcal{B}(K)^* & & \\ {\scriptstyle r}\uparrow & & & \searrow^{s} & \\ V & \xrightarrow{\varphi} & W & \overset{\iota_W}{\hookrightarrow} & W^{**} \end{array} , \qquad (15.2.1)$$

where r and s are completely bounded and $\omega \in \mathcal{B}(H \otimes K)^*$, *but we do not assume that s is weak* continuous.* We define the corresponding *exactly integral norm*

$$\iota^{ex}(\varphi) = \inf\{\|r\|_{cb} \|\omega\| \|s\|_{cb}\},$$

where the infimum is taken over all such factorizations. It is trivial that if $\varphi : V \to W$ is completely integral, then φ is exactly integral and

$$\iota^{ex}(\varphi) \leq \iota(\varphi).$$

Given operator spaces V and W, we let $\mathcal{I}^{ex}(V, W)$ denote the space of all exactly integral mappings from V into W. The fact that ι^{ex} is indeed a norm on $\mathcal{I}^{ex}(V, W)$ can be proved directly, but it is also evident from Theorem 15.4.1. There is a natural operator space matrix norm on $\mathcal{I}^{ex}(V, W)$.

Since we are not going to use it in the following, we leave the details to the reader.

Lemma 15.2.1 *Let V and W be operator spaces. If $\varphi : V \to W$ is completely integral, then $\varphi^* : W^* \to V^*$ is exactly integral with*

$$\iota^{ex}(\varphi^*) \leq \iota(\varphi).$$

Proof We may use (12.3.13) to construct a commutative diagram

$$\begin{array}{ccccc} \mathcal{B}(K) & \xrightarrow{M(\tilde{\omega})} & \mathcal{B}(H)^* & & \\ {\scriptstyle s_*}\uparrow & & & \searrow^{\iota_{V^*} \circ r^*} & , \\ W^* & \xrightarrow{\varphi^*} & V^* & \overset{\iota_{V^*}}{\hookrightarrow} & V^{***} \end{array}$$

where $s = (s_*)^*$, and $\tilde{\omega} : \mathcal{B}(K \otimes H) \to \mathbb{C}$ is the obvious 'flip' of ω. □

It should be noted that in the above proof, ι_{V^*} and $\iota_{V^*} \circ r^*$ need not be weak* continuous, and thus we cannot conclude that $\iota(\varphi^*) \leq \iota(\varphi)$.

Lemma 15.2.2 *If $\mathcal{A}$ is a C^*-algebra and V is an arbitrary operator space, then we have the isometric identification*

$$\mathcal{I}^{ex}(V, \mathcal{A}) = \mathcal{I}(V, \mathcal{A}).$$

Proof Let us assume that $\iota^{ex}(\varphi) \leq 1$. Then we can find a factorization

$$\begin{array}{ccccc} \mathcal{B}(H) & \xrightarrow{M(\omega)} & \mathcal{B}(K)^* & & \\ {\scriptstyle r}\uparrow & & & \searrow^{s} & , \\ V & \xrightarrow{\varphi} & \mathcal{A} & \overset{\iota_{\mathcal{A}}}{\hookrightarrow} & \mathcal{A}^{**} \end{array}$$

where r, s are complete contractions and ω is of norm one. From Theorem 15.1.1, we may approximate s in the point-weak* topology by a net of weak* continuous mappings $s_\gamma : \mathcal{B}(K)^* \to \mathcal{A}$ with $\|s_\gamma\|_{cb} \leq 1$. If we fix γ and let

$$\varphi_\gamma = \iota_{\mathcal{A}} \circ s_\gamma \circ M(\omega) \circ r,$$

we have the commutative diagram

$$\begin{array}{ccccc} \mathcal{B}(H) & \xrightarrow{M(\omega)} & \mathcal{B}(K)^* & & \\ {\scriptstyle r}\uparrow & & \downarrow{\scriptstyle s_\gamma} & \searrow^{\iota_{\mathcal{A}} \circ s_\lambda} & , \\ V & \xrightarrow{\varphi_\gamma} & \mathcal{A} & \overset{\iota_{\mathcal{A}}}{\hookrightarrow} & \mathcal{A}^{**} \end{array}$$

where $\iota_{\mathcal{A}} \circ s_\gamma : \mathcal{B}(K)^* \to \mathcal{A}^{**}$ is a weak* continuous complete contraction. It follows from Corollary 12.3.7 that $\iota(\varphi_\gamma) \leq 1$. Since each of the mappings s_γ and φ has its range in $\mathcal{A}$, the net φ_γ converges to φ in the point-weak topology. It easily follows from the definition of the integral norm that $\iota(\varphi) \leq 1$. □

Although the definition of the exactly integral mappings might seem contrived, such mappings play a natural and important role in operator space theory. We shall substantiate this claim in §15.4.

15.3 STRONG LOCAL REFLEXIVITY FOR VON NEUMANN ALGEBRA PREDUALS

Theorem 15.3.1 *For any C^*-algebra $\mathcal{A}$, $\mathcal{A}^*$ is a locally reflexive operator space.*

Proof From condition (ii$'$) of Theorem 14.3.1, it suffices to show that we have the isometry

$$\mathcal{I}(\mathcal{A}^*, F) = \mathcal{N}(\mathcal{A}^*, F)$$

for all finite-dimensional operator spaces F. Given $\varphi : \mathcal{A}^* \to F$, it is trivial from the definition that $\iota(\varphi) \leq \nu(\varphi)$. On the other hand, if we let $S(\varphi) = \varphi^*$, then we have the mappings

$$\begin{aligned}\mathcal{I}(\mathcal{A}^*, F) \xrightarrow{S} \mathcal{I}^{ex}(F^*, \mathcal{A}^{**}) &\cong \mathcal{I}(F^*, \mathcal{A}^{**}) \\ &\cong \mathcal{N}(F^*, \mathcal{A}^{**}) \xrightarrow{S^{-1}} \mathcal{N}(\mathcal{A}^*, F),\end{aligned}$$

where S is contractive by Lemma 15.2.1, the first identification is proved in Lemma 15.2.2, and the second is trivial. Thus, $\nu(\varphi) = \iota(\varphi)$. □

We have from Corollary 14.3.2 that any subspace of a locally reflexive operator space is again locally reflexive. In particular, if $\mathcal{R}$ is a von Neumann algebra, then we may identify $\mathcal{R}_*$ with an operator subspace of $\mathcal{R}^*$, and we obtain the following result.

Corollary 15.3.2 *For any von Neumann algebra $\mathcal{R}$, the predual $\mathcal{R}_*$ is a locally reflexive operator space.* □

The following result of Pisier is the completely bounded analogue of a well-known theorem in Banach space theory.

Lemma 15.3.3 *Suppose that V is an operator space, and that $v_i \in V$, $f_i \in V^*$ $(i = 1, \ldots, n)$ are biorthogonal; that is, $f_i(v_j) = \delta_{i,j}$. Then given $0 < \epsilon < 1$ and elements $w_i \in V$ such that*

$$\sum_{i=1}^{n} \|f_i\| \, \|v_i - w_i\| < \epsilon, \tag{15.3.1}$$

there is a complete isomorphism $\varphi : V \to V$ such that $\varphi(v_i) = w_i$, where $\|\varphi\|_{cb} < 1 + \epsilon$ and $\left\|\varphi^{-1}\right\|_{cb} < (1 - \epsilon)^{-1}$.

Proof Let $\delta : V \to V$ be the mapping defined by

$$\delta(v) = \sum_{i=1}^{n} f_i(v)(v_i - w_i).$$

Since $\|f_i\|_{cb} = \|f_i\|$, we have

$$\|\delta\|_{cb} \le \sum_{i=1}^{n} \|f_i\| \, \|v_i - w_i\| < \varepsilon.$$

If we let $\varphi = id_V - \delta$, then $\varphi(v_i) = w_i$ $(i = 1, \ldots, n)$ and

$$\|\varphi\|_{cb} < 1 + \|\delta\|_{cb} < 1 + \varepsilon.$$

Let $\psi = \sum_{k=0}^{\infty} \delta^k$, where $\delta^0 = id_V$. Then ψ is a well-defined completely bounded mapping on V with

$$\|\psi\|_{cb} \le \sum_{k=0}^{\infty} \|\delta\|_{cb}^{k} < (1-\varepsilon)^{-1}.$$

It is easy to see that $\psi = \varphi^{-1}$ is the completely bounded inverse of φ since

$$\varphi \circ \psi = (id_V - \delta) \circ \sum_{k=0}^{\infty} \delta^k = id_V$$

and

$$\psi \circ \varphi = \sum_{k=0}^{\infty} \delta^k \circ (id_V - \delta) = id_V.$$

□

An operator space V is called *strongly locally reflexive* if given a finite-dimensional subspace $F \subseteq V^{**}$, a finite-dimensional subspace $N \subseteq V^*$ and $\varepsilon > 0$, there exists a complete isomorphism φ of F onto a subspace E of V such that

(a) $\|\varphi\|_{cb}, \|\varphi^{-1}\|_{cb} < 1 + \epsilon$;
(b) $\langle \varphi(v), f \rangle = \langle v, f \rangle$ for all $v \in F$ and $f \in N$;
(c) $\varphi(v) = v$ for all $v \in F \cap V$.

We note that in contrast to Banach space theory, locally reflexive C^*-algebras need not be strongly locally reflexive. For example, K_∞ is not strongly locally reflexive. To see this, we let F be any finite-dimensional operator subspace of $M_\infty = (K_\infty)^{**}$. If we suppose that K_∞ is strongly locally reflexive, then for any $\varepsilon > 0$ there exists a completely bounded isomorphism φ of F onto a subspace S of some M_n such that

$$\|\varphi\|_{cb} \left\|\varphi^{-1}\right\|_{cb} < 1 + \epsilon.$$

This implies that

$$d_{cb}(F, S) < (1 + \varepsilon)^2,$$

and thus F is exact. This contradicts the result in §14.5 that M_∞ contains non-exact finite-dimensional operator spaces. Therefore, K_∞ is not strongly locally reflexive.

Theorem 15.3.4 *Suppose that V is a locally reflexive operator space for which there exists a completely isometric injection*

$$\theta : V^{**} \to \mathcal{B}(H),$$

which satisfies the W^*AP (see §15.1). Then V is strongly locally reflexive.

Proof In order to simplify the notation, we shall assume that V^{**} is a subspace of $\mathcal{B}(H)$, and that we have a net of weak* continuous finite-rank complete contractions $t_\gamma : V^{**} \to \mathcal{B}(H)$ for which

$$\|t_\gamma(v^{**}) - v^{**}\| \to 0$$

for each $v^{**} \in V^{**}$.

Let us fix $0 < \epsilon < 1$ and finite-dimensional subspaces $F \subseteq V^{**}$ and $N \subseteq V^*$, and let us assume that $0 < \delta < \epsilon/3n^3$, where $n = \dim F$. Then we can choose a complete contraction $t = t_\gamma$ such that

$$\|t(v) - v\| < \delta \|v\| \tag{15.3.2}$$

for all $v \in F$. If we let W be the range of t, we have $t : V^{**} \to W$ and $t^* : W^* \to V^*$.

From Lemma 14.3.3, there exists a mapping $\varphi : F \to V$ such that

$$\|\varphi\|_{cb} < 1 + \delta \tag{15.3.3}$$

and

$$\langle \varphi(v), f\rangle = \langle v, f\rangle \tag{15.3.4}$$

for all $v \in F$ and $f \in t^*(W^*) + N$. We let $C \subseteq \mathcal{CB}(F, V)$ be the convex set of all mappings $\varphi : F \to V$ satisfying (15.3.3) and (15.3.4). We let $F_0 = F \cap V$, and let $\iota_0 : F_0 \to V$ be the inclusion mapping. We let $C_0 \subseteq \mathcal{CB}(F_0, V)$ denote the convex set of all mappings $\varphi \circ \iota_0$, where $\varphi \in C$. We claim that ι_0 is in the point-norm closure of C_0. This is apparent since if we are given an arbitrary finite-dimensional subspace $G \subseteq V^*$, then our previous argument shows that there is a mapping $\varphi' : F \to V$ satisfying

$$\|\varphi'\|_{cb} < 1 + \delta$$

and

$$\langle \varphi'(v), f\rangle = \langle v, f\rangle$$

for all $v \in F$ and $f \in t^*(W^*) + N + G$. Since $\iota_0(F_0) \subseteq V$, we can suitably choose a net of φ' such that $\varphi' \circ \iota_0$ converges to ι_0 in the point-weak topology. We infer from the usual convexity argument that ι_0 is in the point-norm closure of C_0, and since F_0 is finite-dimensional, we may choose a mapping $\varphi \in C$ satisfying (15.3.3) and (15.3.4), for which

$$\|\iota_0 - \varphi \circ \iota_0\| < \delta,$$

and thus by Corollary 2.2.4

$$\|\iota_0 - \varphi \circ \iota_0\|_{cb} < n\delta.$$

For all $v \in F$ and $f \in W^*$,

$$\langle t(\varphi(v)), f\rangle = \langle \varphi(v), t^*(f)\rangle = \langle v, t^*(f)\rangle = \langle t(v), f\rangle,$$

and thus

$$t(\varphi(v)) = t(v) \tag{15.3.5}$$

for all $v \in F$.

We next perturb φ in order to satisfy (c). Since F is n-dimensional, there is a projection P of F onto $F_0 = V \cap F$ with $1 \leq \|P\|_{cb} \leq n$. To see this, it suffices to choose an Auerbach basis $v_1, \ldots, v_r \in F_0$ $(r \leq n)$ and a dual basis $f_j \in F_0^*$, where both the v_j and f_j have norm 1. We may extend each f_j to a functional on F, which we also denote f_j, again with norm 1. The mapping $P = \sum f_j \otimes v_j$ is a projection such that

$$\|P\|_{cb} \leq \sum \|f_j \otimes v_j\| = r \leq n.$$

We have

$$\varphi_1 = (\iota_0 - \varphi)P + \varphi : F \to V$$

is a completely bounded mapping such that $\varphi_1(v_0) = v_0$ for $v_0 \in F_0$. If $v \in F$,

$$\langle \varphi \circ P(v), f\rangle = \langle P(v), f\rangle,$$

and thus

$$\langle \varphi_1(v), f\rangle = \langle \varphi(v), f\rangle = \langle v, f\rangle$$

for $f \in N$, hence φ_1 satisfies (b) and (c). We also have

$$\|\varphi_1\|_{cb} \leq \|\iota_0 - \varphi \circ \iota_0\|_{cb} \|P\|_{cb} + (1+\delta) \leq \delta n^2 + (1+\delta) < 1 + \epsilon. \tag{15.3.6}$$

We let E be the range of φ_1. We must show that φ_1 is almost a complete isometry of F onto E. Let us assume that $v_1, \ldots, v_n$ is an Auerbach basis for F with biorthogonal dual basis f_i. We infer from (15.3.5) that for all i,

$$\begin{aligned}\|v_i - t \circ \varphi_1(v_i)\| &\leq \|v_i - t \circ \varphi(v_i)\| + \|t \circ \varphi(v_i) - t \circ \varphi_1(v_i)\| \\ &\leq \|v_i - t(v_i)\| + \|t\|_{cb} \|\varphi \circ \iota_0 - \iota_0\| \|P\|_{cb} \\ &\leq \delta + \delta \|P\|_{cb} \leq 2\delta \|P\|_{cb} < 2\delta n^2.\end{aligned}$$

Thus,

$$\sum_i \|v_i - t \circ \varphi_1(v_i)\| \|f_i\| \leq 2\delta n^3 < \epsilon.$$

From Lemma 15.3.3, we may find a mapping $s : \mathcal{B}(H) \to \mathcal{B}(H)$ for which

$$s \circ t \circ \varphi_1(v_i) = v_i,$$

and $\|s\|_{cb} \leq (1-\epsilon)^{-1}$. It follows that $\varphi_1^{-1} = s \circ t|_E$, and since t is completely contractive,

$$\|\varphi_1^{-1}\|_{cb} \leq (1-\epsilon)^{-1}. \tag{15.3.7}$$

Then φ_1 will also satisfy (a). □

Theorem 15.3.5 *If $\mathcal{R}$ is a von Neumann algebra, then $\mathcal{R}_*$ is strongly locally reflexive.*

Proof From Theorem 15.1.1, any completely isometric inclusion

$$\mathcal{R}^* \hookrightarrow \mathcal{B}(H)$$

has the W^*AP. Since $\mathcal{R}_*$ is locally reflexive (Corollary 15.3.2), the result follows from Theorem 15.3.4. □

15.4 EXACTLY INTEGRAL MAPPINGS II

Given a finite-dimensional operator space L, we may assume that $L \subseteq M_\infty$. We cannot use the truncation mappings

$$P^n : x \in M_\infty \to x^n \in M_n$$

to approximate L by subspaces of matrix spaces M_n in the completely bounded Banach–Mazur distance. In fact, we have shown in the proof of Theorem 14.4.1 that such an approximation is equivalent to the exactness of L. Nevertheless, as we shall see in the following demonstration, these truncations do provide a certain type of asymptotic approximation. Let us again let ρ_n denote the restriction of P^n to a mapping from L onto $S_n = \rho_n(L)$.

In the proof of the following theorem, we shall employ some delicate tensor product manipulations. For any finite-dimensional operator spaces L and M,

$$L^* \mathbin{\widehat{\otimes}} M^* \cong (L^* \mathbin{\widehat{\otimes}} M^*)^{**} \cong (L \mathbin{\check{\otimes}} M)^*, \tag{15.4.1}$$

since the second dual of a finite-dimensional operator space coincides with itself. In contrast to Banach space theory, this is generally false if only one of the spaces is finite-dimensional. However, if S is a subspace of some matrix space M_n, then for any operator space W,

$$S^* \mathbin{\widehat{\otimes}} W^* \cong (S \mathbin{\check{\otimes}} W)^*. \tag{15.4.2}$$

This can be seen by either applying (iii) of Theorem 14.4.1 or more simply by using the commutative diagram

$$\begin{array}{ccc} T_n \mathbin{\widehat{\otimes}} W^* & \cong & (M_n \mathbin{\check{\otimes}} W)^* \\ \downarrow & & \downarrow \\ S_n^* \mathbin{\widehat{\otimes}} W^* & \cong & (S_n \mathbin{\check{\otimes}} W)^* \end{array} .$$

Theorem 15.4.1 *If $\varphi : V \to W$ is a completely bounded mapping of operator spaces, then the following are equivalent:*

(i) $\iota^{ex}(\varphi) \leq 1$;

(ii) $\nu(\varphi \circ \psi) \leq 1$ *for all complete contractions* $\psi : S \to V$ *with* $S \subseteq M_n$ *and* $n \in \mathbb{N}$;

(iii) $\|id \otimes \varphi : S^* \mathbin{\check{\otimes}} V \to S^* \mathbin{\widehat{\otimes}} W\| \leq 1$ *for all* $S \subseteq M_n$ *and* $n \in \mathbb{N}$;

(iv) *there exists an infinite index set I, a free ultrafilter $\mathcal{U}$ on I, and a commutative diagram*

$$\begin{array}{ccccc} \prod M_\infty & \xrightarrow{\mathcal{M}} & \prod T_\infty/\mathcal{U} & & \\ {\scriptstyle r}\uparrow & & & \searrow^{s} & \\ V & \xrightarrow{\varphi} & W & \xrightarrow{\iota_W} & W^{**} \end{array} \quad , \tag{15.4.3}$$

where r and s are complete contractions, and $\mathcal{M} = \pi_{\mathcal{U}} \circ (M(a_\alpha, b_\alpha))$ is determined by the multiplication operators

$$M(a_\alpha, b_\alpha) : M_\infty \to T_\infty$$

with $\|a_\alpha\|_2, \|b_\alpha\|_2 \leq 1$.

Proof (i)⇒(ii) Let us suppose that φ satisfies (i). Then we may assume that φ has a factorization of the form (15.2.1). Given an operator space $S \subseteq M_n$ and a complete contraction $\psi : S \to V$, $\omega((r \circ \psi) \otimes id)$ is a contractive element of

$$(S \check{\otimes} \mathcal{B}(K))^* \cong S^* \hat{\otimes} \mathcal{B}(K)^* = \mathcal{N}(S, \mathcal{B}(K)^*)$$

(see (15.4.2)). The corresponding element of $\mathcal{N}(S, \mathcal{B}(K)^*)$ is $M(\omega) \circ r \circ \psi$, since if $x \in S$, and $y \in \mathcal{B}(K)$,

$$M(\omega)(r(\psi(x))(y) = \omega(r(\psi(x)) \otimes y) = \omega((r \circ \psi) \otimes id)(x \otimes y).$$

Thus, using the given factorization,

$$\begin{aligned} \nu(\varphi \circ \psi) &= \nu(\iota_W \circ \varphi \circ \psi) \\ &= \nu(s \circ M(\omega) \circ r \circ \psi) \\ &\leq \|s\|_{cb} \nu(M(\omega) \circ r \circ \psi) \leq 1. \end{aligned}$$

(ii)⇔(iii) is immediate from the commutative diagram

$$\begin{array}{ccc} S^* \check{\otimes} V & \xrightarrow{id \otimes \varphi} & S^* \hat{\otimes} W \\ \| & & \| \\ \mathcal{CB}(S, V) & \xrightarrow{\psi \mapsto \varphi \circ \psi} & \mathcal{N}(S, W) \end{array} \quad .$$

(iii)⇒(iv) We let I be the index set of all triples $\alpha = (E, F, k)$, where $E \subseteq V$ is finite-dimensional, $F \subseteq W$ is finite-codimensional and $k \in \mathbb{N}$. Given such a triple α, we shall also use the notation $E = E_\alpha, F = F_\alpha$ and $k = k_\alpha$. We write $\iota_\alpha : E_\alpha \hookrightarrow V$ and $\pi_\alpha : W \to W/F_\alpha$ for the inclusion and quotient mappings. We define a partial order on I by $\alpha \preceq \alpha' = (E', F', k')$ if $E \subseteq E'$, $F' \subseteq F$ and $k \leq k'$. For each $\alpha \in I$, we let $I_\alpha = \{\alpha' \in I : \alpha \preceq \alpha'\}$, we write $\mathcal{F}_\preceq$ for the filter generated by these I_α's, and we fix a free ultrafilter $\mathcal{U}$ on I containing $\mathcal{F}_\preceq$.

For each $\alpha = (E, F, k) \in I$, W/F is a finite-dimensional operator space, and thus we may identify it with a finite-dimensional subspace G_α of M_∞. We let $P^n : M_\infty \to M_n$ be the truncation mapping, and we let

$$\rho_{n,\alpha} : G_\alpha \to S_\alpha = P^n(G_\alpha)$$

be the corresponding restriction. As in the proof of Theorem 14.4.1, we may choose an integer $n_\alpha \in \mathbb{N}$ for which

$$\left\|(\rho_{n_\alpha,\alpha})_{k_\alpha}^{-1}\right\| \leq 1 + \frac{1}{k_\alpha}. \tag{15.4.4}$$

We let $\rho_\alpha = \rho_{n_\alpha,\alpha}$ and $\sigma_\alpha = \rho_{n_\alpha,\alpha}^{-1}$. We have

$$\varphi_\alpha = \rho_\alpha \circ \pi_\alpha \circ \varphi \circ \iota_\alpha$$

is a linear mapping from E_α onto $S_\alpha \subseteq M_{n(\alpha)}$.

From (iii), we have

$$\left\|id \otimes \varphi : S_\alpha^* \check{\otimes} V \to S_\alpha^* \hat{\otimes} W\right\| \leq 1,$$

and thus

$$\left\|id \otimes \varphi_\alpha : S_\alpha^* \check{\otimes} E_\alpha \to S_\alpha^* \hat{\otimes} S_\alpha\right\| \leq 1.$$

For each $\psi \in S_\alpha^* \check{\otimes} E_\alpha$, we have

$$(id \otimes \varphi_\alpha)(\psi) \in S_\alpha^* \hat{\otimes} S_\alpha$$

and thus from (11.2.5) that

$$|\text{trace}\,(\varphi_\alpha \circ \psi) = |\text{trace}\,(id \otimes \varphi_\alpha)(\psi)| \leq \|(id \otimes \varphi_\alpha)(\psi)\|_{S_\alpha^* \hat{\otimes} S_\alpha} \leq \|\psi\|_{S_\alpha^* \check{\otimes} E_\alpha}.$$

We conclude that φ_α is a contractive element in

$$(S_\alpha^* \check{\otimes} E_\alpha)^* = (S_\alpha \hat{\otimes} E_\alpha^*)^{**} = S_\alpha \hat{\otimes} E_\alpha^* = \mathcal{N}(E_\alpha, S_\alpha)$$

(see (15.4.1)) and $\nu(\varphi_\alpha) \leq 1$. Thus, we may find a commutative diagram

$$\begin{array}{ccc} M_\infty & \xrightarrow{M(a_\alpha, b_\alpha)} & T_\infty \\ {\scriptstyle r_\alpha}\uparrow & & \downarrow{\scriptstyle s_\alpha} \\ E_\alpha & \xrightarrow{\varphi_\alpha} & S_\alpha \end{array},$$

where r_α and s_α are complete contractions and $\|a_\alpha\|_2, \|b_\alpha\|_2 \leq 1$. If we let $\tilde{r}_\alpha : V \to M_\infty$ be a completely contractive extension of r_α to V, then we obtain the following commutative diagram

$$\begin{array}{ccccccccc} & & M_\infty & & \xrightarrow{M(a_\alpha, b_\alpha)} & T_\infty & \xrightarrow{\sigma_\alpha \circ s_\alpha} & W/F_\alpha & \\ & {\scriptstyle \tilde{r}_\alpha}\nearrow & \uparrow{\scriptstyle r_\alpha} & & & & \searrow{\scriptstyle s_\alpha} & \uparrow{\scriptstyle \sigma_\alpha} & \\ V & \xleftarrow{\iota_\alpha} & E_\alpha & \xrightarrow{\varphi \circ \iota_\alpha} & W & \xrightarrow{\pi_\alpha} & W/F_\alpha & \xrightarrow{\rho_\alpha} & S_\alpha \end{array}.$$

We let $\tilde{r} = (\tilde{r}_\alpha) : V \to \prod M_\infty$ and $\mathcal{M} = \pi_{\mathcal{U}} \circ (M(a_\alpha, b_\alpha))$. The mappings s_α and σ_α determine corresponding ultraproduct mappings

$$(\sigma_\alpha \circ s_\alpha)_{\mathcal{U}} : \prod T_\infty/\mathcal{U} \xrightarrow{(s_\alpha)_{\mathcal{U}}} \prod S_\alpha/\mathcal{U} \xrightarrow{(\sigma_\alpha)_{\mathcal{U}}} \prod (W/F_\alpha)/\mathcal{U},$$

where $(s_\alpha)_{\mathcal{U}}$ is a complete contraction, and from (15.4.4), it is evident that the same is true for $(\sigma_\alpha)_{\mathcal{U}}$. Finally, given $\pi_{\mathcal{U}}((w_\alpha + F_\alpha)) \in \prod (W/F_\alpha)/\mathcal{U}$, we may assume that (w_α) is a net of bounded elements in W. Then the weak* ultralimit $\lim_{\mathcal{U}} w_\alpha$ exists in W^{**} (the unit ball is weak* compact), and we may define $\tilde{s}(\pi_{\mathcal{U}}((w_\alpha + F_\alpha))) = \lim_{\mathcal{U}} w_\alpha$. It is easy to see that $\tilde{s} : \prod (W/F_\alpha)/\mathcal{U} \to W^{**}$ is a well-defined complete contraction such that the diagram

$$\begin{array}{ccccc} & & \prod (W/F_\alpha)/\mathcal{U} & & \\ & {\scriptstyle \pi_{\mathcal{U}} \circ (\pi_\alpha)} \nearrow & & \searrow {\scriptstyle \tilde{s}} & \\ W & & \xrightarrow{\iota_W} & & W^{**} \end{array}$$

commutes. This gives us the following commutative diagram of complete contractions

$$\begin{array}{ccccc} \prod M_\infty & \xrightarrow{\mathcal{M}} & \prod T_\infty/\mathcal{U} & \xrightarrow{(\sigma_\alpha \circ s_\alpha)_{\mathcal{U}}} & \prod (W/F_\alpha)/\mathcal{U} \\ {\scriptstyle \tilde{r}} \uparrow & & & {\scriptstyle \pi_{\mathcal{U}} \circ (\pi_\alpha)} \nearrow & \downarrow {\scriptstyle \tilde{s}} \\ V & \xrightarrow{\varphi} & W & \xrightarrow{\iota_W} & W^{**} \end{array} .$$

If we let $s = \tilde{s} \circ (\sigma_\alpha \circ s_\alpha)_{\mathcal{U}}$, then we obtain (15.4.3).

(iv)⇒(i) Assuming that we have the factorization (15.4.3), our task is to construct from it a factorization of the form (15.2.1). From (10.3.3), we have a complete isometry

$$\theta : \prod T_\infty/\mathcal{U} \to \mathcal{T} = \left(\prod M_\infty\right)^* e,$$

where e is a central projection in $(\prod M_\infty)^{**}$, and thus we may define a completely contractive projection P_e of $(\prod M_\infty)^*$ onto $\prod T_\infty/\mathcal{U}$ by letting $P_e(F) = Fe$. We may modify (15.4.3) by using the commutative diagram

$$\begin{array}{ccccc} \prod M_\infty & \xrightarrow{\mathcal{M}} & \prod T_\infty/\mathcal{U} & \xrightarrow{\theta} & \left(\prod M_\infty\right)^* \\ {\scriptstyle r} \uparrow & & & \searrow {\scriptstyle s} & \downarrow {\scriptstyle s \circ P_e} \\ V & \xrightarrow{\varphi} & W & \xhookrightarrow{\iota_W} & W^{**} \end{array} .$$

Turning to the left side of this diagram, we may suppose that V is an operator subspace of $\mathcal{B}(H)$ and let $\iota : V \hookrightarrow \mathcal{B}(H)$ be the inclusion mapping. Then $r = (r_\alpha)$, where each $r_\alpha : V \to M_\infty$ is a complete contraction, and using the Arveson–Wittstock–Hahn–Banach theorem, we may extend each r_α to a complete contraction $\tilde{r}_\alpha : \mathcal{B}(H) \to M_\infty$. These determine the complete contraction $\tilde{r} = (\tilde{r}_\alpha) : \mathcal{B}(H) \to \prod M_\infty$, and we have the commutative diagram.

$$\begin{array}{ccccccc}
\mathcal{B}(H) & \xrightarrow{\tilde{r}} & \prod M_\infty & \xrightarrow{\mathcal{M}} & \prod T_\infty/\mathcal{U} & \xrightarrow{\theta} & \left(\prod M_\infty\right)^* \\
{\scriptstyle\iota}\uparrow & {\scriptstyle r}\nearrow & & & & \searrow{\scriptstyle s} & \downarrow{\scriptstyle s\circ P_e} \\
V & & \xrightarrow{\varphi} & W & \xhookrightarrow{\iota_W} & & W^{**}
\end{array} .$$

For each $\alpha \in I$, we let $\omega_\alpha : \mathcal{B}(H) \otimes M_\infty \to \mathbb{C}$ be the linear functional given by

$$\omega_\alpha(x \otimes y) = \langle a_\alpha \tilde{r}_\alpha(x) b_\alpha, y \rangle .$$

Since $\|a_\alpha\|_2, \|b_\alpha\|_2 \le 1$, it is clear that ω_α is a contractive linear functional on $\mathcal{B}(H) \check{\otimes} M_\infty$. Then $\pi_\mathcal{U}((\omega_\alpha))$ is a contractive element in

$$\prod (\mathcal{B}(H) \check{\otimes} M_\infty)^*/\mathcal{U} \hookrightarrow \left(\prod (\mathcal{B}(H) \check{\otimes} M_\infty)\right)^*,$$

where we have used the corresponding identification of the ultraproduct with a subspace of $(\prod(\mathcal{B}(H) \check{\otimes} M_\infty))^*$ (see Corollary 10.3.5). We can identify $\mathcal{B}(H) \check{\otimes} \prod M_\infty$ with an operator subspace of $\prod(\mathcal{B}(H) \check{\otimes} M_\infty)$, and we let ω be the restriction of $\pi_\mathcal{U}((\omega_\alpha))$ to $\mathcal{B}(H) \check{\otimes} \prod M_\infty$. Then ω is a contractive linear functional on $\mathcal{B}(H) \otimes_\vee \prod M_\infty$ such that for every $x \in \mathcal{B}(H)$ and $(y_\alpha) \in \prod M_\infty$,

$$\begin{aligned}
\omega(x \otimes (y_\alpha)) &= \lim_\mathcal{U} \omega_\alpha(x \otimes y_\alpha) \\
&= \lim_\mathcal{U} \langle a_\alpha \tilde{r}_\alpha(x) b_\alpha, y_\alpha \rangle = \langle \theta \circ \mathcal{M} \circ \tilde{r}(x), (y_\alpha) \rangle .
\end{aligned}$$

This shows that $M(\omega) = \theta \circ \mathcal{M} \circ \tilde{r}$.

Finally, we let $J : \prod M_\infty \hookrightarrow \mathcal{B}(K)$ be an identification of $\prod M_\infty$ with a von Neumann subalgebra of $\mathcal{B}(K)$ for some Hilbert space K. Since $\prod M_\infty$ is injective, there is a completely contractive projection π from $\mathcal{B}(K)$ onto $\prod M_\infty$. Taking adjoints, the composition

$$\left(\prod M_\infty\right)^* \xrightarrow{\pi^*} \mathcal{B}(K)^* \xrightarrow{J^*} \left(\prod M_\infty\right)^*$$

is just the identity mapping, and we obtain the commutative diagram

$$\begin{array}{ccccc}
\mathcal{B}(H) & \xrightarrow{M(\omega)} & \left(\prod M_\infty\right)^* & \xrightarrow{\pi^*} & \mathcal{B}(K)^* \\
{\scriptstyle\iota}\uparrow & & & & \downarrow{\scriptstyle s\circ P_e \circ J^*} \\
V & \xrightarrow{\varphi} & W & \xhookrightarrow{\iota_W} & W^{**}
\end{array} .$$

The composition $\omega \circ (id \otimes \pi)$ is a contractive functional on $\mathcal{B}(H) \check{\otimes} \mathcal{B}(K)$, and thus we may extend it to a contractive functional $\tilde{\omega}$ on $\mathcal{B}(H \otimes K)$. For any $z \in \mathcal{B}(K)$,

$$M(\tilde{\omega})(x)(z) = \omega(x \otimes \pi(z)) = (M(\omega)(x))(\pi(z)) = (\pi^* M(\omega)(x))(z),$$

and thus $\pi^* \circ M(\omega) = M(\tilde{\omega})$. We obtain the commutative diagram

$$\begin{array}{ccccc} \mathcal{B}(H) & \xrightarrow{M(\tilde{\omega})} & \mathcal{B}(K)^* & & \\ {\scriptstyle\iota}\uparrow & & & \searrow{\scriptstyle\tau} & \\ V & \xrightarrow{\varphi} & W & \xhookrightarrow{\iota_W} & W^{**}, \end{array}$$

where $\tau = s \circ P_e \circ J^*$ is a complete contraction, and we conclude that φ is exactly integral with $\iota^{ex}(\varphi) \leq 1$. □

It is easy to see from Theorem 15.4.1 that ι^{ex} is a norm on $\mathcal{I}^{ex}(V, W)$, and $\mathcal{I}^{ex}(V, W)$ is a mapping ideal. It is also possible to express ι^{ex} as a dual space norm. We have from (iii) of Theorem 15.4.1 and (15.4.2) that

$$\begin{aligned} \iota^{ex}(\varphi) &= \sup\left\{\left\|id \otimes \varphi : S^* \check{\otimes} V \to S^* \hat{\otimes} W\right\|\right\} \\ &= \sup\left\{\|id \otimes \varphi(u)\|_{S^* \hat{\otimes} W} : \|u\|_{S^* \check{\otimes} V} \leq 1\right\} \\ &= \sup\left\{\|id \otimes \iota_W \circ \varphi(u)\|_{S^* \hat{\otimes} W^{**}} : \|u\|_{S^* \check{\otimes} V} \leq 1\right\} \\ &= \sup\left\{|\langle (id \otimes \varphi)(u), v\rangle| : \|u\|_{S^* \check{\otimes} V} \leq 1, \|v\|_{S \check{\otimes} W^*} \leq 1\right\}, \end{aligned}$$

where the supremum is taken over arbitrary $S \subseteq M_n$ and contractive elements $u \in S^* \check{\otimes} V$ and $v \in S \check{\otimes} W^*$. If we let u and v correspond to the functions $a \in \mathcal{CB}(S, V)$ and $b \in \mathcal{CB}(W, S)$, then a simple calculation with elementary matrices leads to the formula

$$\iota^{ex}(\varphi) = \sup\left\{|\mathrm{trace}\,(\varphi \circ \psi)| : \psi = a \circ b, \|a\|_{cb}, \|b\|_{cb} \leq 1\right\}. \tag{15.4.5}$$

In general, given a finite-rank mapping $\psi : W \to V$, we define

$$\gamma_{SK}(\psi) = \inf\left\{\|a\|_{cb}\,\|b\|_{cb}\right\},$$

where the supremum is taken over all factorizations

$$\begin{array}{ccccc} & & S & & \\ & {\scriptstyle a}\nearrow & & \searrow{\scriptstyle b} & \\ W & & \xrightarrow{\psi} & & V \end{array}$$

with $S \subseteq M_n$ and $n \in \mathbb{N}$. It is easy to see that this determines a norm on $\mathcal{F}(W, V)$, and we let $\gamma^0_{SK}(W, V)$ denote the corresponding normed space. We conclude from (15.4.5) that we have an isometric injection

$$\mathcal{I}^{ex}(V, W) \hookrightarrow \gamma^0_{SK}(W, V)^*, \tag{15.4.6}$$

and in particular, if W is finite-dimensional,

$$\mathcal{I}^{ex}(V, W) = \gamma^0_{SK}(W, V)^*. \tag{15.4.7}$$

We also claim that the exactly integral norm is local.

Proposition 15.4.2 *A linear mapping $\varphi : V \to W$ is exactly integral with $\iota^{ex}(\varphi) \leq 1$ if and only if for every finite-dimensional subspace $L \subseteq V$, we have $\iota^{ex}(\varphi_{|L}) \leq 1$.*

Proof Since $\mathcal{I}^{ex}(V, W)$ is a mapping ideal, it is clear that

$$\iota^{ex}(\varphi_{|L}) \leq \iota^{ex}(\varphi).$$

On the other hand, for any operator subspace S of M_n and any complete contraction $\psi : S \to V$, we have

$$\nu(\varphi \circ \psi) = \nu(\varphi_{|L} \circ \psi) \leq \iota^{ex}(\varphi_{|L}),$$

where we let $L = \psi(S)$. It follows from Theorem 15.4.1 that

$$\begin{aligned}\iota^{ex}(\varphi) &= \sup\{\nu(\varphi \circ \psi) : \|\psi : S \to V\|_{cb} \leq 1,\, S \subseteq M_n, n \in \mathbb{N}\} \\ &\leq \sup\{\iota^{ex}(\varphi_{|L}) : L \subseteq V \text{ finite-dimensional}\}.\end{aligned}$$

□

The following result provides some motivation for our terminology 'exactly integral mappings'.

Proposition 15.4.3 *For any operator space V, the following are equivalent:*

(i) *V is exact;*
(ii) *$\mathcal{I}(V, L) = \mathcal{I}^{ex}(V, L)$ for all finite-dimensional operator spaces L;*
(iii) *$\mathcal{I}(V, W) = \mathcal{I}^{ex}(V, W)$ for all operator spaces W.*

Proof Let $\varphi : V \to W$ be an exactly integral mapping. If V is exact, then for any finite-dimensional subspace $L \subseteq V$ and $\varepsilon > 0$, there exists a linear isomorphism ψ from L onto an operator subspace S of some matrix space M_n such that $\|\psi\|_{cb} < 1 + \varepsilon$ and $\|\psi^{-1}\|_{cb} < 1$. It follows that

$$\iota(\varphi_{|L}) = \nu(\varphi_{|L} \circ \psi^{-1} \circ \psi) \leq \nu(\varphi_{|L} \circ \psi^{-1})\|\psi\|_{cb} \leq \iota^{ex}(\varphi_{|L})(1 + \varepsilon).$$

If we let $\varepsilon \to 0$, we have

$$\iota(\varphi_{|L}) \leq \iota^{ex}(\varphi_{|L}),$$

and thus by the local property of ι and ι^{ex},

$$\iota(\varphi) \leq \iota^{ex}(\varphi).$$

Therefore, we have $\iota(\varphi) = \iota^{ex}(\varphi)$, and this shows that (i) $\Rightarrow$ (iii).

It is clear that (iii)$\Rightarrow$ (ii). Let us suppose that (ii) holds for all finite-dimensional subspaces $L \subseteq V$. Then fixing such a subspace, we have a norm-decreasing linear isomorphism (both sides coincide with the linear space $L^* \otimes V$)

$$\theta : \gamma^0_{SK}(L, V) \to \mathcal{CB}(L, V).$$

But we are given that the adjoint mapping

$$\theta^* : \mathcal{I}(V, L) \to \mathcal{I}^{ex}(V, L)$$

is isometric, and thus θ must itself be an isometry. If $\iota : L \hookrightarrow V$ is the inclusion mapping, then it follows that for any $\epsilon > 0$, we have a commutative diagram

$$\begin{array}{ccccc} & & S & & \\ & {}^{a}\nearrow & & \searrow^{b} & \\ L & & \xrightarrow{\iota} & & V \end{array},$$

where S is an operator subspace of some M_n, and $\|a\|_{cb}\,\|b\|_{cb} < 1+\epsilon$. Thus, by definition, L is exact, and the same follows for V. □

Proposition 15.4.4 *Given operator spaces V and W and a linear mapping $\varphi : V \to W$,*

$$\iota(\varphi) \leq \iota(\varphi^*).$$

Moreover, V is locally reflexive if and only if we have the isometry

$$\iota(\varphi) = \iota(\varphi^*)$$

for all operator spaces W and linear mappings $\varphi : V \to W$.

Proof For any finite-dimensional operator space E

$$E^* \check{\otimes} V^* = (E \hat{\otimes} V)^*,$$

whereas the corresponding mapping

$$E^* \hat{\otimes} V^* \to (E \check{\otimes} V)^*$$

is norm-decreasing. From these we conclude that

$$\begin{aligned}\iota(\varphi) &= \sup\{\|id \otimes \varphi : E^* \check{\otimes} V \to E^* \hat{\otimes} W\| : E \text{ finite-dimensional}\}\\ &= \sup\{\|(id \otimes \varphi)^* : (E^* \hat{\otimes} W)^* \to (E^* \check{\otimes} V)^*\| : E \text{ finite-dimensional}\}\\ &\leq \sup\{\|id \otimes \varphi^* : E \check{\otimes} W^* \to E \hat{\otimes} V^*\| : E \text{ finite-dimensional}\}\\ &= \iota(\varphi^*).\end{aligned}$$

If V is locally reflexive, then $E \hat{\otimes} V^* \to (E^* \check{\otimes} V)^*$ is isometric, and the above calculation shows that $\iota(\varphi) = \iota(\varphi^*)$.

Conversely, if W is a finite-dimensional operator space, we have the isometries

$$\mathcal{CB}(V, W) \cong V^* \check{\otimes} W \cong W \check{\otimes} V^* = \mathcal{CB}(W^*, V^*),$$

and from (12.3.3), we have the isometry

$$\mathcal{I}(W^*, V^*) = \mathcal{N}(W^*, V^*).$$

Therefore, if for every $\varphi : V \to W$, we have $\iota(\varphi) = \iota(\varphi^*)$, then we have the isometries

$$\mathcal{I}(V, W) = \mathcal{I}(W^*, V^*) = \mathcal{N}(W^*, V^*) = \mathcal{N}(V, W),$$

and we conclude from Theorem 14.3.1 that V is locally reflexive. □

15.5 RELATING THE MAPPING SPACES

We have seen that as in Banach space theory, mapping spaces provide a fundamental tool for studying operator spaces. In this regard, it is particularly useful to consider the conditions under which various mapping spaces coincide. In order to emphasize this feature of the theory, we have devoted this section to a summary of our previous results, together with some additional information. For simplicity, we only consider the norm structure on the mapping spaces.

We recall that given operator spaces V and W, we have the contractive inclusions

$$\mathcal{N}(V,W) \subseteq \mathcal{I}(V,W) \subseteq \mathcal{I}^{ex}(V,W) \subseteq \Pi_1(V,W) \subseteq \mathcal{CB}(V,W).$$

Our task is to list conditions under which these are in fact isometric surjections.

The isometry between $\mathcal{N}(V,W)$ and $\mathcal{I}(V,W)$

From (12.3.3), if V is an arbitrary finite-dimensional operator space, then

$$\mathcal{N}(V,W) = \mathcal{I}(V,W)$$

for all operator spaces W. More generally, we have shown in Theorem 14.3.1 that an operator space V is locally reflexive if and only if

$$\mathcal{N}(V,W) = \mathcal{I}(V,W)$$

for all finite-dimensional operator spaces W.

The isometry between $\mathcal{I}(V,W)$ and $\mathcal{I}^{ex}(V,W)$

If we fix an operator space V for the domain, we have from Proposition 15.4.3 that V is exact if and only if we have the isometry

$$\mathcal{I}(V,W) = \mathcal{I}^{ex}(V,W)$$

for all (finite-dimensional) operator space W. On the other hand, from Lemma 15.2.2, if $\mathcal{A}$ is an arbitrary C^*-algebra, then we have the isometry

$$\mathcal{I}(V,\mathcal{A}) = \mathcal{I}^{ex}(V,\mathcal{A})$$

for *all* operator spaces V.

The isometry between $\mathcal{I}(V,W)$ and $\Pi_1(V,W)$

Proposition 15.5.1 *If $\mathcal{A}$ is an injective C^*-algebra, then we have the isometry*

$$\mathcal{I}(V,\mathcal{A}) = \Pi_1(V,\mathcal{A})$$

for all operator spaces V.

Proof Given an arbitrary operator space V, suppose that $\varphi \in \Pi_1(V,\mathcal{A})$ satisfies $\pi_1(\varphi) \leq 1$. Then from Theorem 13.2.3, there exist indexed families

$(a_\alpha)_{\alpha\in I}$ and $(b_\alpha)_{\alpha\in I}$ of Hilbert–Schmidt contractions, a free ultrafilter $\mathcal{U}$ on I, and a commutative diagram

$$\begin{array}{ccc} \prod M_\infty & \xrightarrow{\mathcal{M}} & \prod T_\infty/\mathcal{U} \\ \cup| & & \cup| \\ V_\infty & \longrightarrow & V_1 \\ {\scriptstyle r}\uparrow & & \downarrow{\scriptstyle s} \\ V & \xrightarrow{\varphi} & \mathcal{A} \end{array},$$

where r and s are complete contractions, $\mathcal{M}$ is determined by the complete contractions

$$M(a_\alpha, b_\alpha) : M_\infty \to T_\infty : x \mapsto a_\alpha x b_\alpha,$$

and where $V_\infty = \overline{r(V)}$, and $V_1 = \overline{\mathcal{M}(V_\infty)}$. Since $\mathcal{A}$ is injective, s has a completely contractive extension $\tilde{s} : \prod T_\infty/\mathcal{U} \to \mathcal{A}$. Therefore, we obtain a commutative diagram of complete contractions

$$\begin{array}{ccccc} \prod M_\infty & \xrightarrow{\mathcal{M}} & \prod T_\infty/\mathcal{U} & & \\ {\scriptstyle r}\uparrow & & \downarrow{\scriptstyle \tilde{s}} & \searrow^{\iota_{\mathcal{A}}\circ\tilde{s}} & \\ V & \xrightarrow{\varphi} & \mathcal{A} & \overset{\iota_{\mathcal{A}}}{\hookrightarrow} & \mathcal{A}^{**} \end{array}.$$

This implies by Theorem 15.4.1 that $\varphi : V \to A$ is an exact integral mapping with $\iota^{ex}(\varphi) \leq 1$. Therefore, we have the isometry

$$\mathcal{I}(V, \mathcal{A}) = \mathcal{I}^{ex}(V, \mathcal{A}) = \Pi_1(V, \mathcal{A}).$$

□

An operator space W is said to have the *weak expectation property* (or simply, WEP) if for any completely isometric inclusion $\tau : W \hookrightarrow B(H)$, there exists a completely contractive mapping $P : B(H) \to W^{**}$ such that $P \circ \tau = \iota_W$, where $\iota_W : W \hookrightarrow W^{**}$ is the canonical inclusion.

Corollary 15.5.2 *If W has WEP, then we have the isometry*

$$\mathcal{I}(V, W) = \Pi_1(V, W)$$

for all operator spaces V.

Proof Let us fix a completely isometric inclusion $\tau : W \hookrightarrow B(H)$ and a complete contraction $P : B(H) \to W^{**}$ such that $P \circ \tau = \iota_W$. For any operator space V and $\varphi \in \Pi_1(V, W)$ with $\pi_1(\varphi) \leq 1$, $\tau \circ \varphi$ is an element of $\Pi_1(V, B(H)) = \mathcal{I}(V, B(H))$ with

$$\iota(\tau \circ \varphi) = \pi_1(\tau \circ \varphi) \leq \|\tau\|_{cb}\, \pi_1(\varphi) \leq 1.$$

Therefore, we have

$$\iota(\varphi) = \iota(\iota_W \circ \varphi) \leq \|P\|_{cb}\, \iota(\tau \circ \varphi) \leq 1$$

since $\varphi \in \mathcal{I}(V,W) \to \iota_W \circ \varphi \in \mathcal{I}(V,W^{**})$ is an isometric inclusion by Lemma 12.3.2. This shows that we have the isometry

$$\mathcal{I}(V,W) = \Pi_1(V,W).$$

□

On the other hand, if we fix the domain space V, we may obtain the following result.

Corollary 15.5.3 *If V is a nuclear operator space, then we have the isometry*

$$\mathcal{I}(V,W) = \Pi_1(V,W).$$

for all operator spaces W.

Proof Let W be an arbitrary operator space and $\varphi \in \Pi_1(V,W)$ with $\pi_1(\varphi) \le 1$. For any complete contraction $\psi : M_n \to V$, we may identify ψ with a contractive element in $\mathcal{CB}(M_n, V) = T_n \check{\otimes} V$. Then $\pi_1(\varphi) \le 1$ implies that $\varphi \circ \psi \in \mathcal{N}(M_n, W) = T_n \widehat{\otimes} W$ with $\nu(\varphi \circ \psi) \le 1$.

Since V is a nuclear operator space, there exist two nets of complete contractions $\varphi_\alpha : V \to M_{n(\alpha)}$ and $\psi_\alpha : M_{n(\alpha)} \to V$ such that $\psi_\alpha \circ \varphi_\alpha \to id_V$ in the point-norm topology. Then $\varphi \circ \psi_\alpha \circ \varphi_\alpha$ is a net of finite-rank mappings from V into W such that $\nu(\varphi \circ \psi_\alpha \circ \varphi_\alpha) \le \nu(\varphi \circ \psi_\alpha) \|\varphi_\alpha\|_{cb} \le 1$ and $\varphi \circ \psi_\alpha \circ \varphi_\alpha \to \varphi$ in the point-norm topology. Therefore, we have $\iota(\varphi) \le 1$.

□

It is well known to Banach space specialists that for any Banach space W, we have the isometry

$$\mathcal{N}^B(c_0, W) = \Pi_1^B(c_0, W).$$

The following corollary shows that the operator analogue of this result is also true.

Corollary 15.5.4 *For any operator space W, we have the isometry*

$$\mathcal{N}(K_\infty, W) = \Pi_1(K_\infty, W).$$

Proof To see this, since T_∞ has the operator space approximation property, for any operator space W, the canonical mapping

$$\Phi : T_\infty \widehat{\otimes} W \to T_\infty \check{\otimes} W$$

is a one-to-one mapping. Therefore, we have the isometries

$$\mathcal{N}(K^\infty, W) = T_\infty \widehat{\otimes} W \hookrightarrow T_\infty \widehat{\otimes} W^{**} = (K_\infty \check{\otimes} W^*)^*.$$

On the other hand, we have the isometric inclusion

$$\mathcal{I}(K_\infty, W) \hookrightarrow (K_\infty \check{\otimes} W^*)^*$$

by Lemma 12.3.3. This implies the isometries

$$\mathcal{N}(K_\infty, W) = \mathcal{I}(K_\infty, W) = \Pi_1(K_\infty, W).$$

□

If the range space is a dual operator space W^*, then we may define the *weakly integral* space $\mathcal{I}^w(V, W^*)$ to be the mapping space such that

$$\mathcal{I}^w(V, W^*) \cong (V \check{\otimes} W)^*,$$

or, in other words, we may identify mappings $\varphi \in \mathcal{I}^w(V, W^*)$ with the bounded linear functional $F_\varphi \in (V \check{\otimes} W)^*$ such that

$$F_\varphi(v \otimes w) = \langle \varphi(v), w \rangle$$

for all $v \in V$ and $w \in W$, and we define

$$\iota^w(\varphi) = \|F_\varphi\|_{V \check{\otimes} W}.$$

It follows from (12.3.9) that if $\varphi : V \to W^*$ is an integral mapping, then it is weakly integral with $\iota^w(\varphi) \leq \iota(\varphi)$.

Proposition 15.5.5 *Given operator spaces V and W and a linear mapping $\varphi : V \to W^*$, we have*

$$\iota^{ex}(\varphi) \leq \iota^w(\varphi).$$

Proof Given a mapping $\varphi \in \mathcal{I}^w(V, W^*)$ with $\iota^w(\varphi) \leq 1$, then for any operator space $S \subseteq M_n$ and complete contraction $\psi : S \to V$, the mapping

$$\varphi \circ \psi : S \to W^*$$

corresponds to a contractive linear functional

$$F_\varphi \circ (\psi \otimes id_W) \in (S \check{\otimes} W)^* = S^* \hat{\otimes} W^*.$$

This implies that $\varphi \circ \psi \in \mathcal{N}(S, W^*)$ with $\nu(\varphi \circ \psi) \leq 1$. It follows from Theorem 15.4.1 that $\iota^{ex}(\varphi) \leq 1$. □

Therefore, given operator spaces V and W, we have contractive inclusions

$$\mathcal{I}(V, W^*) \subseteq \mathcal{I}^w(V, W^*) \subseteq \mathcal{I}^{ex}(V, W^*).$$

The isometry between $\mathcal{I}(V, W^*)$ and $\mathcal{I}^w(V, W^*)$

It follows from Corollary 14.2.3 and Theorem 14.3.1 that an operator space W is locally reflexive if and only if we have the isometry

$$\mathcal{I}(V, W^*) = \mathcal{I}^w(V, W^*)$$

for all operator space V. On the other hand, it follows from Corollary 14.2.3 and Theorem 14.4.1 that an operator space V is exact if and only if we have the isometry

$$\mathcal{I}(V, W^*) = \mathcal{I}^w(V, W^*)$$

for all operator space W.

The isometry between $\mathcal{I}^w(V, W^*)$ and $\mathcal{I}^{ex}(V, W^*)$

If $\mathcal{R}$ is an von Neumann algebra and V is an arbitrary operator space, then from Lemma 15.2.2 we have the isometries

$$\mathcal{I}(V, \mathcal{R}) = \mathcal{I}^w(V, \mathcal{R}) = \mathcal{I}^{ex}(V, \mathcal{R}).$$

One cannot replace $\mathcal{R}$ with a general operator space since the corresponding result is false even in the finite-dimensional case.

Proposition 15.5.6 *There exist finite-dimensional operator spaces V and W such that*

$$\mathcal{I}^w(V, W^*) \neq \mathcal{I}^{ex}(V, W^*).$$

Proof Given an operator space V, if for any finite-dimensional operator space W we have the isometry

$$\mathcal{I}^w(V, W^*) = \mathcal{I}^{ex}(V, W^*),$$

then we have the isometries

$$\mathcal{I}(V, W^*) = \mathcal{I}^w(V, W^*) = \mathcal{I}^{ex}(V, W^*).$$

This implies that V is exact by Proposition 15.4.3. Therefore, if V is a finite-dimensional non-exact operator space, then there exists a finite-dimensional operator space W such that

$$\mathcal{I}^w(V, W^*) \neq \mathcal{I}^{ex}(V, W^*).$$

□

15.6 NOTES AND REFERENCES

The material of this chapter appeared in Junge (1996), Effros and Ruan (1997), and Effros *et al.* (2000). Corollary 15.5.4 is due to Junge and Le Merdy (1999). There is additional evidence that the predual of a von Neumann algebra is well-behaved (see Effros and Ruan 1998). It seems possible that one could begin a general structure for von Neumann algebras by studying the manner in which the subspaces of their preduals which are completely isometric to T_n for various $n \in \mathbb{N}$ are interrelated.

Part V

Some Algebraic Applications

16
Non-commutative harmonic analysis

16.1 QUANTIZED BANACH ALGEBRAS

For the convenience of the reader, we shall first recall some basic definitions in Banach algebra theory. A *Banach algebra* is an associative algebra $\mathcal{A}$ (over the complex numbers $\mathbb{C}$) with a Banach space norm $\|\cdot\|$ such that the multiplication

$$m : \mathcal{A} \times \mathcal{A} \to \mathcal{A} : \ m(a, b) = ab$$

is a contractive bilinear mapping, i.e.

$$\|m(a, b)\| \leq \|a\|\|b\|$$

for all $a, b \in \mathcal{A}$. Equivalently, m determines a contraction

$$m : \mathcal{A} \otimes^\gamma \mathcal{A} \to \mathcal{A}.$$

Given a Banach algebra $\mathcal{A}$ and an $\mathcal{A}$-bimodule V, V is called a *Banach $\mathcal{A}$-bimodule* if V is a Banach space and the $\mathcal{A}$-bimodule operations

$$\rho_l : \mathcal{A} \times V \to V : (a, v) \mapsto a \cdot v$$

and

$$\rho_r : V \times \mathcal{A} \to V : (v, a) \mapsto v \cdot a$$

are bounded bilinear mappings. If V is a Banach $\mathcal{A}$-bimodule, then there is a natural Banach $\mathcal{A}$-bimodule structure on the dual space V^* given by

$$\langle a \cdot f, v \rangle = \langle f, v \cdot a \rangle$$

and

$$\langle f \cdot a, v \rangle = \langle f, a \cdot v \rangle$$

for all $a \in \mathcal{A}, f \in V^*$ and $v \in V$.

Let $\mathcal{A}$ be a Banach algebra and V a Banach $\mathcal{A}$-bimodule. A linear mapping $D : \mathcal{A} \to V$ is called a *derivation* if it satisfies

$$D(ab) = D(a) \cdot b + a \cdot D(b)$$

for all $a, b \in \mathcal{A}$. Given $v \in V$, we let $D_v : \mathcal{A} \to V$ denote the linear mapping defined by

$$D(a) = a \cdot v - v \cdot a$$

for all $a \in \mathcal{A}$. It is easy to verify that for every $v \in V$, D_v is a bounded derivation from $\mathcal{A}$ into V, and we call D_v an *inner* derivation. A Banach algebra $\mathcal{A}$ is called *amenable* if for any Banach $\mathcal{A}$-bimodule V, every bounded derivation from $\mathcal{A}$ into V^* is inner. This concept was introduced by B. Johnson in the early 1970s as the natural algebraic analogue of amenability for locally compact groups. One of the major results in the subject states that a locally compact group G is amenable if and only if the convolution algebra $L^1(G)$ is an amenable Banach algebra (see Johnson 1972*a*).

The main purpose of this chapter is to briefly consider the operator analogues of Banach algebras and the corresponding notion of amenability. One of the most important applications of these ideas is to the study of amenability for the Fourier algebras of locally compact groups. As we shall see in §16.2, the Fourier algebra has a natural operator space structure which must be used if one wishes to generalize the theory of convolution algebras.

Let $\mathcal{A}$ be an associative algebra over complex numbers $\mathbb{C}$. We call $\mathcal{A}$ a *completely contractive Banach algebra* if $\mathcal{A}$ is a complete operator space and the multiplication

$$m : \mathcal{A} \times \mathcal{A} \to \mathcal{A}$$

is a completely contractive bilinear mapping, i.e. it determines a completely contractive linear mapping

$$m : \mathcal{A} \hat{\otimes} \mathcal{A} \to \mathcal{A}.$$

Equivalently, we have

$$\|[a_{i,j} b_{k,l}]\| \leq \|a\| \, \|b\|$$

for all $a = [a_{i,j}] \in M_m(\mathcal{A})$, $b = [b_{k,l}] \in M_n(\mathcal{A})$ and arbitrary $m, n \in \mathbb{N}$. It is important to note that although the multiplication on $\mathcal{A}$ determines a corresponding multiplication on $M_n(\mathcal{A})$ in the usual way, it does not follow that $M_n(\mathcal{A})$ is a Banach algebra. We shall see in §17.1 that for unital algebras $\mathcal{A}$, the latter condition implies that $\mathcal{A}$ is an operator algebra.

Let $\mathcal{A}$ be a completely contractive Banach algebra and V be an $\mathcal{A}$-bimodule. Then V is called an *operator $\mathcal{A}$-bimodule* if V is a complete operator space and the left and right $\mathcal{A}$-module operations

$$\rho_l : \mathcal{A} \times V \to V : \ (a, v) \mapsto a \cdot v \tag{16.1.1}$$

and

$$\rho_r : V \times \mathcal{A} \to V : \ (v, a) \mapsto v \cdot a \tag{16.1.2}$$

are completely bounded. Equivalently, we have for some constant C

$$\|[a_{i,j} \cdot v_{k,l}]\| \leq C \, \|a\| \, \|v\|$$

and

$$\|[v_{k,l} \cdot a_{i,j}]\| \leq C \, \|v\| \, \|a\|$$

for all $a = [a_{i,j}] \in M_m(\mathcal{A})$, $v = [v_{k,l}] \in M_n(V)$ and arbitrary $m, n \in \mathbb{N}$.

We let $\mathcal{Z}_{cb}(\mathcal{A}, V)$ denote the space of all completely bounded derivations from $\mathcal{A}$ into V, and we let $\mathcal{I}_{cb}(\mathcal{A}, V)$ denote the space of all inner derivations from $\mathcal{A}$ into V. In this case, we have $\mathcal{I}_{cb}(\mathcal{A}, V) \subseteq \mathcal{Z}_{cb}(\mathcal{A}, V)$ since every inner derivation D_v $(v \in V)$ is automatically completely bounded. The linear space $\mathcal{H}^1_{cb}(\mathcal{A}, V) = \mathcal{Z}_{cb}(\mathcal{A}, V)/\mathcal{I}_{cb}(\mathcal{A}, V)$ is called the *first completely bounded cohomology group* of $\mathcal{A}$ on V.

Let V be an operator $\mathcal{A}$-bimodule. Then (16.1.1) and (16.1.2) determine an $\mathcal{A}$-bimodule structure on the operator space dual V^* of V. These operations are completely bounded since if $a = [a_{i,j}] \in M_m(\mathcal{A})$, $f = [f_{k,l}] \in M_n(V^*)$ and $v = [v_{g,h}] \in M_p(V)$,

$$\left\| [a_{i,j} \cdot f_{k,l}]_p (v) \right\| = |||[\langle a_{i,j} \cdot f_{k,l}, v_{g,h} \rangle]||| = |||[\langle f_{k,l}, v_{g,h} \cdot a_{i,j} \rangle]|||$$
$$\leq C \|f\|_{cb} |||[v_{g,h} \cdot a_{i,j}]||| \leq C \|f\| \|v\| \|a\| .$$

This shows that

$$|||[a_{i,j} \cdot f_{k,l}]||| \leq C \|a\| \|f\| .$$

A similar calculation can be used to prove that

$$|||[f_{k,l} \cdot a_{i,j}]||| \leq C \|a\| \|f\| .$$

We say that V^* is a *dual* operator $\mathcal{A}$-bimodule.

A completely contractive Banach algebra is said to have a *bounded approximate identity* if there exists a net of bounded elements $\{e_\gamma\}$ in $\mathcal{A}$ such that

$$\|e_\gamma a - a\| \to 0 \quad \text{and} \quad \|a e_\gamma - a\| \to 0 \tag{16.1.3}$$

for all $a \in \mathcal{A}$. A completely bounded Banach algebra $\mathcal{A}$ is called *operator amenable* if for every operator $\mathcal{A}$-bimodule V, every completely bounded derivation from $\mathcal{A}$ into the dual operator $\mathcal{A}$-bimodule V^* is inner. Equivalently, we have $\mathcal{H}^1_{cb}(\mathcal{A}, V^*) = 0$ for all dual operator $\mathcal{A}$-bimodules V^*.

To illustrate how operator amenability may be used, we reproduce one of the standard results of classical amenability theory.

Proposition 16.1.1 *If $\mathcal{A}$ is operator amenable, then $\mathcal{A}$ has a bounded approximate identity.*

Proof We first prove that if $\mathcal{A}$ is operator amenable, then $\mathcal{A}$ has a right approximate identity. Since $\mathcal{A}^*$ is an operator $\mathcal{A}$ -bimodule , the completely bounded right $\mathcal{A}$-module operation on $\mathcal{A}^*$ induces a completely bounded left $\mathcal{A}$-module operation on $\mathcal{A}^{**}$. More precisely, the left $\mathcal{A}$-module operation on $\mathcal{A}^{**}$ is given by

$$\langle a \cdot \overline{a}, f \rangle = \langle \overline{a}, f \cdot a \rangle$$

for all $a \in \mathcal{A}$, $f \in \mathcal{A}^*$ and $\overline{a} \in \mathcal{A}^{**}$. On the other hand, we define the right $\mathcal{A}$-module operation on $\mathcal{A}^{**}$ to be the zero map, i.e. $\overline{a} \cdot a = 0$ for all $a \in \mathcal{A}$

and $\overline{a} \in \mathcal{A}^{**}$. It is obvious that $\mathcal{A}^{**}$ with above $\mathcal{A}$-bimodule operations is a dual operator $\mathcal{A}$-bimodule.

The canonical embedding $\iota : \mathcal{A} \hookrightarrow \mathcal{A}^{**}$ is a completely bounded derivation since we have

$$\iota(ab) = a \cdot \iota(b) + \iota(a) \cdot b,$$

where $\iota(a) \cdot b = 0$ for all $a, b \in \mathcal{A}$. If $\mathcal{A}$ is operator amenable, then the completely bounded derivation ι is inner. There exists an element $\overline{e} \in \mathcal{A}^{**}$ such that

$$\iota(a) = a \cdot \overline{e} - \overline{e} \cdot a = a \cdot \overline{e}$$

for all $a \in \mathcal{A}$. This shows that $\overline{e}$ is a right unit of $\mathcal{A}$ in $\mathcal{A}^{**}$. It follows from the Bipolar theorem that there is a net of bounded elements $b_\beta \in \mathcal{A}$ such that $\|b_\beta\| \leq \|\overline{e}\|$ and $b_\beta \to \overline{e}$ in the weak* topology. It follows that for every $a \in \mathcal{A}$, ab_β converges to a in the weak topology in $\mathcal{A}$. The usual convexity argument permits one to replace b_β by a net of elements in the convex hull of $\{b_\beta\}$, which is still denoted by b_β, for which $ab_\beta \to a$ in the norm topology. This shows that $\{b_\beta\}$ is a bounded right approximate identity for $\mathcal{A}$.

Similarly, we can show that $\mathcal{A}$ has a bounded left approximate identity $\{a_\alpha\}$ by considering the zero left $\mathcal{A}$-module operation and the usual right $\mathcal{A}$-module operation on $\mathcal{A}^{**}$. Let

$$e_{(\alpha,\beta)} = a_\alpha + b_\beta - b_\beta a_\alpha.$$

For every $a \in \mathcal{A}$, we have

$$\begin{aligned} \left\|ae_{(\alpha,\beta)} - a\right\| &= \|(ab_\beta - a) + (a - ab_\beta)a_\alpha\| \\ &\leq \|ab_\beta - a\| + \|a - ab_\beta\|\, \|a_\alpha\| \to 0. \end{aligned}$$

Similarly, we can prove that

$$\left\|e_{(\alpha,\beta)}a - a\right\| \to 0.$$

Therefore, $\{e_{(\alpha,\beta)}\}$ is a bounded approximate identity for $\mathcal{A}$. □

Let $\mathcal{A}$ be a completely contractive Banach algebra. Then there is a completely bounded $\mathcal{A}$-bimodule structure on $\mathcal{A} \widehat{\otimes} \mathcal{A}$ given by

$$a \cdot (b \otimes c) = ab \otimes c$$

and

$$(b \otimes c) \cdot a = b \otimes ca$$

for $a \in \mathcal{A}$ and $b \otimes c \in \mathcal{A} \otimes \mathcal{A}$. The operator dual $(\mathcal{A} \widehat{\otimes} \mathcal{A})^*$ and the operator second dual $(\mathcal{A} \widehat{\otimes} \mathcal{A})^{**}$ of $\mathcal{A} \widehat{\otimes} \mathcal{A}$ are dual operator $\mathcal{A}$-bimodules.

We next consider Johnson's notions of virtual and approximate diagonals. These are the Banach algebraic analogues of *means*, and they can be used to 'symmetrize' mappings. In order to motivate this, it is useful to briefly consider the algebraic theory.

Let us suppose that $\mathcal{A}$ is a unital algebra over the complex numbers $\mathbb{C}$. In this algebraic context, we consider only left $\mathcal{A}$-modules which are unital, i.e. which satisfy $1x = x$. The same argument can be applied to right $\mathcal{A}$-modules (respectively, $\mathcal{A}$-bimodules). A left $\mathcal{A}$-module V is said to be *semisimple* if each submodule W of V is a direct summand of V, i.e. there is a submodule X of V such that $V = W + X$ and $W \cap X = \{0\}$. The following is a version of Wedderburn's theorem (see Theorem I.4.2 in Cartan and Eilenberg 1956). We note that it is unnecessary to initially assume that the algebra is finite-dimensional.

Theorem 16.1.2 *Let us suppose that $\mathcal{A}$ is a unital algebra over the complex numbers $\mathbb{C}$. The following are equivalent:*

(i) *any left unital $\mathcal{A}$-module V is semisimple;*
(ii) *$\mathcal{A}$ is finite-dimensional and there exist $n_1, \ldots, n_r \in \mathbb{N}$ such that*

$$\mathcal{A} \cong M_{n_1} \oplus \cdots \oplus M_{n_r}.$$

□

Given a unital algebra $\mathcal{A}$, we can regard $\mathcal{A} \otimes \mathcal{A}$ as a bimodule over $\mathcal{A}$ by letting

$$c\Big(\sum a_i \otimes b_i\Big)d = \sum ca_i \otimes b_i d.$$

Multiplication determines a linear mapping

$$m : \mathcal{A} \otimes \mathcal{A} \to \mathcal{A};\ m(a \otimes b) = ab.$$

We say that $U \in \mathcal{A} \otimes \mathcal{A}$ is a *diagonal* for $\mathcal{A}$ if $cU = Uc$ for all $c \in \mathcal{A}$, and $m(U) = 1$. If, for example, $\mathcal{A} = M_n$, then

$$U = \sum \varepsilon_{i,1} \otimes \varepsilon_{1,i}$$

is a diagonal since

$$m(U) = \sum \varepsilon_{i,i} = 1$$

and for any j, k with $1 \leq j, k \leq n$,

$$\varepsilon_{j,k} U = \varepsilon_{j,1} \otimes \varepsilon_{1,k} = U \varepsilon_{j,k}.$$

If $U = \sum a_i \otimes b_i$ is a diagonal, then for any $c \in \mathcal{A}$, we have

$$\sum ca_i \otimes b_i = \sum a_i \otimes b_i c.$$

Then for arbitrary linear space V, any bilinear mapping $F : \mathcal{A} \times \mathcal{A} \to V$ satisfies

$$\sum F(ca_i, b_i) = \sum F(a_i, b_i c)$$

for all $c \in \mathcal{A}$. The reader may find that this ability to 'move' the variable c in such a sum is reminiscent of what occurs in integrals with respect to invariant measures or means.

Theorem 16.1.3 *If $\mathcal{A}$ is a unital algebra over complex numbers $\mathbb{C}$, then $\mathcal{A}$ has a diagonal if and only if there exist $n_1, \ldots, n_r \in \mathbb{N}$ such that*

$$\mathcal{A} \cong M_{n_1} \oplus \cdots \oplus M_{n_r}.$$

Proof Let us assume that $\mathcal{A}$ has a diagonal

$$U = \sum_{i=1}^{n} a_i \otimes b_i \in \mathcal{A} \otimes \mathcal{A}.$$

Given a left $\mathcal{A}$-module V and a submodule W, we let φ be an arbitrary linear projection of V onto W. To see that such a mapping exists, we fix an algebraic (Hamel) basis $\{e_\gamma\}_{\gamma \in I}$ for V for which there is a subset $I_0 \subseteq I$ with $\{e_\gamma\}_{\gamma \in I_0}$ a basis for W. Each element $v \in V$ has a representation

$$v = \sum_{\gamma \in I} c_\gamma e_\gamma,$$

where only finitely many c_γ are non-zero. We define a linear mapping $\varphi : V \to W$ by letting

$$\varphi(v) = \sum_{\gamma \in I_0} c_\gamma e_\gamma.$$

It is evident that $\varphi^2 = \varphi$ and that φ maps V onto W. We define a linear mapping $\tilde{\varphi} : V \to W$ by

$$\tilde{\varphi}(v) = \sum a_i \varphi(b_i v).$$

Since for each $v \in V$,

$$F_v : \mathcal{A} \times \mathcal{A} \to W : F_v(a, b) = a\varphi(bv)$$

is a bilinear mapping, we have

$$c\tilde{\varphi}(v) = \sum c a_i \varphi(b_i v) = \sum a_i \varphi(b_i c v) = \tilde{\varphi}(cv).$$

This shows that $\tilde{\varphi}$ is a left $\mathcal{A}$-module morphism, and thus $\ker \tilde{\varphi}$ is a left $\mathcal{A}$-submodule of V. Since $b_i w \in W$ for any $w \in W$, we have

$$\tilde{\varphi}(w) = \sum a_i \varphi(b_i w) = \sum a_i b_i w = w,$$

and $\tilde{\varphi}$ is a left $\mathcal{A}$-module projection of V onto W. It is clear that

$$V = W + \ker \tilde{\varphi} \text{ and } W \cap \ker \tilde{\varphi} = \{0\}.$$

Therefore, $\ker \tilde{\varphi}$ is a left $\mathcal{A}$-module complement of W, and V is semisimple. We conclude from Theorem 16.1.2 that $\mathcal{A} \cong M_{n_1} \oplus \cdots \oplus M_{n_r}$.

Conversely, let us assume that

$$\mathcal{A} = M_{n_1} \oplus \cdots \oplus M_{n_r}.$$

We may choose a diagonal

$$U_k \in M_{n_k} \otimes M_{n_k}$$

for each k (see above), and then let

$$U = \sum_k U_k \in \sum (M_{n_k} \otimes M_{n_k}) \subseteq \mathcal{A} \otimes \mathcal{A}.$$

It is easy to see that U is a diagonal in $\mathcal{A} \otimes \mathcal{A}$. □

From the preceding discussion, it is apparent that we cannot expect a Banach algebra to have a diagonal. Nevertheless, Johnson showed that the amenable Banach algebras can be characterized by the fact that they have 'asymptotic' versions of diagonals (see Johnson 1972*b*). In the following, we study the operator analogue of Johnson's results.

A net of bounded elements $\{x_\alpha\}$ in $\mathcal{A} \widehat{\otimes} \mathcal{A}$ is called a *bounded approximate diagonal* if for every $a \in \mathcal{A}$ it satisfies

- **AD1** $\|a \cdot x_\alpha - x_\alpha \cdot a\|_{\mathcal{A} \widehat{\otimes} \mathcal{A}} \to 0$;
- **AD2** $\|m(x_\alpha)a - a\| \to 0$.

An element $U \in (\mathcal{A} \widehat{\otimes} \mathcal{A})^{**}$ is called a *virtual diagonal* if for every $a \in \mathcal{A}$,

- **VD1** $a \cdot U = U \cdot a$;
- **VD2** $m^{**}(U) \cdot a = a$.

The argument for the following result is just a rewording of Johnson's classical argument.

Theorem 16.1.4 *Let $\mathcal{A}$ be a completely contractive Banach algebra. Then the following are equivalent:*

(i) *$\mathcal{A}$ is operator amenable;*
(ii) *$\mathcal{A}$ has a virtual diagonal;*
(iii) *$\mathcal{A}$ has a bounded approximate diagonal.*

Proof (i)⇒(ii) Let $m : \mathcal{A} \widehat{\otimes} \mathcal{A} \to \mathcal{A}$ be the completely contractive multiplication of $\mathcal{A}$. Then the second dual $m^{**} : (\mathcal{A} \widehat{\otimes} \mathcal{A})^{**} \to \mathcal{A}^{**}$ is completely contractive and weak* continuous, and thus the kernel $\ker m^{**}$ is a weak* closed subspace of $(\mathcal{A} \widehat{\otimes} \mathcal{A})^{**}$. Moreover, since $(\mathcal{A} \widehat{\otimes} \mathcal{A})^{**}$ is an operator $\mathcal{A}$-bimodule, and

$$m^{**}(a \cdot \tilde{a}) = a \cdot m^{**}(\tilde{a}) \text{ and } m^{**}(\tilde{a} \cdot a) = m^{**}(\tilde{a}) \cdot a$$

for all $a \in \mathcal{A}$ and $\tilde{a} \in (\mathcal{A} \widehat{\otimes} \mathcal{A})^{**}$, $\ker m^{**}$ is a weak* closed $\mathcal{A}$-subbimodule of $(\mathcal{A} \widehat{\otimes} \mathcal{A})^{**}$. Hence, $\ker m^{**}$ is a dual operator $\mathcal{A}$-bimodule.

Since $\mathcal{A}$ is operator amenable, $\mathcal{A}$ admits a bounded approximate identity $\{e_\gamma\}$. Then $\{e_\gamma \otimes e_\gamma\}$ is a bounded approximate identity for $\mathcal{A} \widehat{\otimes} \mathcal{A}$, and there exists a subnet of $\{e_\gamma \otimes e_\gamma\}$ converging to an element $\tilde{e} \in (\mathcal{A} \widehat{\otimes} \mathcal{A})^{**}$ in the weak* topology. Without loss of generality, we may assume that $\{e_\gamma \otimes e_\gamma\}$ converges to $\tilde{e} \in (\mathcal{A} \widehat{\otimes} \mathcal{A})^{**}$ in the weak* topology. We let

$$D_{\tilde{e}} : \mathcal{A} \to (\mathcal{A} \widehat{\otimes} \mathcal{A})^{**}$$

denote the completely bounded inner derivation defined by

$$D_{\tilde{e}}(a) = a \cdot \tilde{e} - \tilde{e} \cdot a.$$

Since $e_\gamma^2 = m(e_\gamma \otimes e_\gamma)$ is also a bounded approximate identity for $\mathcal{A}$,

$$m^{**}(a \cdot \tilde{e}) = a \cdot m^{**}(\tilde{e}) = a = m^{**}(\tilde{e}) \cdot a = m^{**}(\tilde{e} \cdot a)$$

for all $a \in \mathcal{A}$. It follows that for every $a \in \mathcal{A}$,

$$m^{**}(D_{\tilde{e}}(a)) = m^{**}(a \cdot \tilde{e} - \tilde{e} \cdot a) = a \cdot m^{**}(\tilde{e}) - m^{**}(\tilde{e}) \cdot a = 0.$$

This shows that $D_{\tilde{e}}$ defines a completely bounded derivation from $\mathcal{A}$ into the dual operator $\mathcal{A}$-bimodule $\ker m^{**}$. Since $\mathcal{A}$ is operator amenable, there exists an element $\tilde{b} \in \ker m^{**}$ such that

$$D_{\tilde{e}} = D_{\tilde{b}}.$$

If we let $U = \tilde{e} - \tilde{b} \in (\mathcal{A} \hat{\otimes} \mathcal{A})^{**}$, we have

$$a \cdot U - U \cdot a = D_{\tilde{e}}(a) - D_{\tilde{b}}(a) = 0$$

and

$$m^{**}(U) \cdot a = m^{**}(\tilde{e}) \cdot a - m^{**}(\tilde{b}) \cdot a = a$$

for all $a \in \mathcal{A}$. Therefore, U is a virtual diagonal for $\mathcal{A}$.

(ii)$\Rightarrow$(iii) Assume that $\mathcal{A}$ has a virtual diagonal $U \in (\mathcal{A} \hat{\otimes} \mathcal{A})^{**}$. There exists a net of bounded elements $\{u_\alpha\}$ in $\mathcal{A} \hat{\otimes} \mathcal{A}$ (with $\|u_\alpha\| \le \|U\|$) such that $u_\alpha \to U$ in the weak* topology. Then for every $a \in \mathcal{A}$, $\{a \cdot u_\alpha - u_\alpha \cdot a\}$ converges to 0 in the weak topology in $\mathcal{A} \hat{\otimes} \mathcal{A}$, and $\{m(u_\alpha)a\}$ converges to a in the weak topology in $\mathcal{A}$.

Let $\mathcal{F}$ be the set of all finite subsets of $\mathcal{A}$. For every

$$F = \{a_1, \ldots, a_n\} \in \mathcal{F}$$

and $\varepsilon > 0$,

$$\{(a_1 \cdot u_\alpha - u_\alpha \cdot a_1, \ldots, a_n \cdot u_\alpha - u_\alpha \cdot a_n, m(u_\alpha)a_1, \ldots, m(u_\alpha)a_n)\}$$

is a bounded net of elements in $(\mathcal{A} \hat{\otimes} \mathcal{A})^n \oplus \mathcal{A}^n$ converging to

$$(0, \ldots, 0, a, \ldots, a)$$

in the weak topology. Hence, there exists a convex combination $u_{(F,\varepsilon)}$ of $\{u_\alpha\}$ such that

$$\|a_j \cdot u_{(F,\varepsilon)} - u_{(F,\varepsilon)} \cdot a_j\| < \varepsilon \text{ and } \|m(u_{(F,\varepsilon)})a_j - a_j\| < \varepsilon$$

for $j = 1, \ldots, n$. It follows that the net $\{u_{(F,\varepsilon)}\}$ forms a bounded approximate diagonal for $\mathcal{A}$.

(iii)$\Rightarrow$(ii) Let us suppose that $\mathcal{A}$ has a bounded approximate diagonal $\{u_\alpha\}$. Then we may find a weak* limit U of a subnet of $\{u_\alpha\}$ in $(\mathcal{A} \hat{\otimes} \mathcal{A})^{**}$. It is easy to verify that U is a virtual diagonal of $\mathcal{A}$.

(ii)⇒(i) Assume that U is a virtual diagonal in $(\mathcal{A}\widehat{\otimes}\mathcal{A})^{**}$, and we let $\{u_\alpha\}$ be a net of bounded elements in $\mathcal{A}\widehat{\otimes}\mathcal{A}$ converging to U in the weak* topology and $\{m(u_\alpha)\}$ is a bounded approximate identity for $\mathcal{A}$. Given any operator $\mathcal{A}$-bimodule V, it suffices to show that every completely bounded derivation $D : \mathcal{A} \to \tilde{V}^*$ is inner, where

$$\tilde{V} = \{\tilde{v} = a \cdot v \cdot b : \ a, b \in \mathcal{A}, v \in V\}.$$

Given $\tilde{v} \in \tilde{V}$, we may define a linear functional $F_{\tilde{v}} \in (\mathcal{A}\widehat{\otimes}\mathcal{A})^*$ by letting

$$F_{\tilde{v}}(a \otimes b) = \langle \tilde{v}, aD(b)\rangle$$

for all $a, b \in \mathcal{A}$. It is easy to see that $\|F_{\tilde{v}}\| \leq \|\tilde{v}\|\|D\|_{cb}$. Let $f \in \tilde{V}^*$ be defined by

$$\langle f, \tilde{v}\rangle = \langle F_{\tilde{v}}, U\rangle.$$

We claim that $D = D_f$ is an inner derivation.

To see this, for any $a, b, c \in \mathcal{A}$ and $\tilde{v} \in \tilde{V}$, we have

$$\begin{aligned}
&\langle b \otimes c, \ F_{\tilde{v}} \cdot a - a \cdot F_{\tilde{v}}\rangle + \langle \tilde{v}, \ (bc) \cdot D(a)\rangle \\
&\quad = \langle a \cdot (b \otimes c), \ F_{\tilde{v}}\rangle - \langle (b \otimes c) \cdot a, \ F_{\tilde{v}}\rangle + \langle \tilde{v}, \ (bc) \cdot D(a)\rangle \\
&\quad = \langle ab \otimes c, \ F_{\tilde{v}}\rangle - \langle b \otimes ca, \ F_{\tilde{v}}\rangle + \langle \tilde{v}, \ (bc) \cdot D(a)\rangle \\
&\quad = \langle \tilde{v}, \ (ab) \cdot D(c) - b \cdot D(ca)\rangle + \langle \ \tilde{v}, \ (bc) \cdot D(a)\rangle \\
&\quad = \langle \tilde{v}, \ (ab) \cdot D(c) - b \cdot D(c) \cdot a\rangle \\
&\quad = \langle \tilde{v} \cdot a - a \cdot \tilde{v}, \ b \cdot D(c)\rangle \\
&\quad = \langle b \otimes c, \ F_{\tilde{v} \cdot a - a \cdot \tilde{v}}\rangle.
\end{aligned}$$

Therefore, we have

$$\begin{aligned}
\langle \tilde{v}, a \cdot f - f \cdot a\rangle &= \langle \tilde{v} \cdot a - a \cdot \tilde{v}, f\rangle = \langle F_{\tilde{v} \cdot a - a \cdot \tilde{v}}, U\rangle \\
&= \langle F_{\tilde{v}} \cdot a - a \cdot F_{\tilde{v}}, U\rangle + \lim\langle \ \tilde{v}, m(u_\alpha) \cdot D(a)\rangle \\
&= \langle F_{\tilde{v}}, a \cdot U - U \cdot a\rangle + \lim\langle \tilde{v}, m(u_\alpha) \cdot D(a)\rangle \\
&= \lim\langle \tilde{v} \cdot m(u_\alpha), D(a)\rangle.
\end{aligned}$$

Since every element $\tilde{v} \in \tilde{V}$ can be written as $a \cdot v \cdot b$ for some $a, b \in \mathcal{A}$ and $v \in V$, we infer that

$$\lim \tilde{v} \cdot m(u_\alpha) = \lim a \cdot v \cdot (bm(u_\alpha)) = \tilde{v},$$

and thus

$$D(a) = a \cdot f - f \cdot a$$

for all $a \in \mathcal{A}$. Therefore, $D = D_f$ is inner. □

If $\mathcal{A}$ is a completely contractive Banach algebra, then it is trivially a Banach algebra, and each completely bounded $\mathcal{A}$-bimodule is a bounded $\mathcal{A}$-bimodule. If $\mathcal{A}$ is amenable as a Banach algebra, then it is immediate that $\mathcal{A}$ is operator amenable.

Conversely, if $\mathcal{A}$ is a Banach algebra, then the multiplication mapping $\mathcal{A} \otimes^\gamma \mathcal{A} \to \mathcal{A}$ induces a complete contraction

$$\max \mathcal{A} \widehat{\otimes} \max \mathcal{A} \cong \max(\mathcal{A} \otimes^\gamma \mathcal{A}) \to \max \mathcal{A},$$

and thus $\max \mathcal{A}$ is a completely contractive Banach algebra. Since for any operator $\max \mathcal{A}$-bimodule V, every bounded derivation $D : \max \mathcal{A} \to V^*$ is completely bounded with $\|D\|_{cb} = \|D\|$, we may easily obtain the following result.

Proposition 16.1.5 *A Banach algebra $\mathcal{A}$ is amenable if and only if the completely contractive Banach algebra* $\max \mathcal{A}$ *is operator amenable.*

If G is a locally compact group, then the convolution Banach algebra $L^1(G)$ is the predual of the operator space $L^\infty(G)$. The corresponding matrix norm coincides with $\max L^1(G)$. It follows that the amenability result of Jonhson and Ringrose may be regarded as a contractive Banach algebra result.

Theorem 16.1.6 *Let G be a locally compact group. Then the following are equivalent:*

(i) *G is an amenable group;*
(ii) *$L^1(G)$ is an amenable Banach algebra;*
(iii) $\max L^1(G)$ *is an operator amenable completely contractive Banach algebra.*

16.2 OPERATOR AMENABILITY FOR FOURIER ALGEBRAS

Let G be a locally compact group with left Haar measure μ and modular function $\Delta : G \to \mathbb{R}^+$. For any $s, t \in G$, we have

$$\mathrm{d}\mu(st) = \mathrm{d}\mu(t), \quad \mathrm{d}\mu(ts) = \Delta(s)\mathrm{d}\mu(t) \quad \text{and} \quad \mathrm{d}\mu(t^{-1}) = \Delta(t)^{-1}\mathrm{d}\mu(t).$$

The left regular representation $\lambda : G \to B(L^2(G))$ is defined by

$$\lambda(s)\xi(t) = \xi(s^{-1}t),$$

for all $\xi \in L^2(G)$ and $s, t \in G$.

The regular group von Neumann algebra $VN(G) = \lambda(G)''$ has a *comultiplication*

$$\Gamma : VN(G) \to VN(G) \overline{\otimes} VN(G) = VN(G \times G)$$

given by

$$\Gamma(\lambda(s)) = \lambda(s) \otimes \lambda(s),$$

which satisfies

$$(\Gamma \otimes id)\Gamma = (id \otimes \Gamma)\Gamma.$$

The unitary operator W on $L^2(G \times G)$ defined by

$$W\psi(s,t) = \psi(s, st) \quad (\psi \in L^2(G \times G)) \tag{16.2.1}$$

plays a particularly important role in this theory since we have

$$\Gamma(\lambda(s)) = W^*(\lambda(s) \otimes 1)W. \tag{16.2.2}$$

The *Fourier algebra* $A(G)$ consists of all coefficient functions of the left regular representation λ, i.e. $A(G)$ consists of the functions $s \mapsto \langle \lambda(s)\eta \mid \xi \rangle$ with $\eta, \xi \in L^2(G)$. Eymard proved that $A(G)$ is a commutative Banach algebra with the pointwise multiplication m and the norm

$$\|\omega\| = \inf\{\|\xi\| \, \|\eta\|\},$$

where the infimum is taken over all possible representations

$$\omega(s) = \langle \lambda(s)\xi \mid \eta \rangle.$$

In fact, we may identify $A(G)$ with the operator predual $VN(G)_*$ of $VN(G)$ by identifying $\omega \in A(G)$ with the corresponding function $\omega(s) = \omega(\lambda(s))$ for $s \in G$. Thus, $A(G)$ has a natural operator space structure. With this identification, we have a commutative diagram

$$\begin{array}{ccc} VN(G)_* \widehat{\otimes} VN(G)_* & \xrightarrow{\Gamma_*} & VN(G)_* \\ \| & & \| \\ A(G) \widehat{\otimes} A(G) & \xrightarrow{m} & A(G) \end{array},$$

and thus $A(G)$ is a completely contractive Banach algebra.

If G is an abelian group and $\widehat{G}$ is its dual group, it is known that $VN(G)$ is $*$-isomorphic to $L^\infty(\widehat{G})$ and thus $A(G)$ is completely isometrically isomorphic to $L^1(\widehat{G})$. Therefore, for general non-abelian group G, we may regard $VN(G)$ as the non-commutative analogue of $L^\infty(\widehat{G})$ and $A(G)$ as the non-commutative analogue of $L^1(\widehat{G})$. In contrast to the situation for convolution algebra $L^1(G)$, the Fourier algebra $A(G)$ of an amenable locally compact group G need not be Banach algebra amenable. Johnson (1994) showed that for the compact Lie group $G = SU(2, \mathbb{C})$, $A(G)$ is not an amenable Banach algebra. The situation is more satisfactory in the operator space setting.

Theorem 16.2.1 *A locally compact group G is amenable if and only if its Fourier algebra $A(G)$ is operator amenable.*

This result provides a convincing argument for the importance of operator space techniques in non-commutative harmonic analysis. In order to simplify the argument, we shall confine our attention to the elementary argument for compact groups. Thus, we shall show that if G is a compact group, then $A(G)$ is operator amenable. From Theorem 16.1.4, it suffices to prove that there is an approximate diagonal in $A(G) \widehat{\otimes} A(G)$.

In the rest of this section, we let G be a compact group and let μ be the left Haar measure on G for which $\mu(G) = 1$. In this case, the modular function is the constant function 1 on G, and thus μ is also a right invariant measure. We may define the right regular representation $\rho : G \to \mathcal{B}(L^2(G))$ by letting

$$\rho(s)\xi(t) = \xi(ts)$$

for all $\xi \in L^2(G)$ and $s, t \in G$, and define a self-adjoint unitary operator V on $L^2(G)$ by letting

$$V\xi(s) = \xi(s^{-1})$$

for $\xi \in L^2(G)$. It is easy to see that V satisfies

$$\rho(s) = V^*\lambda(s)V = AdV(\lambda(s))$$

for all $s \in G$.

Lemma 16.2.2 *If G is a compact group, then there exists a net of unit vectors $\{\xi_\alpha\}_{\alpha\in I}$ in $L^2(G)$ satisfying*

(i) $\lambda(s)\rho(s)\xi_\alpha = \xi_\alpha$ *for all* $s \in G$ *and* $\alpha \in I$;
(ii) $\|W(\xi_\alpha \otimes \eta) - (\xi_\alpha \otimes \eta)\| \to 0$ *for all* $\eta \in L^2(G)$.

Proof Since G is a compact group, there exists a base of open neighbourhoods $\{U_\alpha\}_{\alpha\in I}$ at e, the unit element of the group G, with compact closures such that

$$sU_\alpha s^{-1} = U_\alpha$$

for all $s \in G$ (see Hewitt and Ross 1963). It is clear that there is a natural partial order on the index set I determined by

$$\alpha \prec \alpha' \quad \text{if and only if} \quad U_{\alpha'} \subseteq U_\alpha.$$

For each $\alpha \in I$, we let $\xi_\alpha = \chi_{U_\alpha}/\mu(U_\alpha)^{1/2}$, where χ_{U_α} is the characteristic function on U_α. Then $\{\xi_\alpha\}_{\alpha\in I}$ is a net of unit vectors in $L^2(G)$ such that $\xi_\alpha \geq 0$ and

$$\lambda(s)\rho(s)\xi_\alpha(t) = \xi_\alpha(s^{-1}ts) = \xi_\alpha(t)$$

for all $s, t \in G$. This shows that $\lambda(s)\rho(s)\xi_\alpha = \xi_\alpha$ for all $s \in G$ and $\alpha \in I$.

Given any $\eta \in L^2(G)$, since $\{U_\alpha\}_{\alpha\in I}$ is a base of open neighbourhoods at e, it follows from Hewitt and Ross (1963, §20.4) that for any $\varepsilon > 0$, there exists an U_{α_0} such that

$$\int_G |\eta(st) - \eta(t)|^2 \,\mathrm{d}\mu(t) < \varepsilon$$

for all $s \in U_{\alpha_0}$. Therefore, for any $\alpha \in I$ with $\alpha_0 \prec \alpha$, i.e. for which $U_\alpha \subseteq U_{\alpha_0}$,

$$\begin{aligned}\|W(\xi_\alpha \otimes \eta) - (\xi_\alpha \otimes \eta)\|^2 &= \int_G\int_G |W(\xi_\alpha \otimes \eta)(s,t) \\ &\qquad - (\xi_\alpha \otimes \eta)(s,t)|^2 \,\mathrm{d}\mu(t)\mathrm{d}\mu(s) \\ &= \int_G\int_G |\xi_\alpha(s)\eta(st) - \xi_\alpha(s)\eta(t)|^2 \,\mathrm{d}\mu(t)\mathrm{d}\mu(s)\end{aligned}$$

$$= \int_{U_\alpha} |\xi_\alpha(s)|^2 \left(\int_G |\eta(st) - \eta(t)|^2 \,\mathrm{d}\mu(t) \right) \mathrm{d}\mu(s)$$
$$< \varepsilon \int_{U_\alpha} |\xi_\alpha(s)|^2 \,\mathrm{d}\mu(s) \le \varepsilon.$$

This shows that

$$\|W(\xi_\alpha \otimes \eta) - (\xi_\alpha \otimes \eta)\| \to 0.$$

□

A vector $\xi \in L^2(G)$ satisfying condition (i) in Lemma 16.2.2 is called a *central vector* in $L^2(G)$. Since

$$\lambda(s)\rho(t) = \rho(t)\lambda(s)$$

for all $s, t \in G$, the mapping

$$(s,t) \in G \times G \to \lambda(s)\rho(t) \in \mathcal{B}(L^2(G))$$

determines a (strongly) continuous unitary representation of $G \times G$. Therefore, for every unit vector $\xi \in L^2(G)$,

$$M_\xi(s,t) = \langle \lambda(s)\rho(t)\xi \mid \xi \rangle$$

is a contractive and positive definite element in the *Fourier–Stieltjes algebra* $B(G \times G) \cong C^*(G \times G)^*$. Since $G \times G$ is a compact group, we have

$$B(G \times G) = A(G \times G) = A(G)\widehat{\otimes}A(G).$$

Then M_ξ is an contractive and positive definite element contained in $A(G)\widehat{\otimes}A(G)$.

Theorem 16.2.3 *If G is a compact group and $\{\xi_\alpha\}_{\alpha \in I}$ is the net of central unit vectors obtained in Lemma* 16.2.2, *then $\{M_{\xi_\alpha}\}_{\alpha \in I} \subseteq A(G) \widehat{\otimes} A(G)$ is a contractive and positive definite approximate diagonal for $A(G)$. Therefore, $A(G)$ is operator amenable.*

Proof Given a net of central unit vectors $\{\xi_\alpha\}_{\alpha \in I} \in L^2(G)$ as in Lemma 16.2.2, M_{ξ_α} are contractive and positive definite elements in $A(G) \widehat{\otimes} A(G)$. To show that $\{M_{\xi_\alpha}\}_{\alpha \in I}$ is an approximate diagonal for $A(G)$, we need to verify the conditions **AD1** and **AD2**.

Since ξ_α are unit central vectors in $L^2(G)$ satisfying

$$\lambda(s)\rho(s)\xi_\alpha = \xi_\alpha,$$

for any $\eta \in L^2(G)$, we have

$$\begin{aligned} m(M_{\xi_\alpha})\omega_\eta(s) &= M_{\xi_\alpha}(s,s)\omega_\eta(s) \\ &= \langle \lambda(s)\rho(s)\xi_\alpha \mid \xi_\alpha \rangle \langle \lambda(s)\eta \mid \eta \rangle \\ &= \langle \lambda(s)\eta \mid \eta \rangle = \omega_\eta(s) \end{aligned}$$

for all $s \in G$. This shows that

$$m(M_{\xi_\alpha})\omega_\eta = \omega_\eta,$$

and thus $m(M_{\xi_\alpha}) = 1$ in $A(G)$. The condition **AD2** is satisfied.

For any $\eta \in L^2(G)$ and $s, t \in G$, we have

$$\begin{aligned}\omega_\eta \cdot M_{\xi_\alpha}(s,t) &= \langle \lambda(s)\eta \mid \eta\rangle\langle \lambda(s)\rho(t)\xi_\alpha \mid \xi_\alpha\rangle \\ &= \langle (\lambda(s)\rho(t) \otimes \lambda(s))(\xi_\alpha \otimes \eta) \mid (\xi_\alpha \otimes \eta)\rangle.\end{aligned}$$

On the other hand, we have

$$(V^* \otimes I)W^* = W(V^* \otimes I),$$

since for any $\psi \in L^2(G \times G)$,

$$(V^* \otimes I)W^*\psi(s,t) = W^*\psi(s^{-1},t) = \psi(s^{-1}, st)$$

and

$$W(V^* \otimes I)\psi(s,t) = (V^* \otimes I)\psi(s, st) = \psi(s^{-1}, st).$$

It follows that

$$\begin{aligned}M_{\xi_\alpha} \cdot \omega_\eta\,(s,t) &= \langle \lambda(s)\rho(t)\xi_\alpha \mid \xi_\alpha\rangle\langle \lambda(t)\eta \mid \eta\rangle \\ &= \langle (\lambda(s) \otimes I)(AdV(\lambda(t)) \otimes \lambda(t))(\xi_\alpha \otimes \eta) \mid (\xi_\alpha \otimes \eta)\rangle \\ &= \langle (\lambda(s) \otimes I)(V^* \otimes I)W^*(\lambda(t) \otimes I)W(V \otimes I)(\xi_\alpha \otimes \eta) \mid (\xi_\alpha \otimes \eta)\rangle \\ &= \langle W^*(\lambda(s) \otimes I)W(V^* \otimes I) \\ &\qquad (\lambda(t) \otimes I)(V \otimes I)W^*(\xi_\alpha \otimes \eta) \mid W^*(\xi_\alpha \otimes \eta)\rangle \\ &= \langle (\lambda(s) \otimes \lambda(s))(AdV(\lambda(t)) \otimes I)W^*(\xi_\alpha \otimes \eta) \mid W^*(\xi_\alpha \otimes \eta)\rangle \\ &= \langle (\lambda(s)\rho(t) \otimes \lambda(s))W^*(\xi_\alpha \otimes \eta) \mid W^*(\xi_\alpha \otimes \eta)\rangle.\end{aligned}$$

Therefore,

$$\begin{aligned}&\omega_\eta \cdot M_{\xi_\alpha}(s,t) - M_{\xi_\alpha} \cdot \omega_\eta(s,t) \\ &\quad = \langle (\lambda(s)\rho(t) \otimes \lambda(s))(\xi_\alpha \otimes \eta) \mid (\xi_\alpha \otimes \eta)\rangle \\ &\qquad - \langle (\lambda(s)\rho(t) \otimes \lambda(s))W^*(\xi_\alpha \otimes \eta) \mid W^*(\xi_\alpha \otimes \eta)\rangle \\ &\quad = \langle (\lambda(s)\rho(t) \otimes \lambda(s))[(\xi_\alpha \otimes \eta) - W^*(\xi_\alpha \otimes \eta)] \mid (\xi_\alpha \otimes \eta)\rangle \\ &\qquad + \langle (\lambda(s)\rho(t) \otimes \lambda(s))W^*(\xi_\alpha \otimes \eta) \mid [(\xi_\alpha \otimes \eta) - W^*(\xi_\alpha \otimes \eta)]\rangle.\end{aligned}$$

Since

$$(s,t) \in G \times G \mapsto \lambda(s)\rho(t) \otimes \lambda(s) \in \mathcal{B}(L^2(G \times G))$$

is a (strongly) continuous unitary representation from $G \times G$ on $L^2(G \times G)$, the functionals $\omega_\eta \cdot M_{\xi_\alpha} - M_{\xi_\alpha} \cdot \omega_\eta$ are contained in $B(G \times G) = A(G)\widehat{\otimes}A(G)$ and we infer from the inequality

$$\|\omega_\eta \cdot M_{\xi_\alpha} - M_{\xi_\alpha} \cdot \omega_\eta\|_{A(G)\widehat{\otimes}A(G)} \leq 2\|(\xi_\alpha \otimes \eta) - W^*(\xi_\alpha \otimes \eta)\|\,\|\eta\|$$

that the norms on the left converge to 0. Thus, $\{M_{\xi_\alpha}\}$ is an approximate diagonal in $A(G)\widehat{\otimes}A(G)$. □

16.3 NOTES AND REFERENCES

The monograph of Patterson (1988) remains the best source for results and references on the amenability of groups and Banach algebras. Johnson's crucial observation that the Fourier algebra of a compact group does not have to be amenable in the Banach space sense may be found in Johnson (1994). The results of §16.2 were proved in Ruan (1995). This theory has been generalized, in part, to Kac algebras in Ruan (1996).

17

An abstract characterization for non-self-adjoint operator algebras

17.1 QUANTIZED FUNCTION ALGEBRAS

The function algebras provide a particularly well-behaved class of Banach algebras. By definition, a *concrete function algebra* $\mathcal{A}$ is a closed unital subalgebra of $C(\Omega)$ for some compact space Ω , and a Banach algebra is said to be a *function algebra* if it is isometrically isomorphic to such an algebra. If we let $\mathbb{D}$ denote the closed unit disk in $\mathbb{C}$, then the algebra

$$\mathcal{A} = \{f \in C(\mathbb{D}) : f_{|\mathrm{int}\mathbb{D}} \text{ is analytic}\}$$

may be regarded as the pivotal example of a function algebra. In fact, function algebras provide a key functional analytic technique for studying holomorphic functions.

By analogy, we define a *concrete operator algebra* to be a closed subalgebra of $\mathcal{B}(H)$ for some Hilbert space H (we shall not insist that the algebra is unital) and a completely contractive Banach algebra is said to be an *operator algebra* if it is completely isometrically isomorphic to a concrete operator algebra. Perhaps the simplest examples of such algebras are the upper triangular algebras $\mathcal{A}_n = \{[a_{i,j}] \in M_n : a_{i,j} = 0 \text{ for } j < i\}$.

Of course, the C^*-algebras are themselves operator algebras. The theory of C^*-algebra began with Gelfand and Naimark's abstract characterization of these $*$-algebras (see §A.4). In this section we shall show that there is a very simple abstract characterization for the unital operator algebras.

If $\mathcal{A}$ is an associative algebra, then it is easy to see that the same is true for $M_n(\mathcal{A})$ with the multiplication

$$ab = \left[\sum_k a_{i,k} b_{k,j}\right] \tag{17.1.1}$$

for $a = [a_{i,j}], b = [b_{j,k}] \in M_n(\mathcal{A})$. Furthermore, if $\mathcal{A}$ has a multiplicative unit $1_{\mathcal{A}}$, then

$$I_n \otimes 1_{\mathcal{A}} = \begin{bmatrix} 1_{\mathcal{A}} & & 0 \\ & \ddots & \\ 0 & & 1_{\mathcal{A}} \end{bmatrix}$$

is a multiplicative unit for $M_n(\mathcal{A})$.

We begin with a refinement of Corollary 9.4.6 .

Lemma 17.1.1 *Suppose that $\mathcal{A}$ is a unital C^*-algebra and that we have a completely contractive mapping*

$$\varphi : \mathcal{A} \overset{h}{\otimes} \mathcal{A} \to \mathcal{A}.$$

Then there exists a faithful $*$-representation π of $\mathcal{A}$ on a Hilbert space H and contractive operators $R, S, T \in \mathcal{B}(H)$ such that

$$\pi(\varphi(a,b)) = R\pi(a)S\pi(b)T.$$

Proof Let $\mathbf{S}(\mathcal{A})$ be the set of all states p on $\mathcal{A}$. For each $p \in \mathbf{S}(\mathcal{A})$, we let $\pi_p : \mathcal{A} \to \mathcal{B}(H_p)$ be the corresponding GNS $*$-representation. The *universal $*$-representation* of $\mathcal{A}$ is defined on $H_u = \bigoplus_{p \in \mathbf{S}(\mathcal{A})} H_p$ by

$$\pi_u = (\pi_p)_{p \in \mathbf{S}(\mathcal{A})}.$$

This $*$-representation has the property that if π_1 is any $*$-representation of $\mathcal{A}$, then there exists an index set $\mathfrak{s}$ such that π_1 is unitarily equivalent to a $*$-subrepresentation of the direct sum

$$\pi_u^{\mathfrak{s}} = (\pi_u)_{\mathfrak{s}}.$$

It follows from Corollary 9.4.6 that there exist Hilbert spaces K_1 and K_2, non-degenerate $*$-representations π_1 and π_2 of $\mathcal{A}$ on K_1 and K_2, respectively, and a diagram of contractions

$$H_u \xrightarrow{T} K_2 \xrightarrow{S} K_1 \xrightarrow{R} H_u$$

such that

$$\pi_u(\varphi(a,b)) = R\pi_1(a)S\pi_2(b)T$$

for all $a, b \in \mathcal{A}$. We choose an index set $\mathfrak{s}$ such that π_1 and π_2 are unitarily equivalent to $*$-subrepresentations of $\pi = \pi_u^{\mathfrak{s}}$. If, in addition, we suppose that $\mathfrak{s}$ has the same cardinality as $\mathfrak{s} \times \mathfrak{s}$, then we have that $\pi_1^{\mathfrak{s}}$ and $\pi_2^{\mathfrak{s}}$ are also unitarily equivalent to $*$-subrepresentations of π. Thus, changing notation, we may find projections $e_i \in \pi(\mathcal{A})'$ such that $\pi_i^{\mathfrak{s}}(a) = \pi(a)e_i$ for $i = 1, 2$. It follows that

$$\begin{aligned} \pi(\varphi(a,b)) &= R^{\mathfrak{s}}\pi_1^{\mathfrak{s}}(a)S^{\mathfrak{s}}\pi_2^{\mathfrak{s}}(b)T^{\mathfrak{s}} \\ &= R^{\mathfrak{s}}\pi(a)e_1S^{\mathfrak{s}}e_2\pi(b)T^{\mathfrak{s}} \end{aligned}$$

for all $a, b \in \mathcal{A}$, and changing notation again, we have the desired decomposition. □

Theorem 17.1.2 *Suppose that $\mathcal{A}$ is a unital associative algebra with a complete operator space matrix norm such that $\|1_{\mathcal{A}}\| = 1$. Then $\mathcal{A}$ is an* *operator algebra if and only if for each $n \in \mathbb{N}$, $M_n(\mathcal{A})$ is a Banach algebra, i.e. we have*

$$\|ab\| \leq \|a\|\,\|b\| \tag{17.1.2}$$

for all $a, b \in M_n(\mathcal{A})$.

Proof We begin by noting that for any matrices $a, b \in M_n(\mathcal{A})$, we have from (17.1.1)

$$ab = m_n\left(\left[\sum_k a_{i,k} \otimes b_{k,j}\right]\right) = m_n(a \odot b),$$

and thus the condition (17.1.2) is equivalent to the assumption that the mapping

$$m : \mathcal{A} \overset{h}{\otimes} \mathcal{A} \to \mathcal{A} \tag{17.1.3}$$

is completely contractive. If $\mathcal{A}$ is a concrete operator algebra on a Hilbert space H, then this follows from the diagram

$$\begin{array}{ccc} \mathcal{A} \overset{h}{\otimes} \mathcal{A} & \xrightarrow{m} & \mathcal{A} \\ \cap\!\mid & & \cap\!\mid \\ \mathcal{B}(H) \overset{h}{\otimes} \mathcal{B}(H) & \xrightarrow{\mathbf{m}} & \mathcal{B}(H) \end{array}$$

(see Proposition 9.2.5) and the fact that the bottom row is completely contractive (see (9.1.5)).

Conversely, let us suppose that (17.1.3) is completely contractive. We can assume that $\mathcal{A}$ is an operator subspace of $\mathcal{B}(H)$ for some Hilbert space H. We can extend the multiplication mapping

$$m : \mathcal{A} \overset{h}{\otimes} \mathcal{A} \to \mathcal{B}(H)$$

to a completely contractive mapping

$$\tilde{m} : \mathcal{B}(H) \overset{h}{\otimes} \mathcal{B}(H) \to \mathcal{B}(H).$$

We use Proposition 17.1.1 to decompose this mapping. Thus, changing notation, we may assume that $\mathcal{A} \subseteq \mathcal{B}(K)$ for some Hilbert space K and that the mapping $\tilde{m}$ is determined by

$$\tilde{m}(a, b) = RaSbT \tag{17.1.4}$$

for $a, b \in \mathcal{A}$, where R, S, T are contractions on K.

We define a completely positive mapping $\Psi : \mathcal{B}(K) \to \mathcal{B}(K)$ by letting

$$\Psi(a) = T^* aT$$

for all $a \in \mathcal{B}(K)$, and let $\Psi^{(k)}$ be the k-fold composition of Ψ with itself. We define a mapping $\Phi : \mathcal{B}(K) \to \mathcal{B}(K)$ by

$$\langle \Phi(a)\eta \mid \xi\rangle = \mathrm{LIM}\, \langle \Psi^{(k)}(a)\eta \mid \xi\rangle \tag{17.1.5}$$

for all $a \in \mathcal{B}(K)$ and $\xi, \eta \in K$. We recall that a *Banach limit* is a state $\varphi : \ell_\infty \to \mathbb{C}$ such that

$$\varphi((\alpha_k)) = \varphi((\alpha_{k+1})) \qquad ((\alpha_k) \in \ell_\infty).$$

It is a simple consequence of the Hahn–Banach theorem that such states exist. We fix such a state φ and we use the notation $\mathrm{LIM}(\alpha_k) = \varphi((\alpha_k))$. It is easy to see that if a bounded sequence α_k has a limit, then

$$\mathrm{LIM}(\alpha_k) = \lim_{k\to\infty} \alpha_k$$

or in other words, LIM is an extension of the usual limit functional. It readily follows that Φ is a completely contractive and completely positive mapping on $\mathcal{B}(K)$.

Given any elements $a = [a_{ij}], b = [b_{jk}] \in M_n(\mathcal{A})$, we have from (17.1.4) that

$$\begin{aligned} m_n(a \odot b) &= \left[\sum m(a_{ij} \otimes b_{jk})\right] \\ &= (I_n \otimes R)\, a(I_n \otimes S)\, b(I_n \otimes T) \in M_n(\mathcal{B}(K)). \end{aligned}$$

We have

$$\begin{aligned} (\Psi^{(k)})_n(m_n(a \odot b)^* m_n(a \odot b)) &= (I_n \otimes T^*)^{k+1} b^*(I_n \otimes S^*) a^*(I_n \otimes R^*) \\ &\qquad (I_n \otimes R) a(I_n \otimes S) b(I_n \otimes T)^{k+1} \\ &\le \|a\|^2 (I_n \otimes T^*)^{k+1} b^* b(I_n \otimes T)^{k+1} \\ &= \|a\|^2 (\Psi^{(k+1)})_n(b^* b). \end{aligned}$$

We take the Banach limit and thereby obtain

$$\Phi_n(m_n(a \odot b)^* m_n(a \odot b)) \le \|a\|^2 \Phi_n(b^* b). \qquad (17.1.6)$$

On the other hand, since for each $n \in \mathbb{N}$, $1_{\mathcal{A}}^n$ is the contractive multiplicative unit of $M_n(\mathcal{A})$, we have

$$b^* b = m_n(1_{\mathcal{A}}^n \odot b)^* m_n(1_{\mathcal{A}}^n \odot b) \le (T^* \otimes I_n) b^* b(T \otimes I_n) = \Psi_n(b^* b).$$

By an induction for $k = 1, 2, \ldots$,

$$b^* b \le (\Psi^{(k)})_n(b^* b),$$

and thus if we take the Banach limit,

$$b^* b \le \Phi_n(b^* b). \qquad (17.1.7)$$

As in the proof of the Stinespring theorem for completely positive mappings, Φ induces a semi-inner product on $\mathcal{A} \otimes K$ given by

$$\langle a \otimes \eta \mid b \otimes \xi \rangle = \langle \Phi(b^* a)\eta \mid \xi \rangle \qquad (17.1.8)$$

for all $a, b \in \mathcal{A}$ and $\xi, \eta \in K$. Let $X = \{u \in \mathcal{A} \otimes K : \langle u \mid u \rangle = 0\}$. Then (17.1.8) induces an inner product on $(\mathcal{A} \otimes K)/X$, and we let H denote its

Hilbert space completion. We shall let $a \otimes \eta$ also denote the equivalence class of $a \otimes \eta$ in H, and $\mathcal{A} \otimes K$ the corresponding dense subspace in H.

For any $a \in \mathcal{A}$, we define a linear mapping $\theta(a)$ on $\mathcal{A} \otimes K$ by

$$\theta(a)(b \otimes \eta) = m(a, b) \otimes \eta$$

for all $b \otimes \eta \in \mathcal{A} \otimes K$. Given $b_1 \otimes \eta_1, \ldots, b_n \otimes \eta_n \in \mathcal{A} \otimes K$, we let b denote the element in $M_n(\mathcal{A})$ with $b_1, \ldots, b_n$ in the first row and zero otherwise, and we let $\eta = (\eta_i) \in K^n$. It follows from (17.1.6) that

$$\begin{aligned}\left\|\theta(a)\left(\sum b_i \otimes \eta_i\right)\right\|^2 &= \sum_{i,j} \langle \Phi(m(a, b_j)^* m(a, b_i))\eta_i \mid \eta_j\rangle \\ &= \langle \Phi_n(m_n(I_n \otimes a, b)^* m_n(I_n \otimes a, b))\eta \mid \eta\rangle \\ &\leq \|a\|^2 \langle \Phi_n(b^* b)\eta \mid \eta\rangle = \|a\|^2 \left\|\sum b_i \otimes \eta_i\right\|^2 .\end{aligned}$$

Therefore, $\theta(a)$ can be extended to a bounded mapping on the whole Hilbert space H, and $\theta : \mathcal{A} \to \mathcal{B}(H)$ is a well-defined contraction. Using a similar matricial argument, it is easy to see that θ is a complete contraction.

In fact, θ is a complete isometry. To see this we recall that for each $n \in \mathbb{N}$, the multiplicative unit $1^n_{\mathcal{A}}$ of $M_n(\mathcal{A})$ is an element of norm one in $M_n(\mathcal{B}(K))$. Thus, it is clear that $(1^n_{\mathcal{A}})^* 1^n_{\mathcal{A}}$ and $\Phi_n((1^n_{\mathcal{A}})^* 1^n_{\mathcal{A}})$ are contractive and positive elements in $M_n(\mathcal{B}(K))$. For any $\eta = (\eta_j) \in K^n$,

$$\begin{aligned}\|(1_{\mathcal{A}} \otimes \eta_j)\|^2 = \sum \|1_{\mathcal{A}} \otimes \eta_j\|^2 &= \sum \langle \Phi((1_{\mathcal{A}})^* 1_{\mathcal{A}})\eta_j \mid \eta_j\rangle \\ &= \langle \Phi_n((1^n_{\mathcal{A}})^* 1^n_{\mathcal{A}})\eta \mid \eta\rangle \leq \|\eta\|^2 .\end{aligned}$$

Then, for any $a = [a_{ij}] \in M_n(\mathcal{A})$ and $\eta = (\eta_j) \in K^n$,

$$\begin{aligned}\|\theta_n(a)\|^2 \|\eta\|^2 \geq \|\theta_n(a)(1_{\mathcal{A}} \otimes \eta_j)\|^2 &= \sum_i \|\sum_j \theta(a_{ij})(1_{\mathcal{A}} \otimes \eta_j)\|^2 \\ &= \sum_{i,j,k} \langle m(a_{ij} \otimes 1_{\mathcal{A}}) \otimes \eta_j \mid m(a_{ik} \otimes 1_{\mathcal{A}}) \otimes \eta_k\rangle \\ &= \sum_{i,j,k} \langle a_{ij} \otimes \eta_j \mid a_{ik} \otimes \eta_k\rangle = \sum_{i,j,k} \langle \Phi(a^*_{ik} a_{ij})\eta_j \mid \eta_k\rangle \\ &= \langle \Phi_n(a^* a)\eta \mid \eta\rangle \geq \langle a^* a\eta \mid \eta\rangle = \|a\eta\|^2,\end{aligned}$$

where we used the fact that $m(a \otimes 1_{\mathcal{A}}) = a$ in the fourth identity and (17.1.7) in the seventh inequality. This shows that $\|\theta_n(a)\| \geq \|a\|$ for all $a \in M_n(\mathcal{A})$, and thus θ is a complete isometry.

Finally, since m is associative on $\mathcal{A}$, i.e.

$$m(m(a \otimes b) \otimes c) = m(a \otimes m(b \otimes c))$$

and

$$m(1_{\mathcal{A}} \otimes a) = a$$

for all $a, b, c \in \mathcal{A}$, it is easy to verify that θ is a homomorphism from $\mathcal{A}$ into $\mathcal{B}(H)$ such that $\theta(1_{\mathcal{A}}) = I_H$. Therefore, $\theta : \mathcal{A} \to \mathcal{B}(H)$ is a completely isometric unital isomorphism. □

Theorem 17.1.2 led to the first proof that a quotient of an operator algebra by a closed two-sided ideal is again (completely isometric to) an operator algebra. It is easy to generalize Theorem 17.1.2 to operator algebras which have a contractive approximate identity (see Poon and Ruan 1994 and Blecher 1995). A suggestive theorem for the general case has been proved by Blecher. We shall content ourselves with including just the statement of his result.

Theorem 17.1.3 *Let $\mathcal{A}$ be an associative algebra. Then $\mathcal{A}$ is completely boundedly isomorphic to a concrete operator algebra, i.e. there exists a Hilbert space H, a concrete operator algebra $\mathcal{B}$ on H and completely bounded (algebraic) isomorphism*

$$\theta : \mathcal{A} \to \mathcal{B}$$

such that $\|\theta\|_{cb}\|\theta^{-1}\|_{cb} < \infty$ if and only if $\mathcal{A}$ is an operator space such that the multiplication

$$m : \mathcal{A} \otimes_h \mathcal{A} \to \mathcal{A}$$

is completely bounded (or equivalently, completely contractive by a suitable matrix norm modification). □

In operator algebra theory, there is a well-known theorem due to Sakai, which states that if $\mathcal{A}$ is a C^*-algebra and is a dual Banach space, then there exists a Hilbert space H and a faithful weak* continuous $*$-homomorphism $\pi : \mathcal{A} \to \mathcal{B}(H)$ such that $\pi(\mathcal{A})$ is a von Neumann algebra on H. In this case, $\mathcal{A}$ has a unique predual and the multiplication of $\mathcal{A}$ is automatically weak* continuous in each variable. Recently, Le Merdy has proved an analogous result for (not necessarily self-adjoint) operator algebras. We shall state Le Merdy's result in the following theorem without giving the proof.

Theorem 17.1.4 *Let $\mathcal{A}$ be a unital operator algebra. If $\mathcal{A}$ is an operator space dual with an operator space predual V, and the multiplication*

$$m : \mathcal{A} \times \mathcal{A} \to \mathcal{A}$$

is weak continuous in each variable, then there exists a Hilbert space H and a weak* continuous completely isometric homomorphism*

$$\pi : \mathcal{A} \to \mathcal{B}(H)$$

such that $\pi(\mathcal{A})$ is a weak closed unital operator subalgebra of $\mathcal{B}(H)$.* □

17.2 NOTES AND REFERENCES

The characterization of the unital operator algebras proved by Blecher *et al.* in (1990) provided a dramatic justification for the axiomatization of operator spaces. The proofs for Theorem 17.1.3 and Theorem 17.1.4 can be found in Blecher (1995) and Le Merdy (1999), respectively. Sakai's characterization of the von Neumann algebras is presented in Sakai (1971).

There have been a great many further applications of the Haagerup tensor product to other algebraic questions. Examples of this may be found in Blecher *et al.* (1999), and Paulsen (1998). In addition, Blecher and Paulsen (1991*b*) used a notion of multiplicative boundedness to calculate the norms of elements of the full C^*-algebras of free groups, and in various other universal objects. This elegant theory will undoubtedly play a role in modern algebraic analysis.

In a related development, Pisier (1996*a*) has examined one of the oldest questions in operator algebra theory, Kadison's question of whether or not a bounded (non-self-adjoint) unital representation of a unital C^*-algebra is necessarily equivalent to a $*$-representation. Using operator space techniques he has been able to greatly extend the methods of Christensen and Haagerup. In this monograph and subsequent preprints, he has introduced the notion of a 'similarity degree' for an operator algebra, which once again demonstrates the close links between the algebraic and operator space structures of operator algebras.

As we pointed out in Chapter 9.5, one of the first applications of the Haagerup tensor product occurred in a simplified proof that a C^*-algebra is amenable if and only if it is nuclear. This seems to be a suitable place to include a few more words about this application. The result depends directly upon a non-commutative version of Grothendieck's famous inequality due to Pisier (1978) and Haagerup (1985*a*). The latter is perhaps the most remarkable of Grothendieck's results in Banach space theory, and it is fitting that it should arise again in this totally unexpected context. The difficulty point in the argument is to show that if one has a virtual diagonal $M \in (\mathcal{A} \otimes^{\gamma} \mathcal{A})^{**}$, then one can also find a *normal virtual diagonal* in $\mathcal{A}^{**} \otimes^{\sigma h} \mathcal{A}^{**}$ (see Effros 1988). The latter is the notation for the *normal Haagerup tensor product*, a notion we have not considered in this monograph.

We remark that, in contrast to the above situation, Johnson's result shows that if $\mathcal{A}$ is a Fourier algebra of a compact group, then one cannot always 'lift' a virtual diagonal $U \in (\mathcal{A} \widehat{\otimes} \mathcal{A})^{**}$ to a Banach algebra virtual diagonal $\tilde{U} \in (\mathcal{A} \otimes^{\gamma} \mathcal{A})^{**}$.

Finally, we conclude by noting that Pisier has formulated a tensor product which is naturally related to Haagerup's mapping space of decomposable operators. This tensor product, which is generally used for an operator algebra and an operator space, enabled him to give new proofs for some of the early results about the nuclearity of C^*-algebras and semidiscreteness of von Neumann algebras. This result may be found in a monograph which he has recently completed, entitled *An introduction to the theory of operator spaces.* This book contains many other new results, and the reader will find it to be an excellent source for many of the topics not included in this text.

Appendix

Preliminaries

A.1 LINEAR SPACES

We use the terms *linear space* to indicate a real or complex vector space (most of the spaces we consider are complex). By a *pairing* of real or complex linear spaces V and V', we mean a bilinear mapping

$$B : V \times V' \to k,$$

where $k = \mathbb{R}$ or $\mathbb{C}$, respectively, such that $B(v, w) = 0$ for all $w \in V'$ implies that $v = 0$, and similarly $B(v, w) = 0$ for all $v \in V$ implies that $w = 0$. If the pairing is fixed, then we shall often use the notation $\langle v, w\rangle = B(v, w)$.

We say that a linear mapping of linear spaces spaces $\varphi : V \to W$ is *finite-rank* if $\varphi(V)$ is finite-dimensional. We let $\mathbb{L}(V, W)$ (respectively, $\mathbb{FL}(V, W)$) denote the space of all (respectively, finite-rank) linear mappings $\varphi : V \to W$. We let $\mathbb{L}(V) = \mathbb{L}(V, V)$, $\mathbb{FL}(V) = \mathbb{FL}(V, V)$ and $V^d = \mathbb{L}(V, k)$. Given another linear space X, we write $\mathbb{L}(V \times W, X)$ for the bilinear mappings

$$\varphi : V \times W \to X.$$

We assume that the reader is familiar with the direct sum and tensor product constructions, $V \oplus W$ and $V \otimes W$, respectively, as well as the usual identifications

$$k \otimes V \cong V \otimes k \cong V, \tag{A.1.1}$$

$$(V \otimes W) \otimes X \cong V \otimes (W \otimes X), \tag{A.1.2}$$

$$\mathbb{L}(V \otimes W, X) \cong \mathbb{L}(V \times W, X) \cong \mathbb{L}(V, L(W, X)), \tag{A.1.3}$$

$$\mathbb{FL}(V, W) \cong V^d \otimes W. \tag{A.1.4}$$

The bilinear functional

$$V^d \times V \to k : (f, v) \mapsto \langle f, v\rangle = f(v)$$

determines the *trace functional*

$$\text{trace} : \mathbb{FL}(V, V) \to k. \tag{A.1.5}$$

A $*$-*linear space* V is a complex linear space together with a $*$-*operation*, i.e. a conjugate linear mapping $^* : V \to V$ such that $x^{**} = x$ for all $x \in V$.

We say that $x \in V$ is *self-adjoint* if $x^* = x$, and we let V_{sa} be the real subspace of self-adjoint elements in V. If we adopt the usual notation

$$\begin{aligned} \text{Re } x &= (1/2)(x + x^*), \\ \text{Im } x &= (1/2i)(x - x^*), \end{aligned} \tag{A.1.6}$$

we have $x = \text{Re } x + i \,\text{Im } x$ and thus $V = V_{sa} + iV_{sa}$. Given two $*$-spaces V and W, there is also a natural $*$-operation on $\mathbb{L}(V, W)$ defined by

$$\varphi^*(x) = \varphi(x^*)^* . \tag{A.1.7}$$

It is a simple exercise to verify that $\varphi = \varphi^*$ if and only if $\varphi(V_{sa}) \subseteq W_{sa}$.

Given elements $v_1, \ldots, v_n$ in a vector space V and scalars α_j for which $\alpha_j > 0$ and $\sum_{j=1}^{n} \alpha_j = 1$ (respectively, $\sum_{j=1}^{n} |\alpha_j| \leq 1$), we say that the corresponding linear combination $\sum_{j=1}^{n} \alpha_j v_j$ is a *convex* (respectively, an *absolutely convex*) combination of the v_j. We say that a subset K of V is *convex* (respectively, *absolutely convex*) if it is closed under the formation of convex combinations (respectively, absolutely convex combinations). Given a set $D \subseteq V$, we define the *the absolutely convex hull* $|co|\,(D)$ of D to be the smallest absolutely convex set containing D, or equivalently, the set of all absolutely convex combinations of elements in D.

A *cone* C in a vector space V is a subset C such that $C + C \subseteq C$ and $\alpha C \subseteq C$ for all $\alpha \geq 0$. We define an *ordered linear space* to be a complex $*$-linear space V, together with a distinguished cone $V^+ \subseteq V_{sa}$ such that $V_{sa} = V^+ - V^+$. We define by $x \leq y$ if $y - x \in V^+$. If V and W are ordered spaces, then we say that a linear mapping $\varphi : V \to W$ is *positive* if $\varphi(V^+) \subseteq W^+$, and we write $\varphi \geq 0$. It follows that if $\varphi \geq 0$, then $\varphi = \varphi^*$.

A.2 BANACH SPACES

The elements of Banach space theory can be found in many texts, such as Dunford and Schwartz (1958). For simplicity we consider only complex linear spaces in this section. With the obvious changes, the definitions are also valid for real linear spaces. A *normed space* V is a linear space with a function $\|\cdot\| : V \to [0, \infty)$ such that for all $x, y \in V$ and $\alpha \in \mathbb{C}$,

- **N1** $\|x + y\| \leq \|x\| + \|y\|$;
- **N2** $\|\alpha x\| \leq |\alpha| \|x\|$;
- **N3** $\|x\| = 0$ implies that $x = 0$.

V is a *Banach space* if it is complete in the metric $d(x, y) = \|x - y\|$.

We use the notation

$$\begin{aligned} V_{\|\cdot\|<r} &= \{x \in V : \|x\| < r\}, \\ V_{\|\cdot\|\leq r} &= \{x \in V : \|x\| \leq r\}. \end{aligned}$$

Given a linear mapping of normed spaces $\varphi : V \to W$ we let

$$\|\varphi\| = \sup\{\|\varphi(x)\| : \|x\| \leq 1\},$$

and we say that φ is *bounded* if $\|\varphi\| < \infty$. This determines a norm on the linear space $\mathcal{B}(V, W)$ of all bounded linear mappings $\varphi : V \to W$. We write $\mathcal{FB}(V, W)$ for the subspace of finite-rank linear mappings, and we let $\mathcal{B}(V) = \mathcal{B}(V, V)$ and $\mathcal{FB}(V) = \mathcal{FB}(V, V)$. The space $V^* = \mathcal{B}(V, \mathbb{C})$ is called the *dual* of V. A linear mapping $\varphi : V \to W$ is *contractive* (respectively, *isometric*) if $\|\varphi\| \leq 1$ (respectively, $\|\varphi(x)\| = \|x\|$ for all $x \in V$), and $\varphi \in \mathcal{B}(V, W)$ is an *isomorphism* if it is a bijection and $\varphi^{-1} \in \mathcal{B}(W, V)$.

A linear mapping of Banach spaces $\varphi : V \to W$ is said to be *compact* if the norm-closure of $\varphi(V_{\|\cdot\|\leq 1})$ is norm compact in W. We let $\mathcal{K}(V, W)$ be the linear space of compact bounded linear mappings $\varphi : V \to W$ and $\mathcal{K}(V) = \mathcal{K}(V, V)$. We have

$$\mathcal{B}(V, W) \supseteq \mathcal{K}(V, W) \supseteq \mathcal{FB}(V, W),$$

where $\mathcal{K}(V, W)$ is norm closed in $\mathcal{B}(V, W)$. If W is a Hilbert space, then $\mathcal{K}(V, W)$ is the norm closure of $\mathcal{FB}(V, W)$ (see §11.4).

If W is a Banach space, then the same is true for $\mathcal{B}(V, W)$ for any normed space V, hence the dual $V^* = \mathcal{B}(V, \mathbb{C})$ of a normed space V is a Banach space. Each bounded linear mapping $\varphi : V \to W$ determines a corresponding *dual linear mapping* $\varphi^* : W^* \to V^*$ by $\varphi^*(f)(v) = f(\varphi(v))$. From the Hahn–Banach theorem, $\|\varphi^*\| = \|\varphi\|$.

If K is a closed subspace of a normed space V, then the *quotient normed space* is the vector space V/K together with the *quotient norm*

$$\|x + K\| = \inf \{\|x + k\| : k \in K\}.$$

If V is a Banach space, then so is V/K. If $\varphi : V \to W$ is a (non-zero) mapping of normed spaces, and $K = \ker \varphi$, then the induced linear mapping $\tilde{\varphi} : V/K \to W : x + K \mapsto \varphi(x)$ satisfies $\|\tilde{\varphi}\| = \|\varphi\|$.

Lemma A.2.1 *Suppose that V and W are Banach spaces and $\varphi : V \to W$ is a contraction such that*

$$\overline{\varphi(V_{\|\cdot\|<1})} \supseteq W_{\|\cdot\|<r}$$

for some r with $0 < r < 1$. Then if $K = \ker \varphi$, $\tilde{\varphi}$ maps V/K isomorphically onto W and $\left\|\tilde{\varphi}^{-1}\right\| \leq r^{-1}$.

Proof Let us suppose that $y \in W$ satisfies $\|y\| < (1 - \varepsilon)r$. From the assumption, there is an $x_1 \in V_{\|\cdot\|<(1-\varepsilon)}$ such that $y_1 = \varphi(x_1)$ satisfies

$$\|y_1 - y\| < \varepsilon r/2.$$

We may then find an $x_2 \in V_{\|\cdot\|<\varepsilon/2}$ such that $y_2 = \varphi(x_2)$ satisfies

$$\|y - y_1 - y_2\| < \varepsilon r/2^2 .$$

Continuing in this manner, we may find an $x_n \in V_{\|\cdot\|<\varepsilon/2^{n-1}}$ such that $y_n = \varphi(x_n)$ satisfies

$$\|y - y_1 - \cdots - y_n\| < \varepsilon r/2^n .$$

We have that $\sum_n x_n$ converges to an element x of $V_{\|\cdot\|<1}$, and by continuity, $\varphi(x) = y$. We have $\tilde{\varphi}^{-1}(y) = x + K$, where $\|x + K\| \leq \|x\| < 1$, i.e. $\tilde{\varphi}^{-1}(W_{\|\cdot\|<r})$ is contained in the open unit ball of V/K, and we have the desired inequality. □

Given normed spaces V and W, a (non-zero) bounded linear mapping $\varphi : V \to W$ is said to be a *quotient mapping* if the induced mapping $\overline{\varphi} : V/\mathrm{ker}\varphi \to W$ is isometric. It is equivalent to suppose that φ maps $V_{\|\cdot\|<1}$ onto $W_{\|\cdot\|<1}$. From the above result it follows that if V is complete, then it suffices to show that φ maps $V_{\|\cdot\|<1}$ onto a dense subset of $W_{\|\cdot\|<1}$.

We say that a non-zero linear mapping $\varphi : V \to W$ is an *exact quotient mapping* if it maps $V_{\|\cdot\|\leq 1}$ onto $W_{\|\cdot\|\leq 1}$. In general, given a bounded linear mapping $\varphi : V \to W$, it is immediate from the Hahn–Banach Theorem that

$$\begin{aligned} \varphi \text{ is isometric} &\Leftrightarrow \varphi^* \text{ is a quotient mapping} \\ &\Leftrightarrow \varphi^* \text{ is an exact quotient mapping}. \end{aligned} \tag{A.2.1}$$

On the other hand, it is easy to see that

$$\varphi \text{ is a quotient mapping} \Rightarrow \varphi^* \text{ is an isometry}. \tag{A.2.2}$$

If V is complete, then these conditions are equivalent:

$$\varphi \text{ is a quotient mapping} \Leftrightarrow \varphi^* \text{ is an isometry}. \tag{A.2.3}$$

To prove this it suffices to show that if φ^* is an isometry, then $\varphi(V_{\|\cdot\|\leq 1})$ is dense in $W_{\|\cdot\|\leq 1}$ (see Lemma A.2.1). But if this were not the case, we could choose an element $y \in W_{\|\cdot\|\leq 1} \backslash \overline{\varphi(V_{\|\cdot\|\leq 1})}$, and then use the geometric form of the Hahn–Banach theorem to find a function $g \in W^*$ with $|g(\varphi(V_{\|\cdot\|\leq 1}))| < 1$ and $|g(y)| > 1$. This would imply that $\|g\| > 1$, which contradicts the fact that $\|\varphi^*(g)\| \leq 1$.

If V and W are Banach spaces, then we have a natural linear isomorphism

$$V^* \otimes W \cong \mathcal{FB}(V, W),$$

and the bilinear mapping

$$V^* \times V \to \mathbb{C} : (f, v) \mapsto f(v)$$

determines the linear *trace functional*

$$\mathrm{trace} : \mathcal{FB}(V, V) = V^* \otimes V \to \mathbb{C}, \tag{A.2.4}$$

which extends uniquely to a contractive linear functional

$$\mathrm{trace} : V^* \otimes^\gamma V \to \mathbb{C}.$$

Given an indexed family of real or complex normed spaces $(V_s)_{s\in\mathfrak{s}}$, we let

$$\prod_{s\in\mathfrak{s}} V_s = \ell_\infty(\mathfrak{s}; V_s)$$

be the Banach space of *bounded* $\mathfrak{s}$-tuples $(x_s)_{s\in\mathfrak{s}}$ with the norm

$$\|(x_s)\| = \sup_{s\in\mathfrak{s}} \|x_s\| .$$

If $1 \leq p < \infty$, we let $\ell_p(\mathfrak{s}; V_s)$ denote the Banach space of $\mathfrak{s}$-tuples (x_s) for which

$$\|(x_s)\|_p = \left(\sum_s \|x_s\|^p\right)^{1/p} < \infty,$$

together with the norm $\|\cdot\|_p$. We use the abbreviation $\ell_p(\mathfrak{s})$ or $\ell_p^{\mathfrak{s}}$ for $\ell_p(\mathfrak{s};\mathbb{C})$. Given an integer $n \in \mathbb{N}$, we also write n for the index set $\{1,\ldots,n\}$, and thus we let $\ell_p(n;V_i)$ and $\ell_p^n(V)$ denote the linear spaces $V_1 \oplus \cdots \oplus V_n$ and $V \oplus \cdots \oplus V$ with the corresponding p-norms. In particular, $\ell_p^n = \ell_p^n(\mathbb{C})$ is just $\mathbb{C}^n$ with the usual p-norm.

It will be noted that the norms $\|\cdot\|_p$ on $\ell_p^n(V) \cong V^n$ are all equivalent, i.e. they determine the same norm *topology* since for each of them a sequence $v(k) \in \ell_p^n(V)$ converges to an element $v \in \ell_p^n(V)$ if and only if the entries $v(k)_i$ converge to v_i in V. Since this is a recurrent argument in this monograph, it is useful to consider it in greater detail.

If E is an arbitrary finite-dimensional space, then there is a unique vector space topology on E. This may be described by fixing a basis $e_j \in E$ $(1 \leq j \leq n)$, and then using the corresponding isomorphism $E \cong \mathbb{C}^n$ to pull back the usual product topology on $\mathbb{C}^n$. Now let us suppose that V is an arbitrary topological vector space. We may use the corresponding identification

$$E \otimes V \cong \mathbb{C}^n \otimes V \cong V^n$$

to define the *product* topology on $E \otimes V$. This obviously does not depend on the basis e_j. If V is locally convex, then that is also true for $E \otimes V$, and it is immediate that the weak* topology on

$$(E \otimes V)^* \cong E^* \otimes V^* \tag{A.2.5}$$

coincides with the product topology determined by the weak* topology on V^*. If V is a normed space, then we say that a norm on $E \otimes V$ is *natural* if it defines the product topology determined by the norm topology on V. If we fix such a norm $\|\cdot\|$ on $E \otimes V$, then it is easy to see that the dual norm $\|\cdot\|^*$ is natural on (A.2.5).

If L is a subspace of a normed space E, the *annihilator* of L in E^* is the subspace

$$L^\perp = \{f \in V^* : f_{|L} = 0\}.$$

Similarly if M is a subspace of E^*, its *preannihilator* in V is the subspace

$$M_\perp = \{v \in V : f(v) = 0 \text{ for all } f \in M\}.$$

From the bipolar theorem $(L^\perp)_\perp$ is the norm closure of L, whereas $(M_\perp)^\perp$ is the weak* closure of M.

We shall also use some other well-known facts about the weak and weak* topologies. Let us suppose that V and W are Banach spaces.

- If V and W are Banach spaces, then a linear mapping $\varphi : W^* \to V^*$ is weak* continuous if and only if the restriction $\varphi|_{W^*_{\|\cdot\|\leq 1}}$ is continuous in the relative topologies. If that is the case, then there is a unique bounded linear mapping $\psi : V \to W$ for which $\varphi = \psi^*$.
- If X is a closed subspace of V^*, then it is weak* closed if and only if that is the case for $X_{\|\cdot\|\leq 1}$. If that is the case, then there is a subspace W of V and a natural identification $X = W^\perp$. Furthermore, we have a natural weak* homeomorphism and isometry
$$(V/W)^* \cong X.$$
- If f is a weakly continuous functional on V, then it is norm continuous. From this it follows that if K is a convex subset of V, then it is norm closed if and only if it is weakly closed.

Let V be an n-dimensional Banach space. An *Auerbach basis* for V is a vector basis $v_1, \ldots, v_n$ with $\|v_j\| = 1$, for which there exist a basis $f_1, \ldots, f_n$ in V^* with $\|f_j\| = 1$ and $f_i(v_j) = \delta_{i,j}$.

Lemma A.2.2 *Every finite-dimensional Banach space has an Auerbach basis.*

Proof Let V be an n-dimensional Banach space, and let $x_1, \ldots, x_n$ and $g_1, \ldots, g_n$ be arbitrary bases for V and V^*, respectively. The function
$$F(y_1, \ldots, y_n) = \det[g_i(y_j)]$$
is a continuous function on the closed unit ball $\ell^n_\infty(V)_{\|\cdot\|\leq 1}$ of $\ell^n_\infty(V)$. Since $\ell^n_\infty(V)_{\|\cdot\|\leq 1}$ is a compact set, F reaches its maximum value at a point $(v_1, \ldots, v_n)$ in $\ell^n_\infty(V)_{\|\cdot\|\leq 1}$. Clearly, we must have $\|v_j\| = 1$ for $j = 1, \ldots, n$. If we define
$$f_j(x) = F(v_1, \ldots, v_{j-1}, x, v_{j+1}, \ldots, v_n)/F(v_1, \ldots, v_n),$$
then $f_1, \ldots, f_n$ is a basis for V^* with $\|f_j\| = 1$ and $f_i(v_j) = \delta_{i,j}$. □

Finally, we conclude this section with a proof of Helly's lemma.

Lemma A.2.3 *If E is a Banach space, $L \subseteq E^*$ is finite-dimensional and $\tilde{x}_0 \in E^{**}$ is contractive, then for any $\varepsilon > 0$ there exists an $x_0 \in E$ with $\|x_0\| < 1 + \varepsilon$ and*
$$\langle f, x_0 \rangle = \langle \tilde{x}_0, f \rangle$$
for all $f \in L$.

Proof Let $L^\perp$ be the annihilator of L in E^{**} and $L_\perp$ the pre-annihilator of L in E. Then
$$\dim E/L_\perp = \dim L = \dim E^{**}/L^\perp$$

implies that the canonical inclusion

$$E/L_\perp \hookrightarrow (E/L_\perp)^{**} \cong E^{**}/L^\perp$$

is an isometric isomorphism. The contractive element $\tilde{x}_0 + L^\perp$ in $E^{**}/L^\perp$ corresponds to a contractive element $x_0 + L_\perp$ in $E/L_\perp$. Therefore, we may choose $x_0 \in E$ such that $\|x_0\| < 1 + \varepsilon$ and thus

$$\langle f, x_0 \rangle = \langle \tilde{x}_0, f \rangle$$

for all $f \in L$. □

A.3 HILBERT SPACES

A *pre-Hilbert space* is defined to be a linear space H together with a *positive definite sesquilinear form*

$$\langle \cdot \mid \cdot \rangle : H \times H \to \mathbb{C},$$

i.e. a mapping satisfying

- **H1** $\langle \eta \mid \xi \rangle$ is linear in η and conjugate linear in ξ;
- **H2** $\langle \xi \mid \xi \rangle \geq 0$ for all $\xi \in H$;
- **H3** if $\langle \xi \mid \xi \rangle = 0$, then $\xi = 0$.

We often call $\langle \cdot \mid \cdot \rangle$ an *inner product* on H. We have a corresponding norm on H determined by $\|\xi\| = |\langle \xi \mid \xi \rangle|^{1/2}$, and we say that H is a *Hilbert space* if it is complete. Any Hilbert space is isometric to a Banach space of the form $\ell_2(\mathfrak{s})$, where $\mathfrak{s}$ has the cardinality of an orthonormal basis for H. The norm on H determines the sesquilinear form since we have

$$\langle \eta \mid \xi \rangle = \frac{1}{4}\left[\|\eta + \xi\|^2 - \|\eta - \xi\|^2 + \mathrm{i}(\|\eta + \mathrm{i}\xi\|^2 - \|\eta - \mathrm{i}\xi\|^2)\right]. \qquad \text{(A.3.1)}$$

We call bounded linear mappings of Hilbert spaces *bounded operators.*

An important feature of Hilbert spaces is that they are *self-dual.* In order to explain this, it is necessary to introduce the notion of conjugate spaces.

If V is a complex linear space, then we define *a conjugate space* for V to be a pair (W, J) of a linear space W and a conjugate linear isomorphism $J : V \to W$. It is evident that such a pair is essentially unique, i.e. given another conjugate space (W', J'), we have a corresponding commutative diagram

$$\begin{array}{ccc} & V & \\ {}^{J}\swarrow & & \searrow^{J'} \\ W & \xrightarrow{\ \varphi\ } & W' \end{array},$$

where φ is a linear isomorphism. If one has chosen a fixed conjugate space (W, J) for V, it is customary to write $\overline{V}$ for W and $\overline{v} = J(v)$, and to refer to $\overline{V}$ as *the* conjugate of V. Turning to examples, if $V = \mathbb{C}^n$, then we may let $\overline{V} = \mathbb{C}^n$ and $\overline{\alpha} = J(\alpha) = (\overline{\alpha}_1, \ldots \overline{\alpha}_n)$, where $\overline{\alpha}_i$ is the usual complex conjugate of α_i. If V is an arbitrary vector space, then we may define a conjugate space by letting $\overline{V}$ be V with the usual addition and the scalar multiplication

$$(\alpha, v) \mapsto \overline{\alpha}v,$$

and $J : V \to \overline{V}$ be the identity mapping.

If V is a normed space, then there is a natural norm on the conjugate $\overline{V}$ defined by $\|\overline{v}\| = \|v\|$. If H is a Hilbert space, then using (A.3.1), it follows that the sesquilinear form on $\overline{H}$ is given by

$$\langle \overline{v} \mid \overline{w} \rangle = \langle w \mid v \rangle. \tag{A.3.2}$$

If H is a complex Hilbert space, then we define the *canonical duality mapping*

$$\theta : \overline{H} \to H^* : \overline{\xi} \mapsto f_\xi, \tag{A.3.3}$$

where

$$f_\xi(\eta) = \langle \eta \mid \xi \rangle.$$

The Riesz representation theorem of Hilbert space theory states that this is a linear isometry of Hilbert spaces. It follows that the corresponding mapping

$$J : H \to H^* : \xi \mapsto f_\xi$$

is a conjugate linear isometry of Hilbert spaces. The *adjoint* of a bounded operator $S : H \to K$ is the operator $S^\star : K \to H$ defined by

$$S^\star = J_H^{-1} S^* J_K.$$

Following common usage, we shall abandon the 'special star' in this notation. The reader may distinguish between the dual mapping S^* and the adjoint mapping S^* by considering the spaces on which they are defined.

One important application of the fundamental duality mapping may be found in the proof of the following principle, which we shall repeatedly use. If H and K are Hilbert spaces, and

$$B : H \times K \to \mathbb{C}$$

is a sesquilinear functional which is bounded in the sense that

$$|B(\eta, \xi)| \leq C \,\|\eta\|\, \|\xi\|$$

for some constant C, then there exists a linear mapping $T : H \to K$ such that

$$B(\eta, \xi) = \langle T(\eta) \mid \xi \rangle, \tag{A.3.4}$$

where $\|T\| \leq C$.

A.4 C^*-ALGEBRAS AND VON NEUMANN ALGEBRAS

There are a number of excellent introductions to operator algebra theory, including Dixmier (1964, 1981), Kadison and Ringrose (1997*a*, *b*), Pedersen

(1979), and Takesaki (1979). In this section we will summarize some of the results that are used in this text.

We define an *algebra* to be a linear space with an associative multiplication $(a, b) \mapsto ab$. A $*$-*algebra* $\mathcal{A}$ is a $*$-linear space with an associative multiplication for which $(ab)^* = b^*a^*$. We say that an element e in an algebra $\mathcal{A}$ is an *identity* or a *unit* if $ea = ae = a$ for all $a \in \mathcal{A}$, and if $\mathcal{A}$ has a unit we say that $\mathcal{A}$ is a *unital algebra*. A unit must be unique, and it follows that a unit in a $*$-algebra must be self-adjoint. We generally use the symbol 1, or when we are considering operators the symbol I, for a unit. Any $*$-algebra $\mathcal{A}$ has a *unital extension*

$$\mathcal{A}^\dagger = \mathcal{A} + \mathbb{C}1,$$

where the multiplication and the $*$-operation are defined in the usual manner.

If a $*$-algebra $\mathcal{A}$ has a norm, then we say that it is a *Banach $*$-algebra* if we have $\|ab\| \leq \|a\|\,\|b\|$ and $\|a^*\| = \|a\|$. Finally, we say that $\mathcal{A}$ is a C^*-*algebra* if we also have $\|a^*a\| = \|a\|^2$. It is a simple exercise to show that if X is a locally compact Hausdorff space, then the Banach algebra $C_0(X)$ of continuous complex functions vanishing at ∞ on X together with the usual $*$-algebraic operations and the supremum norm is a C^*-algebra. In fact, any commutative C^*-algebra essentially arises in this fashion. If H is a Hilbert space, then $\mathcal{B}(H)$ with the usual $*$-algebraic operations and the operator norm is another example of a C^*-algebra, and thus the same is true for any closed $*$-subalgebra of $\mathcal{B}(H)$. C^*-algebra theory began with the following result of Gelfand and Naimark (1943). This abstract characterization also served as the primary antecedent for the abstract characterizations of the operator systems, operator spaces, and (non-self adjoint) operator algebras.

Theorem A.4.1 (Gelfand–Naimark) *If $\mathcal{A}$ is a C^*-algebra, then there exists a Hilbert space H and an isometric $*$-isomorphism of $\mathcal{A}$ onto a closed $*$-subalgebra of $\mathcal{B}(H)$.* □

The natural ordering on a C^*-algebra $\mathcal{A}$ is algebraically determined since

$$\mathcal{A}^+ = \{a \in \mathcal{A} : a = c^*c \text{ for some } c \in \mathcal{A}\}$$

(it is a non-trivial task to show that this *is* a cone).

The unital extension $\mathcal{A}^\dagger$ of a C^*-algebra $\mathcal{A}$ has a canonical C^*-algebraic structure. We say that a net $\{a_\gamma\}_{\gamma\in\Gamma}$ in $\mathcal{A}$ is an *approximate identity* for $\mathcal{A}$ if $a_\gamma \geq 0$, $\|a_\gamma\| \leq 1$, and $\lim a_\gamma b = \lim ba_\gamma = b$ for all $b \in \mathcal{A}$. A net of self-adjoint elements $\{a_\gamma\}_{\gamma\in\Gamma}$ in a C^*-algebra $\mathcal{A}$ is called *increasing* if $\gamma_1 \leq \gamma_2$ implies that $a_{\gamma_1} \leq a_{\gamma_2}$. Any C^*-algebra has an increasing approximate identity.

The norm on a C^*-algebra $\mathcal{A}$ is also determined by the underlying algebraic structure. Sketching the argument for this, it suffices to consider the case that $\mathcal{A}$ is unital since we may replace $\mathcal{A}$ by its unital extension. If $a \in \mathcal{A}$, then the *spectrum* of a is the non-empty compact set

$$\mathrm{sp}\,(a) = \{\alpha \in \mathbb{C} : a - \alpha 1 \text{ is not invertible}\},$$

and

$$\|a\|^2 = \sup\{|\zeta| : \zeta \in \mathrm{sp}\,(a^*a)\}. \tag{A.4.1}$$

It immediately follows, for example, that any unital $*$-homomorphism of unital C^*-algebras is norm-decreasing (this is also true without the units).

If $\mathcal{A}$ is unital, then the norm and the ordering are closely linked since if $a = a^*$, then

$$\|a\| = \min\{\alpha : -\alpha 1 \leq a \leq \alpha 1\}. \tag{A.4.2}$$

The following result is useful in this context. Owing to the Gelfand–Naimark theorem, it may be applied to elements of arbitrary unital C^*-algebras.

Lemma A.4.2 *Suppose that b is an operator on a Hilbert space H. Then $-I \leq b \leq I$ if and only if*

$$\|b - \mathrm{i}tI\| \leq \sqrt{1 + t^2}$$

for all $t \in \mathbb{R}$.

Proof Given a complex number β, we have $-1 \leq \beta \leq 1$ if and only if

$$|\beta - \mathrm{i}t| \leq \sqrt{1 + t^2}$$

for all $t \in \mathbb{R}$. This is readily verified by inspecting the intersection of all the disks having -1 and 1 in their circumference. From this we see that if f is a continuous function on a compact Hausdorff space Ω, then $-1 \leq f \leq 1$ if and only if

$$\|f - \mathrm{i}t1\| \leq \sqrt{1 + t^2}$$

for all $t \in \mathbb{R}$. It follows that if $b \in \mathcal{B}(H)$ satisfies $-I \leq b \leq I$, then from the spectral theorem,

$$\|b - \mathrm{i}tI\| \leq \sqrt{1 + t^2}$$

for all $t \in \mathbb{R}$.

Conversely, if we have these inequalities, then for any unit vector $\xi \in H$,

$$|\langle b\xi \mid \xi\rangle - \mathrm{i}t| = |\langle (b - \mathrm{i}tI)\xi \mid \xi\rangle| \leq \sqrt{1 + t^2}$$

and

$$-\langle I\xi \mid \xi\rangle = -1 \leq \langle b\xi \mid \xi\rangle \leq 1 = \langle I\xi \mid \xi\rangle.$$

We therefore have $-I \leq b \leq I$. □

The notion of a *state*, or a 'non-commutative probability measure' on a unital C^*-algebra $\mathcal{A}$ provides the key technical tool for proving the Gelfand–Naimark theorem. A state is a functional $p \in \mathcal{A}^*$ such that $p \geq 0$, i.e.

$p(a) \geq 0$ for all $a \in \mathcal{A}^+$ and $p(I) = 1$. It follows that $\|p\| = 1$. We may associate with each state p, a Hilbert space H_p, a unit vector $\xi_p \in H_p$, and an essentially unique GNS *representation* π_p of $\mathcal{A}$ into $\mathcal{B}(H_p)$ such that

$$p(a) = \langle \pi_p(a)\xi_p \mid \xi_p \rangle,$$

and $\pi_p(\mathcal{A})\xi_p$ is norm dense in H_p. The Gelfand–Naimark theorem for a C^*-algebra $\mathcal{A}$ is proved by showing that it has sufficiently many states. If X is a compact space, then the states on $C(X)$ may be identified with the regular Borel probability measures on X.

A *von Neumann algebra* $\mathcal{R}$ on a Hilbert space H is a C^*-algebra on H which is closed in the weak operator topology, i.e. the topology determined by the functionals

$$\mathcal{R} \mapsto \mathbb{C} : r \mapsto \langle r\eta \mid \xi \rangle$$

for $\eta, \xi \in H$. These algebras have the following abstract characterization.

Theorem A.4.3 (Sakai) *A C^*-algebra $\mathcal{R}$ is isometrically $*$-isomorphic to a von Neumann algebra if and only if its underlying Banach space is the Banach dual of another Banach space $\mathcal{R}_*$. If that is the case, then $\mathcal{R}_*$ is essentially unique.* □

If $\mathcal{R}$ is a von Neumann algebra, then the weak* topology determined by $\mathcal{R}_*$ coincides on bounded sets with the weak operator topology.

If $(X, \mathcal{S}, \mu)$ is a measure space, then $\mathcal{R} = L^\infty(X, \mathcal{S}, \mu)$ is a von Neumann algebra with predual $\mathcal{R}_* = L^1(X, \mathcal{S}, \mu)$, whereas for any Hilbert space H, $\mathcal{R} = \mathcal{B}(H)$ is a von Neumann algebra with predual $\mathcal{R}_* = \mathcal{T}(H)$, the trace class operators on H (see §1.3).

Any von Neumann algebra $\mathcal{R}$ is *conditionally order complete.* By this we mean that any increasing net $\{a_\gamma\}_{\gamma\in\Gamma}$ in $\mathcal{R}_{sa}$ which is bounded above has a least upper bound a_0. The net converges to a_0 in the weak* topology.

The first theorem of von Neumann algebra theory is von Neumann's double commutant theorem. If $\mathcal{S}$ is a collection of operators on a Hilbert space H, then its *commutant* is the collection

$$\mathcal{S}' = \{b \in \mathcal{B}(H) : sb = bs \text{ for all s} \in \mathcal{S}\}.$$

Theorem A.4.4 (von Neumann) *If $\mathcal{R}$ is a self-adjoint algebra of operators on a Hilbert space H which contains the identity operator I, then the weak operator and the weak* closures of $\mathcal{R}$ are equal to the double commutant algebra $\mathcal{R}''$.* □

Finally, we shall need an important result regarding commutants of tensor products. The proof of this result represented the simplest major application of the Tomita–Takesaki theory.

Theorem A.4.5 *If $\mathcal{R} \subseteq \mathcal{B}(H)$ and $\mathcal{S} \subseteq \mathcal{B}(K)$ are von Neumann algebras, then*

$$(\mathcal{R} \overline{\otimes} \mathcal{S})' = \mathcal{R}' \overline{\otimes} \mathcal{S}'.$$

□

A.5 A BRIEF LIST OF OPERATOR ALGEBRAS

The study of operator spaces began with the discovery that many of the criteria for distinguishing operator algebras seemed to be invariants of the underlying linear spaces. As an aid to the reader, we shall briefly summarize some of these properties. The definitions necessarily entail notions from operator space theory, and thus the reader may find it necessary to refer to the relevant sections of the monograph.

As was historically case, one usually begins with von Neumann algebras. The finite-dimensional von Neumann algebras coincide with the finite-dimensional C^*-algebras, and a variation on the Wedderburn theorem states that they are necessarily of the form

$$\mathcal{R} = M_{n_1} \oplus \cdots \oplus M_{n_r}.$$

In the global version of the Murray and von Neumann theory (investigated by Segal, Kadison, and Dixmier), the next category of von Neumann algebras are the *type I* von Neumann algebras, which on separable Hilbert spaces are those of the form

$$\mathcal{R} = \prod_{k \in \mathbb{N} \cup \{\infty\}} L^\infty(X_k, \mathcal{S}_k, \mu_k) \overline{\otimes} M_k.$$

At this point, the most general class of von Neumann algebras with a reasonably complete structure theory are the *injective* von Neumann algebras. These were first characterized by the fact that they behave very much like injective Banach spaces. Owing essentially to the outstanding work of Connes (1976), it is now known that these algebras may be internally characterized as those algebras which are suitably generated by finite-dimensional subalgebras. Beginning with the work of Murray and von Neumann, there is a beautiful classification theory for these algebras (see Connes 1976 and Haagerup 1987).

Despite the fact that they naturally arise in both geometry and algebra, little of a general nature is known about the non-injective von Neumann algebras. Perhaps the most important progress that has been made in this direction are results stemming from Voiculescu's theory of free probability.

If $\mathcal{A}$ is a C^*-algebra, then $\mathcal{A}^{**}$ has a natural von Neumann algebraic structure. We have, for example, that $c_0^{**} \cong \ell_\infty$ and $K_\infty^{**} \cong M_\infty$. A C^*-algebra is said to be of type I if $\mathcal{A}^{**}$ is a type I von Neumann algebra. Equivalently, if $\pi : \mathcal{A} \to \mathcal{B}(H)$ is any $*$-representation of $\mathcal{A}$, then the weak operator closure $\overline{\pi(\mathcal{A})}$ is a type I von Neumann algebra. One of the triumphs of the early theory of C^*-algebra was Glimm's proof that a C^*-algebra is type I if and only if it is a GCR algebra in the sense of Kaplansky, or equivalently, it has a 'smooth dual' in the sense of Mackey. If G is a connected semisimple Lie group, then it is known that its reduced group C^*-algebra $C^*_\lambda(G)$ is type I. On the other hand, there are solvable Lie groups G for which $C^*_\lambda(G)$ is not of type I. Furthermore, non-trivially ergodic dynamical systems also lead to non type I C^*-algebras. Finally, inductive limits of C^*-algebras, such as the UHF algebra, are generally not type I. For all of these reasons, the type I algebras do not play a particularly important role in operator algebra theory.

A C^*-algebra $\mathcal{A}$ is *nuclear* if there exist diagrams of completely contractive mappings (or equivalently, completely contractive and completely positive mappings)

$$\begin{array}{ccccc} & & M_n & & \\ & {}^{r}\nearrow & & \searrow^{s} & \\ \mathcal{A} & & \xrightarrow{\;id\;} & & \mathcal{A}\,, \end{array}$$

which approximately commute in the point-norm topology. It is well known in operator algebra theory that a C^*-algebra $\mathcal{A}$ is nuclear if and only if $\mathcal{A}^{**}$ is an injective von Neumann algebra. Owing to the work of Pukanszky and Connes, it is known that if G is any connected second countable locally compact group, then the reduced group C^*-algebra $C^*_\lambda(G)$ is nuclear. On the other hand, if G is a discrete group, then we have Lance's theorem that G is amenable if and only if the full group C^*-algebra $C^*(G)$ (respectively, the reduced group C^*-algebra $C^*_\lambda(G)$) is nuclear. We can conclude from this that $C^*(\mathbb{F}_2)$ and $C^*_\lambda(\mathbb{F}_2)$ are not nuclear since $\mathbb{F}_2$ is not amenable.

Beyond the theory of nuclearity, we find a wide range of properties that are beautifully interrelated. Perhaps the most important of these is Kirchberg's notion of exactness. A C^*-algebra $\mathcal{B}$ is said to be *exact* if for any C^*-algebra $\mathcal{A}$ and closed ideal $\mathcal{I} \subseteq \mathcal{A}$, the sequence

$$0 \longrightarrow \mathcal{I}\check{\otimes}\mathcal{B} \longrightarrow \mathcal{A}\check{\otimes}\mathcal{B} \longrightarrow (\mathcal{A}/\mathcal{J})\check{\otimes}\mathcal{B} \longrightarrow 0$$

is 1-exact (see §14.4). One of the key motivations of operator space theory is the light that it sheds on this property. The deepest theorem about exact C^*-algebras is the fact that a C^*-algebra is exact if and only if it is a C^*-subalgebra of a nuclear C^*-algebra. It is known (see §14.5) that the full group C^*-algebra $C^*(\mathbb{F}_2)$ is not exact. It has been conjectured by Kirchberg that the reduced group C^*-algebra of any discrete group is exact.

There are a host of other C^*-algebraic properties that are currently being investigated, including various approximation properties, local reflexivity, the weak expectation property, the local lifting property, quasi-diagonality, and so on. The most convincing demonstration of the importance of these various notions may be found in Kirchberg's deep paper of (1993).

A.6 ASYMPTOTIC PRODUCTS AND ULTRAPRODUCTS OF BANACH SPACES

Let us suppose that $(V_n)_{n\in\mathbb{N}}$ is a sequence of Banach spaces. The linear space

$$\sum_{n\in\mathbb{N}} V_n = \left\{ (v_n) \in \prod_{n\in\mathbb{N}} V_n : \lim_{n\to\infty} \|v_n\| = 0 \right\}$$

is a closed subspace of the product Banach space $\prod_{n\in\mathbb{N}} V_n$, and we define the *asymptotic product Banach space* to be the quotient Banach space $\prod_{n\in\mathbb{N}} V_n / \sum_{n\in\mathbb{N}} V_n$. We let π_∞ denote the corresponding quotient mapping.

Proposition A.6.1 *Given a sequence of Banach spaces $(V_n)_{n\in\mathbb{N}}$ and an element $v = (v_n) \in \prod_{n\in\mathbb{N}} V_n$, we have*

$$\|\pi_\infty((v_n))\| = \limsup_{n\to\infty} \|v_n\| . \tag{A.6.1}$$

Proof Given $v = (v_n) \in \prod_{n\in\mathbb{N}} V_n$ and $p \in \mathbb{N}$, we define $v^{[p]} \in \prod_{n\in\mathbb{N}} V_n$ by $v_n^{[p]} = 0$ for $1 \le n \le p-1$ and $v_n^{[p]} = v_n$ for $n \ge p$. Since

$$\pi_\infty(v) = \pi_\infty(v^{[p]}),$$

it follows that

$$\|\pi_\infty(v)\| \le \sup \{\|v_n\| : n \ge p\} ,$$

and thus

$$\|\pi_\infty(v)\| \le \limsup_{n\to\infty} \|v_n\| .$$

On the other hand, for every $\varepsilon > 0$, we may find an $h = (h_n) \in \sum_{n\in\mathbb{N}} V_n$ such that

$$\|v + h\| \le \|\pi_\infty(v)\| + \varepsilon.$$

Since $\|h_n\| \to 0$,

$$\begin{aligned} \limsup_{n\to\infty} \|v_n\| &= \limsup_{n\to\infty} \|v_n + h_n\| \\ &\le \sup \{\|v_n + h_n\|\} \le \|v + h\| \le \|\pi_\infty(v)\| + \varepsilon, \end{aligned}$$

and since $\varepsilon > 0$ is arbitrary, we have the desired equality. □

We shall also need the usual notion of ultraproducts, which is quite useful in the 'local theory' of Banach spaces, and seems to be even more important in operator space theory. This technical material is not essential to much of this monograph, and the reader may wish to skip over the various places it occurs. Although filters and ultrafilters are often used in this context, we prefer to use nets. The simplest way to relate these approaches is to introduce the Stone–Čech compactification of the index space, as we indicate below. A detailed presentation of this theory can be found in Kelley (1955).

Let $\mathfrak{s}$ be an arbitrary set. A non-empty collection $\mathcal{F}$ of subset of $\mathfrak{s}$ is a *filter* on $\mathfrak{s}$ if

(i) $\emptyset \notin \mathcal{F}$;
(ii) $A, B \in \mathcal{F}$ implies that $A \cap B \in \mathcal{F}$;
(iii) $A \in \mathcal{F}$ and $A \subseteq B$ implies that $B \in \mathcal{F}$.

A filter $\mathcal{F}$ on $\mathfrak{s}$ is an *ultrafilter* if it is maximal in the sense that whenever $\mathcal{G}$ is a filter on $\mathfrak{s}$ and $\mathcal{F} \subseteq \mathcal{G}$, then $\mathcal{F} = \mathcal{G}$. Given an element $s \in \mathfrak{s}$, it is easy to see that

$$\mathcal{U}_s = \{A \subseteq \mathfrak{s} : \ s \in A\}$$

is an ultrafilter on $\mathfrak{s}$. We call such an ultrafilter $\mathcal{U}_s$ a *fixed* ultrafilter. An ultrafilter $\mathcal{U}$ is said to be *free* if it is not fixed.

Theorem A.6.2 *If $\mathfrak{s}$ is an infinite set, then there is a free ultrafilter on $\mathfrak{s}$.*

Proof The collection of cofinite sets in $\mathfrak{s}$

$$\mathcal{F} = \{F \subseteq \mathfrak{s} : \ \mathfrak{s} \setminus F \text{ finite}\}$$

is a filter. It follows from Zorn's lemma that there is an ultrafilter $\mathcal{U}$ on $\mathfrak{s}$ containing $\mathcal{F}$. Since $\bigcap_{F \in \mathcal{F}} F = \emptyset$, $\mathcal{U}$ is a free ultrafilter on $\mathfrak{s}$. □

Given an infinite set $\mathfrak{s}$, we define the *Stone–Čech compactification* $\beta\mathfrak{s}$ to be the spectrum of the commutative C^*-algebra $\ell_\infty(\mathfrak{s})$, i.e. the set of all non-zero homomorphisms $\omega : \ell_\infty(\mathfrak{s}) \to \mathbb{C}$. The set $\beta\mathfrak{s}$ is compact Hausdorff in the weak* topology. Since $\ell_\infty(\mathfrak{s})$ is order-theoretically conditionally complete, its spectrum $\beta\mathfrak{s}$ is also Stonean, i.e. the closure of an open set is open. Each $s \in \mathfrak{s}$ determines a non-zero homomorphism $\omega_s \in \beta\mathfrak{s}$ given by $\omega_s(f) = f(s)$, and if we let $\mathfrak{s}$ have the discrete topology, we can use the mapping $s \mapsto \omega_s$ to identify $\mathfrak{s}$ with a dense open subset of $\beta\mathfrak{s}$. Each function $f = (f(s)) \in \ell_\infty(\mathfrak{s})$ determines a continuous function $\tilde{f}$ on $\beta\mathfrak{s}$, where we let $\tilde{f}(\omega) = \omega(f)$. Using the usual notion of a function limit, we may write $\tilde{f}(\omega) = \lim_{s \to \omega} f(s)$. The mapping $f \mapsto \tilde{f}$ determines a C^*-algebraic isomorphism of $\ell_\infty(\mathfrak{s})$ onto $C(\beta\mathfrak{s})$. If we let $\mathrm{Proj}\mathcal{A}$ denote the projection in a commutative C^*-algebra $\mathcal{A}$, we have the Boolean algebraic isomorphisms

$$\text{the subsets of } \mathfrak{s} \leftrightarrow \mathrm{Proj}\,\ell_\infty(\mathfrak{s}) \leftrightarrow \mathrm{Proj}\,C(\beta\mathfrak{s}) \leftrightarrow \text{the clopen sets in } \beta\mathfrak{s}\,.$$

Indeed, there is a one-to-one correspondence between a set $A \subseteq \mathfrak{s}$ and the clopen set $\overline{A}$ in $\beta\mathfrak{s}$. From this, we can see that the clopen sets in $\beta\mathfrak{s}$ form a topological basis for $\beta\mathfrak{s}$.

Given an ultrafilter $\mathcal{U}$ in $\mathfrak{s}$, the sets $\overline{A}$ $(A \in \mathcal{U})$ form a maximal filter of compact open sets in $\beta\mathfrak{s}$, and thus there is a unique point $\omega = \omega_{\mathcal{U}} \in \beta\mathfrak{s}$ with

$$\bigcap\{\overline{A} : A \in \mathcal{U}\} = \{\omega\}\,,$$

and conversely, for each $\omega \in \beta\mathfrak{s}$,

$$\mathcal{U} = \mathcal{U}_\omega = \{A \subseteq \mathfrak{s} : \omega \in \overline{A}\}$$

is an ultrafilter in $\mathfrak{s}$. These relations give us a bijection of the ultrafilters (respectively, free ultrafilters) in $\mathfrak{s}$ with the points in $\beta\mathfrak{s}$ (respectively, the points in $\beta\mathfrak{s}\backslash\mathfrak{s}$). Given a bounded mapping $f : \mathfrak{s} \to \mathbb{C}$, we define the *ultralimit* $\lim_{\mathcal{U}} f(s)$ by

$$\lim_{\mathcal{U}} f(s) = \lim_{s\to\omega} f(s) = \tilde{f}(\omega).$$

Equivalently, if we choose a net $(\iota_\gamma)_{\gamma\in\Gamma}$ in $\mathfrak{s}$ converging to ω, we have the net-theoretic limit

$$\lim_{\mathcal{U}} f(s) = \lim_{\gamma} f(s_\gamma).$$

Let us fix an ultrafilter $\mathcal{U}$ on $\mathfrak{s}$, and let ω be the corresponding point in $\beta\mathfrak{s}$. If X is a compact Hausdorff space and $x : \mathfrak{s} \to X$ is an arbitrary function, then there exists a unique point $x_{\mathcal{U}} = \lim_{\mathcal{U}} x(s)$ with the property that

$$\varphi(x_{\mathcal{U}}) = \lim_{\mathcal{U}} \varphi(x(s))$$

for each continuous function $\varphi : X \to [0,1]$. To see this, we let $\mathcal{G}$ denote the collection of all such functions φ. From Urysohn's lemma, the mapping

$$x \mapsto (\varphi(x))_{\varphi\in\mathcal{G}}$$

is a homeomorphism θ of X onto a compact subset $X' \subseteq [0,1]^{\mathcal{G}}$. If we let $s_\gamma \in \mathfrak{s}$ be a net converging to ω, then for each continuous function φ, $\varphi(x_{s_\gamma})$ converges to $\lim_{\mathcal{U}} \varphi(x_{s_\gamma})$, and thus the net $(\varphi(x_{s_\gamma}))_{\varphi\in\mathcal{G}}$ converges in the product topology to some element $t = (t_\varphi)_{\varphi\in\mathcal{G}}$ of X'. The point $x_{\mathcal{U}} = \theta^{-1}(t)$ has the desired property.

Given a family of Banach spaces $(V_s)_{s\in\mathfrak{s}}$, and an element

$$v = (v_s)_{s\in\mathfrak{s}} \in \prod_{s\in\mathfrak{s}} V_s = \ell_\infty(\mathfrak{s}; V_s),$$

the set of non-negative numbers $(\|v_s\|)_{s\in\mathfrak{s}}$ is contained in the compact set $[0, \|(v_s)\|_\infty]$. It follows that the ultralimit $\lim_{\mathcal{U}} \|v_s\|$ exists and satisfies

$$\lim_{\mathcal{U}} \|v_s\| \leq \|(v_s)\|_\infty .$$

We let

$$J_{\mathcal{U}} = \left\{ (v_s)_{s\in\mathfrak{s}} \in \prod_{s\in\mathfrak{s}} V_s : \lim_{\mathcal{U}} \|v_s\| = 0 \right\}.$$

Then $J_{\mathcal{U}}$ is a closed subspace of $\prod_{s\in\mathfrak{s}} V_s$. The quotient space

$$\prod_{s\in\mathfrak{s}} V_s/\mathcal{U} = \left(\prod_{s\in\mathfrak{s}} V_s\right) \Big/ J_{\mathcal{U}}$$

is called the *ultraproduct* of $(V_s)_{s\in\mathfrak{s}}$. If $V_s = V$ for some fixed Banach space V, then we write $V^{\mathcal{U}}$ for $\prod_{s\in\mathfrak{s}} V_s/\mathcal{U}$. We let $\pi_{\mathcal{U}}$ denote the quotient mapping and let $\pi_{\mathcal{U}}((v_s)) \in \prod_{s\in\mathfrak{s}} V_s/\mathcal{U}$ denote the image of an element $(v_s) \in \prod_{s\in\mathfrak{s}} V_s$.

Lemma A.6.3 *Given an indexed family of Banach spaces $(V_s)_{s\in\mathfrak{s}}$ with $\mathfrak{s}$ infinite, a free ultrafilter $\mathcal{U}$ on $\mathfrak{s}$ and an element*

$$v = (v_s) \in \prod_{s\in\mathfrak{s}} V_s,$$

we have

$$\|\pi_{\mathcal{U}}(v)\| = \lim_{\mathcal{U}} \|v_s\|. \tag{A.6.2}$$

Proof If $L = \lim_{\mathcal{U}} \|v_s\|$ and $\varepsilon > 0$, we may choose a set $A \in \mathcal{U}$ such that for all $s \in A$, $|\|v_s\| - L| < \varepsilon$, and thus $\|v_s\| < L + \varepsilon$. If we let $v_s^A = v_s$ for $s \in A$ and $v_s^A = 0$ otherwise, then

$$\|\pi_{\mathcal{U}}(v)\| = \left\|\pi_{\mathcal{U}}(v^A)\right\| \leq \sup\left\{\left\|v_s^A\right\|\right\} \leq L + \varepsilon.$$

On the other hand, we may find an element $h = (h_s) \in J_{\mathcal{U}}$ such that

$$\|v + h\| \leq \|\pi_{\mathcal{U}}(v)\| + \varepsilon,$$

and thus since $\lim_{\mathcal{U}} h(s) = 0$,

$$L = \lim_{\mathcal{U}} \|v_s\| = \lim_{\mathcal{U}} \|v_s + h_s\| \leq \sup\{\|v_s + h_s\| : s \in \mathfrak{s}\} \leq \|\pi_{\mathcal{U}}(v)\| + \varepsilon. \quad \square$$

Proofs of the following additional facts may be found in Heinrich (1980).

Proposition A.6.4 *Let $\mathfrak{s}$ be an infinite index set and $\mathcal{U}$ a free ultrafilter on $\mathfrak{s}$.*

(i) *Given indexed families of Banach spaces $(V_s)_{s\in\mathfrak{s}}$ and $(W_s)_{s\in\mathfrak{s}}$, there is a natural isometric inclusion*

$$\prod_{s\in\mathfrak{s}} \mathcal{B}(V_s, W_s)/\mathcal{U} \hookrightarrow \mathcal{B}\Big(\prod_{s\in\mathfrak{s}} V_s/\mathcal{U}, \prod_{s\in\mathfrak{s}} W_s/\mathcal{U}\Big) : \pi_{\mathcal{U}}((\varphi_s)) \mapsto (\varphi_s)_{\mathcal{U}},$$

which is given by

$$(\varphi_s)_{\mathcal{U}}(\pi_{\mathcal{U}}((v_s))) = \pi_{\mathcal{U}}((\varphi_s(v_s))).$$

(ii) *For any indexed family of Banach spaces $(V_s)_{s\in\mathfrak{s}}$, there is a natural isometric inclusion*

$$\prod_{s\in\mathfrak{s}} V_s^*/\mathcal{U} \hookrightarrow \Big(\prod_{s\in\mathfrak{s}} V_s/\mathcal{U}\Big)^* : \pi_{\mathcal{U}}((f_s)) \mapsto (f_s)_{\mathcal{U}},$$

which is given by

$$\langle (f_s)_{\mathcal{U}}, \pi_{\mathcal{U}}((v_s))\rangle = \lim_{\mathcal{U}} f_s(v_s). \quad \square$$

REFERENCES

[1] Akemann, C.A. and Ostrand, P.A. (1976). Computing norms in group C^*-algebras. *Amer. J. Math.*, **98**, 1015–47.

[2] Alfsen, E.M. and Shultz, F.W. (1998). Orientation in operator algebras. *Proc. Natl. Acad. Sci. USA*, **95**, no.12, 6596–601

[3] Archbold, R. and Batty, C. (1980). C^*-tensor norms and slice maps. *J. London Math. Soc.*, **22**, 127–38.

[4] Arveson, W. (1969). Subalgebras of C^*-algebras. *Acta Math.*, **123**, 141–224.

[5] Arveson, W. (1972). Subalgebras of C^*-algebras II. *Acta Math.*, **128**, 271–308.

[6] Arveson, W. (1977). Notes on extensions of C^*-algebras. *Duke Math. J.*, **44**, 329–55.

[7] Blecher, D. (1992*a*). The standard dual of an operator space. *Pacific J. Math.*, **153**, 15–30.

[8] Blecher, D. (1992*b*). Tensor products of operator spaces II. *Canad. J. Math.*, **44**, 75–90.

[9] Blecher, D. (1995). A completely bounded characterization of operator algebras. *Math. Ann.*, **303**, 227–39.

[10] Blecher, D., Muhly, P. and Paulsen, V. (1999). *Categories of operator modules* (*Morita equivalence and projective modules*). Memoirs of the American Mathematical Society, Providence, Rhode Island, to appear.

[11] Blecher, D. and Paulsen, V. (1991*a*). Tensor products of operator spaces. *J. Funct. Anal.*, **99**, 262–92.

[12] Blecher, D. and Paulsen, V. (1991*b*). Explicit construction of universal operator algebras and applications to polynomial factorization. *Proc. Amer. Math. Soc.*, **112**, 839–50.

[13] Blecher, D., Ruan Z.-J. and Sinclair, A. (1990). A characterization of operator algebras. *J. Funct. Anal.*, **89**, 188–201.

[14] Blecher, D. and Smith, R. (1992). The dual of the Haagerup tensor product. *J. London Math. Soc.*, **45**, 126–44.

[15] Browder, A. (1969). *Introduction to function algebras.* W. A. Benjamin, New York.

[16] Cartan, H. and Eilenberg, S. (1956). *Homological algebra.* Princeton Univ. Press, Princeton, New Jersey.

[17] Choi, M.-D. (1974). A Schwarz inequality for positive linear maps on C^*-algebras. *Illinois J. Math.*, **18**, 565–74.

[18] Choi, M.-D. (1975). Completely positive linear maps on complex matrices. *Linear Algebra and Its Applications*, **10**, 285–90.

[19] Choi, M.-D. and Effros, E.G. (1976). The completely positive lifting problem for C^*-algebras. *Ann. Math.*, **104**, 585–609.

[20] Choi, M.-D. and Effros, E.G. (1977*a*). Injectivity and operator spaces. *J. Funct. Anal.*, **24**, 156–209.

[21] Choi, M.-D. and Effros, E.G. (1977*b*). Nuclear C^*-algebras and injectivity: the general case. *Indiana Univ. Math. J.*, **26**, 443–46.

[22] Choi, M.-D. and Effros, E.G. (1978). Nuclear C^*-algebras and the approximation property. *Amer. J. Math.*, **100**, 61–79.

[23] Christensen, E., Effros, E.G. and Sinclair, A. (1987). Completely bounded multilinear maps and C^*-algebraic cohomology. *Invent. Math.*, **90**, 279–96.

[24] Christensen, E. and Sinclair, A. (1987). Representations of completely bounded multilinear operators. *J. Funct. Anal.*, **72**, 151–81.

[25] Connes, A. (1976). Classification of injective factors. Cases II_1, II_∞, III_λ, $\lambda \neq 1$. *Ann. Math.*, **104**, 73–115.

[26] Connes, A. (1978). On the cohomology of operator algebras. *J. Funct. Anal.*, **28**, 248–53.

[27] Dean, D. (1973). The equation $L(E, X^{**}) = L(E, X)^{**}$ and the principle of local reflexivity. *Proc. Amer. Math Soc.***40**, 146–8.

[28] Defant, A. and Floret, K. (1993). *Tensor norms and operator ideals.* North-Holland, Amsterdam.

[29] Dixmier, J. (1964). *Les C*-algèbres et leurs représentations.* Gauthier-Villars, Paris.

[30] Dixmier, J. (1981). *Von Neumann algebras.* North-Holland, Amsterdam.

[31] Dunford, N. and Schwartz, J. (1958). *Linear operators, Part I.* Interscience, New York.

[32] Effros, E.G. (1987*a*). Advances in quantized functional analysis. *Proceedings of the International Congress of Mathematics* (Berkeley, Calif., 1986), American Mathematical Society, Providence, Rhode Island, 906–16.

[33] Effros, E.G. (1987*b*). On multilinear completely bounded maps. *Contemp. Math.*, **62**, 479-501.

[34] Effros, E.G. (1988). Amenability and virtual diagonals for von Neumann algebras. *J. Funct. Anal.*, **78**, 137–53.

[35] Effros, E.G. and Haagerup, U. (1985). Lifting problems and local reflexivity for C^*-algebras. *Duke Math. J.*, **52**, 103–28.

[36] Effros, E.G., Junge, M. and Ruan, Z-J. (2000). Integral mappings and the principle of local reflexivity for non-commutative L^1-spaces. *Ann. Math.*, **151**, 59–92.

[37] Effros, E.G. and Kishimoto, A. (1987). Module maps and Hochschild-Johnson cohomology. *Indiana Univ. Math. J.*, **36**, 257–76.

[38] Effros, E.G., Kraus, J. and Ruan, Z-J. (1993). On two quantized tensor norms. *Operator algebras, mathematical physics, and low dimensional topology* (Istanbul 1991), Research Notes in Math. 5, A K Peters, Wellesley, Massachusetts, 125–45.

[39] Effros, E.G. and Lance, C. (1977). Tensor products of operator algebras. *Adv. Math.*, **25**, 1–34.

[40] Effros, E.G., Ozawa, N. and Ruan, Z-J. (1999). On injectivity and nuclearity for operator spaces. Preprint.

[41] Effros, E.G. and Ruan, Z-J. (1988*a*). On matricially normed spaces. *Pacific J. Math.*, **132**, 243–64.

[42] Effros, E.G. and Ruan, Z-J. (1988*b*). Representations of operator bimodules and their applications. *J. Operator Theory*, **19**, 137–57.

[43] Effros, E.G. and Ruan, Z-J. (1990). On approximation properties for operator spaces. *International J. Math.*, **1**, 163–87.

[44] Effros, E.G. and Ruan, Z-J. (1991*a*). A new approach to operator spaces. *Canad. Math. Bull.*, **34**, 329–37.

[45] Effros, E.G. and Ruan, Z-J. (1991*b*). Self-duality for the Haagerup tensor product and Hilbert space factorizations. *J. Funct. Anal.*, **100**, 257–84.

[46] Effros, E.G. and Ruan, Z-J. (1993). On the abstract characterization of operator spaces. *Proc. Amer. Math. Soc.*, **119**, 579–84.

[47] Effros, E.G. and Ruan, Z-J. (1994*a*). Mapping spaces and liftings for operator spaces. *Proc. London Math. Soc.*, **69**, 171–97.

[48] Effros, E.G. and Ruan, Z-J. (1994*b*). The Grothendieck-Pietsch and Dvoretzky–Rogers Theorems for operator spaces. *J. Funct. Anal.*, **69**, 171–97.

[49] Effros, E.G. and Ruan, Z-J. (1997). On the analogues of integral mappings and local reflexivity for operator spaces. *Indiana Univ. Math. J.*, **46**, 1289–310.

[50] Effros, E.G. and Ruan, Z-J. (1998). $\mathcal{OL}_p$ spaces. *AMS Contemporary Math.*, **228**, 51–77.

[51] Effros, E.G. and Webster, C. (1997). Operator analogues of locally convex spaces, *Operator algebras and applications* (Samos, 1996), ed. by A. Katavolos, Kluwer, London, 163–207.

[52] Effros, E. G. and Winkler, S. (1997). Matrix convexity: operator analogues of the bipolar and Hahn–Banach theorems. *J. Funct. Anal.*, **144**, 117–52.

[53] Enflo, P. (1973). A counterexample to the approximation property in Banach spaces. *Acta Math*, **130**, 309–17.

[54] Fell, J.M.G. (1962). Weak containment and induced representations of groups. *Canadian J. Math.*, **14**, 237–68.

[55] Gelfand, I.M. and Naimark, M.A. (1943). On the imbedding of normed rings into the ring of operators in Hilbert space. *Mat. Sbornik*, **12**, 197–213.

[56] Groh, U. (1984). Uniform ergodic theorems for identity preserving Schwartz maps on W^*-algebras. *J. Operator Theory*, **11**, 395–404.

[57] Grothendieck, A. (1955). *Products tensoriels topologiques et espaces nuclearies.* Memoirs of the American Mathematical Society, **16**, Providence, Rhode Island.

[58] Haagerup, U. (1979). An example of a non-nuclear C^*-algebra which has the metric approximation property. *Invent. Math.*, **50**, 339–61.

[59] Haagerup, U. (1980). Decomposition of completely bounded maps on operator algebras. Unpublished notes.

[60] Haagerup, U. (1983). All nuclear C^*-algebras are amenable. *Invent. Math.*, **74**, 305–19.

[61] Haagerup, U. (1985*a*). The Grothendieck inequality for bilinear forms on C^*-algebras. *Adv. Math.*, **56**, 93–116.

[62] Haagerup, U. (1985*b*). Injectivity and decomposition of completely bounded maps. *Operator algebras and their connections with topology and ergodic theory* (Buşteni, 1983), Lecture Notes in Math., **1132**, Springer-Verlag, Berlin-New York, 170–222.

[63] Haagerup, U. (1987). Connes' bicentralizer problem and uniqueness of the injective factor of type III_1. *Acta Math.* 158 (1987), no. 1-2, 95–148.

[64] Haagerup, U. and Kraus, J. (1994). Approximation properties for group C^*-algebras and group von Neumann algebras. *Trans. Amer. Math. Soc.*, **344**, 667–99.

[65] Hamana, M. (1979). Injective envelopes of operator systems. *Publ. Res. Inst. Math. Sci. Kyoto Univ.*, **15**, 773–85.

[66] Hamana, M. (1989) Injective envelope of dynamical systems, unpublished manuscript.

[67] Hamana, M. (1992). Injective envelope of dynamical systems, *Operator algebras and operator theory*, Pitman Research Notes in Mathematics Series, No. 271, 69–77, Longman Scientific and Technical, Essex.

[68] Heinrich, S. (1980). Ultraproducts in Banach space theory. *J. Reine Angew Math.*, **313**, 72–104.

[69] Hewitt, E and Ross, K. (1963). *Abstract harmonic analysis I.* Springer-Verlag, New York.

[70] Johnson, B.E. (1972*a*). *Cohomology in Banach algebras.* Memoirs of the American Mathematical Society **127**, Providence, Rhode Island.

[71] Johnson, B.E. (1972*b*). Approximate diagonals and cohomology of certain annihilator Banach algebras. *Amer. J. Math.*, **94**, 685–98.

[72] Johnson, B.E. (1994). Non-amenability of the Fourier algebra of a compact group. *J. London Math. Soc.*, **50**, 361–74.

[73] Junge, M. (1996). *Factorization theory for spaces of operators.* Habilitationsschrift, Universität Kiel.

[74] Junge, M. and Le Merdy, C. (1999). Factorization through matrix spaces for finite rank maps between C^*-algebras, *Duke Math. J.*, **100**, 299–319.

[75] Junge, M. and Pisier, G. (1995). Bilinear forms on exact operator spaces and $B(H) \otimes B(H)$. *Geom. Funct. Anal.*, **5**, 329–63.

[76] Kadison, R.V. (1951). *A representation theory for commutative topological algebra.* Memoirs of the American Mathematical Society **7**, Providence, Rhode Island.

[77] Kadison, R.V. and Ringrose, J.R. (1997*a*). *Fundamentals of the theory of operator algebras. Vol. I. Elementary theory.* Graduate Studies in Mathematics, **15**. American Mathematical Society, Providence, Rhode Island.

[78] Kadison, R.V. and Ringrose, J.R. (1997*b*). *Fundamentals of the theory of operator algebras. Vol II. Advanced theory.* Graduate Studies in Mathematics, **16**. American Mathematical Society, Providence, Rhode Island.

[79] Kadison, R.V. and Singer, I.M. (1960). Triangular operator algebras. Fundamentals and hyperreducible theory. *Amer. J. Math.*, **82**, 227–59.

[80] Kelley, J. (1955). *General toplogy.* Van Nostrand, Princeton, New Jersey.

[81] Kirchberg, E. (1983). The Fubini theorem for exact C^*-algebras. *J. Operator Theory,* **10**, 3–8.

[82] Kirchberg, E. (1993). On nonsemisplit extensions, tensor products and exactness of group C^*-algebras. *Invent. Math.*, **112**, 449–89.

[83] Kirchberg, E. (1995). On subalgebras of the CAR-algebra. *J. Funct. Anal.*, **129**, 35–63.

[84] Kraus, J. (1983). The slice map problem for σ-weakly closed subspaces of von Neumann algebras. *Trans. Amer. Math. Soc.*, **279**, 357–76.

[85] Kraus, J. (1991). The slice map problem and approximation properties. *J. Funct. Anal.*, **102**, 116–55.

[86] Kraus, K. (1983). *States, effects, and operations.* Lecture Notes in Physics **190**, Springer-Verlag, Berlin.

[87] Lance, C. (1973). Nuclear C^*-algebras. *J. Funct. Anal.*, **12**, 157–76.

[88] Le Merdy, C. (1999). An operator space characterization of dual operator algebras. *Amer. J. Math.*, **121**, 55–63.

[89] Lindenstrauss, J. and Rosenthal, H. (1969). The L_p-spaces. *Israel J. Math.*, **7**, 325–49.

[90] Lindenstrauss, J. and Tzafriri, L. (1977). *Classical Banach spaces, I and II.* Springer-Verlag, New York.

[91] Loebl, R. (1975). Contractive linear maps on C^*-algebras. *Michigan Math. J.*, **22**, 361–66.

[92] Murray, F.J. and von Neumann, J. (1936). On rings of operators. *Ann. Math.*, **37**, 116–229.

[93] Murray, F.J. and von Neumann, J. (1937). On rings of operators, II. *Trans. Amer. Math. Soc.*, **41**, 208–48.

[94] Murray, F.J. and von Neumann, J. (1943). On rings of operators, IV. *Ann. Math.*, **44**, 716–808.

[95] Naimark, M.A. (1943). On a representation of additive operator set functions, *C. R. (Doklady), Acad. Sci. URSS* **41**, 359–61.

[96] Von Neumann, J. (1929). Zur algebra der funktionaloperationen und theorie der normalen operatoren. *Math. Ann*, **102**, 370–427.

[97] Von Neumann, J. (1940). On rings of operators, III. *Ann. Math.*, **41**, 94–161.

[98] Von Neumann, J. (1949). On rings of operators. Reduction theory. *Ann. Math.*, **50**, 401–85.

[99] Patterson, A. (1988). *Amenability.* Mathematical Surveys and Monographs **29**. American Mathematical Society, Providence, Rhode Island.

[100] Paulsen, V. (1982). Completely bounded maps on C^*-algebras and invariant operator ranges. *Proc. Amer. Math. Soc.*, **86**, 91–6.

[101] Paulsen, V. (1984). Every completely polynomially bounded operator is similar to a contraction. *J. Funct. Anal.*, **55**, 1–17.

[102] Paulsen, V. (1986). *Completely bounded maps and dilations.* Pitman Res. Notes, Longman Sci. Tech., London.

[103] Paulsen, V. (1992). Representations of function algebras, abstract operator spaces, and Banach space geometry. *J. Funct. Anal.*, **109**, 113–29.

[104] Paulsen, V. (1996). The maximal operator space of a normed space. *Proc. Edinburgh Math. Soc.*, **39**, 309–23.

[105] Paulsen, V. (1998). Relative Yoneda cohomology for operator spaces. *J. Funct. Anal.*, **157**, 358–93.

[106] Paulsen, V. and Smith, R. (1987). Multilinear maps and tensor norms on operator systems. *J. Funct. Anal.* **73**, 258–76.

[107] Pedersen, G.K. (1979). *C^*-algebras and their automorphism groups.* London Mathematical Society Monograph, **14**, Academic Press, London.

[108] Picardello, M. and Pytlik, T. (1988). Norms of free operators. *Proc. Amer. Math. Soc.*, **104**, 257–61.

[109] Pisier, G. (1978). Grothendieck's theorem for non-commutative C^*-algebras with an appendix on Grothendieck's constants. *J. Funct. Anal.*, **29**, 397–415.

[110] Pisier, G. (1986). *Factorization of linear operators and geometry of Banach spaces.* CBMS Regional Conference Series in Math., No. 60.

[111] Pisier, G. (1995). Exact operator spaces. *Recent advances in operator algebras* (Orléans, 1992), Astérisque **232**, 159–86.

[112] Pisier, G. (1996*a*). *Similarity problems and completely bounded maps*, Lecture Notes in Math. **1618**, Springer-Verlag, Berlin.

[113] Pisier, G. (1996*b*). *The operator Hilbert space* OH, *complex interpolation and tensor norms.* Memoirs of the American Mathematical Society, **122**, No. 585, Providence, Rhode Island.

[114] Pisier, G. (1998). *Noncommutative vector valued L_p-spaces and completely p-summing maps.* Astérisque **247**.

[115] Poon, Y-T. and Ruan, Z-J. (1994). Operator algebras with contractive approximate identities. *Canad. J. Math.*, **46**, 397–414.

[116] Raynaud, Y. (1997). On ultrapowers of non-commutative L_1-spaces, *Colloq. del Dept. de Anal. Matematico*, (Univ. Complutense Madrid), to appear.

[117] Ruan, Z-J. (1988). Subspaces of C^*-algebras. *J. Funct. Anal.*, **76**, 217–30.

[118] Ruan, Z-J. (1989). Injectivity of operator spaces. *Trans. Amer. Math. Soc.*, **315**, 89–104.

[119] Ruan, Z-J. (1992). On the predual of dual algebras. *J. Operator Theory*, **27**, 179–92.

[120] Ruan, Z-J. (1994). A characterization of non-unital operator algebras. *Proc. Amer. Math. Soc.*, **121**, 193–8.

[121] Ruan, Z-J. (1995). The operator amenability of A(G). *Amer. J. Math.*, **117**, 1449–74.

[122] Ruan, Z-J. (1996). Amenability of Hopf von Neumann algebras and Kac algebras. *J. Funct. Anal.*, **139**, 466–99.

[123] Sakai, S. (1962). The theory of W^*-algebras, Lecture Notes, Yale University, New Haven.

[124] Sakai, S. (1971). *C^*-algebras and W^*-algebras.* Springer-Verlag.

[125] Schatten, R. (1943). On the direct product of Banach spaces. *Trans. Amer. Math. Soc.*, **53**, 195–217.

[126] Schatten, R. (1950). *A theory of cross-spaces.* Ann. of Math. Studies, **26**, Princeton Univ. Press, Princeton, New Jersey.

[127] Sinclair, A. and Smith, R. (1995). *Hochschild cohomology of von Neumann algebras.* London Mathematical Society Lecture Note Series, 203. Cambridge University Press, Cambridge.

[128] Smith, R. (1983). Completely bounded maps between C^*-algebras. *J. London Math. Soc.*, **27**, 157–66.

[129] Smith, R. (1999). Finite dimensional injective operator spaces. Preprint.

[130] Stinespring, W. (1955). Positive functions on C^*-algebras. *Proc. Amer. Math. Soc.*, **6**, 211–6.

[131] Szankowski, A. (1981). B(H) does not have the approximation property. *Acta Math.*, **147**, 89–108.

[132] Takesaki, M. (1979). *Theory of operator algebras I.* Springer-Verlag, New York.

[133] Tomczak-Jaegermann, N. (1989). *Banach–Mazur distances and finite dimensional operator ideals.* Pitman Monographs and Surveys in Pure and Applied Mathematics **38**.

[134] Tomiyama, J. (1970). *Tensor products and projections of norm one in von Neumann algebras.* Seminar Notes, Univ. of Copenhagen.

[135] Tomiyama, J. (1983). On the transpose map of matrix algebras, *Proc. Amer. Math. Soc.*, **88**, 635–8.

[136] Turumaru, T. (1952). On the direct–product of operator algebras. *Tôhoko. Math. J.*, **4**, 242–51.

[137] Wassermann, S. (1994). *Exact C^*-algebras and related topics.* Lecture Notes Series **19**, Research Institute of Mathematics, Global Analysis Research Center, Seoul National University, Seoul.

[138] Wittstock, G. (1981). Ein operatorvertiger Hahn-Banach Satz. *J. Funct. Anal.*, **40**, 127–50.

[139] Wittstock, G. (1984). Extension of completely bounded C^*-module homomorphism. *Operator algebras and group representations, Vol. II* (Neptun, 1980), Monographs Stud. Math., **18**, 238–50.

[140] Zettl, H. (1983). A characterization of ternary rings of operators. *Adv. Math.*, **48**, 117–43.

Index of Notation

$V^n \cong \mathbb{C}^n \otimes V$	n-tuple space	1
$\varphi^n : V^n \to W^n$	induced mapping of n-tuple spaces	2
$V^{\mathfrak{s}}$	space of $\mathfrak{s}$-tuples	2
$V^\infty = V^{\mathbb{N}}$	space of all sequences	2
$\mathbb{M}_{m,n}(V) \cong \mathbb{M}_{m,n} \otimes V$	m by n matrix space over V	2,3
$\mathbb{M}_n(V) = \mathbb{M}_{n,n}(V)$	n by n matrix space over V	3
$\varepsilon_{i,j} = \varepsilon_{i,j}^{[m,n]}$	matrix units	3
$E_j = E_j^n = \varepsilon_{1,j}^{[1,n]}$	row matrix unit	3
$v \oplus v'$	matrix direct sum	3
$\alpha v \beta$	matrix product	4
$\mathbb{M}_{\mathfrak{s},\mathfrak{t}}(V)$	$\mathfrak{s}$ by $\mathfrak{t}$ matrix space over V	4
$\mathbb{M}_{\mathfrak{s},\mathfrak{t}}^{\text{fin}}(V)$	$v \in \mathbb{M}_{\mathfrak{s},\mathfrak{t}}(V)$ with finitely many non-zero entries	5
$v \circledast w$	Kronecker product of matrices	5
$\varphi_n : \mathbb{M}_n(V) \to \mathbb{M}_n(W)$	induced mapping of matrix spaces	6
$\langle v , w \rangle$	scalar pairing of matrices	6
$\langle\!\langle v , w \rangle\!\rangle$	matrix pairing of matrices	6
trace	trace functional	6, 14, 330, 333
$\|\alpha\|_2$	Hilbert–Schmidt norm of a matrix	7
$\|\alpha\|_1$	trace class norm of a matrix	7
$S_\infty(n) = M_n$	matrix space with operator norm	7
$S_2(n) = HS_n$	Hilbert-Schmidt matrix space	7
$S_1(n) = T_n$	trace class matrix space	7
$\mathbf{t} : M_n \to M_n$	transpose mapping	7
$\mathbf{S}_n$	n by n density matrices	8
$\mathbf{P}_n$	probability measures on $\{1, \ldots, n\}$	8
$H = \bigoplus_{s \in \mathfrak{s}} H_s$	Hilbert space direct sum	10
$H \otimes K$	Hilbert space tensor product	10
$b_1 \otimes b_2$	tensor product of operators	10
$M_n(\mathcal{B}(H)) = \mathcal{B}(H^n)$	matrix norm	11
$\mathcal{FB}(H)$	finite-rank bounded operators on H	14
$\mathcal{S}_1(H) = \mathcal{T}(H)$	trace class operators on H	14
$\mathcal{S}_2(H) = \mathcal{HS}(H)$	Hilbert-Schmidt operators on H	14
$\mathcal{S}_\infty(H) = \mathcal{K}(H)$	compact operators on H	14
$\ell_\infty^n(E)$	Banach n-tuples with supremum norm	19
$\|\cdot\|_n$	matrix norm	19

$\mathbb{M}_{\mathfrak{f}}(V)$	matrix space over a finite index set $\mathfrak{f}$	21
$\|\varphi\|_{cb}$	completely bounded norm	23
$\mathcal{CB}(V,W)$	space of completely bounded mappings	23, 40
$V \hookrightarrow W$	completely isometric injection	23
$V \twoheadrightarrow W$	complete quotient surjection	23
V/N	quotient operator space	37
$\prod_{s\in\mathfrak{s}} V_s = \ell_\infty(\mathfrak{s};V_s)$	product space	38, 333
$\prod_{n\in\mathbb{N}} V_n / \sum_{n\in\mathbb{N}} V_n$	asymptotic product space	39, 343
V_∞	direct limit operator space	40
$\overline{V}$	conjugate space	40, 336
V^*	dual operator space	40
$M_n(V^*) \cong \mathcal{CB}(V, M_n)$	the matrix dual	41
$(\min E, \|x\|_{\min})$	minimal quantization	48
$(\max E, \|x\|_{\max})$	maximal quantization	48
$\mathbb{Z}_n$	free abelian group on n generators	52
$\mathbb{F}_n$	free group on n generators	52
$(H_c, \|\xi\|_c)$	column Hilbert space	54
$(H_r, \|\xi\|_r)$	row Hilbert space	55
$(H, \|\xi\|_o)$	OH operator norm on a Hilbert space	62
$(T_n(V), \|v\|_1)$	the operator space analogue of $\ell_1^n(E)$	67, 129
$T_n(\varphi)$	induced mapping	72
$\varphi \geq_{cp} 0$	φ is complete positivity	78
$\mathcal{L}$	Paulsen's 'off-diagonal' operator system	86
$\mathcal{CP}(\mathcal{A},\mathcal{B})$	cone of completely positive mappings	93
$\|\varphi\|_{dec}$	decomposable norm	95
$L(\mathbb{F}_n)$	left regular von Neumann algebra of $\mathbb{F}_n$	100
$\mathcal{I}(V)$	injective envelope	114
$(E \otimes_\gamma F, \|u\|_\gamma)$	incomplete Banach projective tensor product	123
$E \otimes^\gamma F$	Banach space projective tensor product	123
$(V \otimes_\wedge W, \|u\|_\wedge)$	incomplete operator space projective tensor product	124
$V \widehat{\otimes} W$	operator space projective tensor product	124
$\mathcal{CB}(V \times W, X)$	space of completely bounded bilinear mappings	126
$R(\omega_1) = \omega_1 \otimes id$	right slice mapping	134
$L(\omega_2) = id \otimes \omega_2$	left slice mapping	134
$V^* \overline{\otimes} W^*$	normal spatial tensor product	134
$V^* \overline{\otimes}_{\mathcal{F}} W^*$	normal Fubini tensor product	134
$(E \otimes_\lambda F, \|u\|_\lambda)$	incomplete Banach injective tensor product	137
$E \otimes^\lambda F$	Banach injective tensor product	137
$\Phi : E \otimes^\gamma F \to E \otimes^\lambda F$	canonical contraction	138
$(V \otimes_\vee W, \|u\|_\vee)$	incomplete operator space injective tensor product	139

$V \check{\otimes} W$	operator space injective tensor product	139
$\Phi : V \hat{\otimes} W \to V \check{\otimes} W$	canonical complete contraction	142
$\mathbf{m}$	operator multiplication mapping	148
$\Vert\varphi\Vert_{mb}$	multiplicative bound	149
$\mathcal{MB}(V \times W, X)$	space of multiplicatively bounded mappings	150
$v \odot w$	matrix inner product	150
$(V \otimes_h W, \Vert u\Vert_h)$	incomplete Haagerup tensor product	153
$V \overset{h}{\otimes} W$	Haagerup tensor product	153
$M_\infty(V)$	bounded infinite matrices	175
$M_{\mathfrak{s}}(V)$	bounded matrices with indices in $\mathfrak{s}$	176
$v^{\mathfrak{r}}$, $v^{\mathfrak{f}}$	matrix truncations	176
$K_\infty(V)$	the closure of $\mathbb{M}_\infty^{\mathrm{fin}}(V)$ in $M_\infty(V)$	177
$T_\infty(V)$	infinite trace class matrices over V	180
$\prod_{s\in\mathfrak{s}} V_s/\mathcal{U}$	ultraproduct space	184, 345
$V^{\mathcal{U}}$	ultrapower operator space	185
$c_0(E) = c_0 \otimes^\lambda E$	Banach space of null sequences	197
$\varphi^\infty : c_0(E) \to c_0(E)$	induced mapping	197
$\mathcal{T}_n$	stable point-norm topology on $\mathcal{CB}(V, W)$	197
$\mathcal{T}_w$	stable point-weak topology on $\mathcal{CB}(V, W)$	198
$\mathcal{FCB}(V, W)$	finite-rank completely bounded mappings	200
$\mathcal{F}(X, V; Z \check{\otimes} W)$	relative Fubini product	203
$\mathcal{O}(E, F)$,	Banach space mapping ideal	207
$(\mathcal{N}^B(E, F), \nu^B(\varphi))$	nuclear mappings of Banach spaces	207
$(\mathcal{I}^B(E, F), \iota^B(\varphi))$	integral mappings of Banach spaces	209
$\mathcal{O}(V, W)$	operator space mapping ideal	210
$(\mathcal{N}(V, W), \nu(\varphi))$	completely nuclear mappings	210
$M(a, b)$	multiplication mapping	211
$(\mathcal{I}(V, W), \iota(\varphi))$	completely integral mappings	216
S_{int}	mapping of $\mathcal{I}(V, W^*)$ into $(V \check{\otimes} W)^*$	219
$(\Pi_1^B(E, F), \pi_1^B(\varphi))$	1-summing mappings of Banach spaces	227
$(\Pi_1(V, W), \pi_1(\varphi))$	completely 1-summing mappings	229
$\mathcal{M}$	ultraproduct of $M(a_\alpha, b_\alpha)$	231
$(\Gamma_2^c(V, W), \gamma_2^c(\varphi))$	column Hilbert factorable mappings	234
$(\Gamma_2^r(V, W), \gamma_2^r(\varphi))$	row Hilbert factorable mappings	238
$E \overset{r}{\cong} F$	linear isomoprhism of E and F	243
$d(E, F)$	Banach–Mazur distance	247
$d_{cb}(V, W)$	completely bounded Banach–Mazur distance	247
$\mathcal{OS}_n$	n-dimensional operator spaces	247
$:\check{\otimes}:$, $\check{\otimes}:$, and $:\check{\otimes}$	augmented injective tensor products	248
$d_{ex}(L)$, $d_{ex}(V)$	exact approximation constant	261
$(\mathcal{I}^{ex}(V, W), \iota^{ex}(\varphi))$	exactly integral mappings	285
$\mathcal{Z}_{cb}(\mathcal{A}, V)$	completely bounded derivations	309

$\mathcal{I}_{cb}(\mathcal{A}, V)$	inner derivations	309
$\mathcal{H}^1_{cb}(\mathcal{A}, V)$	first completely bounded cohomology group	309
$A(G)$	Fourier algebra	317
LIM	Banach limit	325
$\mathbb{L}(V, W)$	linear mappings of vector spaces	330
$\mathbb{F}(V, W)$	finite-rank linear mappings of vector spaces	330
$\mathcal{B}(V, W)$	bounded linear mappings of normed spaces	332
$\mathcal{FB}(V, W)$	finite-rank bounded linear mappings of normed spaces	332
$\mathcal{K}(V, W)$	compact linear mappings of Banach spaces	332
$\ell_p(\mathfrak{s}; V_s)$, $\ell_p^{\mathfrak{s}}(V)$	Banach spaces of p-summable sequences	334
$\ell_p^n(V)$, $\ell_p^n = \ell_p^n(\mathbb{C})$	n-tuple spaces with ℓ_p-norm	334
$L^\perp$	annihilator space	334
$M_\perp$	preannihilator space	334
$\langle \cdot \mid \cdot \rangle$	sesquilinear form	336
$\mathcal{A}^\dagger$	unital extension	338
$C_0(X)$	space of continuous complex functions vanishing at ∞ on X	338
$\mathcal{S}'$	commutant	340
$C^*(G)$	full group C^*-algebra	341
$C^*_\lambda(G)$	reduced group C^*-algebra	341
$\mathcal{U}$	ultrafilter	344
$\beta\mathfrak{s}$	Stone–Čech compactification	344
$\lim_{\mathcal{U}} f(s)$	ultralimit	345

Index